GRANITE: FROM SEGREGATION OF MELT TO EMPLACEMENT FABRICS

Petrology and Structural Geology

VOLUME 8

Series Editor:

ADOLPHE NICOLAS

Department of Earth and Space Sciences,
University of Montpellier, France

Granite: From Segregation of Melt to Emplacement Fabrics

Edited by

J. L. BOUCHEZ

Department of Earth Sciences, University of Toulouse, France

D. H. W. HUTTON

Department of Geological Sciences, University of Durham, United Kingdom

and

W. E. STEPHENS

Department of Geology, St. Andrews University, United Kingdom

KLUWER ACADEMIC PUBLISHERS

DORDRECHT / BOSTON / LONDON

A C.I.P. Catalogue record for this book is available from the Library of Congress

ISBN 978-90-481-4812-7

Published by Kluwer Academic Publishers,
P.O. Box 17, 3300 AA Dordrecht, The Netherlands.

Kluwer Academic Publishers incorporates
the publishing programmes of
D. Reidel, Martinus Nijhoff, Dr W. Junk and MTP Press.

Sold and distributed in the U.S.A. and Canada
by Kluwer Academic Publishers,
101 Philip Drive, Norwell, MA 02061, U.S.A.

In all other countries, sold and distributed
by Kluwer Academic Publishers Group,
P.O. Box 322, 3300 AH Dordrecht, The Netherlands.

Printed on acid-free paper

TABLE OF CONTENTS

« Il n'y a guère de royaumes connus qui ne renferment des granits, ou qui ne donnent lieu d'y en soupçonner ».
« There is no known kingdom where granites are not present, or where they may not be suspected ».

Jean-Etienne Guettard (1715-1786), French geologist, member of the Royal Academy of Sciences, initiator in Europe of geological mapping. He was the first to reveal the existence, in Auvergne (Massif Central), of numerous extinct volcanos. J.E. Guettard died one year after James Hutton (1726-1797) proposed that granite formed by consolidation from the fluid state, the beginning of our understanding of the magmatic origin of granites.

PREFACE

The historical development of granite geology, from its roots in the great 18th century debate between Neptunists and Plutonists to the point attained in the present volume, illustrates iterative paths in Progress that are also observed in other fields of research. Strikingly, for an external observer grounded in the study of peridotites and ophiolites, similar lines of evolution appear which can be integrally transposed from one research field to the other. One is the marked shift of granite research from a balance between structural and petrographical studies, before and just after the Second World War, to a geochemistry-dominated science in the late fifties and sixties, until, in the late seventies, studies on granite structures bloomed again. The second striking fact of granite studies is the emergence of a new type of field geology, thanks to this renewed development of structural studies.

That beautiful field oriented structural studies were conducted before the war can be appreciated if we consider, for example, Robert Balk's GSA Memoir in 1937 on the "Structural Behavior of Igneous Rocks" (and, notably, its impressive bibliography on this subject), or Ernst Cloos' 1946 GSA Memoir on lineations. More vividly, the highly developed nature of these studies is attested to by the rather hot debate that grew up between various workers, fuelled by distinct structural observations made in the field, pitting disciples of emplacement of granitic magmas against those favouring transformation or granitisation more-or-less in situ by metamorphic reactions. The debate of those earlier days has been beautifully summarized by H.H. Read in his famous "Granite Controversy" (1957).

Read's formulation of the controversy occurred at the time when geochemistry was emerging as a new and powerful tool. The new techniques opened an era during which granites were considered mainly from this new viewpoint. Geochemical signatures have shown that mantle and crustal origins for granites were both possible, but the debate on how and why granites are emplaced did not progress much. Meanwhile, structural geology was essentially geometrical and mechanistic. In the early 70's, the structural approach began to widen to include solid state physics and fluid dynamics. Detailed structural maps of granitic bodies were again published, mainly in France, and analysed in terms of magmatic and plastic flow. The senior editor of this volume and his students deserve much of the credit for this new development. Via microstructural and petrofabric studies, they were able to discriminate between strain in the presence of residual melt or in the solid-state, and, by systematically measuring magnetic fabrics (AMS), they have been able to map magmatic foliations and lineations in ever finer detail, using the internal markers within granites coming from different tectonic environments. The traditional debate has been shifted anew. The burning question now seems to be how the necessary, large-scale or local, crustal extension required for granite emplacement can be obtained. This question necessarily entails consideration of crustal tectonics, and implicitly suggests the idea of using granites as markers of large scale crustal tectonics. At this stage, an increased collaboration between structuralists and petrologists is required, and this is foreshadowed by the present editorial team.

In the continuum from the microstructural scale to that of crustal structures, field data are literally central. It is because the relevant structures have been precisely observed and systematically mapped that a new discussion has commenced and the problem of granites has taken on a more general significance. These insights can be ascribed to the emergence of this new approach in structural geology, which in my view is a new field geology, because structures represent the first and commonly the main information obtained in the field. This new approach is based on systematic measurements and sampling over a grid whose size depends on the particular objectives to be attained, and that may rely on the use of new techniques. The systematic, accurate, and hopefully unbiased measurements resulting from this field approach should allow us to formulate better interpretations and facilitate physical modelling.

With the additional physical aspects treated at the beginning of the volume, all this is beautifully illustrated here.

Montpellier, 07/11/96
Adolphe Nicolas

INTRODUCTION

This Volume "Granite: from segregation of melt to emplacement fabrics" was conceived at the EUG 8 Meeting held in Strasbourg during April 1995, by gathering, selecting and reviewing the most innovative papers that were presented at the X12 Symposium. It covers a broad range of topics related to physical aspects of granite magmatism, topics largely under-represented in comparison to geochemical approaches to granite study, the term granite, of course, being taken in its broadest sense.

Nineteen papers span the range from physical properties of granitic material to several pluton case studies. ***Part I*** "Melt and magmas, properties and segregation" opens with a stylistic exercise about information theory as applied to the formation of granite (N. Petford, J.D. Clemens and J.L. Vigneresse). Two contributions follow reporting laboratory data on the rheology of granitic magmas (B. Scaillet, F. Holtz and M. Pichavant), and wetting angles and equilibrium melt geometry (D. Laporte, C. Rapaille and A. Provost), and two further contributions are inspired by field studies on magmatic fracturing and segregation of melt in the Adamello massif (B.E. John and H. Stünitz) and the relationship between rock structure and rheology in granites from Spain (C. Fernandez, A. Castro, J.D. De La Rosa and I. Moreno-Ventas). ***Part II*** "Fabrics in granites", develops some lively aspects of present-day granite geology. "Granite is never isotropic" is an introduction to magnetic fabrics in granite (J.L. Bouchez), followed by an illustration of the remarkable fabric homogeneity of granite at all scales (P. Olivier, M. de Saint Blanquat, G. Gleizes and D. Leblanc). Analogue experiments, using a torsion apparatus (L. Arbaret, H. Diot, J.L. Bouchez, P. Lespinasse and M. de Saint Blanquat) or a shear box (A. Fernandez and J. Fernandez-Catuxo), help to reproduce the fabrics of biotite, hence (hopefully) model the magnetic fabric of at least magnetite-free granite. Finally, numerical modelling is used to examine the formation of mineral lineation in terms of strain regime and particle shape ratio (K. Schulmann, J. Jesek and Z. Venera) and the periodic versus steady-state nature of the mineral fabric in simple shear (B. Ildefonse, L. Arbaret and H. Diot). ***Part III*** "Emplacement of granite plutons: case studies" comprises eight contributions. It begins with a general consideration of syntectonic granites, and a proposed solution to the space problem (D.W.H. Hutton), followed by a review paper concerning pluton shape inferred from combined fabric and gravity data (L. Améglio, J-L. Vigneresse and J.L. Bouchez). Three exemplary case studies of plutons emplaced along shear zones are then presented: the huge Los

Pedroches Batholith, Spain (A. Aranguren, F. Larrea, M. Carracedo, J. Cuevas and J.M. Tubia), the majestic Mono Creek granite, Sierra Nevada (M. de Saint Blanquat and B. Tikoff), and the Pan-African granite-charnockite Rahama complex (Nigeria) for which a tentative 3D-reconstruction is proposed (E. Ferré, G. Gleizes, M.T. Djouadi, J.L. Bouchez and F.X.O. Ugodulunwa). Finally, a synthetic view of spatial and temporal associations of magmas emplaced along transcurrent shear zones is given through the example of the Brasilian orogeny in Northeast Brasil (A. Vauchez, S. Pacheco Neves and A. Tommasi). The Volume ends with two associated papers devoted to the emplacement of plutons along subduction zones, one showing a well-documented case study, the Yakushima granite of Japan (R. Anma), and the other, a spectacular experiment of buoyant obliquely rising inclusions in the sense of Ramberg's school, reminding us that diapirism may exist (R. Anma and D. Sokoutis).

Thanks to all the contributors.

The editors are particularly indebted to the external referees: C. Archanjo (University of Natal, Brazil), K. Benn (University of Ottawa), S. Blake (Open University), J. Blencoe (Oak Ridge National Laboratory), G. Borradaile (Lakehead University), R. Caby (University of Montpellier), B. Clarke (Dalhousie University), L. Corriveau (Centre Géologique du Québec), A. Cruden (University of Mississauga), J. Grocott (Kingston University), J.P. Hogan (University of Oklahoma), K. Karlstrom (University of New Mexico), D. Kolhstedt (University of Minnesota), P. Launeau (University of Nantes), R. Law (Virginia Technological Institute), J.P. Lefort (University of Rennes), N. Manktelow (ETH, Zürich), W. Means (University of Albany), B. McNulty (University of Santa Cruz), T. Rushmer (University of Vermont), R. Panozzo-Heilbronner (University of Basel), P.Y. Robin (University of Mississauga), P. Rochette (University of Aix-Marseille), C. Teyssier (University of Minneapolis), R. Trzebski (University of Göttingen), E. Sawyer (University of Quebec at Chicoutimi), and J.M. Tubia (University of Bilbao).

Internal referees are also greatly acknowledged, and last but not least, all the staff of the Laboratory of Petrophysics in Toulouse (UMR CNRS 5563) and particularly Pierre Lespinasse who orchestered the organisation, and Christiane Cavaré-Hester, our Paganini of the computer for many drawings and the editing, are warmly thanked.

Thanks to all of you.

Jean Luc Bouchez, Donny Hutton and Ed Stephens

PART I

MELT AND MAGMAS:

PROPERTIES AND SEGREGATION

APPLICATION OF INFORMATION THEORY TO THE FORMATION OF GRANITIC ROCKS

Nick PETFORD, John D. CLEMENS and Jean-Louis VIGNERESSE*
School of Geological Sciences, Kingston University,
Kingston-upon-Thames, Surrey, KT1 2EE, United Kingdom
**CNRS, BP 23, 54501 Vandoeuvre/Nancy Cedex, France*

ABSTRACT

Information theory, proposed originally by Shannon (1948), has been applied to the formation of granitoid rocks. Application of the theory allows the four main elements involved in granite formation viz.: partial melting (M), melt segregation (S), magma ascent (A) and emplacement (E) to be analysed qualitatively as a single, holistic process. The information content (chemical, isotopic and mineralogical) is contained in the magma, and the received message is the crystallised pluton. Noise added (e.g., by tectonism, weathering or sample collection) can lead to distortion or even irretrievable loss of information. Against this is the energy introduced into the system by the observer. Defining the message to be transmitted during granitoid formation as the initial composition of the partial melt (H_S), the total information or entropy content (H') preserved in an exposed granitic pluton can be expressed symbolically as:

$$H' = [H_S\text{-}N\text{-}\Lambda] + R + \delta$$

where N and Λ represent information loss by noise and weathering, counteracted by redundancy R and input from the investigators δ. The processes involved in granitoid formation, as identified by the application of information theory, are likely to be cyclic. By defining a total assembly time for a pluton, based on the rate of each key granite-forming process, information theory suggests that the rate-limiting step in pluton formation is the process of partial melting of the protolith – ultimately the rate of thermal diffusion necessary to accomplish this. Analysis suggests that total melt volume and melting rate in the source will control both the segregation mechanism and magma ascent rate.

INTRODUCTION

Over the past decade there has been a significant shift in scientific method, away from the reductionist approach, that tends to concentrate on one particular aspect of a phenomenon, to a more open approach, in which systems are considered in their entirety. This holistic approach is becoming increasingly successful in the life sciences and ecology, where systems are commonly greater than the sums of their parts (e.g., Cornwell, 1995).

J.L. Bouchez et al. (eds.) Granite: From Segregation of Melt to Emplacement Fabrics, 3–10.

In contrast, within the geological sciences, many petrologists and geochemists still focus on one particular aspect of the granite phenomenon while ignoring the others. The present authors have not been blameless in this regard. Though this approach has the advantage of maximising the information about process *x*, it may be at odds with what is currently known about process *y* so that, while both appear correct in isolation, they become mutually contradictory when taken together. Recent examples include the apparently irreconcilable difference in the rates of pluton ascent based on fracture transport and fluid dynamical theory (Clemens and Mawer, 1992; Petford et al., 1993) and rates of emplacement based on structural, tectonic and field studies (e.g., Paterson, 1994). Other current examples, chosen at random from discussions on the "granite-research" discussion group on the Internet, include the role of enclave-bearing magmas in granitoid evolution, the shapes of plutons with depth, and the origin and significance of myrmekite.

A recurring problem with the holistic approach is how best to express the various processes that comprise a given system in a coherent and meaningful way. One approach, that has proven successful in a number of disciplines, including geology (e.g., Pelto, 1954; Tasch, 1965; Ferguson, 1980), is the mathematical theory of communications, known as information theory (Shannon, 1948). The theory allows information that is passed from one place to another to be defined and quantified at both its source and destination, thus providing insight into processes occurring during transmission.

The ultimate aim of information theory is to predict the behaviour of the communications system mathematically. Given that the theory in its original form does not require transmitted information to have any meaning, two questions of immediate relevance are: 1) can the concepts of information theory be applied to the genesis of granitic rocks and 2) if applicable, will it tell us anything *new* about the origin and evolution of these rocks? The first question can be answered pragmatically by considering the generation, segregation, ascent and emplacement of granitic (s.l.) magma as a process of information transmission. The final product (a granitic pluton) contains within it information (chemical, mineralogical, isotopic, textural and structural) that is the sum of the various steps involved in its formation. Thus, in principle, each step should be open to analysis by information theory. The second question requires a more philosophical approach. Millions of chemical analyses of granitoid rocks must now exist worldwide and, although this database expands daily, we are still in many ways no closer to understanding the origins of these rocks than in the days of Hutton. Thus, although the application of information theory to the origin of granitic rocks may not reveal new answers immediately, it might provide a framework for re-thinking old problems.

In this paper, we show how information theory can be used to describe holistically the main processes involved in the formation of granitoid rocks. We begin by introducing the basic ideas germane to information theory and provide examples with the use of flow diagrams. A general mathematical and symbolic treatment of the processes involved in granite formation is given. By defining the message to be transmitted during granitoid formation as the composition of the partial melt (H_S), we show how its information or entropy content on emplacement can be expressed symbolically. We conclude by showing how simple application of information theory can help constrain the relative rates of processes involved in granite formation, and argue that these processes are likely to be geologically short-lived, cyclic events.

PRINCIPLES AND DEFINITIONS

The general components of a simple communication system are shown in figure 1. Any communication system contains information in the form of bits (Shannon, 1948), and the processes that lead ultimately to the formation of something observable (e.g., a television picture, fossil, granite pluton, etc.) will be encoded in *n* bits of information. These bits are equal in summation to the total entropy (H') of the system. We consider granitic rocks to be dominantly the products of the crystallisation of hydrous silicate magmas, at high temperatures. Thus, for a granite, every aspect of the rock, from its inception as a melt, its initial chemistry, mode of segregation, ascent, chemical evolution and emplacement are subsumed under H'. A close association exists between information and uncertainty, so that the greater the information content, the more uncertain the answer. Stated differently, the greater the uncertainty of an answer, the greater its information content.

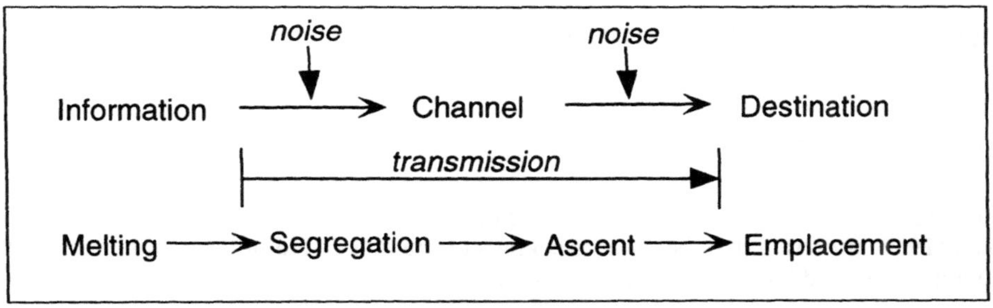

Figure 1. General concept of information theory as defined by Shannon (1948). Information at source is transmitted via a channel to a final destination, where the message is received. Noise may be introduced during transmission. Recast in these terms, granite magmatism is seen as a process whereby information about the source region is transmitted by the melt through an ascent channel to emplacement, where the information stored within a pluton may be decoded by geoscientists.

We stress that while the entropy of Shannon is not strictly speaking the entropy (S) of thermodynamics, the second law can be expressed as:

$$S = k(\ln P) \tag{1}$$

where k is the Boltzmann constant and P is probability. Thus, the less probable an event, the lower its entropy (Schrödinger, 1944, see also Haken, 1983). In a similar way, information theory forces us to assign probabilities to events. For a system of N alternatives, with P_i probabilities from 1 to n, the information or entropy content of the system is:

$$H = -K \sum_{i=1}^{n} P_i \log_2 P_i \tag{2}$$

where K is a unit of choice (Shannon, 1948). As an example, for a case where a system has two alternatives that are both equally likely, $P_i = 0.5$ and $H = -1$. Alternatively, if one or other probability = 0, then $H = 0$. In reality, for a given number of alternatives, H will have a range of values (from 0 to 1) that depends ultimately on the sum of the probabilities P_i (Ferguson, 1988). We are thus faced with the intriguing problem of assigning probabilities to all the events that lead to the formation of a pluton!

Not surprisingly, any theory that attempts to apply probabilities to a finite number of events will have limitations, especially when the number of events becomes very large. In cases where probabilities remain discrete, Ulam's theorem, which predicts an N^2 dependence on interactions between events, should apply (Atlan, 1972). It is also noted that a purely statistical treatment based on probability theory may not be the best way to estimate the sensitivity of a system to external conditions (Haken, 1983). Other technical problems inherent in any application of information theory are uncertainties introduced by redundancy and noise (insignificant information). Redundancy (R) is repeated information, and is expressed as $R = H_{max}\text{-}H/H_{max}$, where H_{max} is the maximum quantity of information in a discrete event N. Where a connection exists between groups (x and y), $H_{max} = H_x + H_y \text{-}(H_{xy})$.

GRANITES AS INFORMATION SYSTEMS

THE GRANITE SERIES

H.H. Read's granite series is a seminal treatment of granitic rocks in a holistic way. Read (1957) proposed that the numerous kinds of granites found in the continental crust are not distributed haphazardly, and that certain types recur in similar parts of orogenic belts and at comparable stages in orogenic evolution. According to Read (op. cit.), the essential idea behind the granite series is that of the mobilisation and transport of material formed in-situ in the crust. Thus, during the course of orogeny, early, deep-level, anatectic migmatites are followed in the mid crust by forcefully intruded granites, which themselves give way finally to high-level, permitted intrusions. It is now recognised that Read's granite series does not provide a reliable account of the evolution of granites during orogenesis. Major problems exist in relating migmatite terranes to large granite plutons, while many high-level ring complexes are related directly to the influx of mantle-derived basaltic magmas. Furthermore, the realisation that granitic melts may be transported large distances through the crust along faults and as self-propagating fissures (Clemens and Mawer, 1992; Petford et al., 1993) challenges the classical ideas of forceful versus permitted ascent and emplacement mechanisms. However, in attempting to relate the various styles of granitoid magmatism in time and space, Read's granite series does provide an orderly and sequential account of granitoid formation that can be examined further using information theory.

Most granite workers (see Brown, 1994 for a review) would accept that pluton formation involves four main steps: 1) melt generation, 2) melt segregation, 3) vertical ascent of magma from the source region and 4) magma emplacement and solidification. Given the prevailing axiom that granites reflect the *P-T-X* conditions of their source regions (e.g., Chappell and White, 1974; Clemens, 1984; Clemens and Vielzeuf, 1987), granite magmatism, expressed in the language of information theory, is seen essentially as the transmission of information, or entropy (H'), about the source region by the magma, from the lower to the middle or upper continental crust, via ascent and emplacement. This sequence of generation and segregation (input or data encoding) ascent (transmission) and emplacement (received message) is set out in figure 1.

Noise is added to the system through the effects of deformation, erosion and sampling error, although these can be overcome to some degree by information input from the observer. An example of redundancy (see previous section) might be the presence of an ancient radiogenic component in the source region as indicated by restitic cores in zircon crystals. However, no extra amount, or repetition, of zircon crystals adds anything new to the message that an inherited component is present.

THE CASE OF A SINGLE EVENT: MAGMA GENERATION

To show how information theory forces us to consider granite formation in a holistic manner, we take as an example the processes involved in the generation (defined here as partial melting and segregation) of melt in the source region. Information theory allows us to consider the generation of melt as a discrete event (Fig. 2). Granites start out in their source regions as partial melts. However, as a number of different rock types can act as protoliths for the magmas, we are faced instantly with a choice of possible source rocks. We also need to know the amount of melting, which is, in turn, related to the pressure, the duration and intensity of heat input, the presence or absence of a fluid phase (and its composition), the porosity, permeability, thermal conductivity, thermal diffusivity and latent heat of fusion of the source, along with the physical properties (e.g., viscosity and density) of the resultant partial melt. These environmental variables, and their possible outcomes during a partial melting event, are summarised in figure 2. An adequate description of the generation process, using information theory, requires the assignment of a probability level to each of the three possible outcomes: 1) extraction with no ponding, 2) ponding with no extraction or 3) some combination of these. We might choose as most realistic $P = 1$ for option 3.

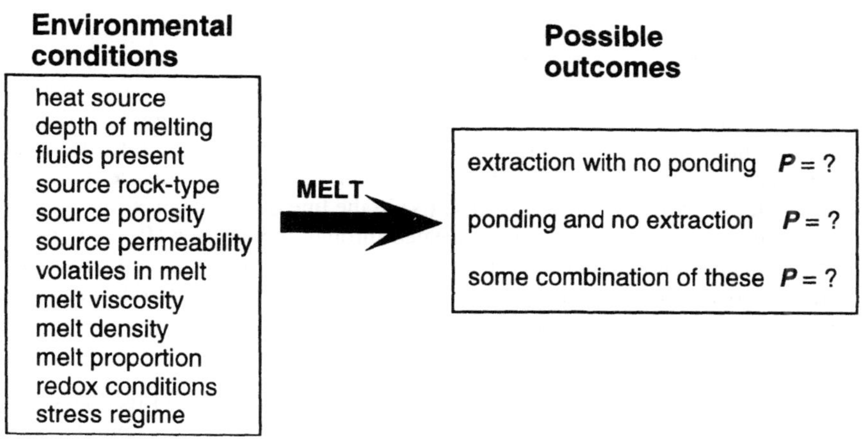

Figure 2. Summary of some of the processes (environmental variables) involved in the generation of granitic melt during a partial melting event. Also shown are some possible outcomes that describe the subsequent fate of the melt. The reader is urged to consider the geological controls that would favour one event having a higher probability (P) that any other.

THE CASE OF MULTIPLE EVENTS: ASCENT AND EMPLACEMENT

The problem becomes more acute when we consider ascent and emplacement as well as generation. We now have to consider not just an individual event, but also the relationships between multiple events. In magmatic systems that involve the generation, movement and solidification of melt, each successive event depends upon the previous one since, without melt, there would be nothing to ascend or be emplaced. For the four dependent events of melt generation (*M*), segregation (*S*), ascent (*A*) and emplacement (*E*) with *n* possibilities for each event, the combined entropy $H_{(M,S,A,E)}$ of the system is:

$$H_{(M,S,A,E)} \leq H_{(M)} + H_{(S)} + H_{(A)} + H_{(E)}. \tag{3}$$

Thus, for two or more dependent events, the total information content is equal to or less than the sum of the information contents of the individual events (Shannon, 1948). The implications of this statement are frustrating yet illuminating. During the process of granitoid formation, some information may be lost irretrievably. Unfortunately, information theory does not tell us from where in the system any potential loss will occur, or indeed whether any information has been lost at all.

GRANITOID MAGMATISM AS A COMMUNICATION SYSTEM

We are now in a position to consider a model describing the total information content involved in granitoid magmatism, from source to final exposure at Earth's surface, in a way similar to that first proposed by Tasch (1965) for the process of fossilisation. We begin by defining a set of variables that take into account the different processes that are likely to occur in the formation of a granite. The total amount of information (H') passed is a combination of the original source information (H_S) less the amount of original geochemical information lost during hybridisation or assimilation and late-stage fluid processes, and information lost during deformation caused by any tectonic activity that may occur during uplift (t_X in figure 3). Data on physical parameters (e.g., density and viscosity) that control melt segregation and ascent are encoded cryptically in the bulk magma composition. Information lost by weathering at Earth's surface is expressed as Λ. The terms together constitute noise (*N*) in the line of communication. Against this irretrievable loss stands information repeated through redundancy (*R*), and energy input from the investigators themselves (δ), defined by Tasch (1965) as derivative data. This final input is crucial since, without an observer, the output information cannot be decoded. These relationships are related symbolically in the expression:

$$H' = [H_S\text{-}N\text{-}\Lambda] + R + \delta \tag{4}$$

Figure 3 shows this relationship schematically, in the form of a delayed feedback loop (Elasser, 1958). Shading denotes those parts of the loop that occur deep within the crust, and are thus normally unobservable. The loop shown here is considered as time-asymmetric, in that processes occurring within the crust may be cyclic (dashed lines). In contrast, the granite body, in its final crystallised state, can only ever make one journey to Earth's surface with the majority of its original source information content intact. Evidence for cyclicity within the crustal loop may be found in the inferred pulsed nature of emplacement of many large plutons (e.g., Pitcher, 1979; Petford, 1996). The fact that different pulses are distinguishable means that each granitic component of a composite pluton has a unique MSAE history. Subsequent reburrial and remelting would erase much of this original source information, and the resultant partial melt would have a new, distinct information content.

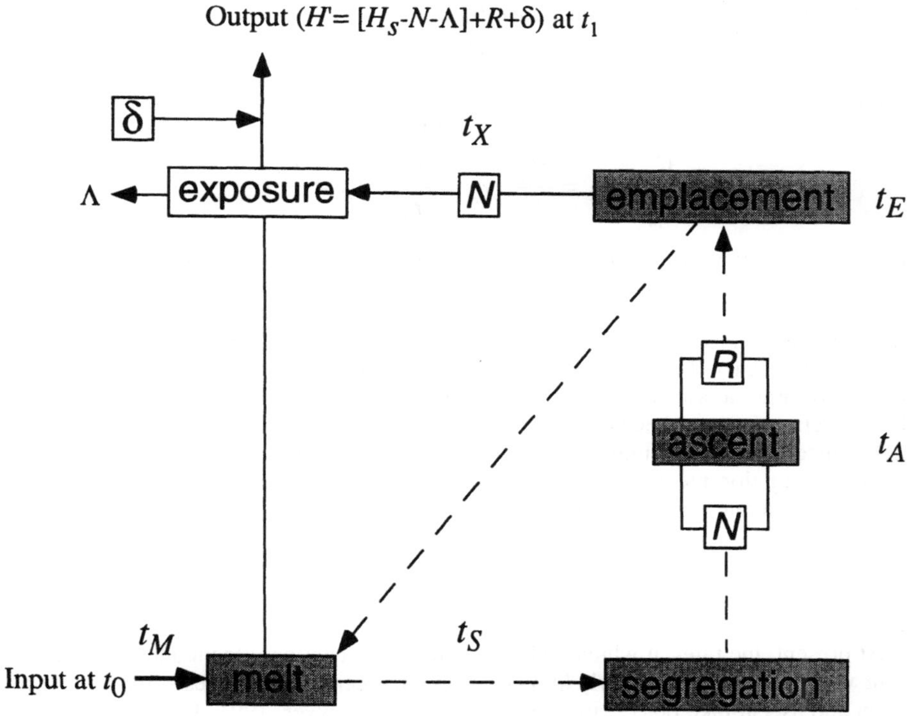

Figure 3. Feedback loop showing the processes involved in granitoid formation during crustal melting. Input at time t_0 leads to the formation of melt in a time t_M that contains primary information about its source (H_S). Segregation is followed by magma ascent and emplacement. These processes occur within the crust (shaded) and although unobservable may be cyclic (dashed line). Post-emplacement uplift and exhumation leads ultimately to exposure, where irretrievable information loss occurs due to weathering (Λ). Further loss of information through noise (N) is counteracted by repetition (R) and energy added by the investigator (δ). Total information output is defined as H'.

PLUTON ASSEMBLY: A HOLISTIC APPROACH

Explicit to the holistic system described here is a temporal component that links the various stages of MSAE together. During assembly of a pluton, the total assembly time (t_a) will be the sum of the times required for generation of partial melt (t_M), segregation of this melt (t_S), ascent (t_A) and emplacement (t_E):

$$t_a = t_M + t_S + t_A + t_E \tag{5}$$

Assembly time (t_a) will also depend upon the volume of the magma pulse, itself dependent upon the amount of partial melt formed in the first place. With the possible exception of diapirism, there is no reason to suppose that ascent and emplacement must occur either by the same mechanism or at the same volumetric rate (see e.g., Clemens et al., 1996). If we assume the volume condition:

$$V_M > V_S > V_A > V_E \tag{6}$$

which takes into account the volume of melt not mobilised during segregation and ascent, the rates of the individual processes are given by:

$$r_M = \frac{V_M}{t_M},\ r_S = \frac{V_S}{t_S},\ r_A = \frac{V_A}{t_A},\ r_E = \frac{V_E}{t_E} \tag{7}$$

These rates assume that the processes are continuous (i.e., there are no hiatuses). From the above analysis, we can identify the following limiting conditions that will disrupt the MSAE process and lead to temporal breaks (pulsing) in pluton formation: 1) $r_M < r_S$ (rate of melting is less than rate of segregation) and 2) $r_S < r_A$ (rate of segregation is less that rate of ascent). There will be positive mechanical feedback between segregation rate and volumetric ascent rate. However, the rate of melting should not exceed the rate of segregation ($r_M > r_S$) because of negative chemical feedback (control of the melting rate) induced by melt accumulation. Total melt volume and melting rate will therefore control segregation mechanism and magma ascent rate. By combining the expressions for volume and rate it is apparent that:

$$t_M > t_S > t_A > t_E \tag{8}$$

suggesting that the rate limiting step in granitoid formation is the time (t_M) required to generate melt in the source region.

At present, the rates at which melt generation, segregation, ascent and emplacement occur within the crust are poorly known, although recent work suggests that melt segregation and ascent take place within months to years (Sawyer, 1991, 1996; Clemens and Mawer, 1992; Petford, 1995). Theoretical and experimental modelling of partial melting above a mafic heat source suggest time-scales of melt generation on the order of 10^2-10^3 years (Huppert and Sparks, 1988), in accordance with the above results. Our analysis of granite formation based on information theory predicts that emplacement should also be a geologically rapid process. It also implies that granitoïd magmatism may be self regulating, with the final volume of plutonic material at emplacement controlled by feedback mechanisms operating *between* the individual stages of melting, segregation ascent and emplacement, rather than a single, dominant MSAE process.

ACKNOWLEDGEMENTS

Jean-Louis Vigneresse (prior to his input as a co-author) and Barrie Clarke are thanked for incisive and constructive reviews.

RHEOLOGICAL PROPERTIES OF GRANITIC MAGMAS IN THEIR CRYSTALLIZATION RANGE

Bruno SCAILLET, François HOLTZ and Michel PICHAVANT
CRSCM-CNRS, 1A rue de la Férollerie,
45071 Orléans Cedex 02, France.

ABSTRACT

Experimental phase equilibria carried out on three natural granites are used to constrain the rheological patterns of natural cooling felsic intrusions with initial H_2O contents between 4.5 wt.% and 7 wt.% and emplacement temperatures between 920°C and 750 °C. Crystallization paths of H_2O-bearing magmas show eutectic melt fraction trends, with only 20-30 wt.% crystals after 90% of cooling. Melt viscosities increase by a factor of 2 or less during crystallization while magma viscosities remain within one order of magnitude of the initial value during 90% of the crystallization interval. This behaviour is due to the build-up in melt H_2O content during crystallization of hydrous granitic magmas that counterbalances the effects of temperature drop and increase in crystallinity on magma viscosity. Crystallization paths of H_2O-CO_2-bearing magmas show that the presence of CO_2 counteracts the lowering effect of H_2O upon viscosity, except at very low fluid/melt mass ratios. As a consequence, magma viscosities of H_2O-CO_2-bearing magmas may increase much more rapidly during crystallization, in some instances by many orders of magnitude relative to H_2O-bearing magmas. Phase equilibrium results show that oxidation may significantly increase granitic magma viscosities, especially when occuring at constant melt H_2O content. Thus, changes in redox state may affect the fluid dynamics of granitic magmas either during their ascent or at their emplacement level. In particular, diapiric granitic magmas are particularly sensitive to oxidation and may have their level of emplacement ultimately controlled by an oxidation event. In contrast, none of the oxidation mechanisms considered (H_2 diffusive loss, H_2-H_2O fluid phase fractionation and Fe^{2+}-rich crystal removal) is fast enough to affect granitic magmas ascending through dykes.

INTRODUCTION

Recently considerable progress has been made in understanding the rheology of silicate melts (see review by Dingwell et al., 1993). Physical properties such as the viscosity or the density of anhydrous superliquidus silicate melts have been extensively measured

J.L. Bouchez et al. (eds.) Granite: From Segregation of Melt to Emplacement Fabrics, 11–29.

and accurate equations for calculating these properties are now available (e.g., Bottinga and Weill, 1972; Lange and Carmichael, 1990; Persikov, 1991). In contrast, the physical properties of hydrous silicate melts are much less well understood, in part because of technical problems associated with measuring rheological properties at the high pressures needed to incorporate H_2O in silicate melts. Nevertheless, the major controlling factors of the viscosity of hydrous magmas have been identified. It is well established that, in addition to temperature, the viscosity of a natural silicate melt is controlled primarily by its SiO_2 content and by the amount of dissolved volatiles which is dominated by H_2O. Although other volatiles such as F and B are known to affect viscosity (Dingwell et al., 1985, 1992; Baker and Vaillancourt, 1995), their initial abundance in magmas is poorly constrained and they are not considered in this work. However, it should be borne in mind that these volatiles play a similar role than H_2O upon viscosity and as such their presence in a magma will enhance any effect due to H_2O. For the magma viscosity to be known, the effect of crystals must be taken into account, which is usually done through the Einstein-Roscoe equation (see Pinkerton and Stevenson, 1992; Lejeune and Richet, 1995). For granitic magmas, the common view is that these relatively SiO_2- and H_2O-rich, low temperature magmas have much higher viscosities than their mafic counterpart (10^3 Pa.s, see Shaw, 1969). Values in the range 10^6-10^7 Pa.s, and sometimes higher, have been reported in the literature. However, determining viscosities for granitic magmas has proven to be a difficult task for two reasons: (1) in contrast with basic magmas (e.g., Shaw, 1969; Lange et al., 1994), liquid lines of descent and melt fraction trends are very poorly defined in granitic systems and, (2) no attempts have been made to directly measure the rheology of granite at hypersolidus temperatures. The classic work of Van der Molen and Paterson (1979) involved the determination of rheology of a partially melted granite under H_2O-saturated conditions only, a situation not representative of many natural granitic magmas. Consequently, current estimates of the viscosity of granitic magmas are entirely based on empirical calculations which require several assumptions regarding melt H_2O content, crystal content, composition, temperature, and the variation of these parameters during ascent, emplacement, and crystallization. For instance, during crystallization melt composition is postulated to evolve toward more silica-rich compositions. This compositional evolution, coupled with decreasing temperature and the complementary increase in crystal content, has led most authors to conclude that the viscosity of granitic magmas increases dramatically during crystallization (e.g., Fernandez and Gasquet, 1994). Implicit in this view is the supposition that the effect of H_2O on melt viscosity during crystallization is negligible relative to that of the other controlling factors (temperature, melt composition and crystal content). However, the incompatible behaviour of H_2O in granitic magmas, with anhydrous phases constituting more than 90% of the total volume of the crystalline phases, implies the H_2O content of the residual melt increases as crystallization proceeds. This must counteract to some extent the effects of temperature, SiO_2 increase and crystal/melt ratio effects on the viscosities of both the melt phase and the crystal-laden magma (e.g., Petford, 1993). It is noteworthy that the extent of the effect of H_2O on melt viscosity has never been quantitatively measured in natural granites.

In this paper we use experimentally determined phase diagrams for natural granitic compositions to critically assess the role of the above mentioned variables on the rheology of granitic melts. After reviewing the available experimental data constraining the P-T-wt.% $H_2O_{\text{in melt}}$ conditions during emplacement of granitic magmas in the crust, crystallization paths obtained for three natural granites with representative compositions (two peraluminous and one metaluminous) are presented. The effects of P, T, fH_2O and fO_2 and melt composition during crystallization on magma rheology are outlined with particular emphasis on the importance of fH_2O and of fO_2. In view of the dramatic effects that fO_2 can have on the phase relations in granitic magmas, some of the mechanisms that affect the redox state of a crystallizing granitic magma are presented and their relative significance are discussed in light of the principal ascent mechanisms of granitic magmas (dykes or diapirs). Although the approach followed in this paper is limited to some case studies, we believe it illustrates the essential features of the rheological evolution of crystallizing granitic magmas.

PHASE EQUILIBRIUM CONSTRAINTS ON THE P-T-wt.% H_2O_{MELT} CONDITIONS DURING THE EMPLACEMENT OF GRANITIC MAGMAS IN THE CRUST

The most powerful method for inferring crystallization conditions (P-T-wt.% H_2O_{melt}) in plutonic rocks is to perform laboratory experiments that constrain phase relations for the particular bulk composition of interest. This technique was pioneered by Maaløe and Wyllie (1975), and has been applied more recently by Clemens and Wall (1981), Clemens et al. (1986) and Whitney (1988), and by many others. Information on conditions during emplacement of the granite magma such as temperature, melt-H_2O content, and pressure, can be obtained simply by locating the portion of the phase diagram that most closely matches the crystallization sequence of the natural rock (e.g., Clemens and Wall, 1981; Webster et al., 1987; Johnson and Rutherford, 1989; Scaillet et al., 1995a). This approach has confirmed that, as for their volcanic equivalent, most plutonic felsic rocks have initial H_2O contents higher than 3-4 wt.% (Clemens and Wall, 1981; Clemens, 1984; Clemens et al., 1986; Webster et al., 1987; Whitney, 1988; Johnson and Rutherford, 1989; Dall'Agnol et al., 1994; Scaillet and Pichavant, 1994; Scaillet et al., 1995a and in prep.). This is shown in figure 1, where the initial H_2O contents, as deduced from phase equilibria constraints, are plotted against pressure and temperature at the time of emplacement of felsic plutonic rocks and some related volcanics. Granitic magmas emplaced at 4 kbar display initial H_2O contents between 5-7 wt.% (e.g., Scaillet et al., 1995a), 4-5 wt.% at 3 kbar (e.g., Clemens and Wall, 1981; Scaillet et al., in prep.) and 2-4 wt.% at 1 kbar (Clemens et al., 1986; Webster et al., 1987).

All the investigated granites are undersaturated with respect to H_2O, with the degree of undersaturation increasing with pressure of emplacement (Fig. 1a). The decrease in H_2O content with decreasing pressure is accompanied by an increase in temperature during emplacement: from 750-800 °C at 4 kbar to temperatures higher than 900 °C at 1 kbar (Fig. 1b), the driest and hotter magmas being the shallowest. An exception to this trend is the Spor Mountain rhyolite, presumably because of its very high F content (Webster et al., 1987). The final important point strongly suggested by the above mentioned studies is the crystal-poor character of the granitic magmas at the time of emplacement. This conclusion is supported by other evidence such as structural and petrographic studies (e.g., Clemens et al., 1986; Clemens and Mawer, 1992; Scaillet et al., 1995b), and indicates that nearly all the crystallization of the investigated granites can be modelled as an isobaric process.

Accordingly, isobaric crystallization paths have been calculated for three natural granites (Fig. 2, Table 1, Appendix 1): a biotite-muscovite leucogranite (hereinafter referred to as DK), a tourmaline-muscovite leucogranite (GB) and an amphibole-biotite granite (RD NNO+2, NNO standing for nickel-nickel oxide oxygen buffer).

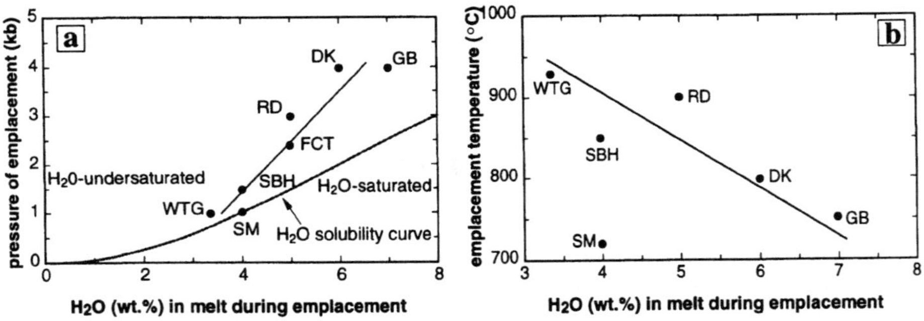

Figure 1. Initial H_2O contents of felsic magmas, as deduced from recent phase equilibrium studies, versus (a) pressure and (b) temperature, at the time of emplacement. In A, the continuous curve represents the H_2O saturation boundary for a haplogranite melt at 850°C (Holtz and Johannes, 1994). DK and GB: biotite-muscovite and tourmaline-muscovite leucogranites (Scaillet et al., 1995a), RD: amphibole-biotite granite (Scaillet et al., 1994); FCT: Fish Canyon Tuff (Jonhson and Rutherford, 1989); SBH: Strathbogie granite (Clemens and Wall, 1981); WTG: Watergums granite (Clemens et al., 1986); SM: Spor Moutain rhyolite (Webster et al., 1987). The straight line is just a manual fit of the experimental data.

The fourth phase diagram (RD FMQ-0.5, FMQ standing for the fayalite-magnetite-quartz buffer, Fig. 2d) also refers to the RD granite but under a lower oxygen fugacity (Fig. 2c-d) indicates the effects of f_{O_2} on the rheological properties of crystallizing granitic magmas. The experimental methods used and the full list of results concerning these phase diagrams can be found elsewhere (Dall'Agnol et al., 1994; Scaillet and Pichavant, 1994; Scaillet et al., 1995a; Scaillet, Pichavant and Dall'Agnol, in prep.). For the peraluminous compositions, melt viscosities have been calculated using the experimental results of Scaillet et al. (1996), whereas for the metaluminous composition, the model of Shaw (1972) was adopted. The viscosities of the melt + crystal suspensions were calculated using the Einstein-Roscoe formula:

$$\eta_{magma} = \eta_{melt} (1 - 1.67 \Phi)^{-2.5} \quad (1)$$

Φ being the volume fraction of crystals. This relation has been shown to hold for a volume fraction of crystals lower than 40% in silicate melts (Lejeune and Richet, 1995) and thus it can be used, as will be shown below, for more than 90% of the crystallization interval of granitic magmas.

EFFECTS OF MELT COMPOSITION AND VOLATILES ON VISCOSITY DURING CRYSTALLIZATION

MELT COMPOSITION

As previously stated, crystal fractionation changes the composition of the residual melt, and this must affect viscosity. Although scarce, available experimental data shows that the silica content of the residual melt during fractional crystallization can either increase (e.g., Scaillet et al., 1995a) or decrease (Pichavant et al., 1987). However, the magnitude of this change remains small, with a correspondingly small effect on viscosity. For instance, the most fractionated melt obtained for the DK composition (140°C below the liquidus with ~ 40 wt.% of crystallization), differs by less than 1 wt.% in SiO_2 relative to the starting bulk composition (on an anhydrous basis), the changes in other element concentrations, except H_2O, being small (see Scaillet et al., 1995a). The effect on viscosity can be evaluated by comparing the viscosity of the initial melt composition and that of the fractionated melt at the same H_2O content of the melt. This compositional change affects the viscosity by no more than 0.1 log unit. The same result has been obtained for the amphibole-biotite metaluminous granite (Scaillet, Pichavant and Dall'Agnol, in prep.). It follows that, from a rheological standpoint, the main compositional variable during the crystallization of these granites is the concentration of H_2O, the variations in other elements having a negligible effect. It is clear, however, that greater compositional changes might occur at higher degrees of crystallisation (> 40-50 wt%) but in this case the effect of crystals on viscosity dominates over that due to chemical changes alone. Thus, for the sake of simplicity, the viscosity changes of granitic melts discussed below have been calculated using a constant bulk composition (that of the granite), the master variables being the changes in H_2O concentration and temperature.

MAGMAS WITH H_2O ONLY

Although most, if not all, natural magmas contain a finite amount of CO_2, the pure H_2O-pure system is a useful limiting case that aids in the understanding of the rheology of crystallizing granitic magmas. For the three compositions used, the starting conditions of crystallization are those inferred from comparison of the natural rock and the phase equilibria diagram, including mineral compositions (Table 1). The initial conditions of crystallization obtained from these studies (initial H_2O contents 7-4.5 wt.%, temperatures 750-920°C) reproduce the range of conditions inferred for most ancient or modern silicic volcanoes, which suggest the results obtained in the present study are applicable to most felsic magmas. The crystallization paths obtained are characterized by small changes in H_2O content during a substantial part of the crystallization interval (Fig. 2). This is due to the small amount of crystallization that takes place over a large temperature range (see also Nekvasil, 1988), as shown by the change in melt fraction with temperature (converted to dimensionless units for the purpose of comparison, see Table 1 to retrieve original values) for the three compositions (Fig. 3a). In all cases, the magma is still 70-80% melt after more than 80% of cooling over the crystallization interval (corresponding to a temperature 15°-

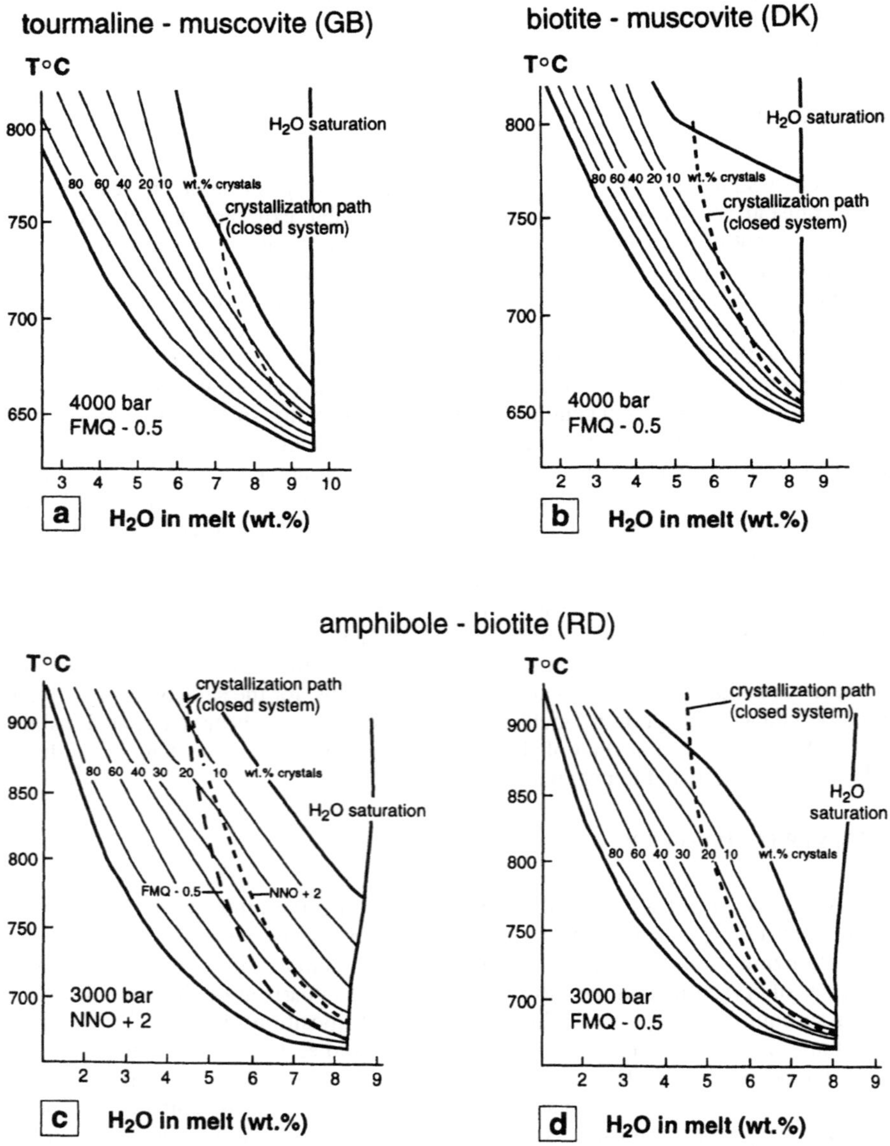

Figure 2. Temperature-wt.% H_2O in melt diagrams for natural granites contoured for wt.% crystals. For the RD composition, the wt.% crystals contours are for oxidized (NNO+2) and reduced (FMQ-0.5) conditions. Note the general depression of melt abundance at a given temperature under reduced conditions compared to oxidized conditions. Calculated crystallization paths are for a closed system with H_2O as the only volatile. The dashed line in C is the crystallization path obtained under reduced conditions, as depicted in D. A reduced magma undergoing oxidation at constant melt H_2O content must remain on the dashed path, along which there is a sharp increase in crystal/melt ratio.

TABLE 1. Composition of granites with initial P, T, wt.% H_2O_{melt}, fO_2, and melt viscosity

	DK[1]	GB[1]	RD[2]
	Bt-Ms	Tu-Ms	Am-Bt
SiO_2	73.04	72.94	71.54
Al_2O_3	15.32	15.57	13.64
Fe_2O_3	0.17	0.52	-
FeO	0.74	0.27	3.42
MgO	0.2	0.14	0.68
CaO	0.85	0.57	2.31
Na_2O	3.85	4.56	3.53
K_2O	4.96	4.14	3.55
TiO_2	0.13	0.06	0.67
P_2O_5	0.14	0.19	-
CO_2	0.05	0.06	-
H_2Ot	0.65	0.64	-
Total	100.20	99.73	99.33
F=O	0.04	0.05	-
Total	100.16	99.68	-
Pressure of emplacement (kbar)	4	4	3
Temperature of emplacement (°C)	800	750	900
initial H_2O (wt.% in melt)	5.5	7	5
fO_2	FMQ-0.5	FMQ-0.5	NNO+2/ FMQ-0.5
T liquidus (°C)	800	750	920
T solidus (°C)	645	630	665
log (η_0) (Pa.s)	4.56[3]	4.14[3]	3.67[4]

[1]wet chemical analysis of bulk rock.
[2]electron probe analysis of fused glass used as starting material in the experiments. All Fe as FeO.
[3]calculated from Scaillet et al. (1996).
[4]calculated from Shaw (1972).
Bt: biotite, Tu: tourmaline, Ms: muscovite, Am: amphibole

20°C above the solidus). The slight departure of the RD trend compared to the trends for the GB and DK granites is due to its higher CaO content (Table 1), which allows plagioclase to crystallize much earlier. Such melt fraction trends are eutectic-like, a fact that is borne out by the near minimum composition of each investigated granite (note that it is also consistent with small changes in composition of the residual melt during most of the crystallization period). Water saturated conditions are achieved only a few degrees above the solidus (Fig. 3a). This indicates that, when H_2O is the only volatile present, bubbles produced by H_2O saturation will have only a minor effect on the rheological properties of crystallizing granitic magmas.

Changes in the viscosity of the residual melt and the magma display virtually the same features for the three compositions. In Fig. 4 all three compositions are plotted using dimensionless variables for the sake of comparison. For the three compositions, melt viscosity changes by no more than a factor of 2 during 90% of the crystallization interval (Fig. 4a). In detail, the changes depend on the initial H_2O content. At 7 wt.% H_2O, the melt viscosity shows a continuous and pronounced decrease toward the end of crystallization, reaching values that are nearly one order of magnitude lower than the initial viscosity.

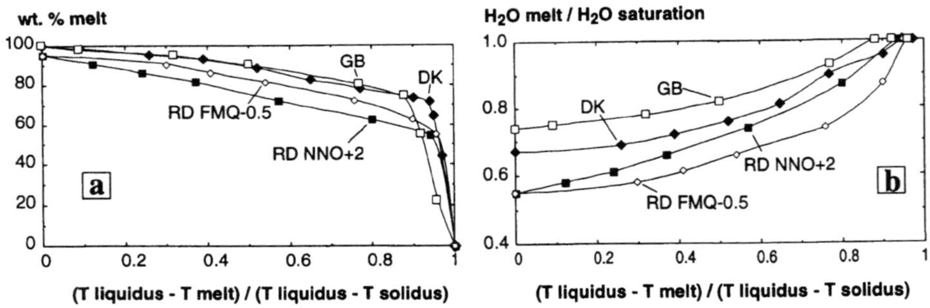

Figure 3. (a) Changes in melt fraction (wt.%) during crystallization of GB, DK and RD at fO_2 = NNO+2 and FMQ-0.5 (see figure 2). Note the eutectic-like behaviour of all melt fraction trends. (b) Changes in the H_2O contents of the corresponding residual melts. Note that, in all cases, H_2O saturation is achieved only after 85 % of cooling at best. To calculate H_2O contents, multiply by the corresponding initial melt H_2O contents listed in Table 1. Table 1 also contains liquidus and solidus temperatures.

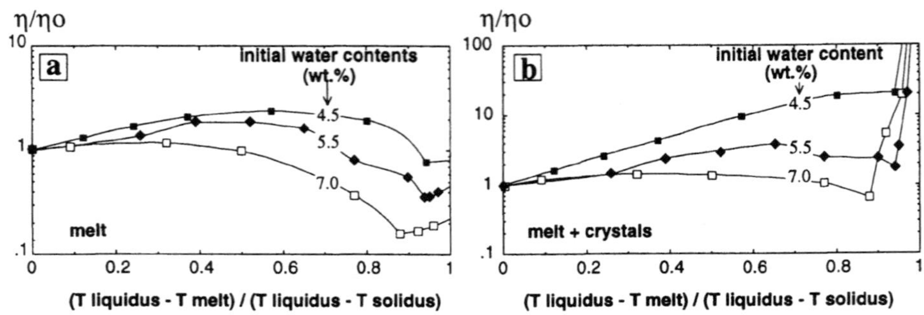

Figure 4. Changes in melt (a) and magma (b) viscosity versus temperature for the GB, DK and RD granites. ηo is the initial viscosity of the melt (Table 1). For the solidus and liquidus temperatures, see Table 1. The H_2O content of the melt at the time of emplacement is indicated on each curve.

This inflexion of viscosity is apparent for all three compositions. However, it clearly decreases as the initial H_2O content decreases. For an initial H_2O content of 4.5 wt.%, it occurs only very late in the crystallization interval. Changes in viscosity of the magma also depend on the initial H_2O content (Fig. 4b). For a H_2O content of 7 wt.%

and 5.5 wt.% H_2O, the viscosity pattern is rather flat, which indicates that the effect of H_2O during crystallization is counterbalanced by the effects of temperature and crystal content. In contrast, for 4.5 wt.% H_2O, the temperature and crystal effects dominate over that of H_2O, and the bulk viscosity steadily increases during crystallization. Therefore, it appears that there is a threshold in initial H_2O content for granitic magmas, at about 6 wt.% H_2O, below which the effect of H_2O on viscosity is dominant over that of temperature and of crystals, and above which the reverse occurs. Unless the granitic magma is particularly H_2O-poor (< 3 wt.% H_2O), the bulk viscosity typically changes by less than one order of magnitude over 90% of the crystallization interval. It is worth noting that such behaviour has far-reaching implications not only for geochemical processes but also for the fluid dynamics of felsic magma chambers where H_2O is the dominant volatile. These topics are beyond the scope of the present paper, but two important points are appropriate here. First, the observed ten-fold increase in magma viscosity, which is commonly cited as a major factor in magma dynamics (i.e., resulting in cessation of convection), takes place after a temperature drop corresponding to 80-90% of the crystallization interval. This implies that, to the extent that this ten-fold increase is indeed a rheological barrier, a large part of the crystallizing boundary layer may well participate in any fluid dynamical process operating deep in the chamber. This is especially true for H_2O-rich magmas like the GB granite. Second, this study emphasizes a point which runs counter to a common belief, namely, that the coldest felsic melts, which are the richest in H_2O, are the least viscous ones. A similar conclusion was reached by Holtz and Johannes (1994) for the haplogranite system (although they restricted their analysis to the case of melts coexisting with quartz and feldspar during the entire period of crystallization, a situation that differs from the cases described here), and by other investigators working on natural granites (e.g., Clemens and Wall, 1981). Nevertheless, such behaviour should enhance melt-crystal fractionation, no matter how it occurs (e.g., crystal settling, boundary layer processes, compaction of crystal mush). As a result, cold magmas will probably undergo fractionation processes to a greater extent than hotter ones, which, in contrast, will have less opportunity to produce large amounts of segregated residual melts.

MAGMAS WITH H_2O AND CO_2

The crystallization paths in H_2O-CO_2-bearing magmas will be detailed for the DK composition only, because GB and RD display qualitatively the same features. Crystallization paths for CO_2-H_2O bearing magmas having the same initial H_2O content and temperature (5.5 wt.% H_2O, 800°C), but initial fluid/melt mass ratios of 1.0, 0.1 and 0.01, are shown on Fig. 5, together with the CO_2-free path, plotted for comparison. For a ratio of 1.0, the crystallization path is essentially vertical with a nearly constant melt H_2O content. The solidus temperature in this case is 40°C higher than that of the CO_2-free system (645°C, Table 1). For a ratio of 0.1, a significant departure from the CO_2-free path is observed once the melt fraction exceeds 20 wt.%, and the solidus is attained at a temperature of 652°C. For a ratio of 0.01, no significant deviations from paths for CO_2-free magmas can be seen. As shown on Fig. 6a, the melt fraction trends obtained in the presence of CO_2 depart significantly from the eutectic-like behaviour of CO_2-free systems, at ratios of 1.0 and 0.1. Similarly, the H_2O contents of melts differ from those for CO_2-free systems (Fig. 6), but only for an initial fluid/melt ratio in excess of 0.1. In all cases, as shown previously by Holloway (1976), H_2O-saturated conditions are never achieved in CO_2-bearing magmas (Fig. 6b).

Variations in melt viscosity are shown in Fig. 7a. A fluid/melt ratio of 0.01 does not affect the viscosity significantly compared to the CO_2-free system. A fluid/melt ratio of 0.1 suppresses the decrease in viscosity caused by H_2O that occurs in CO_2-free systems near the end of crystallization (Fig. 4). For a fluid/melt ratio of 1.0, there is a progressive increase in melt viscosity attributable to decreasing temperature, because the H_2O content of the melt remains constant during crystallization. However, the viscosity of the melt does not change by more than one order of magnitude over the full crystallization interval. Regarding the magma viscosity, fluid/melt mass ratios of 1.0 and 0.1 produce an increase in magma viscosity of 1 to 2 orders of magnitude after 70 % of cooling, whereas this increase occurs only after 95 % of cooling in the CO_2-free magma (Fig. 7b). Clearly, this is related to the greater decrease in melt fraction during cooling when a H_2O-CO_2 fluid phase is present (Fig. 6a). With a mass ratio of 0.01, results are essentially identical to the pure H_2O case.

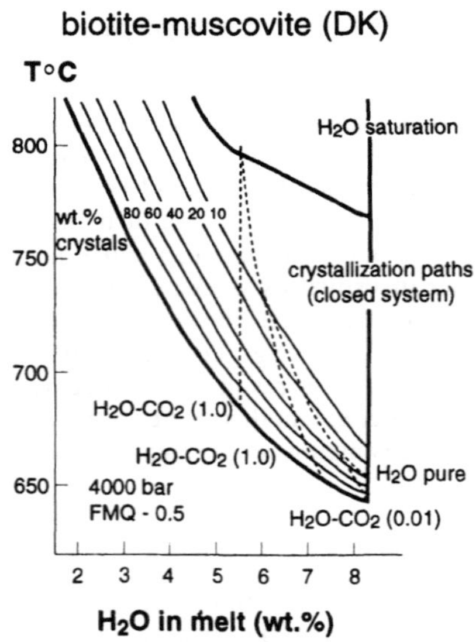

Figure 5. Crystallization paths for a CO_2-H_2O-bearing magma with initial fluid/melt mass ratios of 1.0, 0.1 and 0.01 (mass of CO_2+H_2O in the fluid over the mass of silicate melt including its H_2O content before the magma begins to crystallize). Also shown for comparison is the crystallization path for an H_2O-bearing, CO_2-free magma (see figure 2).

Therefore, depending on its initial fluid/melt mass ratio, a CO_2-H_2O-bearing magma will be able to evolve between two end-members. If the fluid mass is very small compared to that of the melt (ratio < 0.01), the amount of H_2O partitioned into the fluid phase during crystallization will be small. In this case the crystallization path is similar to that of the CO_2-free system. Also, the magma displays the largest possible crystallization interval (Fig. 5), with a temperature of final solidification close to that for an H_2O-bearing, CO_2-free magma.

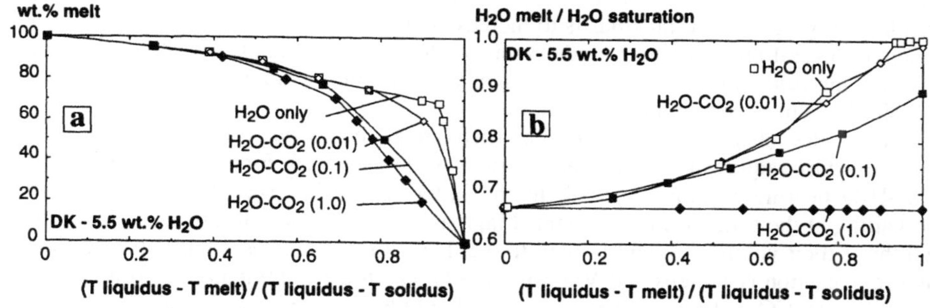

Figure 6. (a) Changes in melt fraction (wt.%) during crystallization of a CO_2-H_2O-bearing magma (DK granite) for different initial fluid/melt mass ratios. Note the departure from eutectic-like behaviour once the mass ratio is > 0.1. (b) Changes in the H_2O contents of corresponding residual melts. Note that, in all cases, H_2O saturation conditions are never achieved. To calculate H_2O contents, multiply by the corresponding initial melt H_2O contents listed in Table 1. Solidus temperatures are 685, 652 and 647 °C for initial fluid/melt mass ratios of 1.0, 0.1, and 0.01, respectively.

On the other hand, if the fluid/melt mass ratio is large (>0.5), the fluid composition does not change with crystallization, buffering the H_2O content of the residual melt to a constant value (by virtue of the small temperature dependence of H_2O activity in silicate melts). In this case, the final solidification temperature is approximatively given by the solidus temperature in the haplogranite system at a H_2O activity corresponding to that of the H_2O/CO_2 ratio in the fluid phase (see Ebadi and Johannes, 1991). In such a scenario, the granite will display the smallest crystallization interval (Fig. 5).

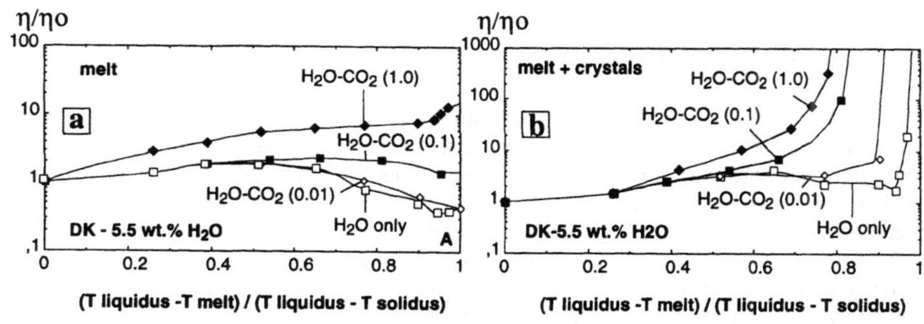

Figure 7. Changes in melt (a) and magma (b) viscosity versus temperature for the crystallization paths of figure 5. η_o is the initial viscosity of the melt (Table 1). Solidus temperatures are 685, 652 and 647 °C for initial fluid/melt mass ratios of 1.0, 0.1, and 0.01, respectively. Note that the presence of CO_2 increases both melt and magma viscosity, especially at high fluid/melt mass ratios.

It should be noted that the results obtained here are in all respects qualitatively similar to those obtained by Holloway (1976), using albite-H_2O-CO_2 as a model system for magmatic crystallization. This agreement validates the approach followed in this study and serves to emphasize the critical influence that CO_2 may have on the rheological properties of granitic magmas. In all cases, the effect of CO_2 on the rheological properties of magmas is to increase bulk viscosity, in some instances by as much as several orders of magnitude compared to hydrous magmas. It should be noted that this increase in bulk viscosity is probably a minimum estimate because the role of bubbles has not been taken into account.

Thus, the characterization of the rheological behaviour of a granite during crystallization requires two essential questions to be answered. First was it a H_2O- or a CO_2-H_2O bearing magma? Second, in the latter case, what was the fluid/melt ratio prevailing at the time of emplacement. Magmas generated under fluid-absent conditions will have their volatile content limited to that locked into the phases participating to the melting reaction (Clemens and Vielzeuf, 1987) which are dominantly hydrous (micas, amphiboles). CO_2 activity in these magmas should be low and conditions close to H_2O saturation should prevail at the end of crystallization (i.e., $aH_2O \sim 1$), as is observed for some peraluminous granites or associated volcanics (e.g., Pichavant et al., 1988). The rheological characteristics of these magmas will probably be similar to those of H_2O-bearing, CO_2-free magmas. On the other hand, fluid-present conditions cannot always be ruled out for the genesis of some crustal magmas (see Johnson et al., 1994). For instance, intrusions displaying mineralogical and textural features consistent with attainment of solidus conditions at $aH_2O < 1$ (e.g., Emslie and Stirling, 1993), are potential examples of magmas having crystallized in equilibrium with a large mass of H_2O-CO_2 fluid. Other examples include magmas generated in overthrust terranes where melting is triggered by large-scale fluid influx in a hot-slab-over-cold slab regime (e.g., Le Fort, 1981). There are numerous processes that can influence fluid compositions and abundances, and this makes it difficult to predict the rheological patterns of the latter category of granites. Nevertheless, they probably display a more viscous behaviour than magmas generated under fluid-absent conditions.

EFFECTS OF PRESSURE

The effects of pressure on melt viscosity can be viewed as either intrinsic or extrinsic. The former are structural changes in the melt produced by a pressure change, such as the change of Al coordination number (e.g., Kushiro, 1976), or the influence of pressure on the FeO/Fe_2O_3 ratio of the melt (Kress and Carmichael, 1992). However, these changes occur over a pressure range on the order of 3-4 GPa, which is far in excess of the pressure drop commonly undergone by most granitic magmas generated in the crust (usually less than 1 GPa). Intrinsic pressure effects on the viscosity of granitic melts are thus not anticipated. Extrinsic pressure effects are mainly exemplified by the increasing solubility of volatile components (H_2O, CO_2) in silicate melts with pressure. High pressure magmas have a greater ability to dissolve H_2O and, consequently, they have lower viscosities as indicated by the GB leucogranite. Another extrinsic influence of pressure, related to that of H_2O, is the effect on phase relations. This effect can be seen by examining isobaric "rock-water" phases diagrams for the same "rock" at various pressures (e.g., Whitney, 1975; Clemens and Wall, 1981). These diagrams show that, for a given melt H_2O content, the stability curves of the major phases rarely exhibit negative dP/dT slopes (Clemens and Wall, 1981). The first impression gained from this observation is that the crystal content of an adiabatically ascending magma should decrease. In detail, however, the amount of crystallization will depend on how fast the magma is transferred upward, i.e., on the efficiency of the mechanisms of heat transfer between the magma and surrounding country rocks (e.g., Marsh, 1982; Sykes and Holloway, 1987; Nekvasil, 1992; Holtz and Johannes, 1994). This problem is in turn related to the mode of magma ascent through the crust (dykes or diapirs), a question still hotly debated today (e.g., Clemens and Mawer, 1992; Weinberg and Podladchikov, 1994; Rubin, 1995). Although this issue cannot be directly addressed with isobaric phase diagrams, it should be noted that the results gathered from phase equilibria studies place strong constraints on the rheological state of the magma at the end of its crustal travel (i.e., at the level of emplacement). This state is characterized, in all the investigated cases, by a low to very low crystal content of the magma which suggests rapid ascent. Thus, although the restricted number of phase equilibria studies precludes any general conclusions about the mechanisms of ascent of granitic magmas, results obtained to date are consistent with dyking as the dominant mode of intrusion.

EFFECTS OF fO_2

Experimental work carried out on the RD granite under two different fO_2 (NNO+2 and FMQ-0.5) shows that fO_2 has a profound influence not only on the stability of ferro-magnesian phases but also on iron-free phases such as the tectosilicates (Dall'Agnol et al., 1994; Scaillet and Pichavant, 1994; Scaillet, Pichavant and Dall'Agnol, in prep.). Under reduced conditions (FMQ-0.5) and for a given melt H_2O content, the thermal stabilities of the tectosilicates are depressed by ca. 50-40°C relative to those under oxidized conditions (NNO+2, Fig. 2). Thus, because the tectosilicates make as much as 90-95% of the volume of the final granitic assemblage, any change in fO_2 conditions will strongly affect the rheological state of the magma. As a rule, for a given temperature and H_2O content, oxidized conditions correspond to a higher crystal/melt ratio than reduced conditions (Fig. 2).

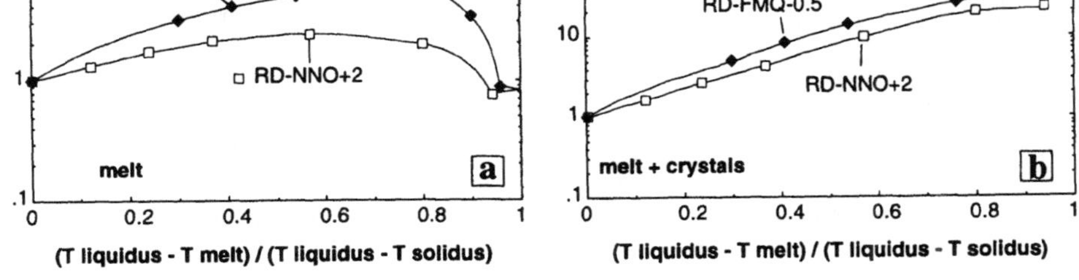

Figure 8. Changes in melt (a) and magma (b) viscosity versus temperature for the crystallization paths of the RD granite under reduced (RD-FMQ-0.5) and oxidized (RD-NNO+2) conditions (Figure 2). Note that magma viscosities are nearly identical for both sets of redox conditions. ηo is the initial viscosity of the melt (Table 1).

The crystallization paths obtained for RD FMQ-0.5 display the same characteristics as its oxidized counterpart (Fig. 2). The RD FMQ-0.5 melt has a slightly higher viscosity than that of RD NNO+2 (Fig. 8a), a fact due to the lower temperature of crystallization prevailing under reduced conditions (50°C, Fig. 2). However, when the crystal effect is taken into account, the two trends of magma viscosities are nearly identical (Fig. 8b).

A case in point is a magma that starts crystallizing under reduced conditions and becomes oxidized at a later stage of its evolution, a circumstance often called upon to account for some chemical and mineralogical trends reported for granites (e.g., Czmanske and Wones, 1973; Linnen et al., 1992; Scaillet et al., 1995a). For example, consider the emplacement of a magma containing 5 wt% H_2O in melt at 920°C under reduced conditions. Its early crystallization path will be that given in Fig. 2d. If, at a given temperature and H_2O content, for example 720°C and 6 wt.% H_2O, the redox state of magma shifts from FMQ-0.5 to NNO+2, then, in order to reequilibrate, the crystal/melt ratio of the magma must also shift, because this parameter depends on fO_2. If oxidation occurs with no gain or loss of H_2O, magma viscosity after oxidation at the same temperature will be identical to that for the magma that crystallized under oxidized conditions. In the example described above, the crystal content at 720°C would change from approximatly 30 wt.% at FMQ-0.5 to approximately 40 wt.% at NNO+2. However, as shown on Fig. 8b, this does not introduce a major change in bulk viscosity, because the effects of increasing crystal and melt H_2O content (from 6 to 7 wt.%) cancel each other. On the other hand, an entirely different result is obtained for a system where the total H_2O content of the melt may vary. The most extreme and interesting case is when the melt H_2O content remains constant during oxidation, being buffered by some external or internal mechanism. In such a situation, fixing both temperature and melt H_2O content during oxidation necessarily imposes a strong crystal fraction increase, from 30 wt.% up to 60 wt.% in the specific example described above, thereby dramatically affecting magma viscosity.

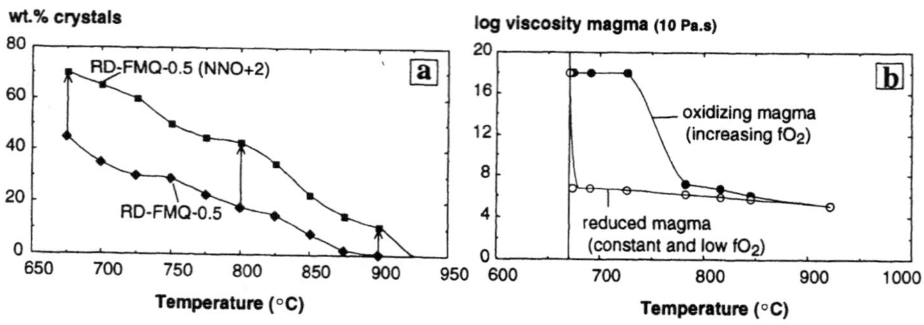

Figure 9. Effects of oxidation at constant melt H_2O content on (a) melt fraction and (b) the magma viscosity, for the RD granite. Note that, at temperatures below ~ 775 °C, oxidation increases the magma viscosity by several orders of magnitude.

The increase in crystal content during oxidation at constant melt H_2O content increases progressively from 15 wt.% at 900°C, up to 30 wt.% at about 725°C with an average increase of 20 wt.% (Fig. 9a). The effect of this change in crystal content in terms of bulk viscosity is shown on Fig. 9b, where the increase in magma viscosity arising from oxidation at constant melt H_2O content is illustrated for different temperatures. It is clear that, under these conditions, a magma undergoing oxidation may undergo huge variations in viscosity, in contrast to closed system behaviour. Significantly, magmas undergoing oxidation can cross the rheological critical melt percentage (RCMP) at much higher temperatures than similar magmas that crystallize under constant redox conditions. As a result, ascending magmas undergoing progressive oxidation may congeal at deeper levels and at various times depending on the rate of the oxidation event.

Although in the preceeding discussion we have considered only instantaneous variations in redox state it is clear that these changes can also occur progressively throughout the crystallization interval. The final result is the same. In view of the important effects of redox state on the rheological properties of granitic magmas on their rheological behaviour, the next section describes the redox mechanisms most commonly invoked for magmatic systems and compares the relative rates of magma ascent, cooling and oxidation.

RATES OF OXIDATION MECHANISMS *vs.* RATES OF ASCENT AND COOLING

The oxidation mechanisms considered are (a) H_2 diffusive loss (Sato, 1977), (b) fractionation of an H_2O-H_2 fluid phase (Candela, 1986), and (c) fractionation of Fe^{2+} rich phases (e.g., Carmichael and Ghiorso, 1990; Carmichael, 1991).

All three mechanisms are based on the assumption that the redox state of the magma is fixed by the following coupled equilibria:

$$2FeO_{melt} + 1/2O_2 \longleftrightarrow Fe_2O_{3\,melt} \quad (3)$$

$$H_2O \longleftrightarrow H_2 + 1/2\,O_2 \quad (4)$$

These reactions show how mechanisms (a), (b) and (c) will affect the redox state of the magma. Mechanisms (a) and (b) imply an open system with loss of either H_2 or H_2O-H_2 which would drive equilibria (3) and (4) toward the right, while mechanism (c) depletes the melt in Fe^{2+} and increases its Fe^{3+}/Fe^{2+} ratio (Carmichael, 1991). The rates of these three mechanisms are significantly different. Mechanism (a) is related to the rate of diffusion of H_2 in silicate melts, the driving force being the difference in chemical potential of H_2 between the magma and its host rock. This hypothesis can be evaluated by calculating the characteristic time needed for H_2 to diffuse out of a 10 m thick (d) dyke or 2 km sized diapir, using the H_2 diffusivity (k) given by Watson (1994) and the following equation:

$$t = d^2/k. \quad (5)$$

Mechanisms (a) and (b) are affected by the fluid-magma or crystal-magma density contrasts. Their rates can be obtained using the Stokes' law:

$$V = 2\,g\,r^2\,(d_{fluid/crystal} - d_{melt})\,/\,9\,\eta \qquad (6)$$

where V is the settling (crystal) or rising (fluid bubble) terminal velocity, g is the gravitational constant, r is the radius of the bubble or crystal, $d_{fluid/crystal}$ is the density of the fluid or crystal, d_{melt} is the melt density, and η is the melt viscosity. Calculations were performed using average values of 10000 Pa.s (4 in log units) and 2.3 g/cm^3 for the melt viscosity and density, respectively. The range of fluid density (0.4 - 1 g/cm^3) corresponds to fluid compositions under normal crustal conditions (1-10 kbar, C-O-H system and an fO_2 between FMQ-0.5 and NNO+2). Crystal sizes (1-2 mm) are those reported for many plutonic intrusions (Brandeis and Jaupart, 1987) with crystal densities varying between 3 and 5 g/cm^3 (e.g., biotite, amphibole, ilmenite). Finally, the rate of dyke propagation was obtained from Clemens and Mawer (1992) and Petford et al. (1993). For granitic diapirs we used the fastest rate determined in the numerical simulations performed by Mahon et al. (1988).

TABLE 2. Comparison between rates of magma ascent and rates of three oxidation processes.

	diapir	dyke	crystal settling	bubble ascent	H_2 loss*	
					1000	10
rate (m/year)	0.3	31000	0.4-1.0	1000	-	-
time (year)	-	-	-	-	31000	3.1

* time needed for H_2 to diffuse out of 10 m thick dyke or out of a 2000 m sized diapir (half thickness of 1000 m) is given.

The results are listed in Table 2. Clearly, a magma ascending along dykes is insensitive to any oxidizing process, rates of dyke propagation being 2 to 5 order of magnitude faster than any of the oxidation mechanisms considered. In contrast, granitic diapirs, which rise through the crust very slowly, are highly sensitive to nearly all oxidation processes. Each of the mechanisms can be shown to be operative under specific circumstances. For instance, H_2 diffusive loss may occur in the case of a reduced magma passing through oxidized levels in the crust (i.e., hematite-bearing rocks). One may then expect to find a reduced halo centered around the intrusion, corresponding to the zone affected by H_2 escape from the granitic magma at the time of H_2 diffusive loss. Given that rates of bubble rise are four orders of magnitude greater than that of diapiric ascent (Table 2), fluid fractionation appears to be the most powerful mechanism for oxidizing a magma. However, this can work only if the melt is fluid-saturated. In a case of an H_2O-bearing magma, such a situation occurs only very late in the crystallization history (Fig. 3b), which considerably limits the mechanical consequences of the oxidation process. In contrast, in CO_2-H_2O-bearing magmas, fluid saturation occurs very early, and oxidation of the magma via upward fluid migration is possible. Although there are no data on Fe^{2+}/Fe^{3+} partitioning between biotite or amphibole phases and granitic melt, the large stability fields reported for both of these phases in most phase equilibria diagrams for felsic systems (e.g., Clemens and Wall, 1981; Naney, 1983; Scaillet et al., 1995a), indicate that crystallization of these minerals may induce oxidation of the magma. Of course, removal of phases such as ilmenite could also have a significant effect on the redox state of the magma. It should be noted that the last two mechanisms (fluid migration and crysatllization of hydrous, iron-bearing minerals) imply that, during the oxidation process, the system undergoes a depletion in its bulk H_2O content. In particular, fractionation of an H_2-H_2O-rich fluid phase would reduce the H_2O content of the residual melt and possibly prevent build-up of H_2O resulting from crystallization of anhydrous phases. Thus oxidation of granitic magma at constant melt H_2O content is possible.

Concerning the relative rates of cooling and oxidation, there is no general answer given the complexities of the fluid dynamics at work in a magma chamber. However, a back-of-the-envelope calculation can be made by considering the time scale of conductive cooling for a normal sized granite (4 km thick), because recent studies have shown that convective motions in granitic magma chambers are probably weak (Brandeis and Jaupart, 1986). The cooling time (t) of such a granite intrusion can be calculated using equation (5), k being the thermal diffusivity of the magma and d its half thickness. With $k = 7\ 10^{-6}\ m^2/s$ (e.g., Brandeis and Jaupart, 1986), the calculated cooling time is about 10^5 years, suggesting that the aforementioned oxidation processes have sufficient time to operate during the solidification of most granitic plutons. Thus, the rheological properties (i.e., the fluid dynamics) of granitic magmas at their emplacement level will also depend on variations in fO_2.

CONCLUSIONS

This study has shown how phase equilibria diagrams can be used to constrain the rheological behaviour of crystallizing granitic magmas. For the first time it is shown that variations in fO_2 can have important effects on the rheological properties of granitic magmas. The results obtained from such an approach clearly demonstrate that changes in the viscosity evolution of a granitic magma are strongly dependent on its initial H_2O content. The abundances of other less soluble volatiles, such as CO_2, are also of major importance in determining the rheology of the crystallizing granite. In particular, melt fraction trends are significantly affected by the introduction of CO_2 into the system. This has important implications for strain analyses carried out on igneous intrusions. Magmas displaying a eutectic-like behaviour (Fig. 3a), will develop most of their mineral fabrics very late in the congealing period. In contrast, H_2O-CO_2-bearing granitic magmas may have melt fraction trends much more linear with respect to temperature, with a significant amount of crystallization occurring over a much wider temperature interval. Thus, the mineral fabric of such intrusions can record a larger fraction of the total amount of strain undergone by the magma than that recorded by CO_2-poor granites. Nevertheless, it is clear that in both cases (H_2O- and H_2O-CO_2 bearing magmas), less than 10-15% of the crystallization occurs during the first 50% of the crystallization interval (Fig. 3a, 6a). Therefore, given the inference of crystal-poor-melts at the time of intrusion, the mineral fabrics preserved in granites probably do not reflect the intrusion mechanism. Similarly, the melt fraction trends obtained in this study show clearly that the release of latent heat from granitic magmas cannot be considered as occurring uniformly across the crystallization interval (e.g., Lange et al., 1994). In addition, the pattern of release of latent heat may vary between magmas of similar composition but with different initial fluid/melt ratios. This implies differences in cooling rates. Accordingly, shape preferred fabrics preserved in granites may indicate crystallization over wide temperature ranges. Interpretation of these data in terms of strain rates in granitic magmas are therefore not straightforward.

AKNOWLEDGMENTS

This study was partly funded by grants of the IDYL research program of the INSU-CNRS. We aknowledge the very thorough reviews of J.P. Hogan and J. Blencoe that greatly improved the quality of the manuscript.

APPENDIX: ISOBARIC CRYSTALLIZATION PATHS IN GRANITIC SYSTEMS

Before calculating isobaric crystallization paths, melt fraction contours for the four phase diagrams in Figure 2 were obtained by mass balance calculations and graphical analysis. Mass balance calculations were performed mainly for the melt-rich field. For the melt-poor region, a graphical analysis was done, based on the following assumptions. First, all along the solidus curve, the modal composition of the solidified granite is considered to be equal to the modal analysis of the natural rock. Second, only quartz and feldspar are taken into account in this calculation, because these phases make up more than 95 wt.% of the total final assemblage. Third, for each of these phases, it was assumed that, at a given temperature, mass fraction varies linearly between 0.01 wt.% at the liquidus and its modal abundance at the solidus. The third assumption was tested whenever possible by mass balance calculation, and both methods agreed within a few wt.%. When H_2O is the sole volatile present, the basic assumption used to determine crystallization paths from phase equilibrium diagrams is that H_2O behaves as a perfect incompatible component during crystallization in a closed system. This is because it can be shown that the amounts of H_2O taken up by the hydrous phases are negligible (very low amount of hydrous crystals at high temperature (< 1wt.%), larger at low temperatures but largely counterbalanced by the effect of anhydrous phases). Therefore, fixing an initial melt H_2O content and temperature permits a crystallization path to be calculated using the mass balance constraint:

$$H_2O_{tot} = H_2O_{melt} \quad (1)$$

When an H_2O-CO_2 fluid phase is present during crystallization, the crystallization paths obtained are as follows: first it is assumed that equilibrium in H_2O activity (aH_2O) between the melt and the fluid phase is maintained throughout the crystallization interval. Second, the H_2O solubility model of Burnham (1979) was used to relate aH_2O and wt.% H_2O in the melt (using the bulk composition), with the assumption that CO_2 does not affect H_2O solubility (see Blank et al., 1993). The fluid phase composition was assumed to be a H_2O-CO_2 binary mixture (a valid assumption for fO_2 higher than FMQ, e.g., Holloway, 1987; Scaillet et al., 1995a) and values for aH_2O in the fluid phase were obtained from the MRK equation of state of Kerrick and Jacobs (1981). Third, it was assumed that all CO_2 remained in the fluid phase, a reasonable assumption considering the very low solubility of CO_2 in granitic melts at pressures prevailing during emplacement of most granitic magmas (< 4 kbar). The calculation was performed in stepwise fashion. First the initial fluid/melt ratio was fixed at liquidus conditions (i.e., no crystals present), for a given wt.% H_2O and temperature. Then assuming $aH_2O_{melt} = aH_2O_{fluid}$ it was possible to calculate the mass of H_2O in both the fluid phase (H_2O_{fluid}) and melt (H_2O_{melt}) which gives the total amount of H_2O (and CO_2) in the system (H_2O_{tot}, CO_{2tot}). Consequently, for any given temperature below the liquidus, the equilibrium melt fraction is that established by $aH_2O_{melt} = aH_2O_{fluid}$, using the mass balance constraint $H_2O_{tot} = H_2O_{melt} + H_2O_{fluid}$ and $CO_{2tot} = CO_{2fluid}$. The starting point of this step is the melt fraction/wt.% H_2O value for the CO_2-free system, with subsequent calculations completed by decreasing the wt.% H_2O of the melt (i.e., by increasing the amount of crystallization). The end point of the crystallization path is the crossing point on the solidus curve. At the solidus all the volatiles are in the fluid phase, so the molar fraction of H_2O (and thus aH_2O) is given by H_2O_2 /(H_2O_{tot} + CO_{2tot}). Therefore, knowing aH_2O fixes the solidus temperature, here taken to be that of the haplogranite system (Ebadi and Johannes, 1991).

WETTING ANGLES, EQUILIBRIUM MELT GEOMETRY, AND THE PERMEABILITY THRESHOLD OF PARTIALLY MOLTEN CRUSTAL PROTOLITHS

Didier LAPORTE, Cédric RAPAILLE and Ariel PROVOST
Université Blaise Pascal and CNRS, URA n° 10,
5, rue Kessler, F-63038 Clermont-Ferrand Cedex, France

ABSTRACT

The permeability of a partially molten rock is one of the primary factors controlling the melt segregation rate. With decreasing melt percentage, permeability becomes increasingly sensitive to the grain-scale geometry of partial melt. At low melt percentages, the ratio of grain-boundary energy to solid-melt interfacial energy, $[\gamma_{ss}]/[\gamma_{sl}]$, is the fundamental physical property that determines the equilibrium melt geometry, including the wetting angle θ at a solid-solid-melt triple junction, the area-to-volume ratio *s/v* of melt pockets at grain corners, and the permeability threshold ϕ_c (ϕ_c is the volume melt percentage at which melt interconnection is established). The trends of increasing θ and ϕ_c or decreasing *s/v* with decreasing $[\gamma_{ss}]/[\gamma_{sl}]$ are well established in the case of a monomineralic system with isotropic interfacial energies. We argue that these general trends must apply as well in natural systems where solid-melt interfacial energies are essentially anisotropic.

Measurements of wetting angles at quartz-quartz-melt, feldspar-feldspar-melt and amphibole-amphibole-melt triple junctions consistently yielded low to very low median values, in the range 10°-60°. These low angles result from high values of $[\gamma_{ss}]/[\gamma_{sl}]$, up to 2.0 for the lowest angles. They indicate that the permeability thresholds of partially molten crustal protoliths should be very low: < 1 vol. % to a few vol. %. This result is confirmed in a series of melt-infiltration experiments in which a hydrous granitic melt was placed in contact with a texturally-equilibrated, polycrystalline aggregate of quartz, at 900 °C-1 GPa. Secondary electron imaging of the quartzite close to the melt reservoir revealed the presence of an interconnected network of grain-edge melt channels at a melt percentage < 0.04 vol. %.

The major implication of the experimental studies is to show that melt segregation may potentially operate at very low degrees of melting (theoretically, any value ≥ ϕ_c). Because of the high viscosities of granitic melts, melt segregation is presumed to be inefficient at such low degrees of melting. There may exist therefore a range of melt percentages above ϕ_c over which the partial melt is interconnected but nearly stagnant.

J.L. Bouchez et al. (eds.) Granite: From Segregation of Melt to Emplacement Fabrics, 31–54.

INTRODUCTION

The dynamics of melt segregation from crustal protoliths is poorly understood (e.g., Brown et al., 1995). The permeability of partially molten rock is, along with the viscosities of partial melt and its crystalline matrix, one of the main physical properties that control the segregation rate. At moderate melt percentages (say > 10 vol. %), the permeability, k, is primarily a function of grain size, a, and volume melt percentage, ϕ (McKenzie, 1989). In the special case of a granular, partially molten rock-type (e.g., a protolith with abundant quartz, feldspar and garnet in the solid residue), the variation of permeability with melt percentage may conveniently be approximated by the permeability-porosity relationships in sandstones. In particular, high values of permeability, ranging from $5\ 10^{-12}\ m^2$ ($a \sim 1$ mm, $\phi \sim 10$ vol. %) to $10^{-9}\ m^2$ ($a \sim 3$ mm, $\phi \sim 30$ vol. %) may be inferred from measurements in Fontainebleau sandstones (see Laporte, 1994). A much more complex behaviour may, however, be anticipated at low melt percentages ($\phi \leq 10$ vol. %):

(1) First, permeability can no longer be expressed as a simple function of ϕ and a. With decreasing melt percentage, permeability becomes increasingly sensitive to the grain-scale geometry, including the frequency of dry grain edges (i.e., direct contact of 3 grains at a triple junction), the tortuosity of melt channels, and the minimum channel cross-sections. Because very different grain-scale melt geometries may prevail depending on the mineralogical make-up of the solid residue, the k-ϕ relationship is expected to be different from one protolith to the other.

(2) Second, the partially molten system may be impermeable below a critical value of the melt percentage, labelled ϕ_c, and thereafter referred to as the permeability threshold (Maaløe, 1985), or the interconnection threshold. ϕ_c is the melt percentage at which melt interconnection is established; below ϕ_c, the partial melt occurs as isolated pockets and is therefore unable to separate from the solid residue. The establishment of an interconnected melt network at ϕ_c results in a dramatic increase of permeability. For example, von Bargen and Waff (1986) calculated that k increases from 0 up to $10^{-13}\ m^2$, as ϕ increases by only a few 0.1 vol. % above ϕ_c.

Experimental and theoretical studies of the grain-scale geometry of melt in partially molten crustal protoliths are few. Early studies by Mehnert et al. (1973) and Busch et al. (1974), and more recent investigations by Hacker (1990), Wolf and Wyllie (1991) and Rubie and Brearley (1991), address the case of disequilibrium melt textures, where the spatial distribution of melt and its geometry are primarily controlled by the location of the reactants and by the volume change on melting. Studies of Jurewicz and Watson (1984, 1985), Laporte (1994) and Laporte and Watson (1995) are, in turn, devoted to equilibrium melt textures, in which the melt geometry at the grain scale is dictated by the principle of interfacial energy minimization. Experiments under non-hydrostatic conditions by van der Molen and Paterson (1979), Dell'Angelo and Tullis (1988), Rushmer (1995) and Rutter and Neumann (1995) provide an insight into the effect of stress on partial melt geometry.

The main purpose of this work is to present a synthesis of equilibrium melt geometries in crustal protoliths. We first summarize the theoretical aspects of textural equilibrium in partially molten systems, then review the experimental studies of the wetting properties of partial melts in crustal protoliths. New experimental results demonstrating that interconnection of a hydrous granitic melt in a polycrystalline aggregate of quartz is established at $\phi < 0.04$ vol. % are summarized. Finally, the implications of the wetting data for melt interconnection at low melt percentage, and the segregation of granitic melts are discussed.

EQUILIBRIUM MELT GEOMETRY AT LOW MELT PERCENTAGE: THEORETICAL BACKGROUND

THE IDEAL MODEL: A SINGLE MINERAL SPECIES WITH ISOTROPIC INTERFACIAL ENERGIES

In a partially molten system, at fixed pressure and temperature, textural adjustments such as rearrangement of grains, interparticle welding, grain growth, and change in grain shape, occur to reduce the total interfacial energy. With time, melt distribution ultimately approaches an equilibrium configuration that corresponds to a minimum total interfacial energy per unit volume. In the special case of a monomineralic, partially molten rock with single-valued solid-solid and solid-liquid interfacial energies, that is subject to hydrostatic stress, the equilibrium melt distribution is characterized by (Smith, 1964; Beere, 1975; Bulau et al., 1979; von Bargen and Waff, 1986; Hunter, 1987):

(1) a constant wetting angle θ (or dihedral angle; see Fig. 1a), whose value is determined by the ratio of solid-solid interfacial energy γ_{ss} (or grain-boundary energy) to solid-liquid interfacial energy γ_{sl} (or surface energy):

$$2\cos\left(\frac{\theta}{2}\right)=\frac{\gamma_{ss}}{\gamma_{sl}} \qquad (1)$$

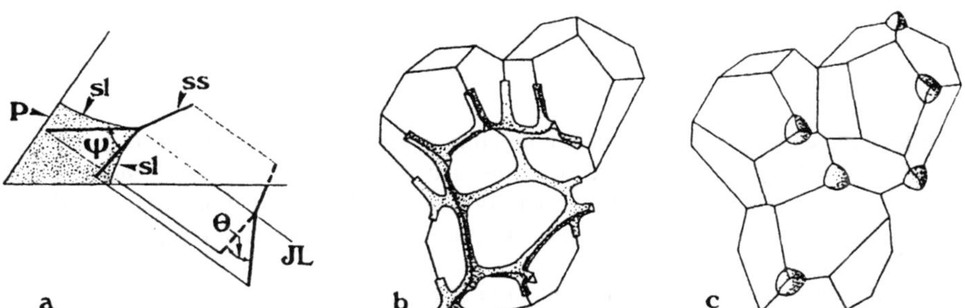

Figure 1. Equilibrium melt distributions in an ideal partially molten system, at low melt percentage (from Laporte, 1994, Fig. 1, p. 487, reproduced here by kind permission of Springer-Verlag, Heidelberg). *a*: Sketch showing melt (stippled) in contact with two crystals at a grain boundary ss; the apparent wetting angle Ψ, in the plane of polished section P, is defined by the tangents to the grain surfaces sl at their point of intersection; the true wetting angle θ is measured in a plane perpendicular to the grain-grain-melt junction line JL. *b*: for $\theta < 60°$, melt forms an interconnected network of grain-edge channels. *c*: for $\theta \geq 60°$ (and melt percentage less than ϕ_c), melt forms isolated pockets at grain corners.

(2) a constant mean curvature of solid-liquid interfaces (the mean curvature is defined at any point of the interface as the quantity $1/2[1/R_{min}+1/R_{max}]$, where R_{min} and R_{max} represent the two principal radii of curvature. In the following, the sign of curvature is considered positive if the center of curvature lies on the liquid side of the interface, and negative in the other case; positively and negatively curved interfaces are referred to as convex and concave, respectively).

Two contrasted types of equilibrium melt geometry at low melt percentage result from the aforementioned requirements: (1) if $0° \leq \theta < 60°$, the melt forms an interconnected network of channels along grain edges (Fig. 1b), and melt inter-connection is theoretically achieved for an infinitely small melt percentage; and (2) if $\theta \geq 60°$, the melt occurs either as isolated pockets at grain corners (Fig. 1c) if the melt percentage ϕ is lower than the critical value ϕ_c, or as an interconnected network of grain-edge channels if $\phi \geq \phi_c$. The permeability threshold ϕ_c increases from ~ 0.6 vol. % at $\theta = 60°$ to ~ 27.8 vol. % at $\theta = 180°$ (Fig. 2; Wray, 1976; Bulau et al., 1979; von Bargen and Waff, 1986).

That two contrasting geometries exist at low melt percentage may be demonstrated by considering the equilibrium shape of a melt pocket at a grain corner in a polycrystalline aggregate (Fig. 3). The equilibrium shape is determined by a balance of surface-tension forces at the point O where the melt is in contact with 3 grains (Fig. 3b). In the special case where all three grains are of the same mineral and interfacial energies are independent of orientation, the equilibrium condition may be derived as follows. Interfacial energy minimization along the four line junctions radiating from O requires that (1) the grain boundaries that meet along the 3-grain line junction (sss in Fig. 3) make dihedral angles of 120° and (2) the wetting angle θ is the same along the 3 solid-solid-melt line junctions (ssl in Fig. 3b) and is given by equation (1). In addition, symmetry considerations require that all 3 ssl junctions make the same angle β with the 3-grain junction (Fig. 3b). These geometrical requirements yield the simple equilibrium condition at O (e.g., von Bargen and Waff, 1986, p. 9263):

$$\cos\beta = -\left(\sqrt{3}\tan\frac{\theta}{2}\right)^{-1} \tag{2}$$

Equation (2) shows that the point junction O can only be stable if $\theta \geq 60°$. For $\theta < 60°$, equation (2) has no real solution, and a melt domain cannot pinch off anywhere along a 3-grain line junction. Therefore, the equilibrium geometry is such that all 3-grain line junctions have been replaced by continuous melt channels, as depicted in Fig. 1b.

For $\theta \geq 60°$, the melt geometry at O is strongly dependent on θ: for θ increasing from 60° to 180° (that is, γ_{ss}/γ_{sl} decreasing from 1.732 to 0), β decreases from 180° to 90°. This results in a dramatic evolution of the shape of an isolated melt pocket at a grain corner (see Bulau et al., 1979). For $60° \leq \theta < 70.53°$, the melt pocket is bounded by concave solid-melt interfaces, at $\theta = 70.53°$ (= arccos1/3), the melt pocket is a regular, plane-faced tetrahedron, and for higher wetting angles, the melt pocket is bounded by convex solid-melt interfaces as depicted in Fig. 3b. The final term of this evolution is a spherical melt pocket at $\theta = 180°$. This trend corresponds to a significant decrease of the area-to-volume ratio of melt pockets with decreasing γ_{ss}/γ_{sl}, as discussed below.

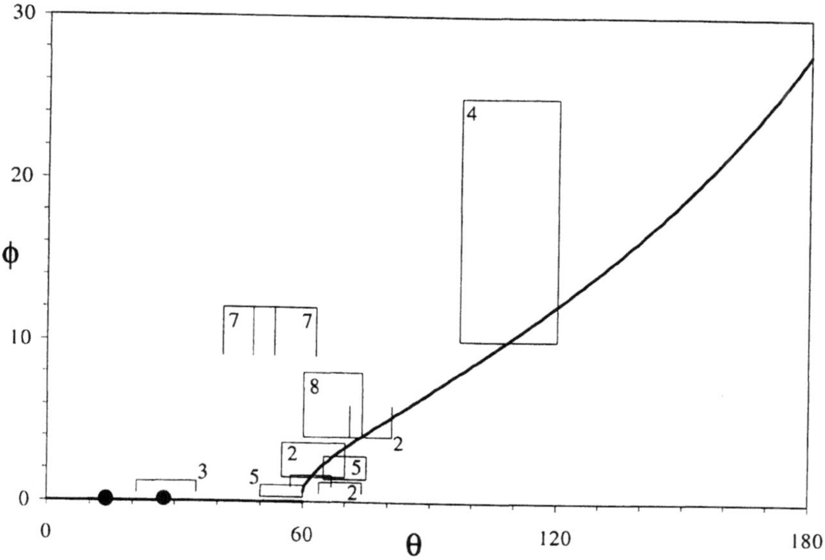

Figure 2. Permeability threshold, ϕ_c (in volume percent), as a function of the wetting angle, θ, in ideal partially molten systems (thick line; corrected from Wray, 1976). Experimentally determined permeability thresholds in fluid-bearing rocks are shown by the boxes and the filled circles (see Table 1 for numerical data; the labels correspond to the reference numbers in Table 1). Box widths correspond to the estimated error in wetting-angle measurements; the upper and lower sides of a box give the upper and lower limits of ϕ_c, respectively; an open box means that only an upper limit or a lower limit has been determined. The filled circles (±5° error in wetting-angle measurements omitted for clarity) are the data for carbonate melt in polycrystalline olivine (Minarik and Watson, 1995), and granitic melt in polycrystalline quartz (this study).

For $\theta \geq 60°$ and $\phi < \phi_c$, a melt pocket may be modelled as a spherical tetrahedron (Fig. 3b; e.g., von Bargen and Waff, 1986). The area-to-volume ratio of a pocket, s/v, may therefore be expressed analytically (see Clemm and Fischer, 1955, and Wray, 1976, though their expressions are incorrect near $\theta = 70.53°$), and is a function of the shape and size of the melt pocket. The size effect may be eliminated by calculating the dimensionless ratio $s/v^{2/3}$. For a given value of $\theta \geq 60°$, the shape of the melt pocket remains constant within the range of melt percentages 0 to ϕ_c, and $s/v^{2/3}$ depends only on θ, that is, on γ_{ss}/γ_{sl}. $s/v^{2/3}$ is plotted as a function of γ_{ss}/γ_{sl} in figure 4 (curves for $\theta < 60°$ in Fig. 4 were obtained from von Bargen and Waff's (1986) numerical calculations; note that, for $\theta < 60°$, $s/v^{2/3}$ is a function of both γ_{ss}/γ_{sl} and the melt percentage). Fig. 4 shows a large increase of $s/v^{2/3}$ with increasing γ_{ss}/γ_{sl} . In other words, with increasing γ_{ss} or decreasing γ_{sl}, a given volume of melt is accommodated by creating more and more solid-melt interfacial area. For a melt percentage of 1 vol. %, for instance, the solid-melt interfacial area at $\gamma_{ss}/\gamma_{sl} = 2$ is larger than at $\gamma_{ss}/\gamma_{sl} = 1$ by a factor of 3.2.

For the ideal model, we conclude that the ratio γ_{ss}/γ_{sl} is the fundamental physical property controlling melt distribution at low melt percentages, including the wetting angle at a solid-solid-melt triple junction (equation [1]), the shape of a melt pocket at a grain corner (equation [2]; Figs. 3-4), and the permeability threshold (Fig. 2). We emphasize in the following section that the ratio of grain-boundary energy to solid-melt interfacial energy is also the primary factor controlling the geometry of partial melts in non-ideal systems.

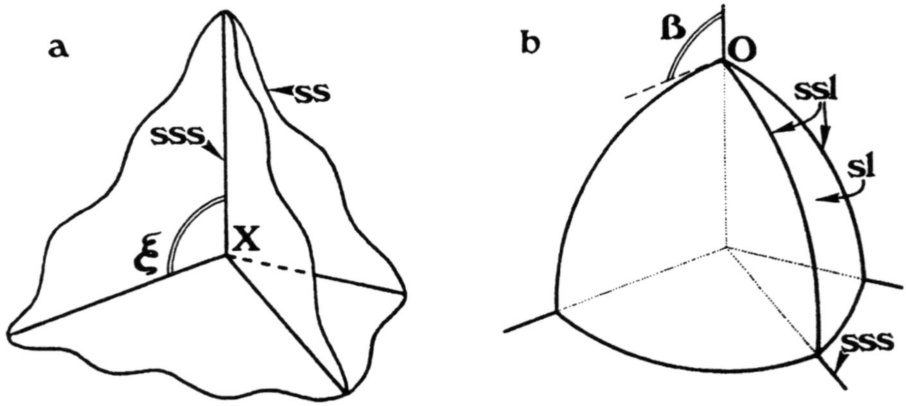

Figure 3. a: Sketch of a grain corner in a polycrystalline aggregate. 4 grains are in contact at the vertex junction X (ss: grain boundary; sss: 3-grain line junction). In the case where all 4 grains are of the same mineral, and grain-boundary energy is isotropic, the vertex junction has a tetrahedral symmetry, and the angle between any two 3-grain line junctions radiating from X is $\xi = 109.47°$. b: Equilibrium shape of a melt pocket at a grain corner (sl: solid-melt interface; ssl: solid-solid-melt line junction); if solid-melt interfacial energy is isotropic, the melt pocket is a spherical tetrahedron: it has a tetrahedral symmetry, and is bounded by four equivalent spherical surfaces, either convex (b, for $\theta > 70.53°$), or concave (not shown, for $60° \leq \theta < 70.53°$).

EFFECTS OF CRYSTALLINE ANISOTROPY ON MELT INTERCONNECTION

The assumptions of a single-valued γ_{sl}, and to a lesser extent, of a single-valued γ_{ss} are inadequate for partially molten systems of geological interest. For most silicates, interfacial energies are significantly anisotropic, that is, the interfacial energy per unit area depends on the orientation of the interface relative to the crystalline lattice (Cooper and Kohlstedt, 1982; Laporte and Provost, 1994). The major effect of anisotropy that one may anticipate is the development of facets parallel to the crystallographic planes of minimum γ_{sl} (e.g., Wortis, 1988). This texture is confirmed by the experimental studies of Waff and Faul (1992), Laporte (1994) and Laporte and Watson (1995). The equilibrium textures in these experiments are characterized by the repeated occurrence of crystallographically-controlled silicate-melt facets. The development of planar solid-melt interfaces is a marked deviation from the ideal model that predicts smoothly curved interfaces with a constant mean curvature.

The effects of anisotropy on the equilibrium melt geometry at low melt percentage were discussed in detail by Laporte and Watson (1995). Their conclusions are as follows:

(1) At a high degree of γ_{sl} anisotropy and a low melt percentage, melt should form isolated, plane-faced pockets at grain corners.

(1) At a high degree of γ_{sl} anisotropy and a low melt percentage, melt should form isolated, plane-faced pockets at grain corners.

(2) The overall shape of these pockets, and as a consequence, the value of the permeability threshold ϕ_c, depend primarily on the ratio of solid-solid to solid-liquid interfacial energies, $[\gamma_{ss}]/[\gamma_{sl}]$, where $[\gamma_{ss}]$ and $[\gamma_{sl}]$ are weighted averages of the anisotropic γ_{ss} and γ_{sl} (see Laporte and Watson, 1995, p. 178, for a mathematical expression of $[\gamma_{sl}]$). Increasing $[\gamma_{ss}]$ or decreasing $[\gamma_{sl}]$ favours the extension of solid-melt interfaces at the expense of grain boundaries, and should therefore result in a higher area-to-volume ratio of melt pockets and a lower permeability threshold.

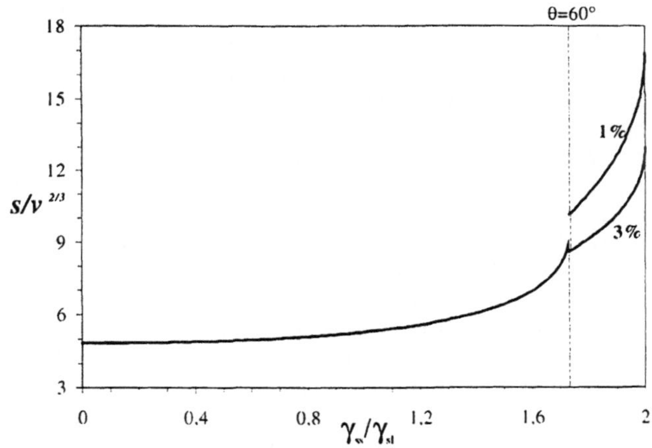

Figure 4. Dimensionless ratio $s/v^{2/3}$ plotted as a function of γ_{ss}/γ_{sl}. s and v are the area and the volume of a melt pocket, respectively. For $\theta < 60°$, $s/v^{2/3}$ depends on both γ_{ss}/γ_{sl} and the melt percentage (curves for $\phi = 1$ vol. % and $\phi = 3$ vol. % are shown for comparison; data for $\theta < 60°$ were obtained from von Bargen and Waff's [1986] numerical calculations). The curve for $\theta \geq 60°$ applies to any melt percentage lower than ϕ_c.

(3) Melt pockets with a high area-to-volume ratio, low but non-constant wetting angles, and a low value of ϕ_c should prevail at high $[\gamma_{ss}]/[\gamma_{sl}]$; and vice-versa, low values of $[\gamma_{ss}]/[\gamma_{sl}]$ should result in a low area-to-volume ratio of melt pockets, high wetting angles, and higher values of ϕ_c.

Most importantly, the ratio of grain-boundary energy to solid-melt interfacial energy is the main parameter that dictates melt geometry at low melt percentage, as in the ideal model (Fig. 4). Specifically, the general trends of increasing θ and ϕ_c or decreasing $s/v^{2/3}$ with decreasing γ_{ss}/γ_{sl} (equations [1-2], Figs. 2-4) should hold in the natural case of anisotropic interfacial energies. Experimental data that may be used to constrain the values of ϕ_c in anisotropic systems are discussed in the next section.

EXPERIMENTAL CONSTRAINTS ON THE PERMEABILITY THRESHOLDS IN ANISOTROPIC SYSTEMS

Experimentally determined permeability thresholds, $\phi_{c,meas}$, for fluid-bearing rocks are listed in Table 1. For each measurement, an ideal value of ϕ_c, $\phi_{c,ideal}$, is given for comparison. This value was computed from the wetting angle in column no. 2, with the assumption that the solid-fluid system behaves ideally (the relationship between ϕ_c and θ in an ideal system is shown in figure 2).

Most measurements in Table 1 were obtained by diffusion experiments. The presence of an interconnected fluid phase (partial melt or supercritical C-O-H mixture) in a rock results in a marked enhancement of diffusive mass transport because fluid diffusion coefficients are several orders of magnitude greater than solid diffusion coefficients (e.g. Watson, 1991). The transition from isolated fluid pockets to a continuous network of fluid-filled channels at ϕ_c should therefore correspond to a dramatic increase of bulk-rock diffusion coefficient, as observed by Watson and Lupulescu (1993) in a series of diffusion experiments in water-bearing clino-pyroxenites at 1500 MPa and 950 °C. In these diffusion-couple experiments, a cylinder of Fe-rich clinopyroxenite was placed against another of Mg-rich clinopyroxenite, and run with various water contents (0, 0.6, 1.3, 2.7, 4.0 and 8.0 vol. %). Only the most fluid-rich couple showed evidence of Fe-Mg interdiffusion, indicating that aqueous fluid interconnection in clinopyroxene-rich rocks is established at somewhere between 4 and 8 vol. % (Table 1).

Three important results emerge from the studies summarized in Table 1:

(1) As a whole, there is good agreement between measured and computed values of ϕ_c (Table 1; Fig. 2). Systems with very low wetting angles (10-30°) yield vanishingly small measurements of ϕ_c: 0.07 vol. % for carbonate melt in dunite (Minarik and Watson, 1995), and <0.04 vol. % for hydrous granitic melt in quartzite (this study). In contrast, most systems with θ close to or above 60° yield distinctly non-zero permeability thresholds that compare well with the ideal values. For example, for a CO_2-H_2O mixture in quartzite, the measured value is 1.4 vol. % < $\phi_{c,meas}$ < 2.8 vol. % (Brenan, 1993), as compared to a computed value of $3.3^{+0.9/-1.0}$ vol. % ($\theta = 70°$; the range of $\phi_{c,ideal}$ accounts for the ±5° error in wetting-angle measurements; see Table 1).

(2) In detail, the behavior at low fluid percentages deviates from the ideal behavior in two respects. First, systems with θ significantly below 60° may have a non-zero permeability threshold: at melt percentages less than ~ 0.07 vol. %, carbonate melt in polycrystalline olivine is not interconnected, despite a median wetting angle of 20-30° (Minarik and Watson, 1995). Second, dry and wet grain edges may coexist at a given fluid percentage, as described by Brenan (1993; Fig. 6a); in addition, the diffusion studies suggest that the frequency of wet grain edges increases with increasing fluid percentage (Watson, 1991). The consequence is thus a progressive development of interconnection and permeability, rather than a sudden establishment of a high-permeability network at ϕ_c, as predicted by the ideal model.

Two factors contribute to the coexistence of dry and wet grain edges at a given fluid percentage. The primary factor is the anisotropy of interfacial energy: in an anisotropic system, the melt configuration at a grain edge depends on the relative orientation of the crystalline lattices in contact, and may therefore vary from one grain edge to the other. In particular, dry grain edges may occur in systems with a median wetting angle much less than 60°; at low fluid percentages, the frequency of dry grain edges may be high enough to prohibit the establishment of an interconnected fluid network at a scale larger than a few grains, resulting in a non-zero permeability threshold. The second factor that may contribute to the coexistence of dry and wet grain edges (but only for systems with $\theta \geq 60°$) is the dispersion of grain sizes. The unique value of ϕ_c in ideal systems with $\theta \geq 60°$ results from the assumptions that all grain edges have the same length and that melt pockets at grain corners are equal-sized (e.g., Wray, 1976): at a given fluid percentage < ϕ_c, the length of the dry portion along a grain edge is equal at all triple junctions, and it decreases with increasing fluid percentage until it reaches 0 at ϕ_c. On the contrary, if grain sizes, and therefore grain-edge lengths, are variable, the shorter grain edges may then be wetted while the longer ones remain partly dry.

(3) The only measured value of ϕ_c that deviates noticeably from the ideal value was obtained in the most distinctly anisotropic system, clinopyroxene-H_2O (Watson and Lupulescu, 1993). The measured value, 4 vol. % < $\phi_{c,meas}$< 8 vol. %, is higher than the ideal value, 0.6 vol. % < $\phi_{c,ideal}$< 4 vol. % (Table 1), although it is not yet possible to tell if it is much higher (by several vol. %), or just marginally higher.

The data set in Table 1 may help constrain the relationship between ϕ_c and θ in anisotropic systems. In the case of a moderate degree of anisotropy (quartz- or olivine-rich rock-types), a reasonable estimate of the permeability threshold may be obtained from the median wetting angle using the ideal relationship between ϕ_c and θ in figure 2 (note that the error on ϕ_c is ~ ± 1 vol. % if an error of ±5° is assumed for the median wetting angle). For θ < 60°, however, the concept of a zero permeability threshold should in general be irrelevant: fluid interconnection will be established at a very small but non-zero fluid percentage (for instance, < 0.1 vol. % for granitic melt in polycrystalline quartz, or carbonate melt in polycrystalline olivine).

In more strongly anisotropic systems (clinopyroxene-, amphibole- or plagioclase-rich rock-types), a more substantial deviation from the ideal value of ϕ_c is suggested by the only available data in the clinopyroxene-H_2O system; at present, there is no evidence, however, for a dramatic effect (for instance, an increase by several vol. % to 10 vol. %) of anisotropy on ϕ_c. Low permeability thresholds (< 1 vol. % to a few vol. %; see below) may therefore be anticipated for anisotropic systems with low median wetting angles, say 10-60°, the range of interest in crustal anatexis.

WETTING PROPERTIES OF CRUSTAL PARTIAL MELTS: A REVIEW OF EXPERIMENTAL DATA

SOLID-SOLID-MELT WETTING ANGLES

Table 2 lists experimentally determined wetting angles of interest for crustal anatexis. Most data are from the literature. We performed 2 additional experiments, Qan22c and QOrH2, to complete the data set (information on experimental techniques and run product analysis are summarized in the Appendix).

Each wetting angle in Table 2 is the median value of fifty to more than one hundred apparent wetting angles (ψ in Fig. 1a). For an ideal system with a single-valued wetting angle θ, the median of the frequency distribution of ψ is indeed equal to θ within ±1° (Riegger and van Vlack, 1960). Apparent wetting angles were measured at magnifications ranging from 1500-2000x (optical microscopy; for instance, Jurewicz and Watson, 1985) to 35000x (back-scattered scanning electron [BSSE] microscopy; Laporte, 1994). For coarse-grained experimental samples, wetting-angle measurements at magnifications of a few 1000x yield satisfactory results. We contend, however, that measurements at higher magnifications are commonly required due to the fine grain-size of many experimental partially molten systems, with an average grain diameter of a few μm to 10 μm.

Quartz-quartz-melt wetting angles

A detailed study of wetting angles at quartz-quartz-melt triple junctions in the quartz-anorthite-water and quartz-albite-orthoclase-water systems, at pressures of 650 to 1000 MPa and temperatures of 800 to 900 °C, has recently been published (Laporte, 1994). The principal results of this study may be summarized as follows: (1) quartz-quartz-melt wetting angles are invariably very low (10° to 19°; Table 2); they seem almost insensitive to melt composition (including water content), as well as to pressure and temperature; (2) in all experiments, smoothly curved quartz-melt interfaces coexisted with planar facets of the form {10$\bar{1}$1} (Fig. 5a). These facets demonstrate that quartz-melt surface energy is not isotropic.

TABLE 1. Measured versus predicted permeability thresholds ($\phi_{c,meas}$ and $\phi_{c,ideal}$, respectively; both in vol. %) in fluid-bearing rocks.

System [a]	wetting angle	$\phi_{c,meas}$ [b]	$\phi_{c,ideal}$ [c]	ref. [d]
ol-carb (1300 °C-1 GPa)	25-30° [e]	~ 0.07	0	1
ol-CO_2 (1200 °C-1 GPa)	76° [e]	> 4[f]	$4.4^{\pm0.9}$	2
ol-COH (1200 °C-1 GPa)	60-65° [e]	$1.5 < \phi_c < 3.6$	$1.7^{+1.6/-1.7}$	2
ol-H_2O (1000 °C-1 GPa)	69° [e]	<1.2[f, g]	$3.1^{+0.9/-1.1}$	2
ol-H_2O (1100 °C-1 GPa)	62° [e]	<1.6[f, g]	$1.5^{+1.2/-1.5}$	2
ol-basalt (1350 °C-1.5 GPa)	26-30°	<1.2[f, g]	0	3
ol-Fe (1150 °C-1 GPa)	107-110°	$10 < \phi_c < 25$	$10.0^{+2.1/-2.0}$	4
qtz-H_2O (1000 °C-1 GPa)	55°	$0.3 < \phi_c < 1.0$	$0^{+0.6}$	5
qtz-COH (1000 °C-1 GPa)	70°	$1.4 < \phi_c < 2.8$	$3.3^{+0.9/-1.0}$	5
qtz-granite (900 °C-1 GPa)	14°	< 0.04	0	6
amph-granite (800 °C-1 GPa)	53-58°	<12[f, g]	$0^{+1.8}$	7
amph-tonalite (800 °C-1 GPa)	46-48°	<12[f, g]	0	7
cpx-H_2O (950 °C-1GPa)	65-69°	$4.0 < \phi_c < 8.0$	$2.7^{+1.3/-2.1}$	8

[a]Rock-type and fluid symbols: ol, dunite; qtz, quartzite; amph, amphibolite; cpx, clinopyroxenite; carb, carbonate melt; COH, H_2O-CO_2 mixture (with X_{CO2}= 0.4 in ref. 5 and 0.5 in ref. 2); basalt, basaltic melt; Fe, solid Fe-metal; granite, granitic melt; tonalite, tonalitic melt.

[b]All measurements were obtained by diffusion experiments, except the one in ref. 4 (electrical conductivity measurement) and our own measurement.

[c] The ideal values of ϕ_c, $\phi_{c,ideal}$, are computed from the wetting angle in column no. 2 assuming that the solid-fluid system behaves ideally (the relationship between ϕ_c and θ in an ideal system is shown in Fig. 2). For θ close to or above 60°, a range of $\phi_{c,ideal}$ is given to account for the estimated error of ±5° in wetting-angle measurements (±10° in ref. 4): considering, for instance, a median wetting angle of 70°, one has $\phi_{c,ideal}$= 2.28, 3.30, and 4.22 vol. % for θ=65, 70, and 75°, respectively, which we summarize as $3.3^{+0.9/-1.0}$ vol. % in column no. 4.

[d] References: 1, Minarik and Watson (1995); 2, Watson (1991); 3, Daines and Richter (1988); 4, Jurewicz and Jones (1995); 5, Brenan (1993); 6, this study; 7, Lupulescu and Watson (1994, 1995); 8, Watson and Lupulescu (1993).

[e] From Watson and Brenan (1987), and Watson et al. (1990).

[f] Only an upper limit or a lower limit of ϕ_c has been determined.

[g] Recalculated to vol. % from original data in wt %.

TABLE 2. Experimentally determined solid-solid-melt wetting angles relevant to crustal anatexis.

Solid phases[a] in contact	Melt composition[b]	P GPa	T °C	median wetting angle	Ref.[c]
qtz-qtz	qtz-ab (0)	0.8	1250	60°	1
qtz-qtz	qtz-ab-or (0)	1.0	1000	59°	2
qtz-qtz	qtz-ab-or (0)	1.0	1150	19°	3
qtz-qtz	qtz-ab-or (6)	1.0	900	14°	3
qtz-qtz	qtz-ab-or (sat.)	1.0	900	10°	3
qtz-qtz	qtz-ab-or (sat.)	1.0	800	16°	3
qtz-qtz	qtz-ab-or (sat.)	0.4	850	49°	4
qtz-qtz	granite	1.0	900	22°	5
qtz-qtz	granite	1.0	950	23°	5
qtz-qtz	qtz-an (sat.)	0.65	900	13°	3
qtz-qtz	qtz-an (sat.)	1.0	900	14°	3
an-an	qtz-an (sat.)	1.0	1000	28°	QAn22c
pl-pl	basaltic andesite	0.8	1050	60°	6
pl-pl	n.d.[d]	1.0	1150	45°	7
sa-sa	qtz-or (sat.)	0.2	850	20°	QOrH2
alkf-alkf	qtz-ab-or (0)	1.0	1000	44°	2
hbl-hbl	intermediate[e]	1.2	975	25°	8
amp-amp	basaltic andesite	0.8	1050	33°	6
amp-amp	granite	1.0	800	53-58°	9
amp-amp	tonalite	1.0	800	46-48°	9
bt-bt	intermediate[e]	1.2	975	23°(//)-90°(⊥)[f]	8
alkf-qtz	qtz-ab-or (0)	1.0	1000	49°	2
amp-pl	basaltic andesite	0.8	1050	54°	6

[a]Mineral symbols: qtz, quartz; an, anorthite; pl, plagioclase; sa, sanidine; alkf, alkali feldspar; hbl, hornblende; amp, amphibole; bt, biotite.
[b]For synthetic systems, we report the components of the system (qtz: SiO_2; ab:$NaAlSi_3O_8$; or: $KAlSi_3O_8$; an: $CaAl_2Si_2O_8$), followed, in parentheses, by the water content in melt (in wt %; "sat" stands for water saturation).
[c]References: 1, Jurewicz and Watson, 1984; 2, Jurewicz and Watson, 1985; 3, Laporte, 1994; 4, Holness, 1995 (many data from ref. 3-4 omitted for conciseness); 5, Laporte and Vielzeuf, 1994 and in preparation; 6, Vicenzi et al., 1988; 7, Longhi and Jurewicz, 1995; 8, Laporte and Watson, 1995; 9, Lupulescu and Watson, 1994, 1995; QAn22c, QOrH2: this study.
[d]The melt composition in the partially molten anorthositic gabbro is not specified.
[e]Silicic melts with SiO_2 in the range 60 to 65 wt %.
[f]// and ⊥ apply to sections parallel and perpendicular, respectively, to the foliation of a biotite-rich rock-type.

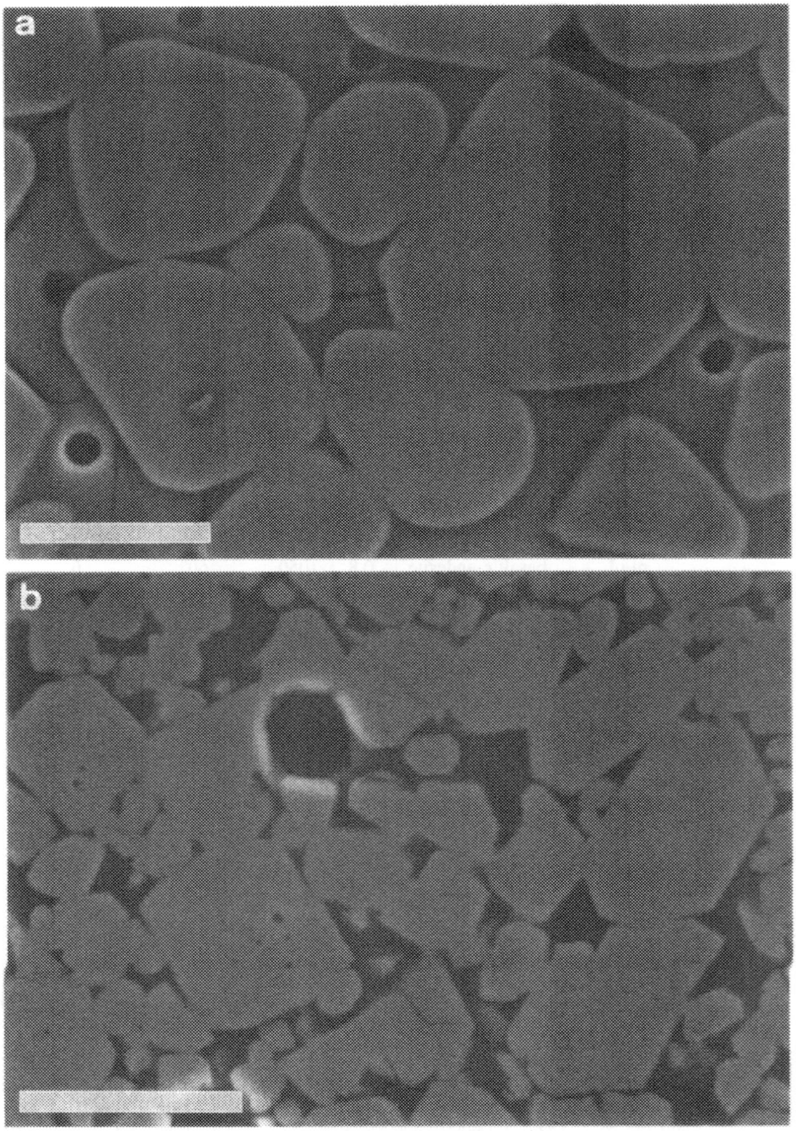

Figure 5. BSSE microphotographs showing the textural relationships of *a*: quartz and *b-c*: anorthite with a H_2O-saturated CAS melt. Quartz and anorthite are light grey, melt is darker; dark voids in *a-b* are bubbles (filled with H_2O at run conditions). Note in *a* the very low apparent wetting angles, and the coexistence of smoothly curved quartz-melt interfaces and planar facets (run QAn21a, equilibrated at 1.0 GPa-900 °C, in Laporte, 1994). *b-c*: Overall texture and a close-up view of sample Qan22c; note the nearly systematic development of planar facets, and the low apparent wetting angles. Scale bars: 20 µm in *a-b*, 5 µm in *c*.

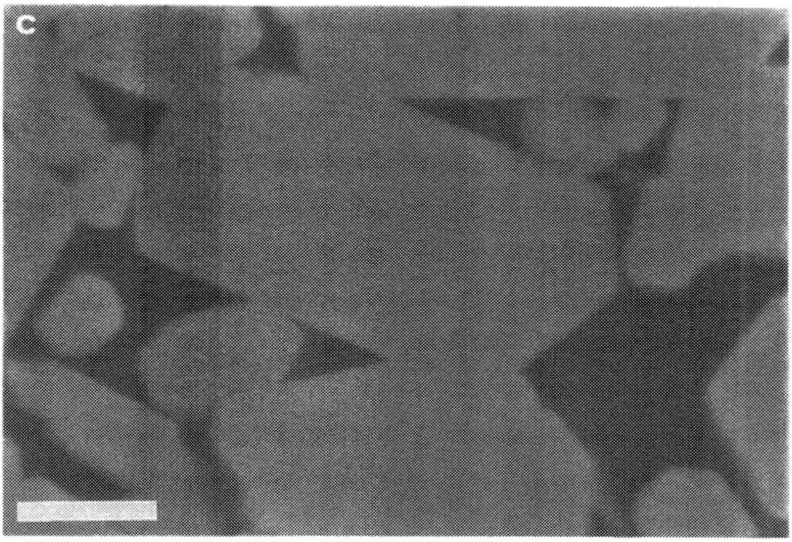

Figure 5 (continued)

Wetting angles at quartz-quartz-melt junctions in the pelitic system are just slightly larger than those measured in the synthetic systems above; namely, 22° at 1000 MPa-900 °C, and 23° at 1000 MPa-950 °C (Laporte and Vielzeuf, 1994). These two measurements are the most directly relevant to natural partially molten crustal sources. They clearly show that, at the pressures and temperatures of crustal anatexis, a very low wetting angle is the rule for quartz in contact with a H_2O-undersaturated granitic melt.

Feldspar-feldspar-melt wetting angles

Melt geometry in feldspar-rich partially molten systems is controlled by the general development of planar, crystallographically-controlled solid-melt interfaces (Longhi and Jurewicz, 1995). This point is best illustrated in figures 5b-c which shows the textural relationships of anorthite with a H_2O-saturated CAS melt (sample Qan22c equilibrated at 1000 MPa-1000 °C for 149 h; CAS stands for CaO-Al_2O_3-SiO_2). The most striking features are the euhedral shape of anorthite grains and the general development of facets, both of which demonstrate that anorthite-melt surface energy is strongly anisotropic.

The frequency distribution of 118 apparent wetting angles measured at anorthite-anorthite-melt triple junctions in sample Qan22c is shown in figure 6a. It displays a main class of low to moderate angles ($\psi \leq 65°$) and a few large angles (up to 127°); the median angle is 28°. The theoretical distribution of ψ expected for a single-valued wetting angle $\theta = 28°$ is shown for comparison (see Jurewicz and Jurewicz, 1986, for the computation of theoretical distributions of apparent wetting angles). The observed distribution is distinctly broader and blunter than the theoretical one, indicating a dispersion of θ's on either side of the median due to anisotropy of anorthite-melt surface energy. The moderate departure from the theoretical distribution in figure 6a, in particular the scarcity of large apparent angles, implies, however, that the dispersion of θs is narrow and that θs at anorthite-anorthite-melt triple junctions have low values close to 28°.

A low median wetting angle (20°, Table 2) was also measured at sanidine-sanidine-melt junctions in our sample QOrH2 (equilibrated at 200 MPa-850 °C; see the Appendix for phase compositions). The frequency distribution of apparent wetting angles in sample QOrH2 is shown in Fig. 6b; it is characterized by a well-defined peak at low angles (≤ 40°) and a long tail of moderate to large angles, up to 139°. We do not know if the elevated number of large apparent wetting angles is due to the anisotropy of surface energy or if it results from a lower degree of textural maturity of sample QOrH2 compared to sample Qan22c (due to the lower pressure of run QOrH2, and the lower water content of the melt). Wetting-angle measurements at quartz-quartz-melt triple junctions in the dry granitic system do indeed suggest that immature textures are characterized by anomalously high frequencies of large apparent angles by comparison with near-equilibrium textures (see Fig. 7d in Laporte, 1994).

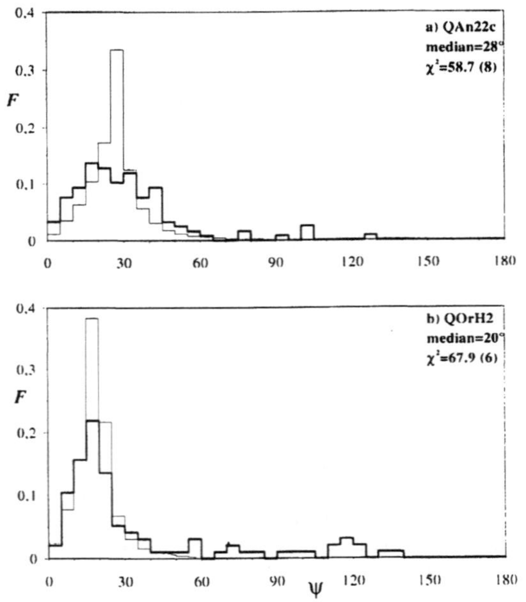

Figure 6. Relative frequency distribution, *F*, of apparent wetting angles at *a*: anorthite-anorthite-melt triple junctions (sample Qan22c, 118 measurements) and *b*: sanidine-sanidine-melt triple junctions (sample QOrH2, 96 measurements). For both samples, the thick line is the observed distribution, and the thin line shows the theoretical distribution computed for a single-valued wetting angle equal to the median angle of the observed distribution (28° in *a*, 20° in *b*). χ^2 is the chi-squared value (the number of degrees of freedom is given in parentheses; see Laporte, 1994, p. 495, for details); for both distributions, χ^2 is larger than the χ^2 variable at the 95% confidence level (=12.6 and 15.5 for 6 and 8 degrees of freedom, respectively).

Other published feldspar-melt wetting angles are higher than our own values, but still ≤ 60°: 44° for alkali feldspar (with Na_2O ~ K_2O in weight) in contact with a dry granitic melt at 1000 MPa-1000 °C (Jurewicz and Watson, 1985); 45° for plagioclase in a partially molten anorthositic gabbro at 1000 MPa-1150 °C (Longhi and Jurewicz, 1995); and 60° for plagioclase in equilibrium with a basaltic andesite in dehydration-melting experiments of a natural, powdered amphibolite, at 800 MPa-1050 °C (Vicenzi et al., 1988).

Biotite-biotite-melt wetting angles

The layered structure of biotite results in a very pronounced anisotropy of γ_{sl} (Spry, 1969). Layer-parallel facets of the form {001} have a much lower interfacial energy per unit area than facets of the forms {110} and {010}. In biotite-rich rock-types, such as melanosomes in migmatites, the combination of crystalline anisotropy and a marked preferred orientation of mica flakes is predicted to lead to very contrasted wetting relationships in sections parallel and perpendicular to the foliation (Laporte and Watson, 1995). In sections parallel to the foliation, apparent wetting angles at biotite-biotite-melt junctions are low, mostly < 60° (with a median value of 23°; Table 2); in contrast, large apparent angles, typically around 90°, dominate in sections perpendicular to the foliation. The ultimate result is a very low permeability to melt flow normal to the foliation in biotite melanosomes.

Amphibole-amphibole-melt wetting angles

In an attempt to estimate the permeability threshold in a partially molten amphibolite, five studies have been focused on the textural relationships of amphibole with melt (Vicenzi et al., 1988; Wolf and Wyllie, 1991; Laporte and Watson, 1995; Lupulescu and Watson, 1994, 1995). All five studies concluded to a strong anisotropy of amphibole-melt interfacial energy, as indicated by the systematic development of planar facets, especially of the form {110}. Median apparent angles at amphibole-amphibole-melt junctions are consistently low to moderate (Table 2); 33° for amphibole in equilibrium with a basaltic andesite at 800 MPa-1050 °C (Vicenzi et al., 1988); 25° for hornblende in equilibrium with a melt of intermediate composition at 1200 MPa-975 °C (Laporte and Watson, 1995); and 46-48° and 53-58° for a calcic amphibole in equilibrium with a tonalitic melt and a granitic melt, respectively, at 1000 MPa-800 °C (Lupulescu and Watson, 1994, 1995).

The frequency distribution of apparent wetting angles at hornblende-hornblende-melt junctions obtained by Laporte and Watson (1995) is quite similar to that shown in figure 6a for anorthite: it is characterized by a broad peak of angles $\psi < 70°$, with a frequency maximum of 16.8% for the class $20° \leq \psi < 25°$, and almost no large angles. This distribution implies that, although hornblende-melt interfacial energy is strongly anisotropic, the dispersion of θs is narrow and that wetting angles at hornblende-hornblende-melt triple junctions have low values close to 25°.

DIRECT INSIGHTS INTO MELT INTERCONNECTION AT LOW MELT PERCENTAGE

Secondary electron imaging of fracture surfaces

Imaging the fracture surfaces of a polycrystalline aggregate may provide a three-dimensional insight into the geometry of any fluid phase present in the aggregate, as exemplified by the study of fluid geometry in synthetic quartzites by Laporte and Watson (1991). In the case of silicate partial melts, this technique was first applied with success to two melt-poor polycrystalline aggregates of quartz containing, respectively, ~ 11 wt % of a water-saturated CAS melt, and ~ 5 wt % of a water-saturated granitic melt (volume melt percentages are just slightly larger than these figures; Laporte, 1994). The two samples were annealed in a piston-cylinder apparatus at, respectively, 1000 MPa-900 °C for 327 h and 1000 MPa-800 °C for 333 h. Cold depressurization at the end of the experiments produced unloading cracks that were observed with a scanning electron microscope. Secondary electron imaging revealed the presence of an interconnected network of melt channels along grain edges in both samples; no dry grain edges, i.e., direct contacts of 3 quartz grains at a triple junction, were observed. We demonstrate in the following section that this type of geometry still holds at much lower melt percentages. For instance, the secondary electron microphotograph in figure 7 shows a fracture surface in a polycrystalline aggregate of quartz containing ~ 0.2 vol. % of a hydrous granitic melt; that the partial melt is interconnected is evident, and the similarity with the network of grain-edge melt channels idealized in figure 1b is striking.

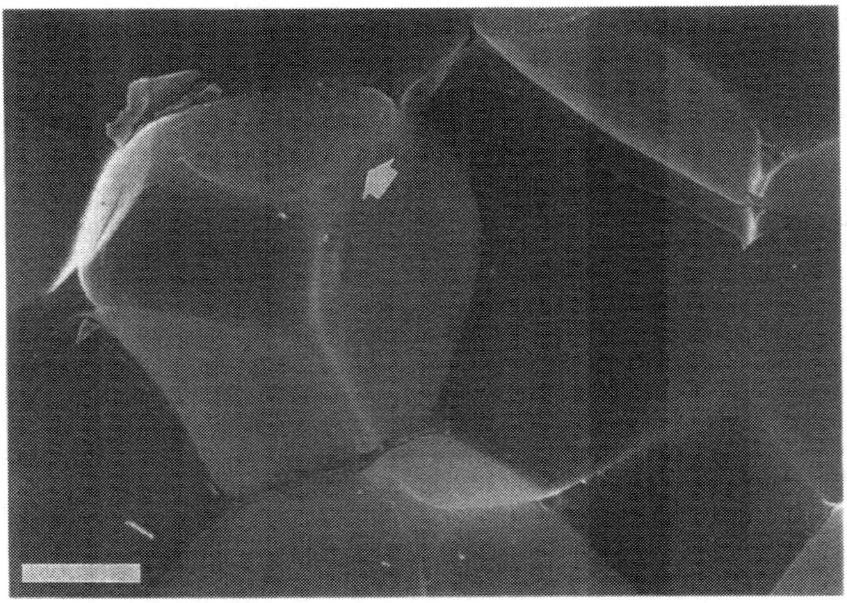

Figure 7. Secondary electron microphotograph of a fracture surface in a polycrystalline aggregate of quartz containing ~ 0.2 vol. % of a hydrous granitic melt (run INFIL#7). Melt forms an interconnected network of channels along grain edges. Melt channels are well preserved in the depressions left behind by grain plucking (center right); elsewhere, the flattened grain edges (arrow) betray the presence of former grain-edge channels that have been destroyed during sample preparation. Scale bar: 5 μm.

Diffusion experiments

The principle of the diffusion experiments was summarized above. The sole diffusion study that addressed the problem of melt interconnection in a crustal protolith is the ongoing work of Lupulescu and Watson (1994, 1995). Lupulescu and Watson's experiments show that tonalitic and granitic melt interconnection in an amphibole-rich residue is achieved at melt contents less than 10 wt % (~ 12 vol. %; Table 1). This result is consistent with the wetting angles measured at amphibole-amphibole-melt triple junctions (53-58° for the granitic melt, 46-48° for the tonalitic melt; Table 2). Because these angles are significantly lower than the wetting angles at clinopyroxene-clinopyroxene-water triple junctions (65-69°; Watson and Lupulescu, 1993), the permeability threshold in partially molten amphibolite should be lower than in a water-bearing clinopyroxenite (4 vol. % $< \phi_c \leq$ 8 vol. %): ≤ 3-4 vol. % according to Laporte and Watson (1995).

SUMMARY

Two general results emerge from the studies summarized in Table 2: (1) wetting angles in partially molten crustal systems are low to very low: all median wetting angles are ≤60°, and most fall in the range 10-30°; and (2) in all systems investigated, silicate-melt surface energy is anisotropic: moderately anisotropic for quartz, strongly anisotropic for feldspar, biotite and amphibole. The degree of anisotropy was determined experimentally for β-quartz in a water-saturated CAS melt at 1000 MPa-900 °C (Laporte and Provost, 1994). Quartz-melt surface energy is found to be maximum parallel to the basal plane (0001) and minimum parallel to facets of the form $\{10\bar{1}1\}$; the ratio of the maximum to the minimum value is ~ 1.4.

Due to anisotropy of surface energy, the interpretation of median wetting angles in terms of melt interconnection is delicate; in particular, interconnection at an infinitely small melt percentage cannot be inferred from the measurement of a median wetting angle < 60°. We argue, however, that the low wetting angles in Table 2 result from high values of the ratio $[\gamma_{ss}]/[\gamma_{sl}]$ and imply very low values of the permeability threshold (presumably, < 1 vol. % to a few vol. %). In the ideal case of single-valued γ_{ss} and γ_{sl}, the exact value of γ_{ss}/γ_{sl} could be obtained from the median wetting angle by using equation (1). Because quartz-melt surface energy is just moderately anisotropic, and γ_{ss} anisotropy is negligible for quartz (Laporte and Watson, 1995), values of $[\gamma_{ss}]/[\gamma_{sl}]$ in the range 1.9 to 2.0 may be inferred from the median wetting angles for quartz in contact with hydrous melts (10 to 23°; Table 2). In the following section, we demonstrate that a hydrous granitic melt in polycrystalline quartz is interconnected at a vanishing melt percentage, in agreement with this high value of $[\gamma_{ss}]/[\gamma_{sl}]$.

PERMEABILITY THRESHOLD OF HYDROUS GRANITIC MELT IN POLYCRYSTALLINE QUARTZ

The evidence for a very low permeability threshold of hydrous granitic melt in polycrystalline quartz comes from an ongoing series of infiltration experiments. The process of melt infiltration that we are referring to was defined by Watson (1982; see also Riley et al., 1990, and Riley and Kohlstedt, 1991). On the basis of interfacial energy considerations, Watson (1982) anticipated that a pool of melt in chemical and thermal equilibrium with its surroundings would tend to infiltrate the dry grain edges of the host material if the wetting angle is < 60°. The driving force for melt infiltration is the reduction of total interfacial energy that results from replacing a dry grain edge by a continuous melt channel when solid-melt interfacial energy per unit area is much lower than grain-boundary energy. Melt infiltration presumably occurs via a process of solution (along grain edges) and precipitation (within the melt pool).

The high values of $[\gamma_{ss}]/[\gamma_{sl}]$ inferred from the wetting-angle measurements in Table 2 suggest that melt infiltration may play an important role in the evolution of crustal source regions. To test this hypothesis, we undertook a series of infiltration experiments in which a cylinder of synthetic, texturally-equilibrated quartzite was placed in contact with a quartz-saturated, hydrous granitic melt, at 1.0 GPa and 900 °C for 3 to 7 days. These experiments are basically similar to those described by Jurewicz and Watson (1984), except that the latter used an anhydrous quartz-albite melt at 0.8 GPa and 1250 °C. A detailed account of our infiltration experiments will be given elsewhere; only the results concerning the grain-scale geometry of melt at very low melt percentages are discussed below.

The synthetic quartzites were prepared from a fine powder of quartz (< 8μm) that was loaded into a platinum container and annealed for 4 hours at 1.0 GPa and 1600 °C in a piston-cylinder apparatus. The melt reservoir was made from a mixture of 84 wt % granitic glass (GminH1, which contains ~ 6.4 wt % H_2O; Laporte, 1994) and 16 wt % powdered quartz. Because the amount of quartz that is dissolved into GminH1 at 1.0 GPa-900 °C is small (Laporte, 1994), the water content of the granitic melt during the infiltration tests was just slightly lower than the initial 6.4 wt %. In preliminary attempts (exemplified below by run INFIL#7), the texturally equilibrated quartzite was first retrieved from its platinum container; a short cylindrical fragment, 0.5 to 1 mm long, was then embedded in ~ 15-20 mg of the reservoir mixture in a 3-mm-OD platinum container. In subsequent experiments (exemplified below by run INFIL#10), the quartzite was kept within its precious metal container whose upper part had been first ground and polished to expose the sample; the decapitated container was then fitted into a 3-mm-OD platinum capsule along with 8-10 mg of the reservoir mixture. After being welded, the infiltration couple was placed horizontally into a piston-cylinder assembly of the type described by Perkins and Vielzeuf (1992). Both runs INFIL#7 and INFIL#10 were conducted at 1.0 GPa and 900 °C for a duration of 159 hr and 151 hr, respectively.

The principal results of interest may be summarized as follows.

(1) Back-scattered electron imaging of a polished, longitudinal section of sample INFIL#10 revealed no quantitative infiltration of melt (Fig. 8a); by quantitative, we mean the infiltration of a significant amount of melt, at least a few vol. %. The abundance of plucked grains in the quartzite is also indicative of the absence of a significant amount of penetrating melt (Jurewicz and Watson, 1984).

(2) "Infiltration" was more extensive in our preliminary experiments (e.g., INFIL#7). The process involved in these experiments is, however, predominantly melt injection into open fractures, rather than surface-energy-driven infiltration. For instance, injection of melt into two dilational fractures was observed in sample INFIL#7. Our interpretation is that, in these earlier experiments, removing the quartzite from its platinum container favoured the opening of unloading cracks and grain boundaries; these open features were subsequently injected by melt at the beginning of the infiltration experiment. In particular, melt injection into open grain boundaries produced domains with a low melt percentage in which textural equilibrium was approached at the end of the experiment. The 3-dimensional geometry of melt in such a domain is characterized by a pervasive network of grain-edge channels (Fig. 7). Using the formula derived below, an estimate of ~ 0.2 vol. % for the melt percentage in figure 7 may be inferred from the channel thickness (~ 0.65 μm) and the grain size of quartz (~ 20 μm).

(3) The most striking result was obtained by imaging a fracture surface of quartzite INFIL#10 close to the melt reservoir (within 500 μm from the reservoir). At high magnifications (Fig. 8b), secondary electron imaging revealed the presence of an interconnected network of narrow melt channels, typically 0.3-μm wide, along grain edges: locally, these channels are well preserved, specially in the depressions left behind by

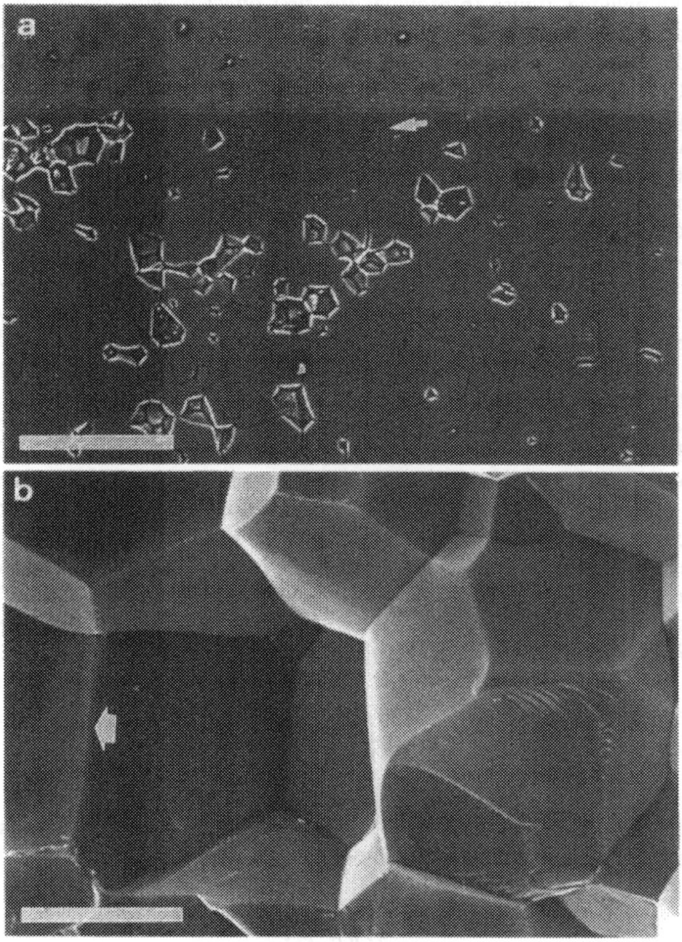

Figure 8. (a) BSSE microphotograph of a polished section of sample INFIL#10 showing the quartz-saturated melt reservoir (top) and the quartzite (bottom); except for a short pod of melt ~ 30 μm long (arrow), there is no evidence for melt infiltration in the quartzite. Systematic opening of grain boundaries and grain plucking in the quartzite result from the $\beta \rightarrow \alpha$ transition at the end of the experiment; (b) Secondary electron microphotograph of a fracture surface in sample INFIL#10 showing the quartzite texture in the vicinity of the melt reservoir. This photograph reveals distinct melt channels along grain edges (or flattened grain edges where the melt channels have broken away during sample preparation; arrow); melt percentage is 0.04 vol. %. Scale bars: 100 μm in *a*; 10 μm in *b*.

grain plucking; more commonly, the channels have been partly or totally destroyed during sample preparation, yet the presence of a former channel is still betrayed by the flattened grain edges of quartz (arrow in Fig. 8b; a dry grain edge would appear as a sharp crest). We emphasize that the texture displayed in figure 8b is not restricted to a few grain diameters from the melt reservoir but that it extends to at least 500 µm from the reservoir.

To our knowledge, figure 8b provides the first direct illustration of the interconnection of a partial melt at a vanishing melt percentage. It displays a topology very similar to that predicted by the ideal model (for $\theta < 60°$; Bulau et al., 1979, p. 6105). In particular, almost all the melt resides in grain-edge tubules that commonly achieve a nearly prismatic geometry, as shown by the constant width of flattened grain edges in figure 8b. In many areas, the channel width is quite constant from one grain edge to the other, with an average value of 0.3 µm (from 12 measurements of the width of near-horizontal flattened grain edges); wider channels (~ 0.5 µm) and thinner channels (0.1-0.2 µm) are present locally, however, as well as channels whose width decreases from the grain corners toward the middle of the grain edge. One puzzling feature of melt topology in figure 8b is that it does not show compelling evidence of interfacial energy anisotropy, whereas similar systems with a higher melt percentage are characterized by the general development of facets (for instance, Fig. 5a). Rare dry grain edges observed in quartzite INFIL#10 in the vicinity of the melt reservoir could, however, be ascribed to crystalline anisotropy.

Because the grain-edge channels can be approximated by prisms (an approximation that is only valid for melt percentages ≤ 1 vol. %; von Bargen and Waff, 1986), an estimate of the melt percentage in figure 8b may be inferred from the average channel width and grain size. The channel width, w, is defined in figure 9, and is related to the radius of curvature, r, by $w = 2r\sin(30-\theta/2)$, where θ is in degrees.

Let the quartzite be modelled as a packing of equal-sized, regular tetrakaidecahedra, of diameter a (between square faces). The volume of a tetrakaidecahedron is $a^3/2$, and it has 36 edges of equal length: $l = a/\sqrt{8}$. The expression relating the cross-sectional area of a channel, S_t, to its width and to the wetting angle θ is

$$S_t = w^2\left[\frac{\sqrt{3}}{4} - \frac{\pi(30-\theta/2)}{240\sin^2(30-\theta/2)} + \frac{3}{4\tan(30-\theta/2)}\right] \tag{3}$$

Every tetrakaidecahedron is edged by 36 channels, and every channel is shared by 3 adjacent tetrakaidecahedra. The volume melt fraction f is therefore given by

$$f = \frac{36l(S_t/3)}{a^3/2} \tag{4}$$

$$f = 6\sqrt{2}\left[\frac{w}{a}\right]^2\left[\frac{\sqrt{3}}{4} - \frac{\pi(30-\theta/2)}{240\sin^2(30-\theta/2)} + \frac{3}{4\tan(30-\theta/2)}\right] \tag{5}$$

Grain sizes in quartzite INFIL#10 were measured on BSSE micrographs of the polished sample (after 10 s etching with HF), using the linear intercept method; the average grain size a was obtained by multiplying the mean linear intercept length by 3/2 (e.g., Olgaard and Evans, 1988), yielding $a = 21$ μm. Setting $a = 21$ μm, $w = 0.3$ μm and $\theta = 14°$ (the median wetting angle measured for quartz equilibrated with glass GminH1 at 1.0 GPa and 900 °C; Laporte, 1994) in equation (5) yields $f = 3.9\ 10^{-4}$, that is a melt percentage of $\phi \sim 0.04$ vol. %. Note that this is a conservative estimate. For instance, Fig. 8b shows an interconnected melt network with $w \sim 0.3$ μm (the channel width at the arrow is 290 nm), but a grain size closer to 30 μm, corresponding to a melt percentage of ~ 0.02 vol. %.

Textures in the infiltration experiments demonstrate that the interconnection of a hydrous granitic melt in a polycrystalline aggregate of quartz is established at an extremely low melt percentage: $\phi_c < 0.04$ vol. %. At present, however, the possibility of a non-zero permeability threshold cannot be discarded because of the observation of a few dry grain edges in INFIL#10. We may therefore conclude that the permeability threshold lies somewhere within the range $0 \leq \phi_c < 0.04$ vol. %. This result is in very good agreement with a median wetting angle < 60°. On the contrary, the failure to infiltrate significant amounts of melt in the quartzites is in contradiction with the predictions at such low θ. For $\theta = 14°$, melt infiltration could theoretically proceed until the melt percentage equals ~ 30 vol. %, at which point a minimum state of total interfacial energy per unit volume would be reached (Laporte and Watson, 1995). The absence of extensive penetration of melt within the quartzite in our experiments is presumably due to a very poor kinetics of the solution-precipitation process by which melt infiltration progresses. The dry quartz-albite melt in Jurewicz and Watson's (1984) study also showed no tendency to infiltrate the grain edges of a quartzite.

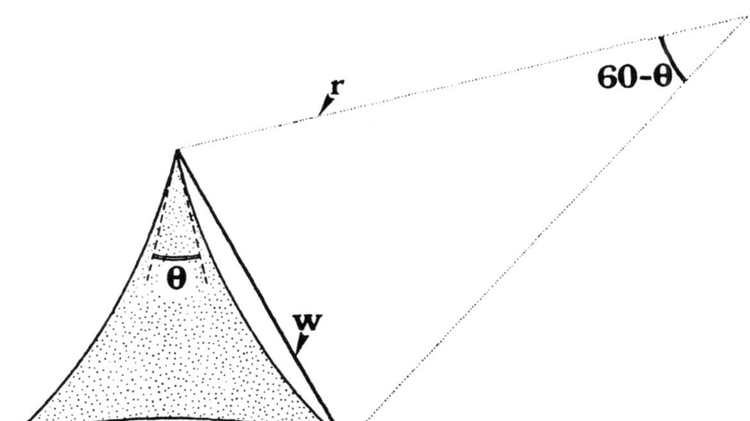

Figure 9. Cross-section of an idealized grain-edge melt channel (stippled); w is the channel width, r is the radius of curvature of solid-melt interfaces, θ is the wetting angle.

DISCUSSION AND GEOLOGICAL IMPLICATIONS

TO WHAT EXTENT IS TEXTURAL EQUILIBRIUM ACHIEVED DURING CRUSTAL ANATEXIS?

Experimental data on equilibrium melt geometries in crustal protoliths were reviewed above. The very long duration of crustal melting events (typically $>10^6$ years; Brown et al., 1995) suggests that textural equilibrium may commonly be approached during crustal anatexis (see Laporte and Watson, 1995, for a discussion). The possibility of disequilibrium melt geometries resulting from either deformation at a high strain rate or from a high rate of melt production cannot be ruled out, however. In any case, the following considerations must be kept in mind:

(1) Because an interconnected melt network provides a rapid pathway for diffusive mass transport, the time-scale for textural equilibration should be much shorter in a partially molten rock than in a subsolidus rock. In this context, the observation that near-equilibrium textures are common in subsolidus, high-grade metamorphic rocks (Kretz, 1966; Vernon, 1968) is a strong argument for the attainment of textural equilibrium during crustal anatexis.

(2) At low melt percentages and slow strain rates, the dominant deformation mechanisms are expected to be dislocation creep and/or diffusion creep, not cataclastic flow (Dell'Angelo and Tullis, 1988). Under these conditions, melt distribution at the grain scale may still be governed by the principle of interfacial energy minimization, as exemplified by the experimental studies of Dell'Angelo et al. (1987) for crustal protoliths, Cooper and Kohlstedt (1984) and Hirth and Kohlstedt (1995) for olivine-basalt systems.

(3) Wetting angles are small-scale features that require only the operation of a very localized dissolution-precipitation process and short-distance transport to establish the equilibrium value.

(4) Melting experiments using solid-rock cores as starting materials (Mehnert et al., 1973; Busch et al., 1974; Rubie and Brearley, 1991; Wolf and Wyllie, 1991) show partial melt textures that are similar to the textures described at the contact of shallow intrusions of basic to intermediate magmas (e.g., Kaczor et al., 1988) or in xenoliths embedded in lava flows (Maury and Bizouard, 1974; Harris and Bell, 1982). Because the grain size is basically equal to that in crustal protoliths and the run durations are typically a few days or a few weeks, the disequilibrium textures in these experiments cannot resemble the partial melt textures established even in short-lived crustal melting events (10^2-10^3 years for roof melting above basaltic sills in the deep crust; Huppert and Sparks, 1988).

From these considerations, the assumption that partial melt geometries are primarily dictated by the principle of interfacial energy minimization is considered a reasonable one. Some important implications for the permeability threshold of crustal protoliths, melt segregation and source evolution may therefore be inferred from the equilibrium studies summarized above.

PERMEABILITY THRESHOLDS IN PARTIALLY MOLTEN CRUSTAL PROTOLITHS

The principal contribution of the experimental studies is to show that melt interconnection in crustal protoliths is established at a low melt percentage. The only direct measurement of ϕ_c is our value of < 0.04 vol. % for a hydrous granitic melt in polycrystalline quartz. Precise measurements of ϕ_c in protoliths more realistic than pure quartzite are presently not available. For most crustal rock-types, a very low value of ϕ_c may, however, be inferred from the median wetting angles summarized in Table 2 for the following reasons:

(1) The ratio of grain-boundary energy to solid-melt interfacial energy is the primary factor controlling melt distribution at low melt percentage. The general trends of increasing θ and ϕ_c or decreasing $s/v^{2/3}$ with decreasing γ_{ss}/γ_{sl}, that are well established in ideal systems with isotropic interfacial energies, must apply as well in natural systems where solid-melt interfacial energy may be strongly anisotropic. In particular, a low median wetting angle results from a high ratio of grain-boundary energy to solid-melt interfacial energy and implies a low permeability threshold.

(2) Permeability thresholds measured in systems where solid-melt interfacial energy is just moderately anisotropic, as in quartz- or olivine-fluid systems, are basically equal to those predicted by the isotropic model (Fig. 2 and Table 1).

(3) The only detailed study of ϕ_c in a strongly anisotropic system (clinopyroxene-H_2O; Watson and Lupulescu, 1993) suggests that the deviation from the ideal value of ϕ_c is only moderate, at most a few vol. % (Fig. 2 and Table 1).

Therefore, very low values of ϕ_c, < 1 vol. % to a few vol. %, may be inferred from the low median wetting angles summarized in Table 2 (10°-60°). Waiting for precise measurements, a value of 3.4 vol. % is considered to provide a reasonable upper limit for the permeability threshold of most crustal protoliths. $\phi_c = 3.4$ vol. % is the permeability threshold for an ideal system with $\theta = 70.53°$, in which melt pockets at grain corners are regular, plane-faced tetrahedra. Because wetting angles in synthetic crustal protoliths are much lower than 70.53°, and melt pockets are bounded by concave solid-melt interfaces, ϕ_c should indeed be < 3.4 vol. %. A much lower value, close to zero, is even probable in quartz-rich protoliths, such as greywackes, due to the very low wetting angles at quartz-quartz-melt junctions and to the low degree of anisotropy of quartz-melt interfacial energy.

IMPLICATIONS FOR THE SEGREGATION OF GRANITIC MELTS

The major implication of the experimental studies is to show that melt segregation may potentially operate at very low degrees of melting (theoretically, at any melt percentage $\geq \phi_c$). Simple compaction calculations for melt percentages in the range 10-30 vol. % indicate that the segregation of granitic melts may occur within time-scales of $\leq 10^5$ to 10^6 years (Laporte, 1994); deformation-enhanced melt extraction could yield much shorter time-scales. Because of the relatively high viscosities of typical crustal melts (10^3-10^4 Pa s for partial melts produced by dehydration-melting of greywackes; Montel and Vielzeuf, 1996), it is not clear, however, whether the separation of melt from the solid residue may be efficient at melt percentages below 5-10 vol. %. Three general conclusions may be drawn from these remarks: (1) melt viscosity, not melt interconnection, should in general be the limiting factor in the segregation of granitic partial melts; in particular, the amount of melt that is ultimately left behind within the solid residue at the end of a segregation event is likely to be dictated by the high melt viscosities, and may be significantly larger than ϕ_c; (2) there may exist a range of melt percentages above ϕ_c over which the partial melt is interconnected but nearly stagnant; and (3) the concept that granitic melts in source regions with melt percentages below 20-30 vol. % are immobile and cannot lead to the generation of large plutons (Wickham, 1987) is questionable. Because large volumes of the continental crust in collision zones and other tectonic settings should experience incipient melting (with melt percentages never exceeding a few vol. % to 10 vol. %) for durations of 10^6 to several 10^7 years, the process of low-melt fraction segregation, and the possibility to generate some form of granite magmatism at melt percentages < 10 vol. %, deserve further investigation.

ACKNOWLEDGMENTS

This manuscript benefitted from discussions with Estelle Rignault, Pierre Boivin, Jean-Marc Montel and Daniel Vielzeuf. Assistance with the SEM by A.M. Mafille (CRMP, Université Blaise Pascal) was greatly appreciated. The experimental and analytical work were supported by a financial support from the Institut National des Sciences de l'Univers (contract no. 92-DBT-4.29; thème "Fluides, Minéraux, Cinétique"). We would like to thank J.L. Bouchez, D.L. Kohlstedt, T. Rushmer and J.-L. Vigneresse for their critical comments and suggestions.

APPENDIX: TECHNICAL INFORMATION ON RUNS QAN22C AND QORH2

For run Qan22c, the experimental techniques and starting materials are the same as those described by Laporte (1994, p. 487-489) for the experiments in the quartz-anorthite system. 8 mg of a mixture of 95 wt % anorthite + 5 wt % quartz were loaded in a platinum container, along with ~ 0.3 µl distilled water to ensure water saturation. The container was then welded shut, placed in a piston-cylinder assembly, and subjected to a pressure of 1000 MPa and a temperature of 1000 °C for 149 hours. The starting composition and the P-T conditions were chosen to produce about 10-20 wt % melt in equilibrium with anorthite (and excess water; Stewart, 1967). Sample QAn22c shows the expected three-phase paragenesis anorthite + melt + vapor. 9 electron microprobe analyses of glass (quenched melt) yielded an average composition of 65.9 (0.8) wt % SiO_2, 16.1 (1.0) wt % Al_2O_3, and 9.7 (0.5) wt % CaO (numbers in parentheses are standard deviations; see Laporte, 1994, p. 489, for analytical conditions).

For run QOrH2, the starting composition was a mixture of 88 wt % K-feldspar + 12 wt % quartz, and yielded ~ 35 wt % melt at 850 °C and a water pressure of 200 MPa (Shaw, 1963). The quartz powder (≤ 20µm) was that prepared by Laporte (1994); the K-feldspar powder (≤ 22µm) was obtained by grinding and sieving a gem-quality monocrystal from Madagascar (whose yellow color indicates the presence of Fe^{3+} in substitution for Al^{3+}). 28 mg of the mixture were loaded in a platinum container, along with 0.9 µl distilled water. After welding, the container was subjected to the P-T conditions of 200 MPa-850 °C for 328 hours in a cold-seal pressure vessel. The resulting paragenesis is K-feldspar (sanidine) + melt + vapor. K-feldspar forms euhedral grains that show a faint zoning; e.g. in BSSE microphotographs, rims, a few µm thick, appear slightly darker than cores, presumably because of a lower iron content. The average phase compositions in wt % are as follows: SiO_2= 68.5 (0.3), Al_2O_3= 11.5 (0.1), K_2O = 9.9 (0.2), Na_2O = 1.0 (0.1) for 9 glass analyses (analytical conditions for the electron microprobe are 15 kV, 5 nA, a counting time of 10 s, and a beam size of 8x8 µm; FeO was not analysed); and SiO_2= 64.4 (0.4), Al_2O_3= 16.8 (0.1), K_2O = 16.4 (0.3), Na_2O = 0.4 (0.0), FeO = 1.2 (0.1) for the cores of K-feldspar (9 analyses at the following condidions: 15 kV, 10 nA, a counting time of 10 s, and a beam size of 5x5 µm).

For both samples, the frequency distribution of apparent wetting angles was characterized by measuring with a protractor about 100 angles from BSSE microphotographs of polished sections (see Laporte, 1994, for details); magnifications ranged from 5000x to 14000x for sample QOrH2, and from 7000x to 15000x for sample Qan22c.

MAGMATIC FRACTURING AND SMALL-SCALE MELT SEGREGATION DURING PLUTON EMPLACEMENT: EVIDENCE FROM THE ADAMELLO MASSIF (ITALY)

Barbara E. JOHN
Department of Geology and Geophysics
University of Wyoming
Laramie, Wyoming 82071, United States of America
and Holger STÜNITZ
Geologisches-Paläontologisches Institut
Universität Basel
CH-4056 Basel, Switzerland

ABSTRACT

Foliated sheet-like structures in a quartz dioritic to granodioritic pluton complex (Lago della Vacca suite) in the southern Adamello Massif, Italy, provide insight into the process of magmatic fracturing and melt segregation during crystallization of a pluton. These structures occur as anastomosing, sinuous circumferential zones, which vary in width from millimetres to hundreds of metres, extending for one to hundreds of metres along strike. The composition of the zones is very similar to their lesser deformed host rocks. Microscopic evidence for melt migration and segregation in these zones comes from intra- and intergranular fractures and veinlets. With increasing melt volume, intergranular fractures coalesce and form through-going sheets (melt segregations). Larger melt segregations show a strong shape preferred orientation, which indicates substantial shortening normal to the sheet walls. The shortening direction remained constant throughout emplacement.

If, at low melt fractions, melt pressure exceeds cohesion along the foliation, fractures can open normal to the shortening direction. Alternatively, dikes may also form as shear fractures parallel to the plane of anisotropy (at σ_1 = 30-45° to the foliation). At higher melt fractions, segregated melt may flow down a pressure gradient (along the extension direction) into the foliation plane. Deformation will in any case be localized in zones of weakness, which consist of partially solidified material, and leads to deformation-assisted melt segregation, a potentially important deformation process during pluton emplacement.

J.L. Bouchez et al. (eds.) Granite: From Segregation of Melt to Emplacement Fabrics, 55–74.

INTRODUCTION

Geologic evidence for the generation, segregation, ascent and emplacement of silicic magmas is fragmentary at best. Much progress has been made in understanding the petrogenesis (source and evolution) of silicic magmas through experimental, isotopic and geochemical studies. There is much controversy however, regarding the mechanism(s) of migration of silicic magma from its source, through significant thicknesses of the crust, to the final site of emplacement. The mechanism of melt segregation during the crystallization of melt has received little attention to date. Experiments and/or natural examples documenting the processes of silicic melt segregation during emplacement are rarely observed. Good natural examples of inhomogeneous melt distribution in a silicic magmatic system are preserved in the southern Adamello Massif, Italy, and provide insight into the processes of magmatic fracturing and melt segregation during pluton emplacement.

The term 'melt segregation' is used to refer to the separation of a melt fraction from its residuum and/or source or crystallization products, and may occur during either melting or crystallization. Melt segregation is usually considered a phenomenon responsible for the extraction of melt from a rock undergoing partial melting (e.g. Spera, 1980; Wickham, 1987; Sawyer, 1994; Brown, 1994). The conditions for melt segregation during crystallization differ from those associated with anatexis, in that the preferred orientation of crystals, the melt composition, and crystal/melt ratios of the framework crystals are different (Brown et al., 1995; Vigneresse et al., 1996). Melt segregation during crystallization produces an inhomogeneous distribution of melt and crystals, potentially leading to deformation concentrated in the mechanically weakest parts of a material, that is in the zones of highest melt-fraction. This inhomogeneous distribution of melt is of great interest for the potential localization of deformation in melt-rich sections of the crust.

The purpose of the present study is to document the geometry and microstructures of naturally occurring silicic- to intermediate composition melt in granitoid rocks frozen in the process of segregation. The migration and segregation of melt in this instance occurred during magma emplacement and crystallization rather than anatexis, and resulted from horizontal compaction segregation into dilatant fractures. Such melt segregations appear to be common features in granitoid plutons during emplacement (Cuney et al., 1990; Brun et al., 1990; Nicolas, 1992; Pons et al., 1995), and their study may help reveal mechanisms of magmatic deformation (fracturing and fabric development) related to melt distribution.

GEOLOGIC FRAMEWORK OF THE ADAMELLO MASSIF

The Adamello Massif, in the Italian Alps, is the largest of the Tertiary intrusions emplaced during the Alpine orogeny. The batholith is exposed over an area greater than 670 km^2 with up to two kilometres of vertical relief, and is elongate in a northeast-southwest direction with a maximum length of about 80 km (Fig. 1). The southern Adamello Massif was emplaced at intermediate to shallow crustal levels into the southern Alpine crystalline basement and cover rocks. The cover rocks were deformed during the Orobic phase of the Alpine orogeny into a series of variably tight, upright folds striking roughly WSW-NNE, and subsequently intruded by the massif (Brack, 1983; 1984). Intrusive relations within the Massif indicate that it evolved as a series of discrete magmatic pulses, emplaced progressively from the southwest to the northeast, resulting in a composite batholith of at least twelve distinct plutons of quartz diorite, tonalite and granodiorite composition (Bianchi et al., 1970). Existing estimates of the emplacement pressure vary from 1.5 to more than 3.5 kbar over the southern part of the Massif (Riklin, 1983; John and Blundy, 1993). This study focuses on the Lago della Vacca Suite (LVS), a small pluton in the southernmost part of the Ré di Castello Group of the southern Adamello Massif (Fig. 1; Bianchi et al., 1970; Brack, 1983; Ulmer et al., 1983).

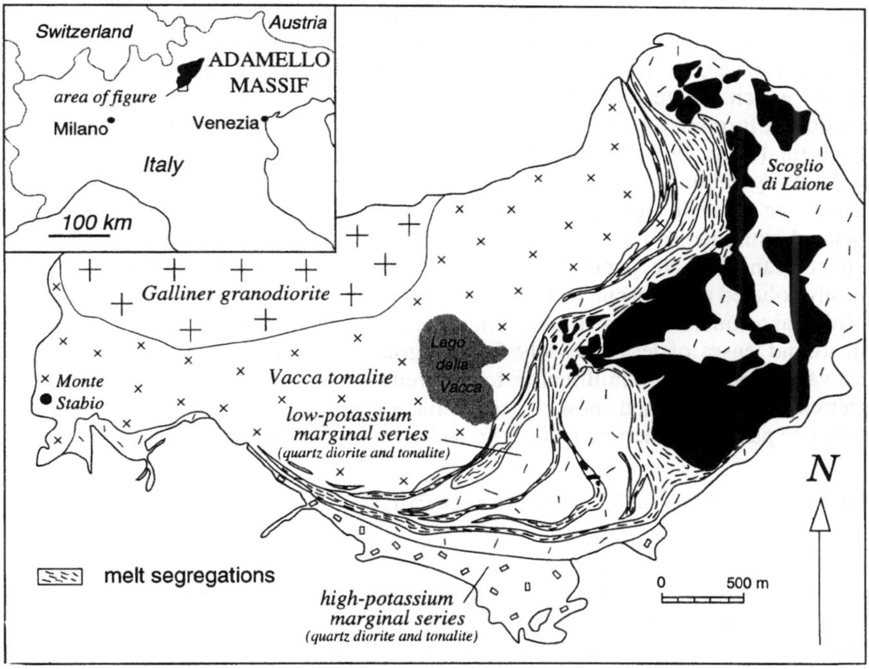

Figure 1. Lago della Vacca suite (southern Adamello Massif, Italy): simplified geologic map showing the distribution of rock types, and melt segregations within the marginal series.

STRUCTURAL AND PETROLOGIC FEATURES OF THE LAGO DELLA VACCA SUITE

The Lago della Vacca Suite forms a small semi-circular intrusion, roughly 4.5 km by 4.7 km, emplaced into folded Mesozoic sedimentary rocks and precursor Tertiary intrusive rocks, including coarse-grained mafic and silicic granitoids (Blundy, 1989; Blundy and Sparks, 1992), and a layered mafic/ultramafic cumulate sequence, the

Blumone complex (Ulmer, 1986). Biotite Rb-Sr (Del Moro et al., 1983), and U-Pb zircon geochronology (Hansmann, 1986; Hansmann and Oberli, 1991) data imply that the LVS was emplaced and cooled to temperatures of less than 400°C between ~43 and 40 Ma. The suite is truncated to the north and west by younger intrusive rocks (Fig. 1). Structural considerations imply that emplacement of these later intrusive bodies, as well as subsequent Alpine deformation have not affected the original geometry and orientation of the Lago della Vacca Suite (John and Blundy, 1993).

The Lago della Vacca Suite exhibits crude concentric compositional and temporal zonation from hornblende quartz diorite, through tonalite to granodiorite. Intrusive relations indicate that age decreases and silica content increases inward from the margins (Fig. 1). The suite was divided by John and Blundy (1993) into four members based on texture and mineralogy, including the outer or marginal units comprising heterogeneous quartz diorite, tonalite, and their sinuous zones of segregated melt. The marginal units are intruded inward by homogeneous, equigranular tonalite and granodiorite, the Vacca tonalite, which in turn is cut by a medium-grained granodiorite, the Galliner granodiorite, forming the core of the pluton. Irregular, penetrative contacts between core and marginal units, though only rarely observed, indicate that the core units are younger than the marginal units.

Evidence for melt segregation within the suite comes from small sheet-like features associated with intense foliation development. The long axes of oblate mafic enclaves in the host granitoids are parallel in orientation to margins of the dikelets. The most intensely foliated rocks occur in the older marginal units (tonalites and quartz diorites), and form anastomosing and sinuous ring-like zones of quartz diorite and tonalite that wrap around lenses of less intensely foliated granitoid of the same composition. These zones vary in width from millimetres to hundreds of metres, extend for tens to hundreds of metres along strike, and show a strong shape preferred orientation of plagioclase and hornblende parallel to their margins. The wider zones contain mafic enclaves.

CONTACT RELATIONS AND FABRIC DEVELOPMENT

Contacts between the four major units in the LVS are sharp in some places while in others they are marked by mineralogical and textural changes that may be diffuse over a distance of several metres. This implies that successive magma pulses or units of the LVS were emplaced before the precursor pulse had fully solidified. The outer contact of the LVS with surrounding country rocks is variable in nature and orientation, depending on host rock type. Intrusion of the marginal units, the earliest pulses of magmatism, generated dikes and thin sheets in adjacent granitoid rocks and stoped fragments of country rock (Fig. 1). Deformation of host rocks by intrusion of the LVS is almost exclusively brittle in character. Subvertical dikes and apophyses of the marginal quartz diorite exploited pre-existing fractures in the host rocks, producing stoped blocks of coarse-grained gabbro now widespread in the southern and eastern parts of the suite (Fig. 1). The arcuate trend of these trains of stoped blocks suggests that they were displaced and rotated by the later, forcefully emplaced inner units. In contrast to external relations, contacts between the marginal and core units are typically vertical or steeply inward dipping (>60°) toward the center of the suite, and are associated with well-developed foliations (Fig. 2).

The dominant fabric in the intrusive rocks of the LVS is a foliation (Fig. 2). Magmatic foliation, defined by the planar alignment of igneous plagioclase, amphibole and biotite, and by mafic enclaves, is developed to varying degrees in all units of the LVS. Linear mineral fabrics are rarely observed: where present, they are subvertical. Foliations define steeply inward-dipping (or funnel shape) concentric trajectories that roughly parallel the marginal contact and, in general tend to be less well defined towards the core of the pluton. A similar trend is observed in the axial ratios of mafic enclaves which tend to decrease towards the pluton core (John and Blundy, 1993). Adjacent to the southern and eastern borders of the Vacca tonalite, foliation trajectories are oblique to the contact between the tonalite and the surrounding marginal units. This relationship is particularly evident where the contact is irregular (north west of Cornone di Blumone), and suggests that fabric development and magma intrusion occurred synchronously during emplacement. In other words, early formed fabrics were repeatedly truncated and reoriented by subsequent magma pulses.

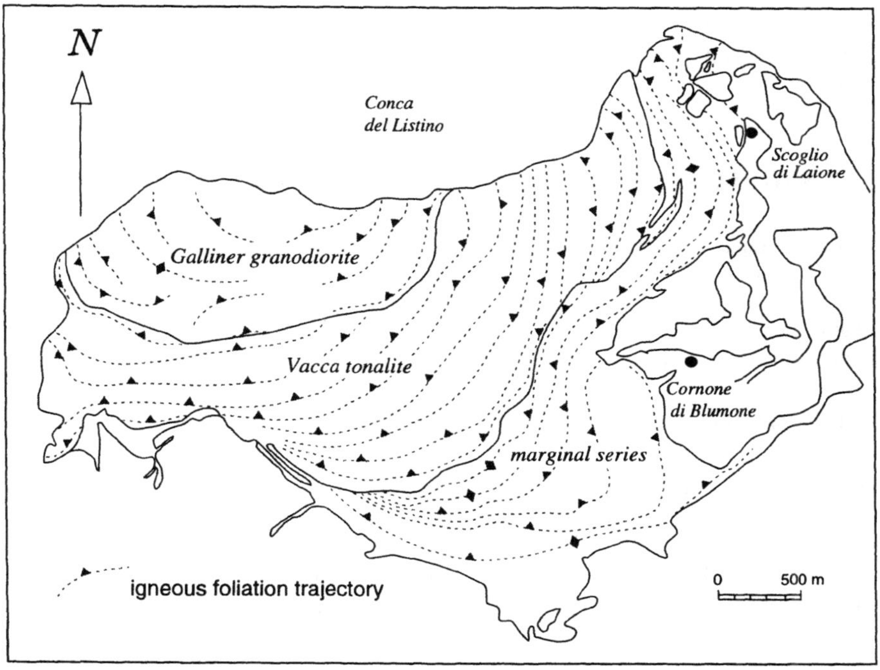

Figure 2. Lago della Vacca suite (southern Adamello Massif, Italy): generalized foliation trajectory map (after John and Blundy, 1993).

PETROLOGY AND GEOCHEMISTRY

Units of the LVS form a chemically related suite of granitoid rocks which define smooth linear trends on Harker variation diagrams (Kagami et al., 1991; John and Blundy, 1993). Quartz diorite and tonalite from the marginal units range in silica from 50 to 64 wt.%, and were divided into low- and high-potassium members (Fig. 1), based on biotite content. Those with more than 10 vol.% biotite form the high-potassium units (K_2O greater than 1.3 wt.%), while those with amphibole in greater proportion than biotite form the low-potassium units (K_2O less than 1.3 wt.%). The younger Vacca and Galliner units form the core of the suite, with a small range of 60 to 65 wt.% silica, high potassium concentrations, and biotite in excess of hornblende. The fine-grained marginal units described as melt segregations show large variations in SiO_2-content (52-64 wt.%) and modal mineralogy (John and Blundy, 1993). They were mapped on the basis of their relatively fine grain size and intensity of foliation development, rather than on mineralogy or chemistry. Consequently, the marginal units represent a textural facies, rather than a facies based on composition. Major and trace element concentrations indicate a complete overlap in composition between strongly and less foliated samples of both low and high-K marginal units. This implies a local origin for the melt segregations, rather than new magma input (John and Blundy, 1993).

The LVS was emplaced at ~350 MPa pressure, with estimated temperatures ranging from 580° to 850°C, with the vast majority clustering around 680°±50°C, at or close to the water-saturated granitoid solidus at pressures of 250 to 350 MPa (John and Blundy, 1993).

MELT SEGREGATION IN THE LAGO DELLA VACCA SUITE

The most intensely foliated rocks in the LVS occur as sinuous, anastomosed, concentric lenses, sheets or dikelets of fine-grained diorite, quartz diorite, and tonalite within the marginal units. These zones are labeled "melt segregations" in Figure 2, and are traceable in a 200° arc around the margin of the suite; they do not intrude into the surrounding country rocks.

MACROSCOPIC FEATURES OF MELT SEGREGATIONS

The melt segregations are characterized by aligned grains of plagioclase, hornblende and, rarely, biotite in a fine-grained foliated matrix of the same mineralogy. These zones vary in width from centimetres to hundreds of metres and extend for tens to hundreds of metres along strike. The melt segregations are characterized by relatively higher concentrations of mafic enclaves: 11-35 inclusions/m^2 compared to 2-10/m^2 in adjacent less intensely foliated rocks. Contacts between the fine-grained zones and their host rocks are remarkably sharp with no intermediate grain sizes on the scale of one to two millimetres (Fig. 3). Some of these foliated rocks have textures similar to synplutonic mafic dikes, whereas others resemble fine-grained porphyritic equivalents of adjacent marginal rocks.

Three-dimensional strain analysis using mafic enclaves from throughout the pluton indicates that these bands represent zones of very high flattening strain (John and Blundy, 1993). The enclaves are weakly prolate in the core of the pluton, and become increasingly oblate as strain increases towards the pluton margin. Their short axis (Z) is uniformly sub-normal to the foliation, implying radial flattening. Within the marginal units themselves considerable local variations in the amount of shortening over short distances are recorded, with the highest percentages of shortening (up to 84%) corresponding to the zones of melt segregation and intense foliation development. These zones therefore represent loci of very high strain separating lenses of relatively low-strain. The oblate nature of enclaves in the most deformed rocks, and the ring-like geometry of the foliated high-strain zones suggests that deformation involved dominantly radial flattening, associated with successive pulses of magma intrusion.

In the marginal units, widespread, subvertical, aplite dikes (1-20 cm wide) intersect the foliation and melt segregations at high angles. This late distribution of melt is consistent with expulsion of evolved interstitial melt along radially propagating fractures during the final stage of magma emplacement.

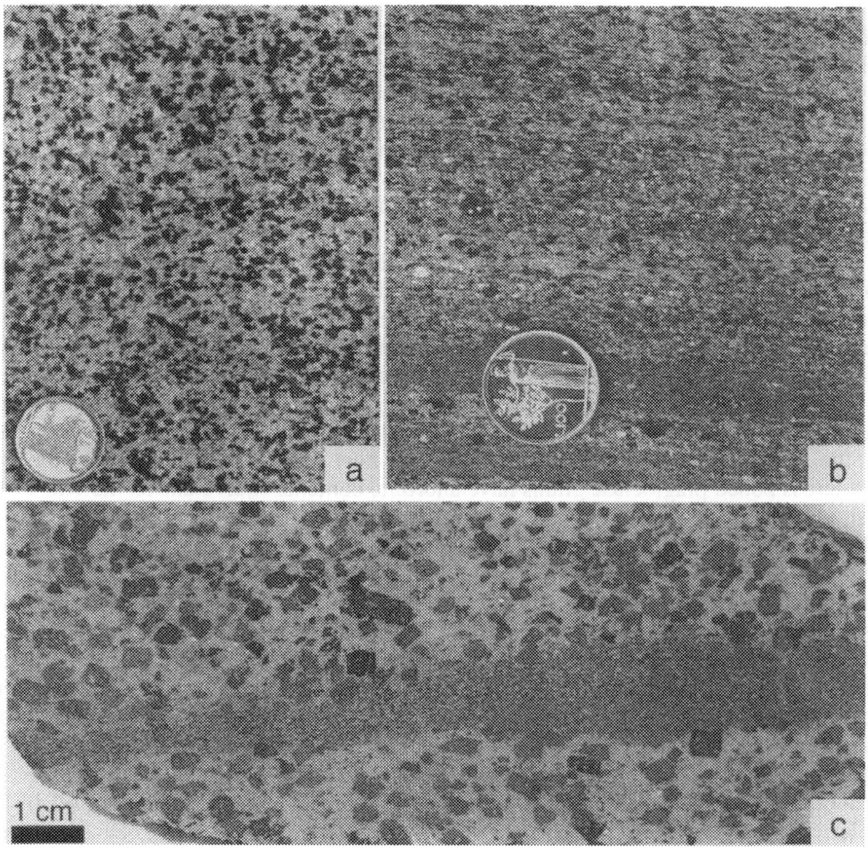

Figure 3. Macroscopic character of the Lago della Vacca Suite marginal series rocks and associated melt segregations. (a) Weakly foliated low-potassium marginal quartz diorite, with poorly aligned amphibole grains in matrix of plagioclase and quartz. Coin is 1.8 cm in diameter. (b) Within the same host rock type as (a) intensely foliated melt segregation and deformed mafic enclaves. 2.8 cm coin. (c) Slab cut normal to thin foliated melt segregation (dark fine-grained zone) cutting low-potassium marginal quartz diorite. Host rock type equivalent to Fig. 2a.

MICROSTRUCTURE OF MELT SEGREGATIONS

In the least foliated marginal rocks the groundmass has a grain size greater than 200 μm, and comprises an assemblage of amphibole (often showing reaction to biotite), biotite, plagioclase, quartz, alkali feldspar, oxides and apatite. Plagioclase, biotite and amphibole show a strong grain shape preferred orientation. Quartz and alkali-feldspar are interstitial and show no sign of plastic strain. Groundmass plagioclase is zoned from calcic cores (An_{51-85}) to sodic rims (An_{40-48}). This zoning pattern is observed in both coarse-grained and groundmass plagioclase in undeformed marginal rocks, and is considered to be a primary magmatic feature.

The melt segregations are characterized by anastomosing zones of fine-grained (50-200 μm) quartz diorite and tonalite that cut the original coarser-grained matrix. Biotite and hornblende within these zones define a strong foliation, though plagioclase and quartz (± alkali feldspar) are polygonal. The groundmass comprises the same mineralogy as the host, although biotite is more abundant, typically at the expense of amphibole, plagioclase is more homogenous (An_{32-54}), and the grain size is substantially finer (50-100 μm). Quartz in particular shows no evidence of grain flattening or other intracrystalline deformation features such as subgrain formation, indicating that all deformation has occurred in the magmatic state.

Microscopic evidence for melt migration and segregation associated with late-stage emplacement of the LVS comes from intra- and intercrystalline fractures and veinlets that characterize the marginal units. In some samples, fractures are confined to a few plagioclase grains, while in others the thin fracture linings coalesce to form through-going dikelets, or anastomosing bands, of fine-grained material.

The smallest traces of melt are present as transgranular fractures through plagioclase and hornblende grains (Fig. 4a). The material filling these small intragranular fractures has the same crystallographic orientation as the host grain (epitaxial growth). Fracture fills across zoned plagioclase grains have anorthite contents equivalent to the rim of the host plagioclase (Fig. 4a). These microstructures are similar to those described by Hibbard (1987), and Bouchez et al. (1992). Some, usually wider, transgranular fractures have rim overgrowths of plagioclase, while the center portion of the fracture fills are composed of small grains (Fig. 4b). Transgranular fractures are typically apophysical cracks extending from wider, polyphase, fine-grained melt segregations (Fig. 4e). These wider melt segregations almost exclusively fill *intergranular* fractures. The fact that these larger melt segregation features have a fracture character is shown by the transition from the major melt-filled zones into their extremities, where the crystallized material is transitional between epitactic growth and polyphase grain aggregates (Fig. 4e). The filling melt forms networks through the already solid framework of crystals (Fig. 4f). Generally, these networks are parallel to the foliation in the host rock defined by the preferred orientation of hornblende and plagioclase.

With increasing melt abundance, melt segregated into layers, a few grain diameters to ~3 cm in width. The average grain size ranges from 30 to 60 μm in the melt networks and thin layers, to 0.2 to 1.2 mm in the wider layers. Isolated large grains (3-4 mm) of hornblende and plagioclase removed from the already solidified host, are commonly suspended in the fine-grained melt segregations (Figs. 4a, b, e, and f). These larger grains or porphyroclasts are characterized by either plagioclase inclusions in hornblende or cores of high-An-content in plagioclase grains, and make up roughly 0-20 modal percent of the thicker segregations. Mafic enclaves are common in the wider melt

Figure 4. Magmatic fractures and melt segregations in the Lago della Vacca Suite. Sequence by increasing melt percentage. (a) Healed intergranular fracture (~10-30 μm wide) in a normally zoned plagioclase grain. The fracture fill comprises epitaxially grown plagioclase of equivalent composition to that of the grain rim. Scale bar =250 μm; (b) Transgranular fracture (~30 μm) of plagioclase grain. The fracture fill consists of plagioclase grown against the fracture walls and individual plagioclase grains with a different orientation in the crack center. Scale bar = 50 μm; (c) Transgranular fractures constitute the extremities of the larger, vertically oriented, intergranular fracture filled with fine-grained hornblende, biotite, plagioclase and quartz (center). Transgranular fractures of plagioclase grains (left and right of the center material) are filled with plagioclase of the same composition as the rim of the clast (right). Scale bar =250 μm; (d) Overview of the melt segregation in (c). The intergranular fractures form networks along grain boundaries of the host. Scale bar =1 mm; (e) Overview of the central part of a thick melt segregation with inclusions of host rock plagioclase (arrows) removed from the dike wall and fully surrounded by the fine-grained material of the segregated melt. Scale bar =250 μm; (f) Close-up view of aligned, zoned plagioclase cores and overgrowths (coming from the surrounding host rock) incorporated in subequigranular matrix of plagioclase, biotite, amphibole and quartz forming the melt segregation. Scale bar =250 μm.

segregation veins. These enclaves have their long axis parallel to the foliation of the veins. Their grain size is smaller than that of the surrounding melt segregations (40 - 200 μm), and composition more mafic than that of their host matrix (Table 1).

Grain shapes in the segregation networks and narrow melt segregation layers are typically equigranular; in wider segregation layers, both hornblende and plagioclase crystals have elongate and prismatic shapes. Generally, the hornblende grains occur between the idiomorphic plagioclase grains, whereas in the surrounding host rock of similar composition, both plagioclase and hornblende have hypidiomorphic shapes with short prismatic hornblende crystals and more elongated plagioclase crystals (Figs 4e, 5).

TABLE 1. Modal composition of adjacent host rock, melt segregations and included mafic enclave in one sample of the Lago della Vacca suite marginal series diorite.

	melt segregation	host rock	mafic enclave
plagioclase	62	61.5	50.4
hornblende	31.3	29.5	45
magnetite	3.4	3	2
biotite	2.3	3.5	2.4
sphene	0.3	0.6	0
quartz	0.8	1.6	0.2

Microstructural analysis by Autocorrelation Function

The preferred orientation of minerals, mainly plagioclase and hornblende, in both the melt segregations and surrounding host rocks were analyzed using the autocorrelation function (ACF) described in detail by Panozzo-Heilbronner (1992). The fabrics were studied in thin sections cut normal to the foliation and parallel to any weak lineation visible in hand specimen.

The autocorrelation function can be envisaged as a brightness distribution, obtained when an image is superimposed on itself at all possible x and y translations (Panozzo-Heilbronner, 1992). The brightness distribution reflects the spatial frequencies of the fabric elements in all directions, that is the ACF reflects the fabric as a whole without distinguishing between individual fabric elements. The ACF is represented as a stepped and grey-level-contoured diagram of the brightness distribution (Figs. 5 and 6). Images used for the analysis are digitally recorded optical micrographs taken under crossed nicols with both the polarizer and analyzer oriented at 45° with respect to the foliation. Under these conditions, the polysynthetic twinning in plagioclase, and hornblende pleochroism stand out most ostensibly. As both hornblende and plagioclase display a crystallographic as well as shape preferred orientation in the magma, this fabric is analyzed most favorably under these conditions. The ACF uses the integral image without separation of individual features such as grain outlines or twin lamellae. The crystallographic fabric of plagioclase, visible through the subparallel alignment of its twin lamellae of grains, and the shape fabric of hornblende and plagioclase, visible as parallel alignment of grain boundaries and long axes, both contribute to the ACF analysis. The result of the ACF analysis is therefore a composite signal, reflecting both shape and crystallographic fabric. This combination of different types of data is justified in the case of magmatic fabrics, because the crystallographic fabric of plagioclase and hornblende arises from the

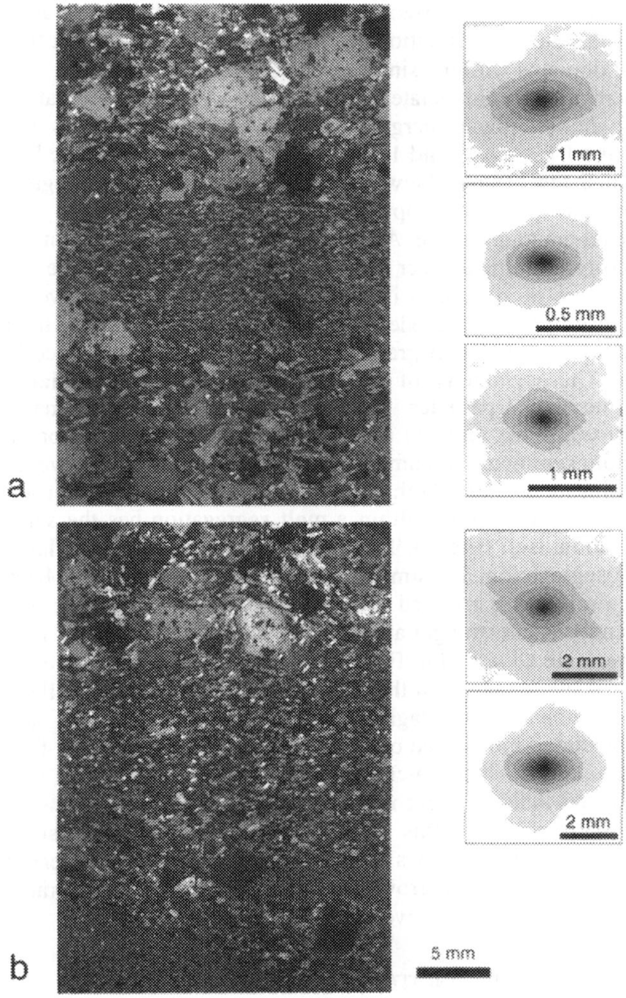

Figure 5. Microstructure and ACF analysis of a melt segregation within the marginal series, coarse grained quartz diorite. (a) The section shows parts of coarse grained host rock (top and bottom) with fine grained melt segregation in the middle. The ACF fabric diagrams on the right refer to the adjacent parts of the section (top and bottom to the coarse grained part, middle to the melt segregation). The analysed region corresponds approximately to half of the area of the individual parts of the section. The fabric anisotropy is greater in the melt segregation than in the coarse grained host rock. This suggests that the fabric in the melt segregation developed when the host rock was nearly solidified. (b) Another section of coarse grained host rock (top) and finer grained melt segregation (below). The very fine grained portions represent mafic enclaves. The upper ACF fabric diagram corresponds to the coarse grained host rock, the lower to the melt segregation. The fabric anisotropy is greater in the melt segregation than in the host rock. For fabric analysis of the mafic enclave see Fig. 6.

alignment of these grains as rigid particles in a melt, and therefore depends on their shape anisotropy. Technically, the half-peak-height of the ACF as a good measure of the grain size, and anisotropy (Panozzo-Heilbronner, 1992), has been used to obtain the aspect ratio, grain size and inclination of the anisotropy, with respect to a reference line identical for all domains within a single thin section.

Typical microstructures associated with the zones of melt segregation were analyzed (Fig. 5). Host diorite (low-K-marginal series) shows a moderately to well-developed fabric anisotropy of 1.39, 1.44 and 1.28. The inclination of the fabric long axis does not vary more than a few degrees between samples. Included melt segregations show a significantly greater fabric anisotropy of 1.53 and 1.68.

The grey-level-contours of the ACF are directly related to grain size, the smaller inner ellipses corresponding to finer grain size, the larger outer ellipses to coarser grains. A systematic feature of all fabrics is that the melt segregations have a constant fabric symmetry (ACF ellipses) independent of their grain size, whereas in the surrounding host rock the ellipse of the larger grains is more inclined to a reference line than that of smaller grains. The asymmetry of the coarser grained host material is due to poor alignment of the larger particles relative to the finer-grained material. The melt segregations have a rather uniform and finer grain size, and therefore tend to show a stronger and more symmetric alignment of particles. However, the overall orientation of alignment remains identical for both melt segregations and host material.

The fabric of a mafic enclave within a melt segregation has the same orientation as the melt segregation itself (Fig. 6). The fabric anisotropy of the enclave is weaker than that of the melt segregation, and similar to that of the coarse grained host rock. As with the host rock, the fabric is inclined to the reference line. The fabric is weaker near the margin of the enclave and stronger at its centre, where the grain size is larger; note that the ACF diagrams are all scaled in Fig. 6, and thus record the differences in grain size. The weaker fabric at the margin of the enclave is in part due to smaller aspect ratios of the individual hornblende and plagioclase grains. The finer grains at the margin are interpreted to be due to more rapid cooling, in turn time was lacking for the particles to develop a shape fabric. Such microstructures are often observed in mafic enclaves (Vernon, 1990), and indicate quenching of the mafic melt against the host rock with a lower solidus temperature (in this case the melt segregation). Distribution of fabric anisotropy in the enclave shows a trend similar to that observed in the melt segregations: the fine-grained, narrow networks show smaller aspect ratios of grains than wider layers and therefore do not develop a preferred orientation.

PHYSICAL AND CHEMICAL CONDITIONS OF MELT SEGREGATION

Geochemical and modal analyses of the host and adjacent melt segregations in the marginal series rocks were carried out to document both the composition and conditions of melt segregation. Care was taken to analyze segregations as much as possible without large porphyroclast or xenocrystic grains of plagioclase and hornblende. Melt segregations in the LVS vary from quartz diorite to tonalite, and have the same mineralogy as their *local* host rocks. Data in Table 1 indicate that the adjacent fine-grained melt segregations and surrounding diorite of the low-K marginal series have identical modal abundance of mineral phases. Similarly, the major and trace-element geochemistry (Fig. 7 and Table 2), shows that the composition of the melt segregations are virtually identical to their adjacent host. Chemical equilibrium between the host and

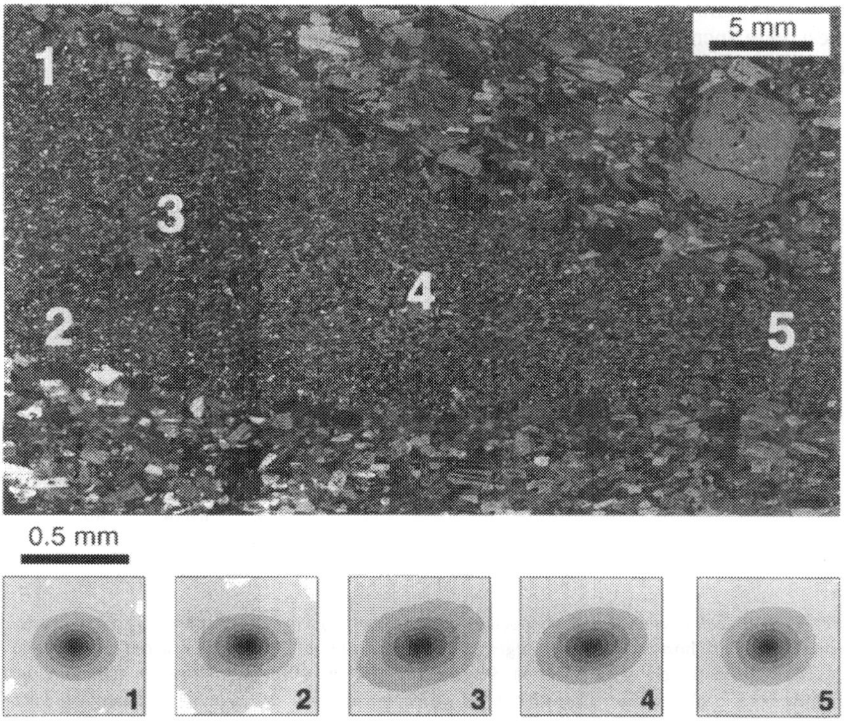

Figure 6. Microstructure and ACF analysis of a mafic enclave within a melt segregation dike, whole rock and plagioclase chemistry: The mafic enclave is from the lower portion of figure 4. The numbers of the ACF fabric diagrams correspond to the location of the analysed regions of the enclave. The fabric anisotropy is weaker at the margins (1, 2, 5) than in the center (3, 4) of the enclave, where the grain size is larger. The fine grained margins are interpreted as more rapidly cooled against the host rock, which has a lower solidus temperature.

melt segregations may have been enhanced by high diffusion rates due to the presence of melt. Yet these relations imply that the melts must have been derived from the same melt, and are of non-eutectic composition. Variations in TiO_2, MgO, and K_2O, and also Ni, Cr not shown in Figure 7, are likely due to the incorporation of small, xenocrystic amphibole grains that were incorporated during stoping of the nearby Blumone Complex.

Mineral chemistry

Amphibole compositions in the melt segregations have relatively high concentrations of potassium, with consistently lower titanium contents and Mg number than in their adjacent host (John and Blundy, 1993). The increase in K_2O and decrease in Mg number is consistent with petrographic evidence of amphibole breakdown to biotite, common in natural and experimental systems of igneous hornblende. At temperatures between 800°C and 900°C evolved interstitial melt reacts with hornblende to produce biotite

(Naney, 1983). Therefore during segregation, migration and emplacement of late-stage, slightly peraluminous silicic melts, there is a continuous reaction: hornblende shows progressively increasing K_2O-contents and decreasing Mg number, and is eventually replaced by biotite (John and Blundy, 1993).

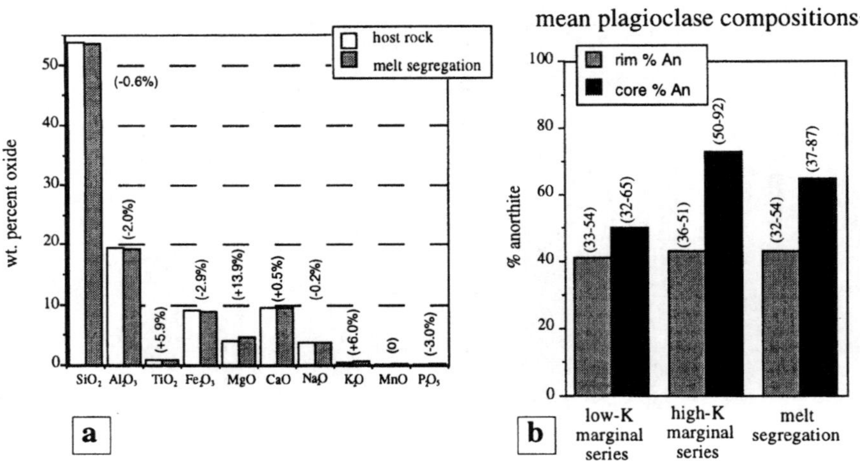

Figure 7. Whole rock chemistry and plagioclase composition. (a) Histogram comparing whole-rock major element chemistry of host rock and melt segregations in a sample of the marginal series of the Lago della Vacca Suite (same sample as in Table 1). Percentages give the differences between host and melt. (b) Mean and range in plagioclase compositions (% anorthite) for melt segregations and associated marginal series rocks of the Lago della Vacca Suite. Note the high anorthite content and nearly complete overlap in composition (both core and rim) between the segregations and their surrounding host implying non-eutectic conditions in their formation.

Small (50-100 μm), subequant grains of plagioclase in the melt segregations have a range in anorthite content (An_{32-54}) that overlaps completely with plagioclase rim compositions (An_{33-54}) in the adjacent host (Fig. 7). Captured porphyroclasts of plagioclase in the segregations locally show myrmekitic intergrowths along their grain boundaries, and show the full range in anorthite content noted in their surrounding host. All of these observations indicate that the melt segregations are of very similar composition to their host rock.

Temperature

Emplacement temperatures for melt segregations were calculated using the amphibole-plagioclase thermometer of Holland and Blundy (1994). All temperature estimates for each rock type of the LVS given in John and Blundy (1993), were made on adjacent amphibole and plagioclase grains assuming an estimated emplacement pressure of 3.5 kbar. Detailed amphibole-plagioclase thermometry on melt segregations from two samples overlap within error with their adjacent host rock temperatures: host 1, 762°±50°C; melt segregation 1, 736°±50°C, and host 2, 748°±50°C; melt segregation 2, 758°±50°C, and indicate that the segregation process occurred in the magmatic or "sub-magmatic" state.

TABLE 2. Comparison of whole-rock major and trace-element geochemistry for both host rock and melt segregation from one outcrop of the low-potassium marginal phase of the Lago della Vacca suite.

SAMPLE	JJ13A melt	JJ13C melt	JJ13D host	JJ13E host
SiO_2	53.16	53.62	53.90	53.56
Al_2O_3	19.25	19.01	19.53	19.54
TiO_2	0.91	0.88	0.78	0.91
$Fe_2O_{3(tot)}$	8.86	8.72	8.99	8.96
MgO	4.54	4.60	3.99	4.03
CaO	9.63	9.46	9.41	9.58
Na_2O	3.88	3.82	3.92	3.80
K_2O	0.59	0.59	0.55	0.54
MnO	0.22	0.22	0.21	0.22
P_2O_5	0.31	0.30	0.33	0.30
Total	101.35	101.23	101.62	101.44
Ba	218	239	215	200
Co	22	20	21	21
Cr	53	60	29	14
Cu	40	26	67	29
Ga	19	19	20	20
Ni	19	21	7	6
Nb	8	6	6	8
Pb	3	6	4	4
Rb	12	11	10	9
Sr	514	500	527	517
V	220	215	227	226
Y	28	27	23	27
Zn	82	79	79	81
Zr	73	76	75	61

DISCUSSION

The following observations have to be reconciled for an emplacement model of the melt segregations in the Lago della Vacca suite:

1. The melt segregations fill fractures parallel to igneous foliation in their host rocks.

2. The fabric in the melt segregations is parallel to the walls of the dykes and foliation of the surrounding host rock. The fabric in the melt segregations is stronger than that of their host.

3. The composition of the melt segregations is only slightly more evolved than that of the host rock.

4. Mafic enclaves occur in the melt segregations, also with a fabric in the same orientation as the segregations.

ORIENTATION OF MELT SEGREGATIONS AND FABRIC DEVELOPMENT

The fact that the fabric of the melt segregations is parallel to that of the igneous host rock suggests that both the shortening and flow directions remained relatively constant during solidification of the host rock, and during fracture and emplacement of the melt segregations. The fabric in both the host and melt segregations indicates bulk shortening normal to fracture orientation, requiring the greatest compressive normal stress (σ_1) to be oriented at a large angle or normal to the fracture walls. If the intrusion of segregated melt is considered to have filled extension fractures under high melt pressure, the inferred stress field for the initiation of such fractures would be such that σ_1 be oriented parallel to the fracture walls (e.g. Anderson, 1951; Shaw, 1980). The stress field for initiation of fractures, and that implied for the magmatic fabrics therefore appear inconsistent. Several possibilities are suggested to reconcile these apparently contradictory observations:

1. A difference in tensile fracture strength exists, due to the anisotropy of the material resulting from the fabrics of hornblende and plagioclase. If the tensile strength parallel to the foliation exceeds the strength normal to the foliation, and if the difference of effective normal stresses does not exceed that difference, fractures parallel to foliation may form at magma overpressures greater than σ_1. This explanation has been proposed for foliation-parallel veins by Gratier (1987) and Wickham (1987), and for larger scale intrusions by Lucas and St. Onge (1995). This explanation requires that the difference in effective stress (σ_1 - σ_3) be very small, as the difference in tensile fracture strength in different directions of the anisotropic fabric will be very small and fractures do not open parallel to σ_1.

On a crustal scale, alternate models have been proposed for sill-dike emplacement at high angles to the regional σ_1 direction, principally to explain the development of horizontal crustal layering during lithospheric extension (McCarthy and Thompson, 1988; Parsons et al., 1992). These models suggest that differences in deviatoric stress supported by layers of contrasting rheology, coupled with high magma pressure can result in principal stress interchanges in the weaker layers that enable extension fracturing normal to σ_1. In this orientation, the well developed crystallographic fabric may enhance the seismic velocity anisotropy character of the layering.

2. The shear strength of anisotropic material is the smallest at an angle of 30 to 45° between the plane of anisotropy and σ_1 (Donath, 1964), and fracture occurs parallel to the plane of anisotropy. The local stress field in the host rock may have produced such shear fractures parallel to foliation, and flow of the melt inside the segregations may have aligned the particles parallel to the fracture surfaces.

The first explanation assumes both a minimum tensile strength and limited strength differences in the host rock during emplacement of the segregations. This assumption may only be valid for nearly solidified rocks, that is for very small melt fractions, as the tensile strength of the grain boundaries (the melt segregations mainly form intergranular fractures) in such grain aggregates must not be zero. A similar argument applies to shear fracture development, because a finite strength between grain boundaries is needed to constitute an anisotropy. These two explanations may only apply to the development of melt segregations in almost completely solidified rock.

A certain width (and presumably cooling time) of the melt segregation is necessary to develop grains with high aspect ratios, and more appropriate to give rise to a preferred orientation. In mafic enclaves, the ACF analysis implies a minimum width for development of a strong fabric of approximately 4-5 mm in mafic enclaves. Planar melt segregation layers do not show marginal quench zones, as both their composition and solidus temperature are equal to those of the host rock, and develop a strong fabric even at minimal thicknesses. Where mafic enclaves occur in the melt segregations, they have a finer grain size (Fig. 6), and a hornblende/plagioclase ratio higher than the melt segregations and adjacent host rock (Table 1).

ROLE OF INTERNAL ANISOTROPY DURING PLUTON EMPLACEMENT

An interesting aspect of repeated emplacement of the melt segregations in the pluton is that fracturing took place parallel to an existing foliation. The fact that the fabric in the melt segregations is better developed than that of the host rock can be explained by intrusion of the melt segregations into a largely solidified host. The host in turn acted as a mechanically rigid carapace, affording stronger fabric development in the melt segregations due to continued shortening normal to the dike walls. The development of anisotropy or foliation is therefore a self-stabilizing process through the formation of sheet-like melt segregations. Thus, inhomogeneous distribution of melt due to crystallization and repeated melt segregation and re-injection into the more solidified host rock is not only controlled by, but also enhances the anisotropy, and therefore appears to be an important process for fabric development in plutonic rocks during emplacement.

Layer and foliation parallel emplacement of granitoid magma is an important process in the regional development of compositional layering in the middle and lower crust (Lucas and St. Onge, 1995). Syntectonic intrusion of dikelets into extension fractures sub-parallel to layering is interpreted to have occurred due to the inherent anisotropy and high magma pressure (see Gratier, 1987; Wickham, 1987; Holliger and Lavander, 1994; McCarthy and Thompson, 1988; Parsons et al., 1992). The geologic observations outlined above from the marginal series of the LVS support a model for the origin of foliation parallel melt segregations emplaced along extension or shear fractures (cf. Etheridge, 1983; Clemens and Mawer, 1992), subparallel to igneous layering/foliation formed during the emplacement and deformation. In addition, syntectonic intrusion may in this way focus deformation into transiently weak zones (ie. melt-rich, cf. Hollister and Crawford, 1986; Davidson et al., 1994) by partitioning deformation into partially solidified granitic rocks/magma and/or pluton margins.

SOURCE OF THE MELT SEGREGATIONS

The melt segregations inject almost completely solidified host rock and are chemically only slightly more evolved than their host material. Thus, they must be of local origin, i.e. do not result from a new magmatic pulse. By the very nature of the problem (only the intrusion of segregated melt into the host rock is preserved), inferences about the source conditions remain speculative and are based on experiments or modeling. Models involving percolation theory, suggest that crystals in a crystallizing melt form a framework capable of supporting hydrostatic stress at approximately 55% crystals (Vigneresse et al., 1996). At this third or "rigid percolation threshold" of Vigneresse et al. (1996) melt may segregate from the framework during compaction. Such melt segregation is enhanced by deformation (Rutter and Neumann, 1995; Brown et al., 1995). Thus, two types of domain may have existed in the pluton: one, where crystallization of the melt is almost complete (the host to the melt segregations), and another domain, where melt segregated from a deformation-assisted compacting crystal framework that contained a high melt fraction. The exact melt fraction, of course, is impossible to determine, but that the melt fraction was rather high is shown by the only slightly more evolved composition of the melt segregations (Table 2). This scenario is also supported by the fact that the melt segregations contain mafic enclaves, which may inject into or be present in larger melt fractions, but are difficult to explain by only locally segregated melt. The exact spatial distribution of these domains is difficult to reconstruct in the LVS, but it is likely that the high melt fraction domains were located below the more solidified domains.

Once melt has segregated, it will flow down the pressure gradient, i.e. parallel to the σ_3 direction, as observed in the experiments by Dell' Angelo and Tullis (1988), in natural examples (Stevenson, 1989), and in melt percolation models (Vigneresse et al., in press). The σ_3 direction coincides roughly with the preferred orientation of the grain boundaries of the already existing anisotropy, approximately normal to σ_1. Segregated melt along axial planes of folds and other flattening planes have been observed by Ramsay (1958), McLellan (1988), Stevenson (1989), Hand and Dirks (1992), and Williams et al. (1995). This flow of segregated melt, associated with fracturing, produces the intrusion of the dikes into the largely solidified domains of the pluton. The rate of flow of the melt has to be faster or equal to the strain rate of the material as a whole for this mechanism to occur (Dell'Angelo and Tullis, 1988). This condition appears to be satisfied, as the melt segregations developed a stronger fabric than their host, indicating higher strain.

The processes of melt transfer from large-volume melt domains to nearly solidified ones is similar to models of filter pressing (Robin 1979, Sawyer and Robin, 1986, Brown et al. 1995), but involves fracturing and flow rather than diffusion. Squeezing melt down a pressure gradient along grain boundaries has been observed in experiments with melt percentages of at least 10-15% (Dell'Angelo and Tullis, 1988). The importance of fracturing for the segregation of melt has been emphasized by Wickham (1987), Clemens and Mawer (1992), Davidson et al. (1994), and Rutter and Neumann (1995). Although the actual mechanism(s) for filter pressing by fracture and flow are not clear, the injection of such melt segregations into the largely solidified host rock by fracturing along the existing anisotropy is observed. Repeated segregation of melt and injection into the partly solidified material by fracturing have been proposed as a mechanism of magmatic differentiation by Mahood and Cornejo (1992). A similar mechanism is envisaged here, although very little chemical differentiation is observed, in part due to melt segregation at higher melt fractions. As a consequence, the melt should be distributed inhomogeneously throughout the pluton during emplacement.

RELATIONSHIP OF MELT SEGREGATIONS AND MAFIC ENCLAVES

Extremely elongate mafic enclaves commonly resemble melt segregations of the LVS in the field, as both have are finer grained and show similar fabric development. Enclaves can only be distinguished from melt segregations by their more mafic composition (Table 1). Mafic enclaves within melt segregations show finer grained margins and weaker shape fabrics than the cores (Fig. 6). These features are difficult to interpret, because quenching of the margin against a lower-solidus-temperature medium would have occurred during the emplacement of the enclave into the melt segregation. Since the melt segregations have moved from their source, the enclaves would then be transported and injected together with the melt segregations as a an object with an outer, more solid shell. The feasibility of this explanation is supported by the fact that the melt segregations have developed a stronger fabric than the enclaves (Figs. 5, 6), indicating a mechanically weaker matrix and stronger enclave.

As the segregations continue to flow into the plane of flattening after the enclaves developed quenched margins, melt from the core of the enclave may inject its host rock along the plane of flattening by fracturing the enclave margin (Vernon et al., 1988). In such cases, the shapes of the enclaves cannot be used as strain markers, as they have not deformed as passive markers, but by fracturing. However, qualitatively they indicate the orientation of the flattening plane and perhaps the relative magnitude of strain.

MELT SEGREGATION AND CRITICAL MELT FRACTION

Natural examples of magmatic systems where evidence is preserved for crystallization process(es) are rare. Bryon and others (1994) have outlined evidence from natural examples in the Peruvian Cordillera, and discuss three crystallization intervals and their respective rheological character. Their conclusions, similar to those drawn from melting experiments (van der Molen and Paterson, 1979), argue that melt volumes below the 'critical melt fraction' cannot escape, whereas volumes of melt above the 'critical melt fraction' can aggregate and migrate to form plutons. Evidence outlined above, indicates that the idea of a 'critical melt fraction' applied to *naturally occurring silicic plutonic rocks* may not be appropriate. That is, even at small percentages of melt (≤1% to 10-15%), migration and segregation are possible processes under conditions of non-hydrostatic stress.

It is essential to distinguish between melt 'creation' and melt introduction when dealing with the melt segregation process, that is whether the melts are due to 'in situ' melting or are crystallizing during or after transport or migration (Nicolas, 1986, Brown et al. 1995, Vigneresse et al., in press). The melt segregations described from the LVS have formed during transport and crystallization of melt once a touching framework of crystals capable of transmitting stress was established. Two spatially adjacent domains have to be distinguished; one, probably at the rigid percolation threshold (more than 55% crystals), where the crystal framework was compacted under differential stress and deformation, and acted as a source for the segregating melt, and the second, certainly below the particles locking threshold, at less than 5-10% melt (Vigneresse et al., in press), where the grain boundaries have some tensile strength. Melt is transferred from the source region to the nearly solidified region, intruding the latter. Deformation is concentrated in the melt segregation dikelets, so that it is difficult to define a "rheologically critical melt percentage" on the scale of the melt segregations, as the melt is continuously redistributed by the process above. As little as 5% melt can significantly affect rock strength (Cooper and Kohlstedt, 1984) by enhancing diffusion of material along fluid- or melt-filled triple points (Pharr and Ashby, 1983). Therefore, the melt injected along grain boundaries of the almost solidified rock leads to intense deformation in those regions of the pluton where the melt segregations intrude.

Our geologic observations from the marginal series of the LVS support a model for the origin of foliation parallel melt segregations emplaced along extension fractures (see Etheridge, 1983; Clemens and Mawer, 1992), subparallel to the igneous foliation formed during the emplacement of consecutive magmatic pulses. If correct, our model for syn-emplacement layer-parallel diking may have important implications for the layered middle-lower crust, especially where identified indirectly from seismic reflection data (McCarthy and Thompson, 1988; Holliger and Levander, 1994).

CONCLUSIONS

From our detailed study of the emplacement history of the Lago della Vacca suite, the following conclusions are drawn:

1. Melt segregation may occur *during* the emplacement and crystallization of magma potentially leading to development of foliated plutons. Their composition is similar or only slightly more evolved than that of the solidified crystal mush into which the melt segregations intrude,

2. The melt intrudes the nearly solidified host by fracturing; the fractures are mostly intergranular and follow the pre-existing foliation of the host rock; this foliation is a purely magmatic fabric, defined by preferred orientations of plagioclase and hornblende,

3. Melt segregation veins formed by fracturing develop a fabric which has the same orientation, but is often stronger than that of their host rock; thus the shortening direction has remained the same (normal to foliation) during emplacement of the veins or dikes.

4. The fractures have probably formed by flow of segregated melt from regions with large melt fractions (more than 55% crystals) in the direction of the least compressive normal stress; the large number of melt segregation veins found suggests that the segregation of melt from crystal frameworks and fracturing parallel to the foliation is a repeated process. The fractures may generate as extension or shear fractures due to high melt pressure. The nearly solidified host rock (with a low melt fraction) has a tensile strength equivalent to that of the prismatic grain boundaries of plagioclase and hornblende.

5. Mafic enclaves within the melt segregation veins have less anisotropic fabrics than the surrounding material of the veins with the same orientation. Fine-grained rims of the mafic enclaves are interpreted as quenched margins, with still lesser fabric anisotropy observed,

6. As the melt fractures tend to follow the already existing foliation, and the melt segregation dikes typically have a stronger fabric than their surrounding host rock. Fabric development in granitoid plutons may be strengthened through repeated melt segregation and injection of melt into portions of the almost solidified rock. This is a way by which foliation development may be self-stabilizing.

ACKNOWLEDGEMENTS

Funds were provided to B. John for this work through the European Scientific Exchange Programme of the Royal Society (UK) and for H. Stünitz by the Swiss Nationalfonds (grant nr. 21-36008.92 and 2000-042134.94), and are gratefully acknowledged. This work has benefited from numerous discussions with Alfons Berger, Renee Heilbronner, Claudio Rosenberg, and Stefan Schmid. Renee Heilbronner provided significant help with the ACF analysis and figure preparation. Comments and careful, constructive reviews by Bruno Lafrance, Jean Luc Bouchez, Edward Sawyer, Jean-Louis Vigneresse, and Donny Hutton have substantially improved the manuscript. We would like to thank all of them very much.

RHEOLOGICAL ASPECTS OF MAGMA TRANSPORT INFERRED FROM ROCK STRUCTURES

Carlos FERNÁNDEZ, Antonio CASTRO, J. D. DE LA ROSA
and I. MORENO-VENTAS
Departamento de Geología, Universidad de Huelva,
La Rábida 21819-Huelva, Spain.

ABSTRACT

Field relationships from plutonic complexes are used to infer rheological properties of magmas, mainly of granite composition. The interpretation of structures acquired during magmatic flow as well as the contact relationships observed between magmatic rocks of contrasted compositions is based on both theoretical and experimental constraints on the rheology of partially melted rocks. Exceptional examples of structures in granitoid complexes from the Hercynian massif of Spain are shown. They support many critical phenomena on the rheology of granite magmas, such as Bingham behaviour, unstable flow, mixing and mingling in conduits.

INTRODUCTION

Magmas are a mix of crystals and silicate melts. Accordingly, magmatic flow consists of relative displacements within the melt together with the consequent rigid-body rotation of suspended crystals, with no-intracrystalline plastic deformation (Paterson et al., 1989). Magmatic flow may produce fabrics very similar to those originated in metamorphic rocks as a consequence of tectonic deformation in the solid state (Nicolas, 1992). Classical kinematic criteria, based on the study of structures in igneous rocks, may be used to know the flow regime, the flow plane and direction of flow, as well as the shear sense in the case of non-coaxial flow (Blumenfeld and Bouchez, 1988; Nicolas, 1992), in a way similar to the study of metamorphic and magmatic rocks deformed in the solid state.

Furthermore, theoretical and experimental forward modeling on magma rheology, in which a given magmatic structure is produced under well constrained conditions, has supplied important advances in our understanding of magma flow in the Earth crust and mantle. This understanding coming from forward modeling is crucial to apply the inverse problem in magma rheology; that is, to reconstruct the original flow regime and rheological evolution from the study of structures in magmatic rocks. Consequently, a

J.L. Bouchez et al. (eds.) Granite: From Segregation of Melt to Emplacement Fabrics, 75–91.

comprehensive knowledge of magma rheology is necessary to infer rheological aspects of magma transport from rock structures. Although many uncertainties remain about the rheological behaviour and the physical properties of silicate magmas, the following features may be stated on the rheology of magmas.

THE RHEOLOGICAL BEHAVIOUR OF CRYSTALLIZING MAGMAS UNDERGOES QUALITATIVE AND QUANTITATIVE CHANGES DURING CRYSTALLIZATION

It seems to be well established from theoretical and experimental works the fact that if the volume fraction of crystals (Φ) is lower than 0.3, the behaviour is close to a Newtonian flow model. If Φ is higher than 0.3, the first rheological threshold (Fernandez and Barbarin, 1991) is reached, and a very narrow transition to visco-plastic behaviour (e.g. Bingham body) is predicted (Shaw, 1969). This kind of rheology, though not wholly supported by the available experimental data to date, may explain the presence of mafic microgranular enclaves within large masses of plutonic rocks of granite to granodiorite compositions (e.g. Fernandez and Barbarin, 1991). This special behaviour imposes a particular kinematics to the magma in such a way that large portions of magma may flow as a plug, bounded by narrow bands of high strain rates and flow velocity gradients.

For a value of Φ higher than 0.65 to 0.7, the melt fraction is within the rheological critical melt percentage (RCMP, Arzi, 1978; van der Molen and Paterson, 1979; Marsh, 1984; Wickham, 1987), also called second rheological threshold by Fernandez and Barbarin (1991). Above the RCMP, magma behaves as a solid body capable of being fractured and brecciated. However, the experimental work of Rutter and Neumann (1995) on deformation of partially molten Westerly granite at strain rates of $10^{-5}s^{-1}$ shows a smooth decrease in the flow stress with increasing melt fraction, without any evidence for a dramatic loss of strength as implied by the concept of the RCMP. These results come from experiments carried out on partially molten rocks and not on crystallizing magmas. In this case, the melt is homogeneously distributed at crystal intersites even for very low melt percentages. Therefore, these experimental results may not be directly extrapolated to describe magmatic systems. Furthermore, Brown et al. (1995) provide a comprehensive review of the distinct geological conditions (volume fraction melt, strain rate, fluid- absent or fluid-present deformation, composition of the system) under which a dramatic weakening is expected in a mix of crystals and silicate melts. Thus, the RCMP still provides a rough, qualitative way to describe the critical transition from melt dominated to crystal dominated behaviour.

THE RELATIVE RHEOLOGY OF COEVAL MAGMAS OF CONTRASTED COMPOSITIONS SHOWS CRITICAL PHENOMENA DURING CRYSTALLIZATION

The study of the structural patterns resulting from magma interactions in nature constitutes a powerful tool to clarify some important features in magma rheology. This is due to the fact that mingling/mixing processes provide a great number of good kinematic markers. In this sense, the works of Fernandez and Barbarin (1991) and Fernandez and Gasquet (1994) emphasize the influence of the contrasted rheological evolution in the coeval magmas over the resulting structures. Several authors (Sparks and Marshall, 1986; Frost and Mahood, 1987; Williams and Tobisch, 1994) unequivocally establish that the interaction between both magmas depends upon their initial temperatures, the mass fraction of mafic magma, the variation in the effective viscosities and the magmatic flow velocities.

These points are used in this paper as the conceptual base to infer magma flow processes from the observation of structures in the rocks. For that, we have studied exceptional outcrops of plutonic rocks on which we carried out comprehensive geological and geochemical works published elsewhere.

GEOLOGICAL SETTING OF SELECTED CASES

The Iberian Massif constitutes one of the largest exposures within the Variscan belt in Western Europe. At the end of the Paleozoic it was affected by important metamorphic and deformational processes; and at the late stages of the orogeny it was intruded by large masses of granitic rocks, with subordinate basic and intermediate rocks (Fig.1, Corretgé, 1983).

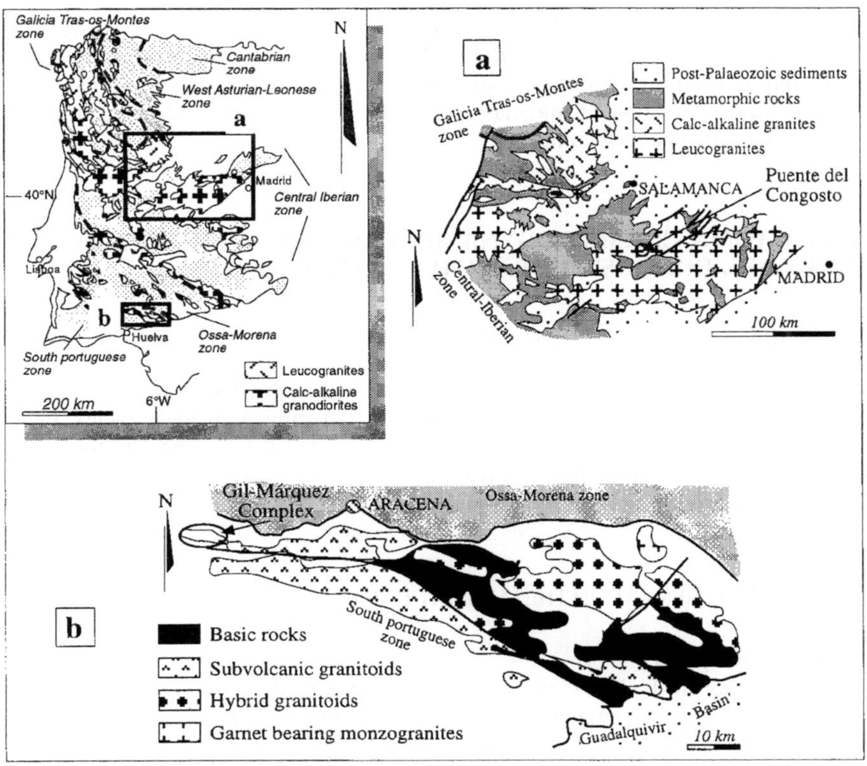

Figure 1. Geological location of the studied selected areas.

One of the best outcrops of granitic rocks in the Iberian Massif is the Central System batholith, with an areal extent of more than 15000 km^2 (Fig. 1a). Migmatites and anatectic granites coexist with basic rocks, and with the intermediate products of hybridization between basic and felsic magmas (Castro et al., 1990, 1991; Moreno-Ventas et al., 1995). An extensional tectonic regime has been proposed to account for the emplacement of this large batholith (e.g., Doblas, 1991). Up to now, there are no studies about the influence of the bulk tectonic regime upon the magmatic fabrics of the igneous rocks in the Central System. However, it is broadly accepted that these rocks were affected by late brittle-ductile shear zones after their complete consolidation (Doblas, 1991).

For this study we have chosen magmatic flow structures appearing in the granodiorites and granites of Puente del Congosto, south of Salamanca (Fig. 1a), in a zone near the contact with the metamorphic host. This zone offers excellent outcrops on the bed of the river Tormes which are used to study the structures originated by the relative movement of granitic masses with slightly different compositions. Flow processes that may account for the disruption of synplutonic dikes and magma mingling and mixing will be also discussed on the basis of the observed structures.

Other cases are located in the Seville Range batholith (De la Rosa, 1992) at the southern part of the Iberian massif (South portuguese zone). This batholith is mainly composed of basic and intermediate (hybrid) rocks disposed in large, elongated bodies parallel to the trend of regional structures. The selected area in this batholith (Fig. 1b) supplies important information about distinct rheological processes during magma movement and the interaction of magmas of contrasted compositions flowing through an ascent conduit (Gil Márquez complex).

INTRA-MAGMA CHAMBER DYNAMIC PROCESSES. THE EXAMPLE OF PUENTE DEL CONGOSTO (SPANISH CENTRAL SYSTEM)

The outcrop of Puente del Congosto, in the Spanish Central system, offers a good scenario to study the influence of magma flow patterns on the extreme heterogeneity and anisotropy of plutonic rocks. The granodiorites of Puente del Congosto are normally peraluminous rocks with biotite and cordierite. In transition with these facies, there are metaluminous granodiorites with biotite and amphibole. All the facies are more or less rich in magmatic microgranular enclaves of tonalite to quartz-diorite composition. A magma mixing origin has been proposed for these heterogeneities (Castro et al., 1991).

FLOW COMPLEXITIES WITHIN THE MAGMA CHAMBER

Structural description

Several granodioritic bodies may be distinguished according to slight compositional differences and relative contents of mafic microgranular enclaves and schlieren. The contacts between these bodies are orientated along an E-W direction, coincident with the orientation of the main contact with the country rocks.

Figure 2a shows one of these contacts between two granodiorite bodies. The contact is marked by a schliere band cross-cutting a set of older schlieren previously developed in the intruded granodiorite body. At both sides of the contact, the granodiorite shows differences in the content and size of K-feldspar megacrysts and in the biotite content of the matrix. The preferred orientation of the K-feldspar megacrysts in the intruding body is nearly parallel to the contact. A similar fabric is truncated in the intruded body. In the sketch drawn from this photograph (Fig. 3), discrete shear zones are depicted. They displace the schlieren of the intruded body but without affecting the granodiorite. They are interpreted as intramagmatic shear zones developed in the flowing magma before its complete consolidation.

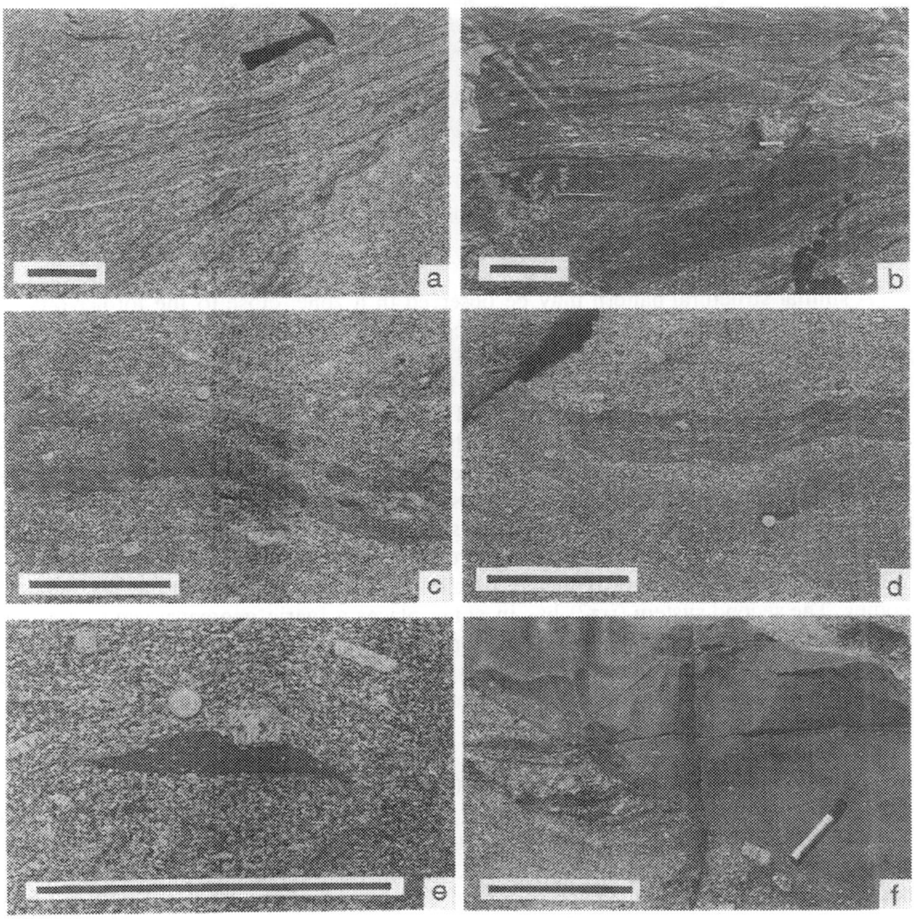

Figure 2. Field aspects of relevant magmatic flow structures in the Puente del Congosto area (Central System batholith). Scale bars: 25 cm. (a) Cross-cutting relationship between schlieren in the contact of two major granite masses.(b, c, d) Magmatic flow structures in elongated mafic microgranular enclaves and their felsic host. (e) Magmatic foliation in a mafic enclave; a feldspar xenocryst indent the mafic enclave. (f) Mafic synplutonic dike showing a hybridized contact with incorporation of feldspar xenocrysts.

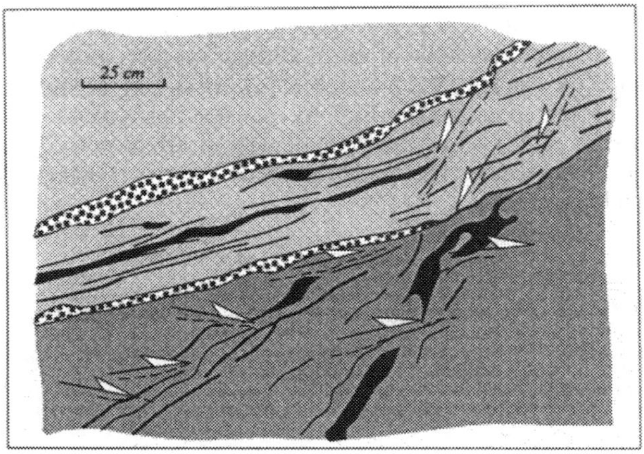

Figure 3. Sketch drawn from figure. 2a, highlighting the intramagmatic shear zones affecting the mafic schliere.

A similar structural pattern may be observed in a zone, close to the former one, in which a body of homogeneous and isotropic granodiorites intruded a very heterogeneous mass with abundant elongated enclaves and schlieren, both included in a matrix with a well developed magmatic foliation defined by biotite and K-feldspar megacrysts (Figs. 2b-e). The enclaves show distinct degrees of hybridization, and their magmatic foliation is parallel both to the long axis of the enclaves and to the foliation in the granodiorite (Fig. 2e). Figure 4 shows a geological sketch of this zone. The intramagmatic shear zones (*isz*) affect the microgranular enclaves, mainly near the contact with the homogeneous granodiorite body. Two conjugate systems of *isz* may be identified: *isz1* (sinistral) and *isz2* (dextral). The first system is more abundant than the second one and it has a WNW-ESE orientation, approximately parallel to the contact between the two igneous bodies. The second system (*isz2*) has in some places a listric geometry and a N-S average orientation. According to kinematic criteria, the *isz* induced (1) local rotations of the planar fabric of the enclaves, (2) local displacements near the boundaries between enclave and host granodiorite and (3) the reorientation of the K-feldspar megacrysts of the granodiorite (Fig. 2b-d), but their effect in the granodiorite matrix is not appreciable. In places, a single *isz* affects two adjacent enclaves but no fabric is developed in the granodiorite band separating the enclaves (Fig. 2b and Fig. 4). The rose diagram of figure 4 shows the relative distribution of the two *isz* systems, as well as the average orientation of the magmatic foliation (*Smg*). From these observations a simple, two-dimensional kinematic model is proposed, attempting to explain the development of such structures related with a band of simple shearing which is associated with the relative movement of two granodiorite masses.

Rheological inferences

The structures described above suggest that several pulses of granodiorite magma intruded into other granodiorite magmas of similar composition within the same magma chamber. This process is inferred to have occurred in the following stages:

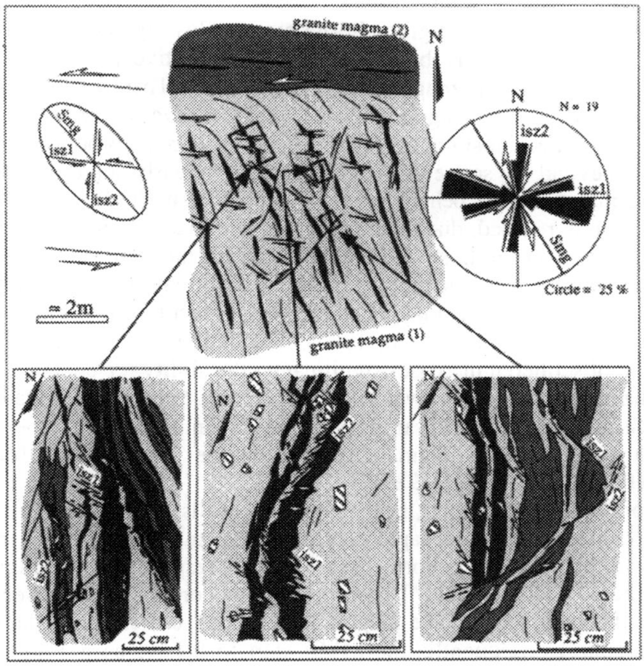

Figure 4. Geological sketch of a contact between two granite masses in the Puente del Congosto area. Insets represent sketches drawn from figures 2b, c and d. Intensity of grey patterns qualitatively reflect the percentage of mafic magma in hybridized enclaves. *Smg*: Magmatic foliation. *isz*: intramagmatic shear zones. Asymmetric arrows indicate the apparent sense of shear in the *isz*. Right of the field sketch: rose diagram of traces of *isz* (19 measurements). Left of the field sketch: kinematic interpretation of the contact. See text for further explanation.

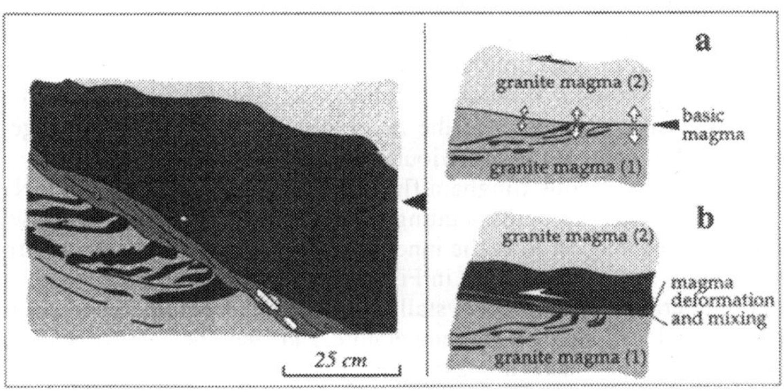

Figure 5. Sketch based on figure 2f. (a) First stage: a granite magma with schlieren is intruded by a second granite magma; in the contact, a mafic magma starts to open a synplutonic dike. (b) Second stage: the mafic dike is emplaced, and the relative movement between both granite magmas is concentrated in one of the margins of the basic dike. See text for details.

1) The process may start with a granodiorite magma flowing near the walls of a large magma chamber. This flow may be due either to convective movements within the chamber or to the ascent of the magma through the crust. There are no data to discern between both hypotheses. The magma included *mafic microgranular enclaves* (MME) during its ascent. The mechanism by which the enclaves were incorporated to the granodiorite magma is beyond the scope of this paper. At some early stage, both granodiorite and enclaves are magmas that behave as nearly-Newtonian fluids. Assuming that (1) thermal equilibrium is reached, due to the fact that the mass of enclave magma is less than the mass of granodiorite host, (2) that this equilibrium temperature may be near 1000° C, and (3) that the flow rate may be high, the deformation of the whole magmatic body, granodiorite plus enclaves, may be strong with aspect ratios of elongated enclaves exceeding 15:1 (Williams and Tobisch, 1994). The extreme distortion of the enclaves and development of schlieren increases significantly the ability for chemical diffusion to produce hybridization between enclaves and host magma (Blake and Koyaguchi, 1991). The effective viscosity of the MME at the equilibrium temperature may be either higher or lower than that of the host magma depending on the local volume fraction of the mafic magma (Frost and Mahood, 1987).

2) As the process is developing near one of the walls of the magma chamber, the high thermal gradient induces the rapid cooling of the magma chamber, making both magmas, enclave and host granodiorite, increase their respective crystallinities (Φ) and reach the first rheological threshold. The differences in composition between both magmas make possible that the enclave magma reaches this critical point before the host granodiorite magma. Consequently, the MME start to behave as a Bingham body. In zones with a high content of MME (Fig. 2b-e and Fig. 4), where a wide diversity of hybrids is developed, the viscosity of the granodiorite may be for a short time higher than that of the enclave magma (Frost and Mahood, 1987). At this stage, fabrics related with the shape preferred orientation of the K-feldspar megacrysts floating within the granodiorite magma may develop (field II of Fernandez and Gasquet, 1994), as well as planar fabrics in the elongated enclaves. When the viscosity inversion temperature is reached, the enclaves do not deform plastically and may fracture if the flow rate is larger than several m/year (Williams and Tobisch, 1994). The absence of this kind of structure indicates that the flow rate may have been strongly reduced with time along with a high increase in the crystallinity of both magmas.

3) At a later stage, the crystallinity of the granodiorite magma and consequently its viscosity, is high enough to be near the rheological critical melt percentage (RCMP). Although it is still a magma, the behaviour of the whole body, granodiorite plus enclaves, is that of a highly viscous Bingham fluid that may be easily fractured. Newly developed intramagmatic fractures, cross-cutting the pre-existing structures, allow the intrusion of new pulses of magma from the inner parts of the magma chamber carrying more magmatic enclaves (granite magma 2 in Fig. 4). This new pulse may be Newtonian due to its higher temperature and lower crystallinity. At the contact, the nearly-solid intruded body undergoes a local increase in temperature and may behave as a Bingham fluid again. Consequently, the intruded granodiorite magma (magma 1 in Fig. 4) behaves as a fluid in a narrow band in which the shear stress is greater than the yield strength. The width of this band depends mainly on magma viscosity and strain rate, and is on the order of several metres (Fig. 4). The magmatic foliation of magma 1 is reoriented toward parallelism with the contact (Fig. 4). This reorientation is partly due to the development of intramagmatic shear zones (Fig. 4: *isz*). Close to this flowing band, to the south, magma 1 behaves as a solid plug and does not develop any visible structure.

DISRUPTION AND DEFORMATION OF SYNPLUTONIC DIKES

Structural description

In places, the contacts between distinct bodies of granodiorites are intruded by quartz-diorite magma forming narrow dikes with synplutonic signatures. These dikes have a composition, texture and grain-size, identical to those of the mafic microgranular enclaves. Figure 2f shows one of these dikes cross-cutting one of the pre-existing schliere of the intruded granodiorite. This granodiorite is slightly different in composition and structures from the granodiorite appearing on the opposite wall of the same dike (see sketch in Fig. 5). Furthermore, the shapes of each contact are different: one of the contacts is lobate as a response to the intrusion of a magma into another magma; the other one is more planar and is marked by the presence of a mesocratic, narrow band very rich in feldspar crystals similar to those of the host granodiorite.

In other places, the mafic rocks appear forming enclave swarms. A good example outcrops in the vicinity of Puente del Congosto, under the Roman bridge. This is a 5 m wide band and more than 20 m long, in which abundant mafic enclaves, different in size, colour, composition, etc, are enclosed in a granodiorite matrix slightly darker than the normal granodiorite outside of this band (Fig. 6a-d and sketches in Fig. 7). Some relevant features of this band are:

1) The composition of the enclaves is varied. Some of them are the typical mafic microgranular enclaves (MME) with a composition of diorite to quartz-diorite, with some amphibole clots, with no feldspar phenocrysts and with a rounded shape. Another type is characterized by the presence of feldspar phenocrysts (mainly plagioclase) and rounded quartz crystals included in a dark matrix nearly identical to that of the typical MME described above. A third type is characterized by a more elongated shape and a composition intermediate between that of the host granodiorite and the typical MME. These are disposed parallel to the main foliation of the host granodiorite, which is marked by the orientation of the K-feldspar megacrysts, coinciding with the orientation of the whole band.

2) The boundaries between the feldspar-bearing MME and the host granodiorite are sharp. They occasionally show angular shapes when they are in contact with the leucocratic facies of the matrix granodiorite. By contrast, the contacts are diffuse and rounded if they are enclosed in the intermediate facies. In these enclaves, feldspars and amphiboles are oriented defining a magmatic fabric parallel to the long axis of the enclave. The magmatic foliation in the host granodiorite is defined by the orientation of K-feldspar megacrysts and by the orientation of extremely elongated bands with a composition intermediate between the normal granodiorite and the MME. The enclaves are embraced by the external foliation indicating that they are more viscous than the host. The long axis of the enclaves and their internal foliation are oriented at 60° with respect to the external foliation (Fig. 6a), suggesting that these enclaves were rotated into a magma matrix, less viscous than the enclaves, during the waning stages of the magmatic flow in the magma chamber.

3) Other kind of enclaves are mafic bodies with lobate, crenulated contacts (Fig. 6b). The contacts are interpenetrated by the host granodiorite. These relationships may result from the flow deformation of pre-existing structures of back-veining. Any intermediate relationship between those described above and sharp fractures in the enclave filled by granodiorite magma may be observed in this outcrop (Figs. 6b and 7). Internally, these enclaves show a local foliation defined by the orientation of amphibole clots. This folia-

tion is preferably developed near the margins. The magmatic foliation of the host granodiorite surrounds the enclaves indicating, as in the former case, the existence of a viscosity contrast during the late stages of magmatic flow.

4) Some MME, with a globular shape, show zones very rich in feldspar and quartz phenocrysts (Fig. 6c). These zones are disposed near the margins of the enclave and are transitional to the typical MME facies. These enclaves are, as in the former cases, embraced by the external foliation of the host granodiorite.

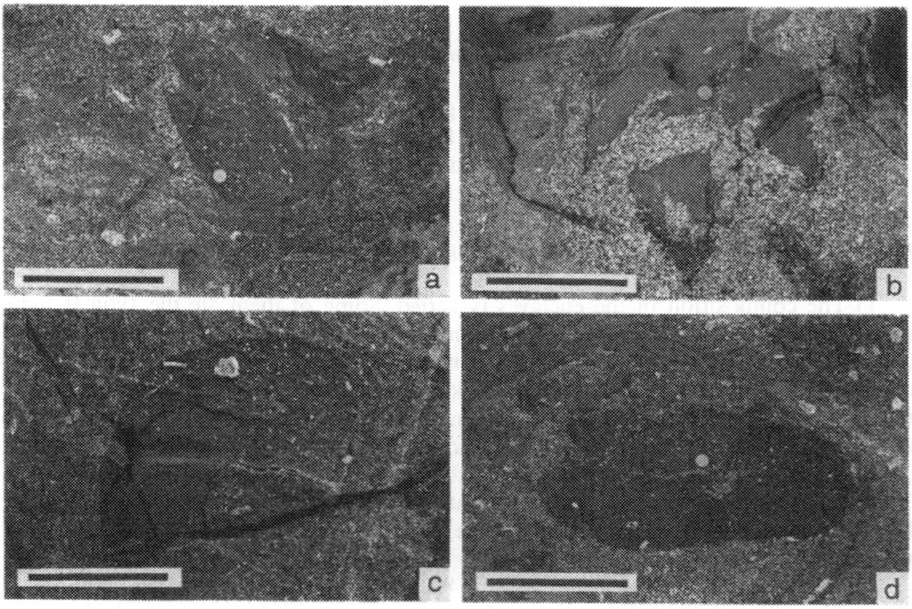

Figure 6. Field aspects of mafic enclaves in the Puente del Congosto area. They belong to the same enclave swarm. Scale bars: 25 cm. (a) Rotated mafic enclave, with an internal magmatic foliation; the external magmatic foliation embraces the mafic enclave; see also the extremely elongated enclaves with an intermediate composition. (b) Partly disrupted mafic enclave, showing crenulated contacts with respect to the host. (c) Rounded mafic enclave; an external, curved band with feldspar megacrysts partly encloses a microgranular inner facies. (d) Rectangular mafic enclave with rounded vertices; the contact with the host is sharp in the bottom and diffuse in the top of the photograph, where incipient mixing takes place; note the grey, elongated enclaves of intermediate composition.

Rheological inferences

The sequence of processes inferred from the observed features in figure 2f may be as follows:

1) A dike of mafic magma is injected at the boundary between two distinct granodiorite magmas that in turn are flowing one, with respect to the other (Fig. 5a). This process requires a transitory brittle behaviour of the granodiorite host, possibly enhanced by a locally high fluid pressure near the tip of the intrusion. This special behaviour may take place in magmas with crystal fractions under the RCMP, within field II of Fernan-

dez and Barbarin (1991), accompanied by locally-high flow rates (Williams and Tobisch, 1994).

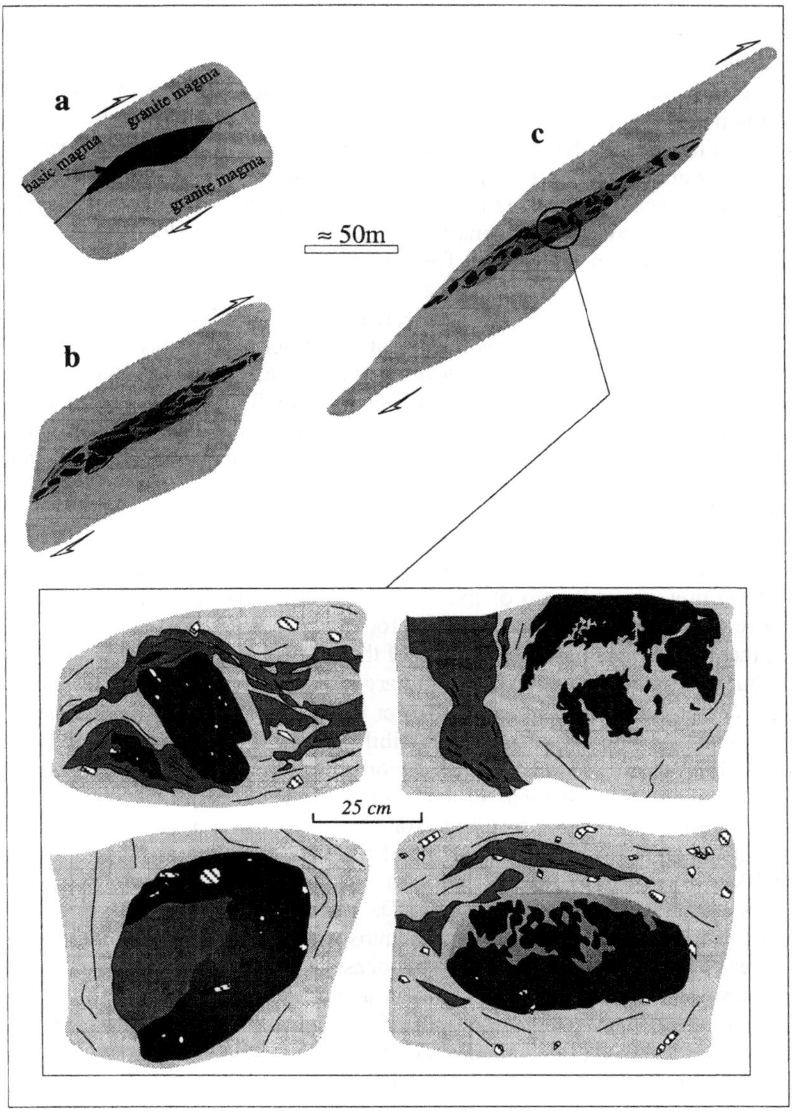

Figure 7. Model explaining the genesis of mafic enclave swarms in the Puente del Congosto area. (a) A local transtension zone develops in-between two independently flowing granite magmas; the gap is filled by mafic magma giving a synplutonic dike. (b and c) Magmatic flow favours the progressive disruption of the dike and the genesis of distinct types of mafic and hybrid enclaves. Inset: sketches drawn from figure 6 are presented as snapshots of a long-lived process preserved in the final frozen stage. Grey pattern intensities reflect the percentage of mafic magma preserved in hybridized enclaves.

2) The injection of the mafic magma may locally increase the temperature of the host granodiorite magma, inducing a slight decrease of its crystallinity, and consequently of its viscosity. However, as the volume of the injected mafic magma is reduced in comparison with the volume of host magma, the effect of the intrusion will be the quenching of the mafic magma, which quickly reaches the viscosity inversion temperature. At this point, the viscosity of the mafic magma may be close to that of the host granodiorite, a favourable rheological situation for mechanical mixing between dike and host (Fernandez and Barbarin, 1991). For a short period of time, it may be possible that the movement between the two granodiorite masses may be transfered to the dike. Due to the viscoplastic behaviour of the mafic magma, the imposed velocity gradient is only able to deform the margin of the dike. This margin is hybridized with the host granodiorite incorporating feldspar xenocrysts (Fig. 5b).

3) Finally, the temperature of the dike decreases under the RCMP and the dike becomes rigid. The flow between the two adjacent granodiorite magmas may continue but not affecting the structures previously developed in the interior of the dike.

In order to explain the enclave swarm illustrated in figure 6, a similar mechanism may be invoked; the main difference is in the volume of injected magma. The relative movement of two adjacent granite masses, with a rheological behaviour of field II of Fernandez and Gasquet (1994), may locally induce the opening of large cavities by transitory brittle fracture of the interface separating the two masses of magma (Fig. 7a). This fracturing may be a consequence of the following factors: (a) local instabilities in the flow with the development of transtension zones; (b) hydraulic fracturing assisted by the high fluid pressure at the tip of the ascending mafic dike (e.g., Shaw, 1980; Spera, 1980), or (c) a combination of these two factors.

Once the dike is injected, the dynamics of the two granodiorite pulses starts to affect the dike, but due to its relatively large size (larger than the above described dike), it may maintain over a larger period of time its lower viscosity, as compared with the viscosity of the host granodiorite, favouring the possibility of mixing with the host felsic magma (Fig. 7b-c). The margins of the dike may become enriched in feldspars coming from the host granodiorite by a mechanism similar to that depicted in the former case. A pervasive back-veining is developed resulting from the propagation of shear zones and tensile fractures related to the relative movement of the two granodiorite masses as they have rotational kinematics as shown in figure 6a. For simplicity, we have considered these kinematics as due to simple shearing (Fig. 7).

By this mechanism, granitic magma is introduced into the mafic dike, still hot and less viscous, allowing the magma mixing process to occur repeatedly in space and time with the development of new structures. As a consequence, a wide variety of enclave types is developed. The examples of figure 6 are consistent with the theoretical results of Williams and Tobisch (1994) predicting that the enclaves of intermediate composition are much more deformable than the mafic enclaves.

MAGMA TRANSPORT IN CONDUITS: THE EXAMPLE OF THE GIL-MARQUEZ COMPLEX (SEVILLA RANGE BATHOLITH)

The Gil-Márquez complex is located in the westernmost part of the Sevilla Range batholith (Fig. 1b). It was described in a recent paper (Castro et al., 1995), where we suggest that it represents a notable field example of magma mixing within a deep-seated conduit.

Comparison with the experimental models conducted by Blake and Campbell (1986) and Koyaguchi (1987) seems to be supported on the basis of the petrography, structure and geochemistry of the Gil-Márquez complex. The Gil-Márquez intrusion has an elongated shape. It is essentially composed of biotite granites, intermediate (hybrid) granodiorites and biotite-hornblende tonalites, and hornblende quartz-diorites. The quartz-diorites appear as a synplutonic dike intruded within partially melted tonalite. A remarkable feature of the complex is the presence of a conspicuous zoning in rock types and structures, with basic rocks in the center. Castro et al. (1995) consider the distinct rock types of the Gil-Márquez complex as a result of the interaction between two end-member magmas: a quartz-diorite basic magma and a granite, felsic magma. The outcropping structures in the consolidated conduit represent frozen, snapshots belonging to different stages of the magma-mixing evolution, because the same process of intrusion and granodiorite digestion by basic magmas occurred repeatedly in space and time. The study of such structures, presently appearing telescoped in space, is essential to understand the intermediate stages by which both end-members interact within the conduit. The structures recognized within the tonalites and granodiorites, as well as in the contact between quartz-diorites and tonalites (Fig. 8), led Castro et al. (1995) to propose the following model of dike evolution.

1) A body of magma of quartz-diorite composition intrudes as a synplutonic dike within a silicic magma. This can be either of granite, granodiorite or tonalite composition depending on the maturity of the mixing process within the conduit. Cooling of the external boundaries of the mafic dike results in the development of chilled margins (Figs. 8a and 9a). Furthermore, the transfer of heat from the mafic magma allows the granitoid (or tonalite) magma to become remobilized; as a result, the chilled margins are disrupted by the relative movement between the mafic dike and its felsic host (Blundy and Sparks, 1992).

2) Tension cracks, oblique to the intrusion direction, develop in the chilled margins. As a consequence, back-veining favours the forced intrusion of leucocratic magma within the mafic dike (Figs. 8b and 9b).

3) Within the dike, back-veining generates elongated septa with lobate magma-magma contacts, as felsic melt interacts with mafic magma of the dike interior (Figs. 8c and 9c).

4) Deformation of felsic septa by flow within the dike may lead eventually to their disruption and the isolation of rounded enclaves of granite magma. Mafic magma still has a lower viscosity than felsic enclaves, as evidenced by the pattern of the magmatic foliation in the quartz-diorite, that embraces granite enclaves (Figs. 8d and 9d). This apparently abnormal viscosity contrast can be established even after the equilibrium temperature has been reached, but in this case, this viscosity contrast is only possible when the mass fraction of the mafic member is large (Frost and Mahood, 1987).

5) Progressive deformation continues by the stretching and folding of enclaves (Figs. 8e-f and 9e-f). A slow decrease in temperature allows the system to reach the inversion point at which mafic and felsic magmas have identical viscosities (Fernandez and Gasquet, 1994). During this stage the mixing between the strained enclaves and the mafic magma is highly enhanced (Figs. 8g and 9g).

6) Finally, the enclaves evolve from irregular bodies with diffuse contacts to elongated patches resulting from complete disaggregation of the felsic magma (Figs. 8h and 9h). Mixing proceeds to produce hybrid tonalites by matrix homogenization and incorporation of feldspar megacrysts within the mafic magma.

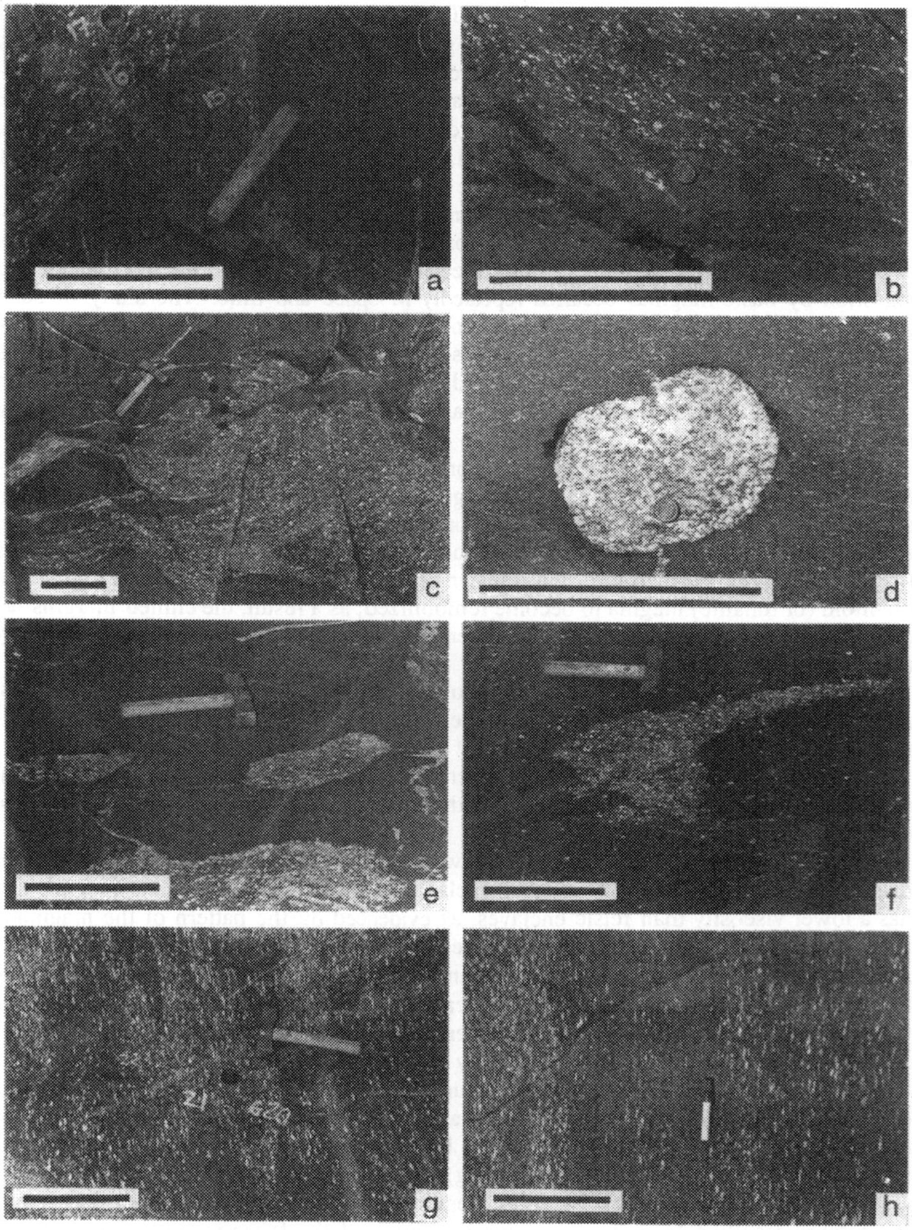
a
b
c
d
e
f
g
h

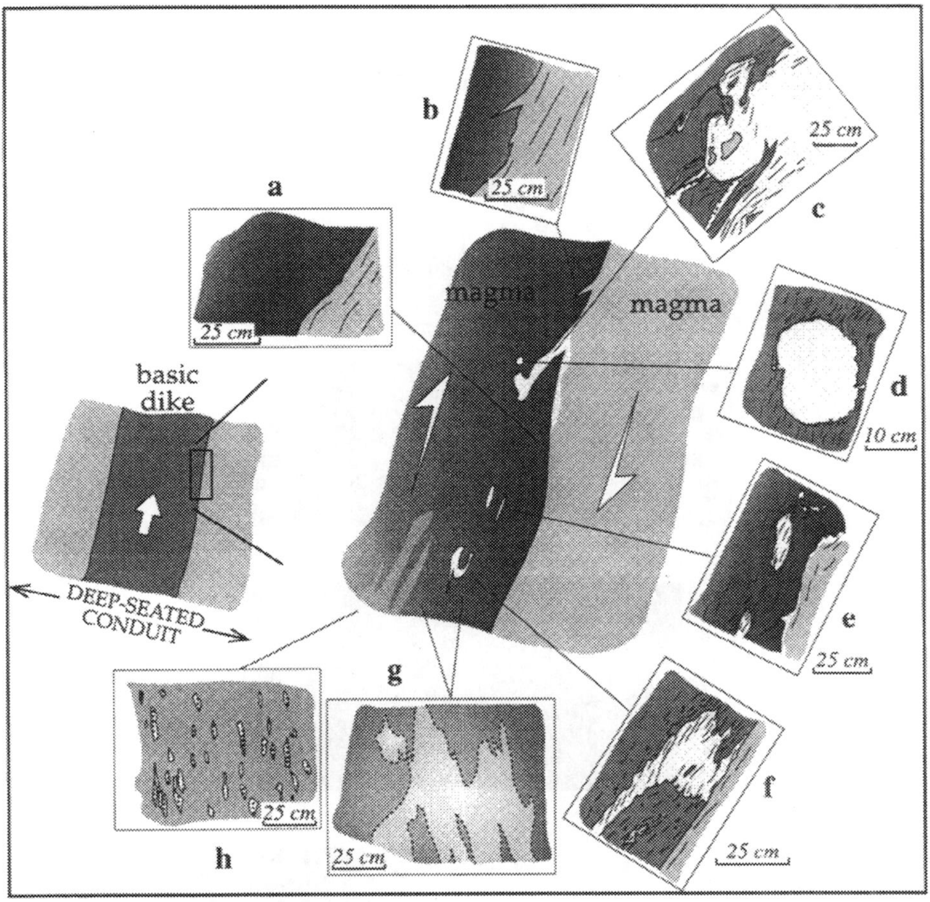

Figure 9. (a to h) sketches drawn respectively from photographs a to h of figure 8. The distinct structures are presented as stages of a complete cycle of magma interaction and mixing between a mafic magma (dark grey) and a felsic magma (light grey). This process takes place in the interior of a deep-seated conduit, and it is triggered by the intrusion of a mafic dike within felsic or intermediate (hybrid) magmas.

Figure 8. (opposite) Field aspects of magmatic flow structures in the Gil-Márquez complex (Sevilla Range batholith). Scale bars: 25 cm. (a) Chilled margin developed at the contact between the quartz-diorite magma and the host hybrid tonalite. (b) Back-veining of the mafic dike by the remobilized felsic magma. (a) Lobate contacts between quartz-diorites and granodiorites; highly deformed contacts lead to partial individualization of granodiorite blobs. (d) Granitic enclaves included within the quartz-diorites; note the magmatic foliation in the mafic rock embracing the felsic enclave. (e) Granitic enclaves strongly elongated. (f) Granitic enclaves folded, disrupted and partly disaggregated. (g) Isoclinally folded enclaves with diffuse contacts with respect to the hosting mafic magma. (h) Heterogeneous hybrid produced by disaggregation of felsic enclaves, matrix homogenization and incorporation of feldspar xenocrysts into the mafic magma.

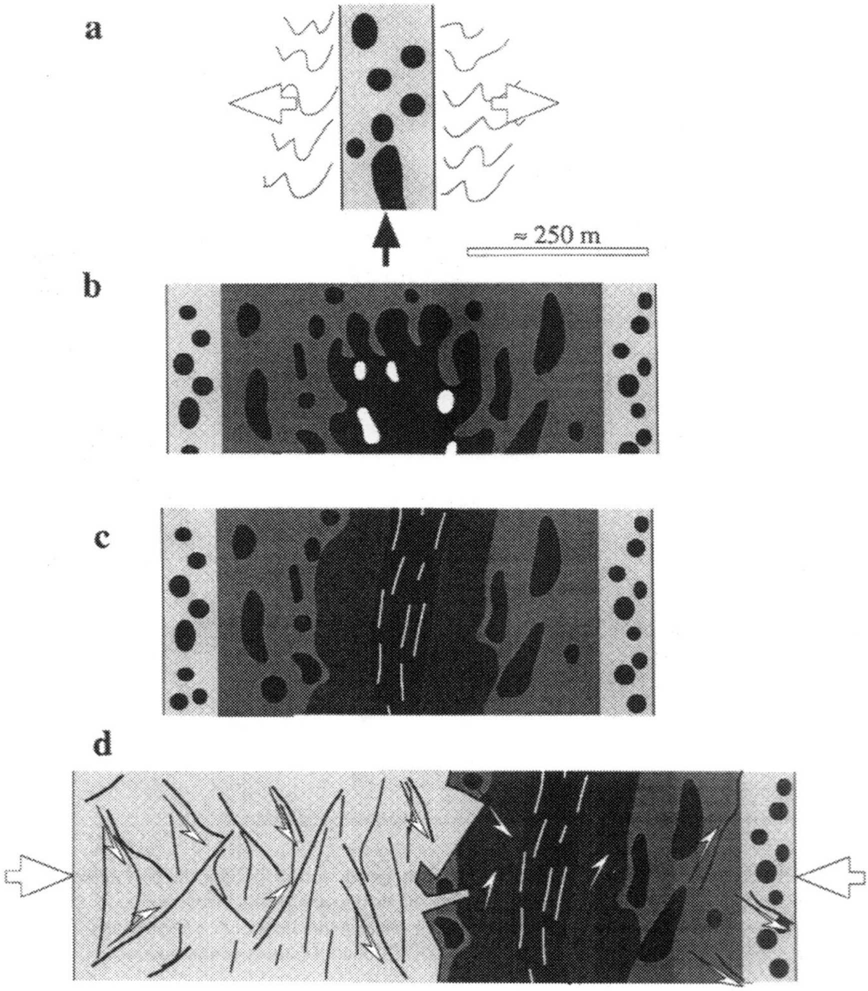

Figure 10. Schematic model of the process of flow and mixing in the Gil-Márquez complex. (a) A tensional fracture is opened and a composite dike injected through the crust. (b) Injection of new mafic dikes initiates the hybridization of the host felsic magma (hybrid granodiorite with enclaves). (c) Late mafic dikes hybridize with the previous magmas, resulting in hybrid tonalites. (d) The process stops by cooling of the whole complex; new pulses of granite magma intrude the conduit. After complete consolidation of the complex, the external tectonic regime starts to deform it, developing heterogenous discrete shear zones. Solid-state foliations appear in the felsic rocks.

This process, repeatedly occurring in space and time, gave rise to a complex facies pattern within the deep-seated conduit. Figure 10, modified from Castro et al. (1995), represents an idealized sketch of the proposed evolution model. Once the RCMP is reached in the felsic facies, external tectonic stresses start to deform the dike. Granites and granodiorites are less competent than tonalites and quartz-diorites. As a consequence, plastic intracrystalline deformation structures appear in quartz and phyllosilicates, and a solid-state foliation develops in the felsic facies. The basic rocks suffer a more heterogeneous, solid-state deformation, generating discrete conjugate ductile-brittle shear zones (Fig. 10d).

CONCLUDING REMARKS

The natural examples of magmatic structures described in this paper and interpreted in terms of magma rheology, clearly indicate that the study of structures in magmatic rocks is a useful tool to infer flow processes during magma transport. Magma dynamics can therefore be reconstructed using the field relationships as the main reasoning base. However, to address this complex inverse problem it is necessary to have a clear understanding of the physical properties of silicate magmas obtained from numerical and experimental forward modeling.

ACKNOWLEDGEMENTS

Financial support by a grant from the National Commission for Science and Technology (CICYT, Project PB91-0600) and the Junta de Andalucía (PAI 4018 and PAI 4108) is gratefully acknowledged. The comments of Sonia H. Santano led to considerable improvements of the English text. D. S. Blake, W.E. Stephens and an anonymous referee considerably improved an early version of the paper.

PART II

FABRICS IN GRANITE

GRANITE IS NEVER ISOTROPIC : AN INTRODUCTION TO AMS STUDIES OF GRANITIC ROCKS

Jean Luc BOUCHEZ
Université Paul-Sabatier, CNRS UMR n° 5563
Laboratoire de Pétrophysique et Tectonique,
38 rue des Trente-Six-Ponts, F-31400 Toulouse, France

ABSTRACT

This paper develops the author's and co-workers' findings that well-organized crystalline fabrics are ubiquitous in granitic rocks. Traditional structural and microstructural measurements of foliation and lineation in granites, performed directly in the field, in the laboratory from macroscopic oriented samples, and applied to entire plutons, reveal that magmatic structures may be homogeneous over huge areas. Hence structural mapping of granitoid massifs is of great interest in the understanding of magma emplacement and deformation in the crust. The low-field magnetic fabric measurement (AMS), a low cost, quick and easy technique, gives a quantitative description of the crystalline fabric and is now used systematically for structural mapping of granites. The present state of the « art » on the magnetic fabrics in granites is reviewed, based on the distinction between the paramagnetic, or magnetite-free, and the ferromagnetic, or magnetite-bearing granites. Magnetic fabrics in granitic rocks are discussed in relation to their variability at various scales and their kinematic and rheological significance. Finally, the relationship between fabrics and modes of emplacement is discussed.

INTRODUCTION

The presence, or absence, of a macroscopically visible planar structure in a granitic rock is the source of the old distinction between « oriented » or « structured », and « equant » or « massive » rock types. The former, with structures in the granite generally parallel to those of the country rocks, was often considered to characterize syntectonic plutons. In contrast, the latter rock type, with no conspicuous internal structures, was generally considered to be post-tectonic. Furthermore, the syn- or post-tectonic character of a pluton is often deduced from regional, geochemical and isotope chronology considerations that have nothing to do with the structure within the magma body. A closer look at the rock with an experienced eye, depending on the quality of the outcrop, often reveals that anisotropic features are ubiquitous. The various forms these features take (elongate

J.L. Bouchez et al. (eds.) Granite: From Segregation of Melt to Emplacement Fabrics, 95–112.

enclaves and schlieren, aligned megacrysts..), the intensities and distributions they have with respect to contacts, among other things, are worth studying.

In a given outcrop, the structural anisotropy, or fabric, of a granite is marked by the preferential grain-shape orientation, or shape fabric of the populations of crystals (Fig. 1). The term subfabric refers to the preferred orientation of a given category of crystals, principally K-feldspar, plagioclase or biotite. A planar fabric defines a *magmatic foliation*, usually marked by polyhedral K-feldspar megacrysts and/or by tabular plagioclase grains or biotite flakes. A major step forward in magmatic fabric studies was to realize that a linear fabric, or *magmatic lineation*, although difficult to evidence, is always associated with the foliation.

Measurements of magmatic fabrics were first determined by « traditional » means (see next heading). They revealed that the magmatic structure that is locally recorded in a pluton may be fairly constant in orientation, or at least may vary gradually, over the entire pluton or batholith. The kinematic and the rheological significance of the remarkable consistency of the magmatic structure will be, at the end of this contribution, briefly hinted at in relation to fabric acquisition mechanisms. Traditional fabric measurements were tedious and time consuming, not easily reproducible from one observer to the other and, above all, did not provide information easy to handle quantitatively.

Magnetic fabric determination, a quick and easy method which provides a semi-quantitative fabric description both in fabric orientation and in fabric intensity, has experienced a rapid development during the last decade. Its potential is considerable in structural mapping of granitoid domains, a geological field which remains under-developed. A statement of the current knowledge of magnetic fabrics of granitoids is the object of the second part of this contribution.

TRADITIONAL STRUCTURAL MEASUREMENTS

The magmatic fabric orientation of a granite may be worked out by detailed examination of favourable outcrops in 3D. The foliation plane is determined first. This is done by measuring the orientation of the average plane intersecting at least two natural section planes parallel to the biotite subfabric trace, and/or the K-feldspar subfabric trace, determined on these natural section planes (Oertel, 1955, p. 21). Then, a natural surface of the outcrop that happens to be parallel to the foliation plane is looked for, in order to find out whether crystals define a lineation by their statistical orientation in one direction of this plane. The lineation may be easy to trace by using elongate markers such as the long axis of the K-feldspar megacrysts or amphiboles. When biotite is the only fabric marker present in the rock, deciphering the lineation may prove difficult since biotite is not elongate. In fact, the lineation marked by the biotite grains, and more generally by tabular and platy crystals, is a « zone axis », or axis around which the largest faces of crystals are statistically disposed (Fig. 1). In other words, the zone axis is the statistically defined axis of rotation of the population of crystals.

The zone axis is generally found to be parallel, or very close within the measurement uncertainties, to the lineation directly measured from the K-feldspar, plagioclase or amphibole crystals (Fig. 1a: lineation). 3D analogue experiments of biotite fabric evolutions indicate that the zone axis is likely to be close to the maximum principal finite stretch (Arbaret et al., this volume). This is apperently due to a predominance, at the grain-scale, of vorticity parallel to the flow line over vorticity perpendicular to it.

This also attests that the overall strain regime is close to simple shear, otherwise clearly defined angular obliquities between the zone axis and the stretch axis would be observed (see Schulmann et al., this volume).

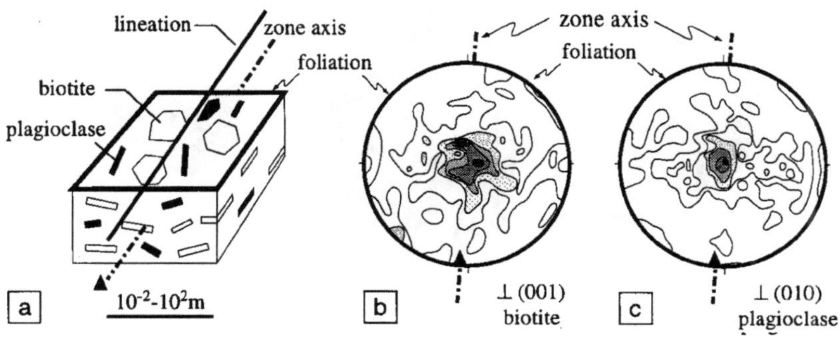

Figure 1. Magmatic foliation and lineation defined from the subfabrics of biotite and plagioclase. a: the lineation is defined by both the linear fabric of the plagioclase crystals and the zone axis of the biotite crystals. b and c: equal area diagrams from the Guérande granite (200 Universal-stage measurements projected into the foliation plane; *in* Bouchez et al., 1981) illustrating a magmatic lineation defined by a zone axis represented by the black triangle, and calculated as the pole to the best fit plane; the fabric maximum, or short axis of the shape fabric ellipsoid, defines the normal to the foliation plane. b: poles to the (001) cleavage of biotite; contours: 0.5, 1.5, 2.5, 3.5 and > 4.5%. c: γ optical index of plagioclase-oligoclase (An 5-10), very close to the (010) normal; (010), the polysynthetic twin plane, is parallel to the morphological flattening plane of plagioclase; contours: 0.5, 1.5, 2.5, 5.0 and > 7%.

Inappropriate orientations of the observation planes, low quality of the outcrop, or particularly weak fabrics may prevent one from obtaining a fabric measurement from a given outcrop. The measurement itself may be long and tedious to obtain, inaccurate and not necessarily reproducible. Preferably, oriented specimens may be collected in the field and their fabric determined in the laboratory using statistical countings on several saw-cut sections (Fig. 2). To obtain the structural map of the Brame-St Sylvestre-St Goussaud granite complex of Limousin (Fig. 3), Mollier (1984) performed such systematic measurements of biotite and K-feldspar subfabrics. He collected 256 specimens of about five kilograms each. Hence rather more than one (metric) ton of rocks was handled in the laboratory! Note that this granite complex of the French Massif Central was formerly almost everywhere considered as « massive ». To the best of our knowledge, the resulting structural map, covering hundreds of square kilometres (Fig. 3; see also Améglio et al., this volume: their figure 7), was the first publication concerning a whole granite massif and for which both the planar and the linear crystalline fabrics were determined. At that time, this contribution suggested that structural mapping of entire massifs, or batholiths, could provide keys to the understanding of the emplacement modes of granites, and to the determination of syn-magmatic regional kinematics.

In this granite complex, it was discovered (i) that the whole massif underwent SE-NW subhorizontal shear deformation during magma emplacement (Fig. 3: stereoplot of the magmatic lineations), and (ii) that a N20°E striking dextral shear zone had deformed the early horizontal foliation in the magmatic state (Fig. 3: shaded and foliation trajectories).

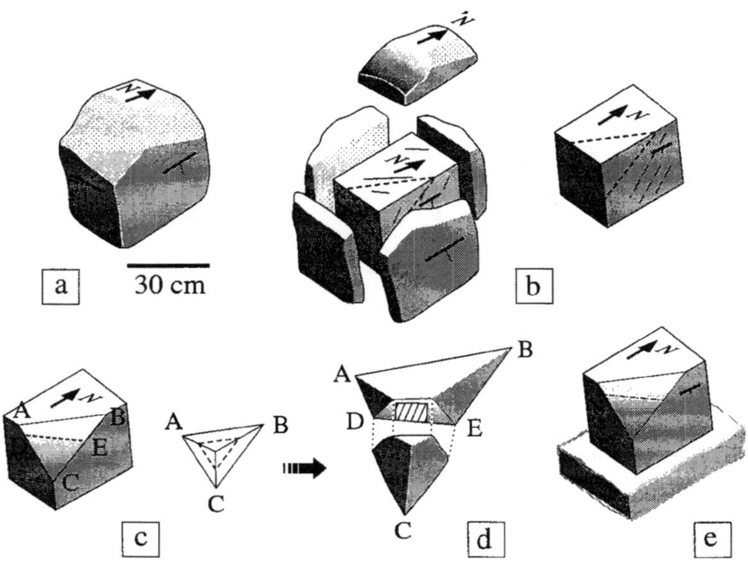

Figure 2. Mesoscopic « traditional » fabric measurements in a granite (*after* Mollier, 1984). a: oriented sample from the field. b: saw-cut sections for determination of foliation traces (dotted lines). c: saw-cut section(s) parallel to foliation for determination of lineation. d: preparation of a thin section (section normal to foliation and parallel to lineation). e: orientation measurement of foliation and lineation with respect to geographical frame, sample resting on modelling clay.

These « traditional » fabric measurements, based on two-dimentional orientation measurements of populations of crystals, are now greatly facilitated by the recent developments in numerical image analysis techniques (Panozzo-Heilbronner, 1992; Allard and Benn, 1986; Launeau et al., 1990, 1994). They allow specific fabric quantification such as the comparison between the subfabrics of biotite and magnetite (see below), both in orientation and intensity. These techniques do not solve, however, the problem of time consumption and cost when mapping large areas, and the difficult problem of 3D fabric reconstruction from 2D-based data (see Fernandez and Fernandez-Catuxo, this volume).

MAGNETIC FABRIC MEASUREMENTS

The magnetic susceptibility K (SI units = 4π CGS units) of a body relates the induced magnetisation M to the magnetic field H into which the body is immersed by the relation M=[K] x H (Fig. 4a). In low magnetic field (Collinson, 1983), [K] is a second rank tensor geometrically represented by a triaxial ellipsoid with axes $K_1 \geq K_2 \geq K_3$, whose mean value, $K=1/3\ (K_1+K_2+K_3)$ is the bulk susceptibility magnitude. Graham (1953) first predicted that the anisotropy of magnetic susceptibility (AMS), or magnetic fabric, held great potential for imaging in 3D the internal fabric of a rock. A wealth of papers have been published on this subject, applying the AMS to granites, among which Daly (1959), Stacey (1960), King (1966), Birch (1979), and Ellwood and Whitney (1980) are some of the milestones.

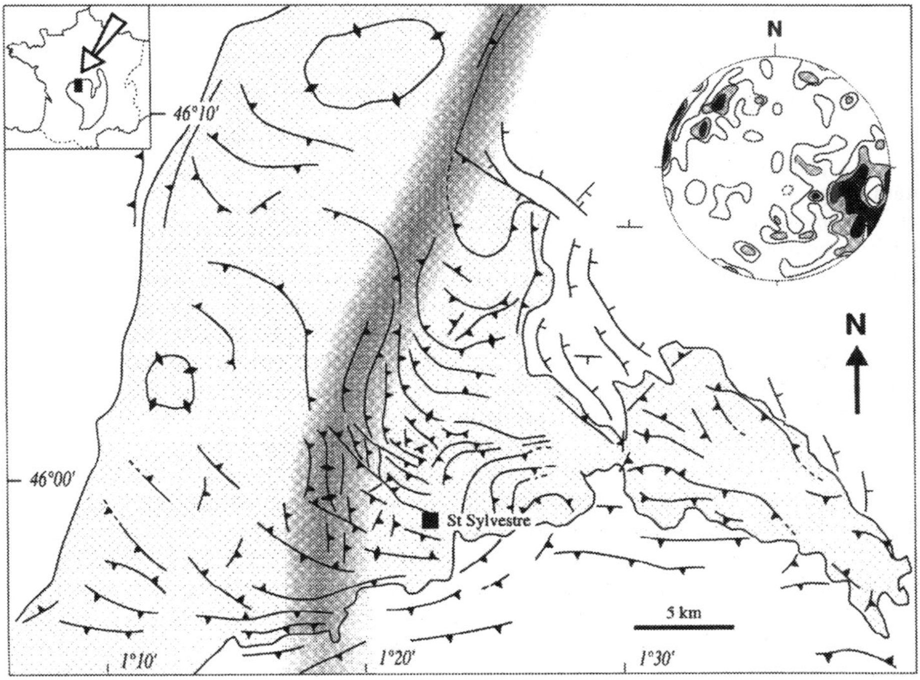

Figure 3. The Brâme-St Sylvestre-St Goussaud (BR-SS-SG) granite complex in the western French Massif Central: map of the magmatic foliation trajectories (determined from 256 different stations) and orientation diagram (equal area, lower hemisphere) of the magmatic lineations (141 measurements). Foliations and lineations determined as indicated in figure 2. Grey shading: magmatic shear zone discovered through this study. *After* Mollier and Bouchez (1982) and Mollier (1984).

Numerous papers have discussed, or reviewed AMS as applied to rocks (Hrouda, 1982; Rochette, 1987; Borradaile, 1988; Rochette et al., 1992, Tarling and Hrouda, 1993, among others). In the following, only the elements that closely relate with the current structural study of granites will be presented, but applications to lavas and sediments bear on the same principles. Other magnetic techniques, such as the anisotropies of magnetic remanences (Jackson, 1991), because they have not yet been used routinely, will not be discussed. Our instrumental experience for the measurement of AMS includes the « Minisep » apparatus of Molspin Ltd (Newcastle-upon-Tyne, UK) based on a specimen rotating in an alternative field ($7x10^{-4}$ T; 10 kHz), with a sensitivity of about $5x10^{-7}$ SI, and the KLY2 « Kappabridge » apparatus of Agico (Brno, Czech Republic) measuring axial susceptibility with an alternating current ($4x10^{-4}$ T; 920 Hz), the inductance variation being measured with a bridge (Jelinek, 1977). The high sensitivity of the Kappabridge instrument, of the order of 5×10^{-8} SI, is particularly well adapted to the study of the anisotropy of low susceptibility rocks, such as paramagnetic granites (see below). These low-field instruments, although very practical, have the drawback of giving a measurement of all magnetic contributions (see below). Therefore, they do not allow separation of the paramagnetic and ferromagnetic subfabrics. The ideal instrument, which would be based on a variable field, does not yet exist for routine measurement.

ELEMENTS OF MAGNETIC MINERALOGY

Iron, the principal element responsible for the magnetic signal, confers to the rock two main magnetic behaviours (Fig. 4a), depending on the lattices into which its atoms enter. *Paramagnetism* is the slightly positive magnetic susceptibility, K_{para}, generated by the iron-bearing silicates. In granites, those minerals are mainly biotite, iron-bearing muscovite, amphibole, and accessorily pyroxene, garnet, cordierite, epidote and tourmaline; although being an oxide, ilmenite belongs to this category. *Ferromagnetism* s.l. (i.e. including ferromagnetism stricto sensu, weak ferromagnetism and ferrimagnetism), K_{ferro}, has a positive susceptibility becoming null at high field strengths. The residual magnetisation, or remanence, at zero inducing field is at the base of palaeomagnetic studies. In granites, magnetite is the main mineral having this property, along with accessory monoclinic pyrrhotite. Strictly speaking these are ferrimagnetic. At low field, they have a strongly positive susceptibility magnitude. In contrast, hematite and goethite are antiferromagnetic, with susceptibility magnitudes lower than K_{para} (Fig. 4). Finally, the *diamagnetic* susceptibility, K_{dia}, is always slightly negative and concerns every mineral. In granites, more than 80% of the minerals, principally quartz and feldspar, display a diamagnetic behaviour only. The bulk magnetic susceptibility of a granite is the addition of all the contributions: $K = K_{para}+K_{ferro}+K_{antiferro}+K_{dia} \approx K_{para}+K_{ferro}$, since $K_{antiferro}$ and $|K_{dia}|$ are almost negligible. Note that K_{ferro} becomes null at high field (Fig. 4a), with a saturation threshold depending on the mineral species (around 1 Tesla for magnetite).

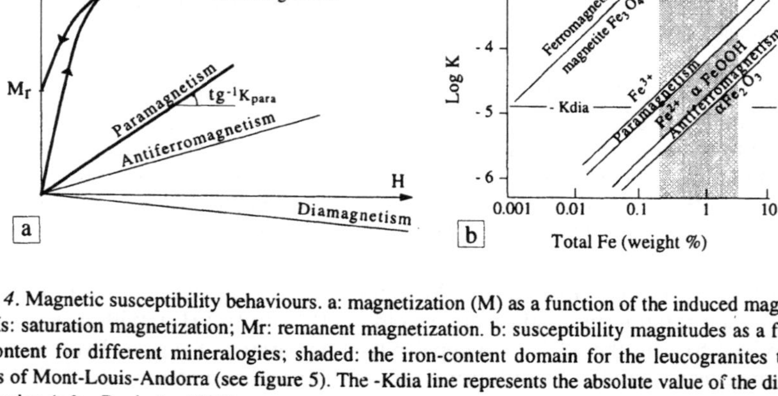

Figure 4. Magnetic susceptibility behaviours. a: magnetization (M) as a function of the induced magnetic field (H); Ms: saturation magnetization; Mr: remanent magnetization. b: susceptibility magnitudes as a function of iron-content for different mineralogies; shaded: the iron-content domain for the leucogranites to quartz-diorites of Mont-Louis-Andorra (see figure 5). The -Kdia line represents the absolute value of the diamagnetic contribution (*after* Rochette, 1987).

Therefore, a magnetite-free granite, or « paramagnetic granite », has a bulk susceptibility $K \approx K_{para}$ whose magnitude is generally below 500×10^{-6} SI, corresponding to the total susceptibility magnitude of the paramagnetic species. To a first approximation K_{para} is proportional to the iron content of the rock. More precisely, a theoretical K value may be calculated using the formula of Rochette et al. (1992) for « matrix » susceptibility, from known Fe^{2+}, Fe^{3+} and Mn^{2+} weight contents (Fig. 5a). This proportionality between iron-content and K, hence between the biotite (± amphibole) fraction and K, allowed Gleizes et al. (1993) to map out the different paramagnetic granite-types of the Mont-Louis-Andorra pluton, Pyrenees. The susceptibility magnitudes of the different rock-types are given (in 10^{-5} SI) in the frequency histogram of figure 5b.

Magnetite has a specific susceptibility about three orders of magnitude higher than that of biotite. Thus, for a given total iron content, the bulk susceptibility of a magnetite-bearing granite will be much higher than for a paramagnetic granite (Fig. 4b). This explains why magnetic susceptibility magnitudes of granitoids are bimodal, which fits with the classification of Ishihara (1981) into « magnetite-series » and « ilmenite-series » granites.

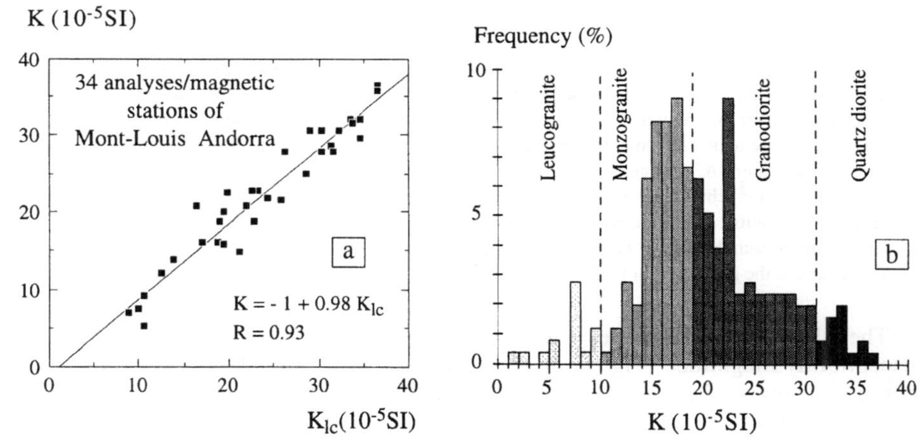

Figure 5. Paramagnetic-type Mont-Louis Andorra pluton (Pyrenees). a: correlation between the measured magnetic susceptibility (K) and the calculated paramagnetic susceptibility (K_{lc}), using chemical analyses, for 34 samples. b: frequency histogram of the susceptibility magnitudes (254 sampling stations) and correspondence with rock-types. *From* Gleizes et al. (1993).

MAGNETIC FABRIC IN PRACTICE

The long axis of the AMS ellipsoid is the magnetic lineation, and the short axis is the normal to the foliation plane. Determining the orientation of this ellipsoid with respect to the geographical frame constitutes a chain of operations (Fig. 6) which is at the base of the structural mapping of granitoids.

The planar and linear magnetic structures are spatially well-described over a pluton by using a regular grid of sampling of about one-kilometre spacing for the usual scales of mapping. This is illustrated by a number of recently published studies, such as in this volume, those of the Los Pedroches batholith (Southern Spain; Aranguren et al.), Sierra

Nevada batholith (California; Saint Blanquat and Tikoff) and Rahama Complex (Nigeria, Ferré et al.). The striking homogeneity, at least in orientation, of the magnetic fabrics and therefore shape fabrics, over considerable areas (often >>10^3 km^2), constitutes a major result of AMS mapping in granitoids. The specific question of the magnetic fabric variability at various scales is addressed in Olivier et al. (this volume).

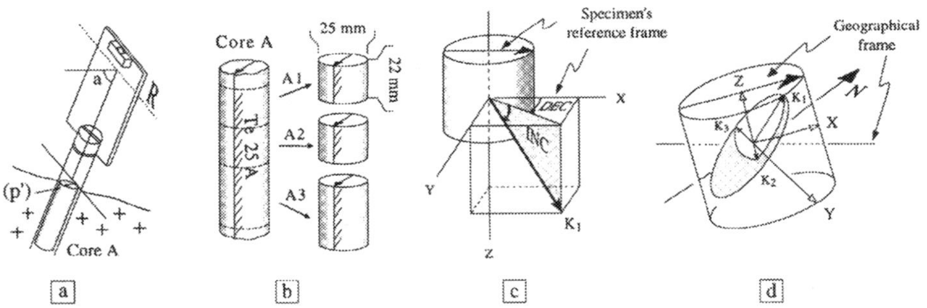

Figure 6. Sample collection procedure for AMS measurement. a: after drilling with a portable machine, the core is oriented with a compass and a clinometer; p is the strike of the plate perpendicular to the core axis; p' (= p ± 90°) and α are the trend and plunge of the core axis; line along core materializes the vertical plane passing through core-axis, and top-arrow of core points down-plunge parallel to p'. b: specimens A1 and A2 are collected from core A; the same is done from core B, thus giving 4 specimens per station, i.e., a rock volume of 4x10.8 cm^3; chunks A3 and B3 may eventuallyprovide additional AMS specimens and are used for thin section preparation and microstructural determination. c: the AMS measurement yields the declination and inclination of each AMS axis with respect to the specimen's axes. d: using p' and α the AMS ellipsoid is calculated back to the geographical frame.

The AMS ellipsoid provides a wealth of scalar information. In addition to the bulk susceptibility magnitude (Fig. 5), it gives the relative magnitudes between axes. The most commonly used parameters are (see Jelinek, 1981; Ellwood et al., 1988):

$P = K_1/K_3$, or total anisotropy degree;

$L = K_1/K_2$, or linear anisotropy degree;

$F = K_2/K_3$, or planar anisotropy degree; and

$T = [Log(K_2/K_3)-Log(K_1/K_3)]/[Log(K_2/K_3)+Log(K_1/K_2)]$,the shape parameter of Jelinek (1981) which varies from T=-1 (cigar shaped ellipsoid) to T=+1 (pancake). It describes the shape with no reference to anisotropy (Borradaile and Werner, 1994). It is easier to handle than the famous parameter of Flinn (1962), P_{Flinn}= (L-1)/(F-1), which varies from zero to infinity as in figure 10b, and involves both shape and anisotropy. Note that plotting T and P together allows shape and anisotropy to be described by two independent parameters. Note also the frequent use of $P_{Jelinek} = \exp[2 \sum (\ln K_i/K)^2]^{1/2}$ (i = 1 to 3), which takes the shape of the ellipsoid into account in the calculation of the anisotropy degree. $P_{Jelinek}$ is slightly larger that P for prolate ellipsoids, smaller than P for oblate ellipsoids, and very close to P for $K_2 = 1/2\ (K_1+K_3)$. Finally, note the widespread use of the parameters given as percentages, such as P% = (P-1)x100, L% and F%.

The degree of paramagnetic anisotropy « P_{para} » is sometimes used for rocks having very low susceptibility magnitudes. This parameter avoids the « artificial » increase of the anisotropy parameter due to K_3 tending to zero, by adding the diamagnetic contribution K_{dia}, considered to be isotropic, to each axis of the ellipsoid (Rochette, 1987; Bouchez et al., 1987). In a granite, K_{dia} is estimated to be around -14×10^{-6} SI, the magnitude of quartz diamagnetism (Hrouda, 1986). Therefore, $P_{para}=(K_1-K_{dia})/(K_3-K_{dia})$. Calculation of L_{para} and F_{para} is similar.

MAGNETIC FABRICS OF PARAMAGNETIC GRANITES

A granite containing biotite as the only iron-bearing phase constitutes the simplest case of paramagnetic fabric. Although less frequent, this is also the case for a granite containing iron-bearing muscovite in addition to biotite (Bernier et al., 1987), or containing only amphibole. In a biotite-only granite, the *orientation* of the magnetic fabric of a population of biotites is similar to that of its petrofabric, or lattice fabric, since biotite has a magneto-crystalline anisotropy. The magnetic axes of biotite are parallel to its crystallographic axes, and therefore to its shape axes. The minimum specific anisotropy magnitude k_3 for biotite, and more generally phyllosilicates, is normal to the cleavage plane: $k_3<k_2=k_1$; $k_3 \perp (001)$. Thus K_3, the short axis of the rock magnetic ellipsoid, is coaxial with the long axis of the fabric ellipsoid of $\perp$ (001), the poles to the cleavage planes of biotite (Fig. 1b). Hence K_3 represents the statistical normal to the biotite foliation. Conversely, K_1, the long axis of the rock magnetic fabric, equates with the minimum of the fabric of $\perp$ (001), which, in essence, is also the zone axis of (001) (see Fig. 1a, and related discussion concerning the zone axis).

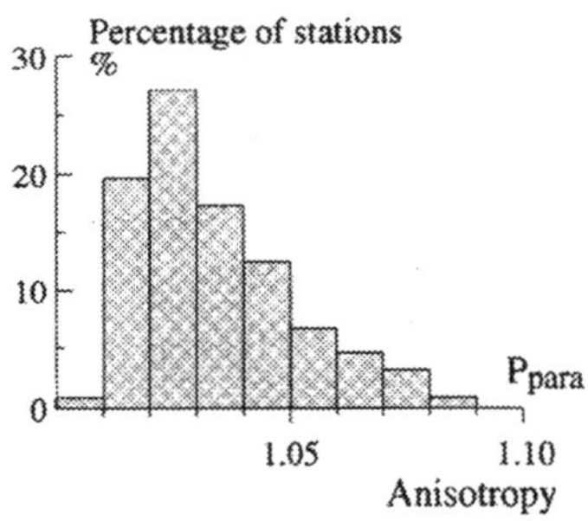

Figure 7. Anisotropy degrees of paramagnetic granites of the Pyrenees (France and Spain). Frequency histogram of 774 stations compiled from Gleizes (1992), including the granite massifs of Ercé, Lacourt, Maladeta, Mont-Louis-Andorra and Trois-Seigneurs; 1% of the data have $P_{para} \geq 10\%$. Note that, in the calculation of P_{para}, the diamagnetic contribution has been subtracted (see text).

The *magnitude* of the magnetic fabric varies between $P=1$ for a random orientation of biotite (isotropy), and $P=k_1/k_3=P_c$, which is around 1.35 (Borradaile et al., 1987; Zapletal, 1990), the intrinsic magneto-crystalline anisotropy of biotite, for a perfect planar orientation of all biotite grains. The anisotropy degree of biotite-bearing granites hardly exceeds P=1.10 (Fig. 7). This ideal paramagnetic fabric behaviour must be tempered, however, since minute inclusions or exsolutions of ferromagnetic material may be present in biotite (Borradaile and Werner, 1994). Traces of ferromagnetic material, easy to detect with the Micromag apparatus (see Borradaile and Werner, 1994), seem to affect mainly the shape parameters of the resulting fabric.

If not perfect, this simple interpretation for the magnetic fabric in the biotite-granites, and more generally for rocks whose magnetic behaviour is entirely controlled by phyllosilicates, is particularly favourable for numerical modelling of magnetic fabrics (e.g., leucogneisses, Siegesmund et al., 1995). In a two-mica granite, the role of muscovite, if iron-bearing, is exactly the same as the role of biotite, except that it has a lower intrinsic susceptibility. Modelling considers the susceptibility tensor of the magnetic marker (biotite), which is introduced into the orientation model of March (1932) for slates and schist, or into the model of Jeffery (1922) for biphasic materials such as magmas (Arbaret et al., and Schulmann et al., this volume). Therefore, the sensitivity of the magnetic fabric to strain intensity, initial fabric, strain overprints, and departures from simple shear can be explored (Hrouda, 1987; 1993; Vergne and Fernandez, 1990; Richter et al., 1993; Benn, 1994; Schulmann et al., this volume).

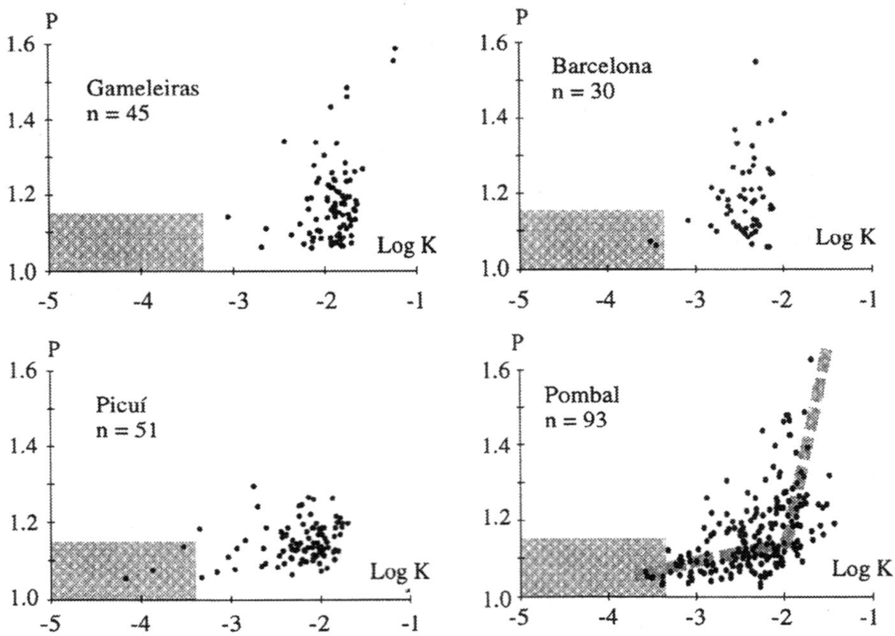

Figure 8. Anisotropy degree versus magnetic susceptibility K, of ferromagnetic granites of Northeast Brazil (Gameleiras, Barcelona, Picui and Pombal); shaded: maximum paramagnetic domain of granitoids; n: number of different stations; *after* Archanjo (1993). Abrupt steepening of the broken line in Pombal underlines onset of magnetic interactions in magnetite-bearing granites.

Other iron-bearing silicates in the granite, such as amphibole, yield composite fabrics that are less easy to model. The simplest case is the biotite-amphibole granite. Although specific magnetocrystalline data on amphibole are incomplete, this mineral is thought to have its magnetic long axis parallel to its shape long axis, or at a moderate angle to it. This is suggested by the numerous magnetic studies of amphibolites showing that K_1 is parallel to the prominent amphibole lineation (Pearce and Fueten, 1989). The experiments of Arbaret (1995) also show that in suspensions of amphibole deformed in torsion, similar to the biotite experiments of Arbaret et al. (this volume), K_1 strongly concentrates parallel to the finite strech direction as soon as $\gamma>1$. This suggests that K_1 is parallel to the mean elongation of amphibole. With a specific anisotropy magnitude of $Pc\approx1.6$ (Borradaile et al., 1987), and supposed to be more linear than planar, amphibole in a biotite-granite is therefore expected to amplify the prolateness of the magnetic ellipsoid.

Other iron-bearing silicates may occur in addition to biotite and amphibole. The addition of minerals having an inverse magnetocrystalline fabric (i.e. magnetic short axis of mineral is parallel to its shape long axis), may lead to anomalous magnetic fabrics that may be difficult or impossible to analyse (Rochette et al., 1992). One such uncommon case, the tourmaline-biotite-leucogranites of the Higher Himalayas, has been documented in detail by Rochette et al. (1994) and Scaillet et al. (1995). Another case with cordierite has been reported by Amice (1990).

MAGNETIC FABRICS OF FERROMAGNETIC GRANITES

This concerns granites for which the oxygen fugacity of the magma favoured magnetite crystallizing with ferro-magnesian silicates. The sizes of magmatic magnetite grains range from a few tens of microns to 1 or 2 millimetres. These large grain sizes result in the multidomain magnetic behaviour of magnetite. Magnetic mineralogy techniques, such as alternating field and thermal demagnetisation of Isothermal Remanent and Anhysteretic Remanent Magnetisation, confirm the predominant multidomain nature of magnetite in many granites (Benn et al., 1993; Archanjo et al., 1995). Multidomain magnetite has a specific magnetic anisotropy degree which, to the first order, is proportional to the grain shape ratio (Uyeda et al., 1963). Therefore, the shape fabric of magnetite is expected to control the magnetic fabric.

In terms of magnetic fabric *magnitudes*, the large range of anisotropy values of magnetite-bearing granites, up to P=1.6 for Log K > -2.5 (Fig. 8), cannot be entirely explained by the magnitude of the shape fabric of magnetite (Archanjo et al., 1995). Rochette et al. (1992) stressed that the positive correlation between P and K (Fig. 8) indicates that a composite magnetic fabric exists. In a granite in which biotite and magnetite are present, both mineral species contribute to the anisotropy degree up to about P=1.15, i.e., the maximum value observed in paramagnetic granites.

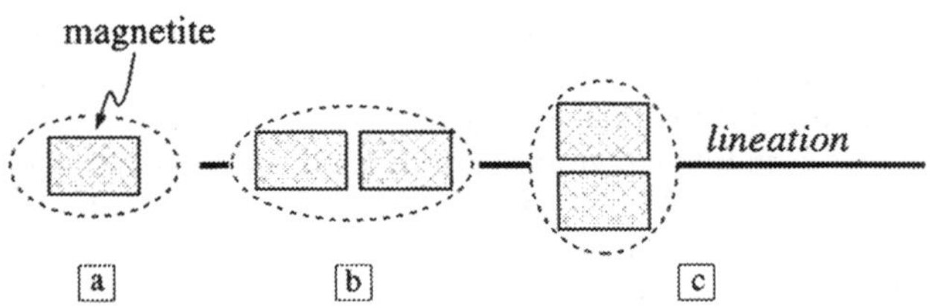

Figure 9. Role of magnetic interaction between elongate grains of magnetite (after Grégoire et al., 1995). a: susceptibility ellipse of a grain of magnetite. b: in the « aligned » configuration, the resulting susceptibility ellipse is stretched parallel to lineation. c: in the « side by side » configuration, the resulting susceptibility ellipse may become elongate perpendicularly to lineation, thus lowering the bulk anisotropy degree of the rock.

The sudden increase of P at « ferromagnetic » values of K (Log K > -2.5) can be explained only by the complementary effect of distribution anisotropy. This effect, based on the fact that an elongated cluster of equant grains produces a similar anisotropy as a single elongated grain, was suggested to be the principal source of anisotropy in basalts by Hargraves *et al.* (1991). It has been studied experimentally by Grégoire *et al.* (1995) for different arrangements of slightly elongate magnetite grains (Fig. 9), on the basis of Stephenson's analysis (1994). Due to magnetic interactions, for a distance between grain centres less than twice the grain size, the susceptibility magnitude increases parallel to grain alignment (Fig. 9b). Note that, for a given concentration of magnetite, the probability of magnetic interaction is larger for small grain sizes of magnetite, as in basalt, than for large grains, as in granite. Similarly, for a given average grain size, the magnetic interactions increase with the amount of magnetite: this explains the strong increase of P with K in the high susceptibility domain (Log K > -2.5; Fig. 8). Note that the magnetite grains in inclusion within silicates, such as in the clinopyroxenes of the syenite of Triunfo in Brasil (Archanjo, 1993), may preferentially interact in the « aligned » configuration (Fig. 9b), hence increase the suscetibility magnitude parallel to the lineation. In contrast, grain interactions in the "side-by-side" configuration may lower the susceptibility magnitude parallel to lineation (Fig. 9c). This may explain the frequent scattered values of P within a given specimen of ferromagnetic granite, and the common presence of a wide range of P values in the P vs. K diagrams of figure 8.

COMPARISON BETWEEN PARA- AND FERROMAGNETIC FABRICS

This question is essential for the use of AMS in the structural analysis of granites. The *orientations* of the subfabrics of biotite and of magnetite are more-or-less parallel. This is exemplified by ferromagnetic granites in which the magnetite content, as estimated by the susceptibility magnitude, varies considerably from place to place without a significant variation in the magnetic ellipsoid orientation (Archanjo, 1993). The precise measurement of biotite and magnetite subfabrics, with an automatic shape fabric analysis procedure from carefully oriented 2D sections of the Gameleiras ferromagnetic granite in Brazil, also supports the coaxiality of both fabrics (Archanjo et al., 1995). Quite surprisingly, in the latter study where the contribution of magnetite to AMS predominates, the statistically measured angle between K_1 and the subfabric of biotite, and variation of this angle, are smaller (7°±8°) than between K_1 and the subfabric of magnetite (12°±18°). This is attributed to the small number of magnetite crystals used in the statistics, to the poor shape fabric of magnetite, which has grain shape ratios barely reaching 1.5, and to the fact that image analysis is not sensitive to distribution anisotropy. Magnetite, either as individual grains or as elongate clusters, has a tendency to align preferentially parallel to the other crystals in the magma such as biotite. This kind of mimetic orientation could be responsible for the parallelism between the subfabrics of biotite and magnetite.

Since several factors, namely grain shape, shape fabric and distribution of magnetite, play a role in the AMS of the non-paramagnetic rocks, the use of the magnetic *anisotropy degree* (P) to quantify the fabric strength, hence the strain intensity, should be restricted to the paramagnetic granites, say for Log K (SI) < -3.5 (shaded in Fig. 8), and if possible to biotite-alone (or amphibole-alone) bearing granites. Indeed in the latter granites, the variations of P may be attributed to strain gradients such as those observed to increase towards a pluton's border (Leblanc et al., 1994; Aranguren et al., this volume), or towards late magmatic shear zones (Bouchez and Gleizes, 1995). In ferromagnetic granites, the contribution of strain intensity to anisotropy becomes secondary with regard to other factors such as magnetic mineralogy (Benn et al., 1993) and interaction anisotropy (Grégoire et al., 1995). Variations in K and P, however, may be used as magnetic mineralogy variation indicators, hence help to track p_{O_2} variations, solid-liquid partitions, traces of hydrothermal activity , etc...

FABRICS IN GRANITIC ROCKS: DISCUSSION

LOCAL AND REGIONAL VARIABILITIES OF AMS MEASUREMENTS

In contrast to the palaeomagnetic technique, which requires a dozen specimens per site, the AMS technique needs only a few specimens, four in the procedure described in figure 6, i.e. about 120 grams of rock, to correctly characterise a given site. This is due to the remarkable homogeneity of grain shape fabrics in granites, down to volumes as small as a few tens of cubic centimetres.

	PARAMAGNETIC						FERROMAGNETIC			
	(a) Foix 69 stations		(b) Trois-Seigneurs 51 stations		(c) Maladeta 253 stations		(d) Pombal 93 stations		(e) Gameleiras 45 stations	
Variability	K_1	K_3	K_1	K_3	K_1	K_3	K_1	K_3	K_1	K_3
$\alpha K<10°$ in%	9	15	40	38	26	38	49	51	62	47
- < 20° -	59	63	51	64	51	64	80	82	82	89
- < 30° -	79	85	68	77	68	77	95	88	90	99

Table I. Compilation of within-site variabilities of the AMS measurements in the paramagnetic plutons of Foix, Trois-Seigneurs and Maladeta (Pyrenees), and ferromagnetic plutons of Pombal and Gameleiras (North-East Brazil). αK_3 (respectively αK_3) is the maximum angular departure between the mean direction of K_1 (resp. K_3) of the four specimens per sampling site. The figures give the percentages of sites having αK_1 and αK_3 less than 10°, 20° and 30°. Origin of data: a: Bouchez et al., 1990; b: Gleizes, 1992; c: Leblanc et al., 1994; d: Archanjo et al., 1994; e: Archanjo et al., 1995.

At the scale of a sampling site (where a pair of specimens is a few meters apart from the other pair), the variability may be measured by αK_1 (and αK_3), the maximum angular departures between the mean direction of K_1 (and K_3) and the direction of the individual K_1 and K_3 of the four specimens per site. Table I gives the sampling site percentages that have αK_1 and αK_3 less than 10°, 20° and 30°. Local variabilities are generally rather small since more than 30% of sites (except for the Foix pluton) have maximum departures of less than 10°, and more than 50% have maximum departures of less than 20°. Note that, in Table I, variabilities are much smaller in ferromagnetic granites (>80% of sites <20°) than in the paramagnetic granites (> 50% of sites <20°). Strong within-site grouping of orientation measurements seems to be common in calc-alkaline magnetite-bearing granites (Saint Blanquat et al., this volume). As expected, at least in paramagnetic granites, the fabric variability decreases with increasing anisotropy (Fig. 10a), and K_1 is better defined than K_3 (αK_1-αK_3<0) in prolate magnetic ellipsoids than in oblate ones (Fig. 10b).

Fabric variability from the scale of the sampling site to the scale of the whole pluton is merely represented by the maps of the magnetic foliations and lineations, and by the stereoplots of their orientations. Several examples of remarkably smooth regional variabilities in fabric orientation, particularly in lineation trends can be cited. This was the topic of the study of Darrozes et al. (1994), who compared the subfabric of the K-feldspar megacrysts with the magnetic fabric in a quarry of the Sidobre pluton. This is also the object of Olivier *et al.* (this volume) who examine the fabric variability at intermediate scales in different granite massifs.

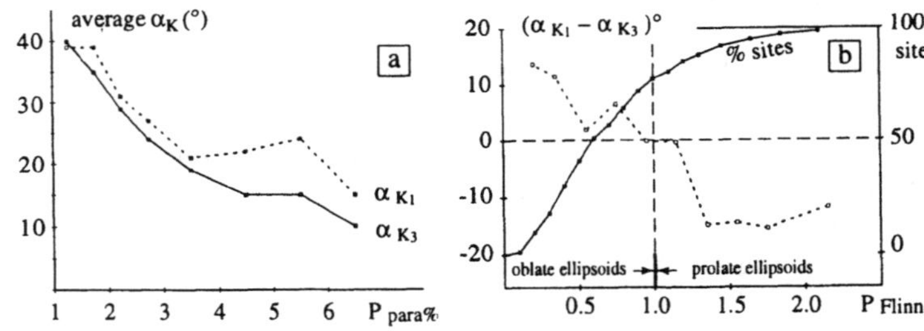

Figure 10. Within-site directional variability of K_1 and K_3 in the paramagnetic granite of Maladeta (253 different sites). a: rôle of the total anisotropy. b: rôle of the shape of the magnetic ellipsoid: dashed line; $\alpha K_1 > \alpha K_3$ in oblate ellipsoids; and $\alpha K_1 < \alpha K_3$ in prolate ellipsoids for which the lineation is best defined; plain line: cumulative percentage of sampling sites. *After* Leblanc et al. (1994).

ORIGIN AND KINEMATIC SIGNIFICANCE OF FABRICS

The different rotation rates of crystals having different shapes within the magma is the source of shape fabric development. Based on Jeffery (1922), fabric development was modelled numerically by Willis (1977), Fernandez et al. (1983) and Jezek et al. (1994). When the axes of the magnetic ellipsoid correspond to the morphological axes of the crystals whose subfabric is measured, these numerical models also apply to magnetic fabrics. Hence, the application is straightforward, and it is « exact » for biotite granites (e.g., Housen and van der Pluijm, 1990; Vergne and Fernandez, 1990; Richter et al., 1993), but only approximate when the paramagnetic mineralogy becomes more complex (Housen et al., 1993). Numerical modelling fails in magnetite-bearing granites, due to the complex interplay of the fabric of magnetite having various intrinsic shape ratios, and anisotropic magnetic interactions related to magnetic grain distribution.

3D analogue experiments in simple shear of suspensions of biotite (Arbaret et al., this volume) show that the biotite subfabrics are much more stable with increasing strain, both in orientation and magnitude, than predicted by numerical models. In a well-sorted suspension, i.e. where particles have consistent shape ratios, fabric stabilization is attributed to the mechanical interactions between crystals, even at low solid fractions (Arbaret et al., 1996), that slow down the crystal rotation rates. In a real magma where the crystals more-or-less form a continuum of shape ratios, the fabric is also efficiently stabilized by the averaging effect of the various rotation rates (see Ildefonse et al., this volume). Returning to the experiments of Arbaret et al. (this volume) the fabric patterns are observed to become stable at γ around 2 or slightly more, and remain more-or-less identical up to $\gamma = 20$. K_1 is observed to be very close to the maximum principal finite stretching axis, and K_3 very close to the shortening axis.

This parallelism between the finite stretch (at least during the final increments of magmatic strain) and the magnetic lineation is demonstrated, in the pluton of Monte Cappanne, Italy (Bouillin et al., 1992), and in the Sidobre pluton (Fig. 11), southern French Massif Central (Darrozes et al., 1994), by the remarkable orthogonal trajectories, in maps views, between the independently measured K_1 axes and strike directions of the late magmatic dikes.

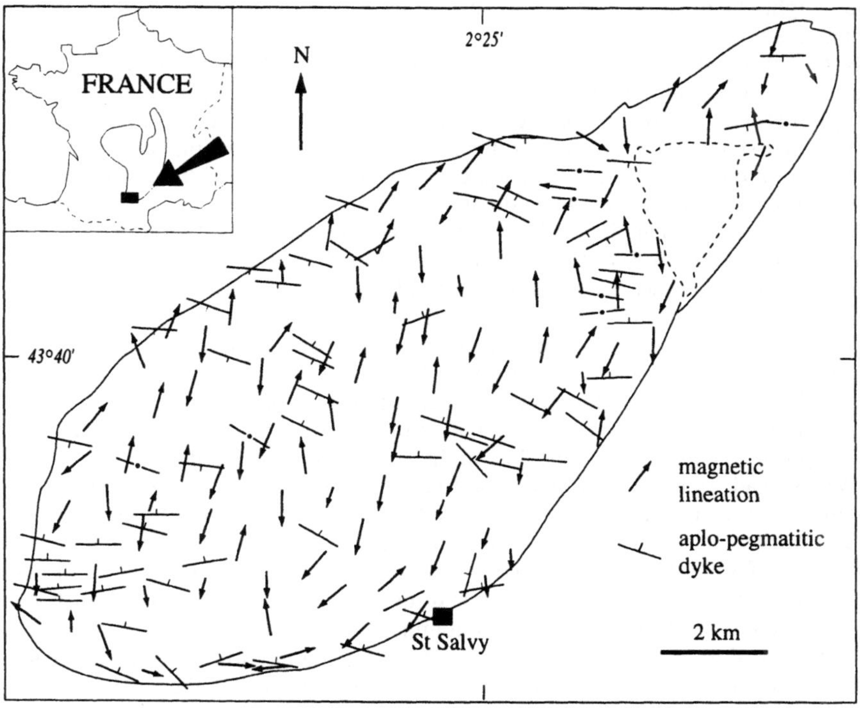

Figure 11. The Sidobre granite pluton (Southern French Massif Central): the magnetic lineations almost everywhere are normal to the aplo-pegmatitic dikes. The overall sigmoidal trend of the lineations, mapped from 103 sampling sites, is interpreted as resulting from dextral shear parallel to thelong sides of the pluton during emplacement. Adapted from Darrozes et al. (1994).

More detailed kinematic analyses, such as the determination of the sense of shear and amount of an eventual coaxial contribution to the progressive strain history, are still the object of sophisticated petrofabric studies in 2D. These must be based on the orientation distributions of the crystals, taken either as individual directions (Benn and Allard, 1989) or globally (Panozzo-Helbronner, 1992; Launeau and Bouchez, 1992), eventually in combinaison with their size distributions (Launeau et al., 1995; Schulmann et al., this volume).

RELATIONS BETWEEN MAGMATIC FABRICS AND EMPLACEMENT MODES

Crystalline fabric axes therefore represent rather exact and stable indicators of the principal axes of the strain undergone by the magma at the end of its emplacement, but the magnitude of the strain is poorly recorded. For example, the existence of a magmatic lineation indicates that, say, the last 2γ of the magmatic shear have been undergone parallel to the extension direction represented by the lineation, but no indication is given concerning the overall strain magnitude itself which may be very large. Since the kinematic memory of a fabric appears to be quite short, other strain intensity markers have to be sought, such as the shape ratios of enclaves (John and Blundy, 1993; John and Stünitz, this volume).

Note that, up to now, we have implicitly considered the basic mechanism of magma deformation to be simple shear, but a component of flattening may be present, for example perpendicularly to the shear plane when the late-magmatic melt is expelled from its matrix into aplitic dikes. In this situation the crystalline planar fabric may get stronger in intensity, more-or-less parallel to the shear plane, and the linear fabric may weaken (see Schulmann et al., this volume).

We can now try to apply our knowledge concernic magmatic fabrics to the analysis of the emplacement modes of plutons. If the magmatic lineations are vertical, as in the whole pluton of Zaër, Morocco (Bouchez and Diot, 1990), and more locally in the plutons of Plouaret, France (Guillet et al., 1985) and Cabeza de Araya, Spain (Amice et al., 1991), it will be concluded that the increments of vertical displacement gradients recorded by the fabric were much larger than the horizontal ones. This situation is suggested at level 2 of figure 12. At this level of magma transfer, the fabric pattern gives no information on what happened closer to the surface, within the more brittle crust, nor on what happened deeper in the crust that might have been more ductile.

Widespread subhorizontal magmatic lineations parallel to the large-scale shear zones where regional strains are concentrated, as in Northeast Brazil (Archanjo et al., 1992; Vauchez et al., this volume), indicate that the magma has been highly strained along theses fault-zones, as suggested by level 1 in figure 12. Large volumes of magma may have been transferred to upper crustal levels but we have no record of this stage since horizontal displacement gradients dominated over the vertical ones in the magma body, at least during the last 2γ of strain. These granite massifs are often elongate and narrow, with subvertical foliations parallel to the shear zones. Other bodies, with also low-plunge lineation, may occupy rather large areas and point to sill-like emplacement, or occupy several thousands of square kilometres in area and form layers parallel to which crustal scale decoupling may have occurred, as in the Variscan granitoids of the French Massif Central (Faure, 1995) or in the PanAfrican stratoid granites of Madagascar (Nédélec et al., 1994).

Finally the largest fabric variability in plutons is found in the upper section of the crust (Fig. 12: level 3). Near a pluton's roof, the fabric patterns depend on the local differential mouvements between the magma and its country rocks, and at a distance, on the irregular geometry of the overlying brittle crust. In the Maladetta pluton, Pyrenees (Leblanc et al., 1994), several more-or-less interpenetrated cupolas have been demonstrated, called polydiapirs, and the fact that these were preserved in the fabrics is interpreted as due to the high-level section of observation in this high-emplacement- level pluton. A common case is provided when the lineation pattern in the pluton is oblique (30°- 40°) almost everywhere with respect to the pluton elongation, as in the Sidobre

(Fig. 11), or, as suggested in figure 12 (level 3), a case inspired by the study of the Tesnou Complex (Djouadi and Bouchez, 1992). In both the latter examples, the dominant structures in the country rocks that are likely to be coeval and related to the pluton's emplacement, are dextral shear zones parallel to the pluton's elongation. Using the late-magmatic dikelets (sketched in Fig. 12) as normal-to-stretch indicators of the very late increments of strain, this may well be a case where the strain magnitude recorded by the fabric could be very low, say $\gamma < 2$.

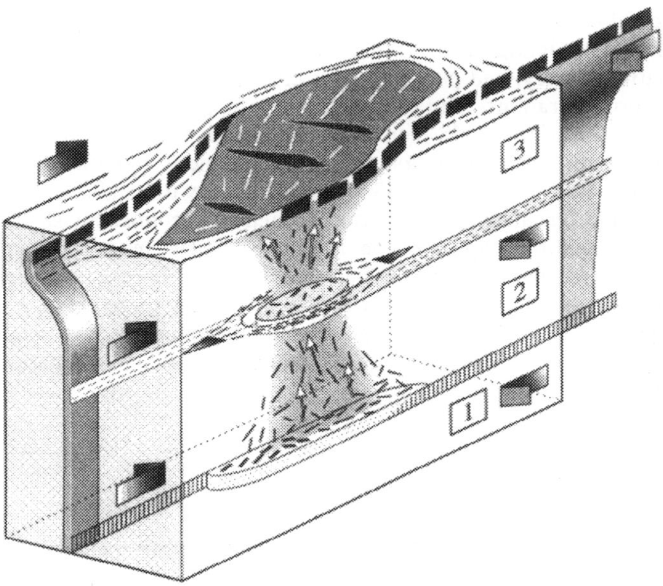

Figure 12. Relationship between magmatic fabrics and emplacement kinematics of granite plutons associated with transcurrent fault-zones, *after* Djouadi (1994). The sketches correspond either to three levels of pluton emplacement, or to three sections of a high-level emplaced pluton. (1): the shear zone is localized, and the highly deformed granitic magma forms elongate bodies whose fabrics are parallel to elongation. (3): the transcurrent shear zone is distributed heterogeneously and dilatant volumes may appear, within which the gathered granitic magma is slightly deformed hence its magmatic fabric is at an angle with the fault trace; sigmoidal fabric patterns may form depending on the relative rates of lateral displacement and magma upwell; note the late magmatic tension gashes infilled by aplitic magma (see figure 10). The intermediate stage (2) may represent the basal part of a pluton where the vertical shear of the ascending magma may result in dominant vertical lineations (see text).

In conclusion, these case studies, which will be complemented by many others in the next decade, clearly show that the fabric pattern of a pluton in map section does not provide, by itself, the key to the emplacement mode. Since the crystalline fabric has a short memory and the magma flow pattern may completely change depending on the section level in the crust, only partial conclusions may be drawn from the examination from the fabric patterns. The other main pieces of information that are required to complement any analysis of emplacement mode are (i) a description of the microstruc-tural state (fully magmatic, subsolidus, HT solid-state and LT solid-state) at every location in the pluton where a structural measurement has been performed, and its comparison with the fabric pattern (see for example de Saint-Blanquat and Tikof, this volume), and (ii) geophysics, mainly gravity modelling, which helps in inferring the extent at depth, the dips of the contact planes, and the location of eventual root-zones (see Améglio et al., this volume).

CONCLUSION

The scientific value of studying crystalline fabrics in magmatic rocks, and particularly in granites, has been highlighted. Information they contain relative to a part of the pluton's deformation history, and possibly to its emplacement mode, are obtained by applying simple considerations. The structural study of granite plutons has been greatly improved by the use of the AMS technique. An introductory review of this technique has been given, with particular emphasis on the magnetic mineralogy point of view which must not be neglected. The AMS technique is easy to handle and efficient since, for example, a little less than 12 kilograms of oriented specimens is sufficient to construct a foliation-lineation map based on 100 stations. The most spectacular result of structural mapping in granite plutons remains the remarkable structural homogeneity of the fabric patterns, and their smooth variability, over very large areas, a phenomenon totally unsuspected less than a decade ago. These fabric patterns now require more numerous data and better modelling to improve the interpretations of the modes of emplacement of granite plutons.

ACKNOWLEDGEMENTS

This study has benefited from the work (sampling, measurements, discussions) of many students and collegues, during many years. I am greated indebted to all of them. Thanks to Keith Benn, Eric Ferré, Pierre Rochette, Michel de Saint-Blanquat and Ed Stephens, who improved the manuscript, and to Christiane Cavaré-Hester, our Paganini of the Illustrator, for the quality of the figures. Under the overall support of the University of Toulouse, funds were provided by the CNRS and INSU, the French Ministry of Education, and complementary funding for bilateral exchanges, such as the CAPES-COFECUB for Brasil.

HOMOGENEITY OF GRANITE FABRICS AT THE METRE AND DEKAMETRE SCALES

Philippe OLIVIER, Michel de SAINT BLANQUAT,
Gérard GLEIZES and Denis LEBLANC
Equipe de Pétrophysique et Tectonique, UMR 5563 CNRS
Université Paul-Sabatier, 38 rue des Trente-Six-Ponts,
31400-Toulouse, France

ABSTRACT

Magnetic fabrics of biotite-bearing granites were systematically determined, at the metre and dekametre scales, in three plutons previously studied for their overall magnetic structures, in order to characterize the spatial homogeneity and variability of the fabrics. These granites, with typically magmatic microstructures, have different mean magnetic anisotropies (P%): Sidobre (southwest Massif Central of France; P%=2.3), Bassiès and Trois-Seigneurs (French Pyrenees; respectively P%= 3.3 and P%=5.6). In each site, two grids of 50 oriented specimens each, respectively one and ten metres apart from each other, have been studied in detail. The directional data, especially the lineations, strongly cluster around their means and have similar orientations on both scales. In map view, the fluctuations of these data are generally gradual and tend to form sigmoids but no clearly defined pattern, such as a C/S system was observed. The magnetic anisotropy and the bulk susceptibility are homogeneous as a whole, and display spatial organizations with no simple relationships with the structures. These results confirm, however, the validity of the homogeneous structural patterns obtained from entire plutons.

INTRODUCTION

Magmatic structure of granitic plutons is the topic of many recent studies, based either on direct measurements (e.g., Guineberteau et al., 1987; Pesquera and Pons, 1990; Gasquet, 1992; John and Blundy, 1993), or on the Anisotropy of Magnetic Susceptibility (AMS) measurements (see Bouchez, this volume). Many studies based on AMS point out the homogeneity of magmatic structures over wide areas and therefore permit better constraining of the modes of emplacement proposed for the studied plutons. In all these studies, the density of sampling is rather low, usually one site per km^2, which raises several questions, such as: i) does the homogeneity observed at the kilometre scale also exist at the lower scales and with which characteristics? ii) how does the fabric varies from one site to the next? iii) how representative is a site characterized by a few samples? To answer these questions, the magnetic fabrics of three different granites, chosen

J.L. Bouchez et al. (eds.) Granite: From Segregation of Melt to Emplacement Fabrics, 113–127.

for their different fabric intensities and grain sizes, have been studied on the scales of metre to several tens of metres, particularly focusing on the characterization of the spatial homogeneity and variability of the fabrics.

METHODOLOGY

MAGNETIC FABRIC MEASUREMENTS

The granites studied here belong to the group of the ilmenite-bearing granitoids (Ishihara, 1977; Bouchez, this volume), for which the iron-bearing silicates, in this case biotite, induce a dominantly paramagnetic behaviour, as confirmed by the low magnitudes of magnetic susceptibility, ranging from 5 to 50 x 10^{-5} S.I. In these rocks a variation of one 10^{-5} S.I. unit corresponds to a variation of about 0.15% in iron content (Gleizes et al., 1993).

Magnetic susceptibility measurements performed at low field (3.8 x 10^{-4} T, 920 Hz, Kappabridge KLY-2 apparatus) on ≈11 cm^3 cylindrical rock samples, allow us to define the intensity and direction of the three principal axes $k1 \geq k2 \geq k3$ of the ellipsoid of susceptibility anisotropy. Tensor means of the axes corresponding to the n samples measured at a site are $K1 \geq K2 \geq K3$. Km, the mean susceptibility, is the arithmetic mean of Ki (i=1, 3). K1, the long axis of the anisotropy ellipsoid, is the magnetic lineation, K3, the short axis, is the normal to the magnetic foliation. Uncertainty on susceptibility measurement may be estimated at a maximum of 3% (2% due to imprecisions in sample volumes and 1% due to the apparatus).

The total anisotropy percentage is given by $P\% = ((K1/K3)-1) \times 100$, or, to isolate the paramagnetic component, $Pp\% = ((K1\text{-}D)/(K3\text{-}D)-1) \times 100$, which is the ratio corrected for the diamagnetic component D, considered as constant and isotropic ($D = -1.4 \times 10^{-5}$ S.I.). This assumes almost constant modal compositions and an extremely weak intrinsic anisotropy of quartz and feldspar (Bouchez et al., 1987). The linear and planar anisotropy ratios are respectively denoted $Lp\% = ((K1\text{-}D)/(K2\text{-}D)-1) \times 100$ and $Fp\% = ((K2\text{-}D)/(K3\text{-}D)-1) \times 100$.

RELATIONSHIP BETWEEN MAGNETIC FABRIC AND MINERAL FABRIC

Various studies using independent measurement techniques (e.g. Heller, 1973; Guillet et al., 1983; Darrozes et al., 1994) have demonstrated that in dominantly paramagnetic granites, the magnetic fabric is parallel to the mineral fabric. In biotite-bearing granites, the magnetic foliation and lineation correspond respectively to the orientation mean plane and zone axis of the biotite flakes.

As the granites studied here have entirely magmatic microstructures, i.e. display no solid-state deformation features (Paterson et al., 1989; Bouchez et al., 1992), the foliation and lineation respectively correspond to the flattening plane and stretching direction acquired by the end of crystallization of the magma. The relationship between the magnetic anisotropy ratio and the finite strain suffered by the granite depends on the strain path and strain intensity (see Arbaret et al., this volume) and several parameters intrinsic to the rock such as the shape of the iron-bearing minerals, their magneto-crystalline anisotropy, the relative abundance of mineral species and the grain size homogeneity. A quantitative evaluation of the finite strain undergone by the crystallizing magma is therefore difficult, but a qualitative evaluation may be proposed: Pp < 3% corresponds to weakly oriented granites, 3%< Pp< 7% corresponds to a clearly visible planar orientation of the rock, and Pp >7% characterizes a strongly oriented granite.

SAMPLING

Three sites in fresh granites displaying different mean bulk magnetic anisotropies were sampled (Fig. 1). The least anisotropic site (Pp = 2.3%) is located in a quarry in the Sidobre (SI) monzogranite pluton (French Massif Central); the second site (Pp = 3.3%) is on a recently eroded glacial surface in the monzogranitic part of the Bassiès (BA) pluton (French Pyrenees); and the most anisotropic (Pp = 5.6%) is on a recently eroded glacial surface in the Trois-Seigneurs (TS) granodiorite (French Pyrenees). These three Variscan plutons were recently studied for their magnetic structures (Moisy, 1993; Darrozes et al., 1994; Gleizes et al., 1991; Leblanc et al., in press).

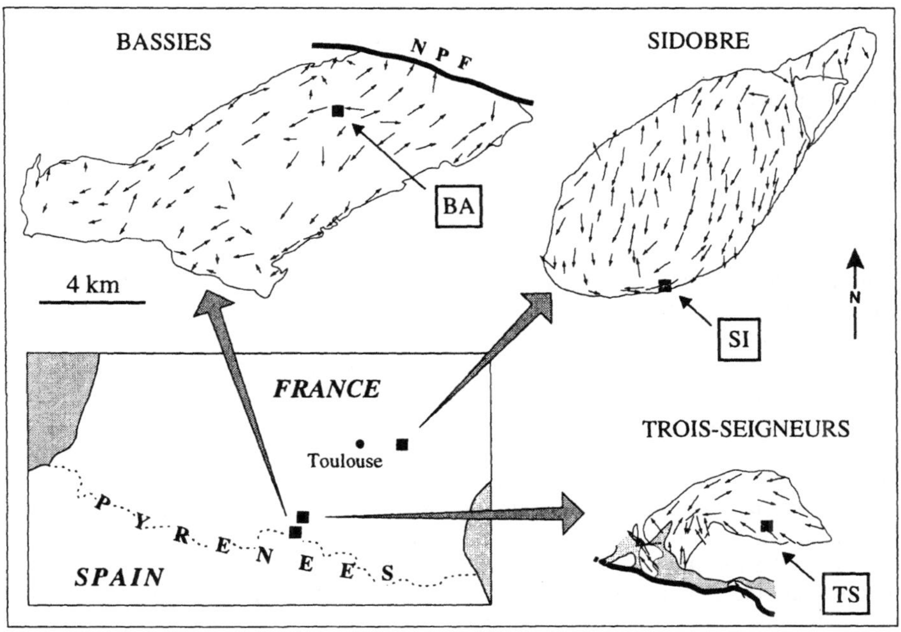

Figure 1. The Sidobre, Bassiès and Trois-Seigneurs plutons. The studied sites (SI, BA and TS) are located on the corresponding maps of magnetic lineations. NPF: North Pyrenean fault. All three plutons.are at the same scale.

At each site, two square grids with 7 x 7 sampling points, plus one outside (Fig. 2), i.e. 50 sampling points, were core-drilled. There were a dekametric grid (dam) for which the cores are ten metres apart, and a metric grid (m) included in the large grid, for which the cores were taken one metre apart. Each core yielded two specimens, about 11 cm^3 in volume each, which were separately measured. The mean of both measurements characterizes each core. The mean grain areas, determined on thin sections (Fig. 3), are the following: SI: 0.9 mm^2 (standard deviation σ = 2.6 mm^2); BA: 0.5 mm^2 (σ = 1.2 mm^2); TS: 0.2 mm^2 (σ = 0.3 mm^2).

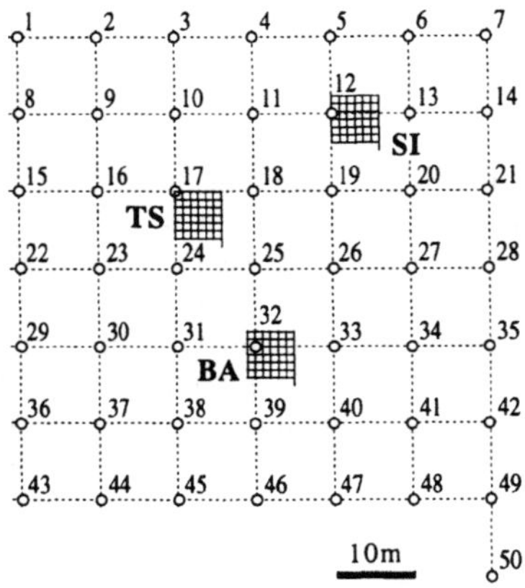

Figure 2. Sampling grids. The small grid is located close to point 12 of the large grid for SI, point 32 for BA and point 17 for TS.

DATA

The raw data, calculated from the sets of 50 cores, are given for each grid in Tables 1 and 2 for the mean orientations of the lineations and foliations, Table 3 for the mean susceptibility magnitudes, and Table 4 for the mean anisotropy percentages. An elementary statistical analysis is given for each sample set, i.e. standard deviation (σ) and coefficient of variation (cv1 = σ/mean), providing an estimate of the within-grid variability. The within-core variability is estimated by the mean (mdsp) and standard deviation (σsp) of the departures between both specimens of the cores, and either, for the directional data, by the percentage of cores displaying a departure between both specimens from 0° to 20°, or, for the scalar data, by a coefficient of variation cv2 = σsp/mean. The variability of the directional data is also represented by orientation in equal-area diagrams (Fig. 4). Spatial organization of the data is depicted on maps of figures 5 to 8 respectively corresponding to the lineations and foliations orientations, susceptibility and anisotropy magnitudes. On these maps, the fiftieth sample is omitted for the sake of simplification. Each data point represents the mean of the two specimens (22cm^3) of the core drilled at this point.

LINEATIONS AND FOLIATIONS

The main results arising from this study are i) the good clustering of the magnetic lineations, and to a lesser extent foliations, around a mean direction; ii) the similar orientations of these structures, for a given site, on both metre and dekametre scales.

Magnetic lineations K1 are the directional data which display the most homogeneous groupings (Table 1; Fig. 4). Their variability is weakest for TS and BA (Table 1, σ),

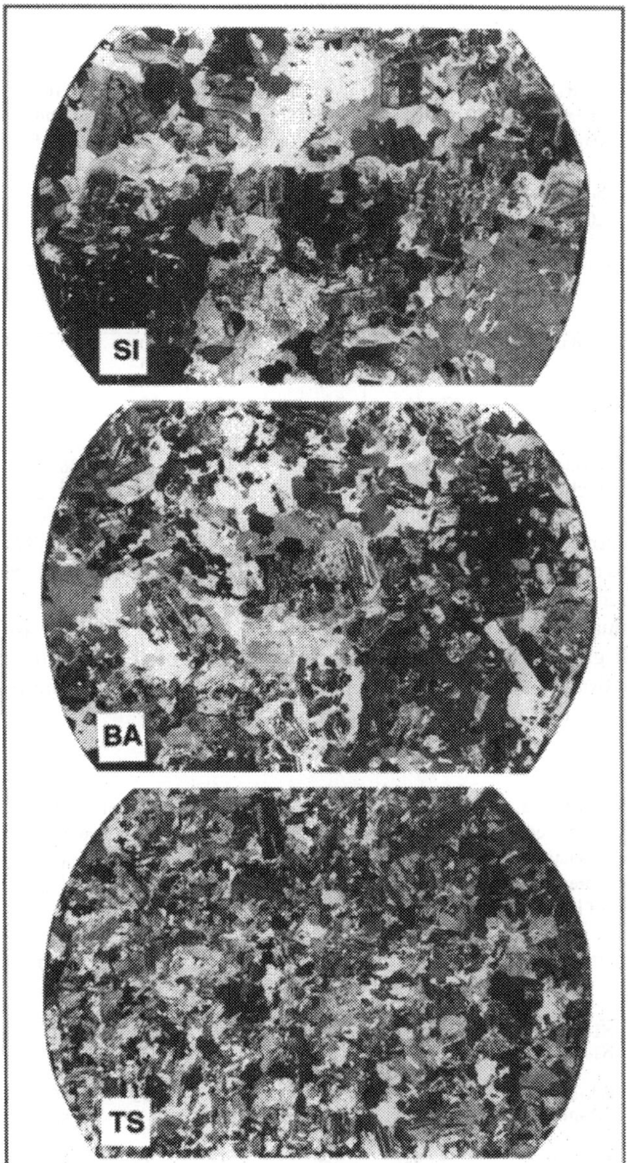

Figure 3. Thin sections perpendicular to core axis (core diameter ≈ 25 mm), of representative textures of SI, BA and TS.

slightly higher for SI, particularly on the large grid (σ = 25°). The angular dispersion on the small grid is lower than on the large grid for all three sites. The angular departure between the mean lineations of both grids of a site is lowest for TS (2°: 290° /19° against 290° / 17°), highest for BA (9°: 47° / 41° against 56° / 35°).

TABLE 1. mean K1 and σ: mean and standard deviation of the lineations of 50 cores (100 specimens); mdsp, σsp: mean and standard deviation of the departures between the two specimens of a same core; 0°-20°: percentage of cores for which departure between lineations of individual specimens lies in the range 0° to 20°.

	mean K1 (%)	σ (°)	mdsp (°)	σsp (°)	0°-20° (%)
SI-dam	248 / 9	25	28	20	46
SI-m	243 / 9	17	22	15	50
BA-dam	47 / 41	21	16	12	70
BA-m	56 / 35	14	12	6	90
TS-dam	290 / 19	19	11	7	88
TS-m	290 / 17	12	9	6	94

The within-core variability of the lineations (Table 1) is low for TS (σsp = 7°-6°; 88%-94% of the cores have an angular departure between individual k1 ranging from 0° to 20°), and, to a lesser extent, for BA (σsp = 12°-6°; 70%-90% of the cores have well-grouped individual specimens), clearly higher for SI (σsp = 20°-15°; 46%-50% of the cores have well-grouped individual specimens). In spite of these intrinsic variabilities, the overall lineation trends and plunges appear to be well organized, the transitions from one direction to another generally being gradual (Fig. 5).

TABLE 2. mean K3 and σ: mean and standard deviation of the K3 of 50 cores (100 specimens); mdsp, σsp.: mean and standard deviation of the departures between the two specimens of a same core; 0-20°: percentage of cores for which departure between k3 of individual specimens lies in the range 0° to 20°.

	mean K3 (%)	σ (°)	mdsp (°)	σsp (°)	0-20° (%)
SI-dam	146 / 52	39	35	23	32
SI-m	149 / 22	52	41	25	28
BA-dam	179 / 35	23	14	15	86
BA-m	173 / 30	15	11	7	88
TS-dam	34 / 41	34	12	12	86
TS-m	32 / 42	26	15	18	84

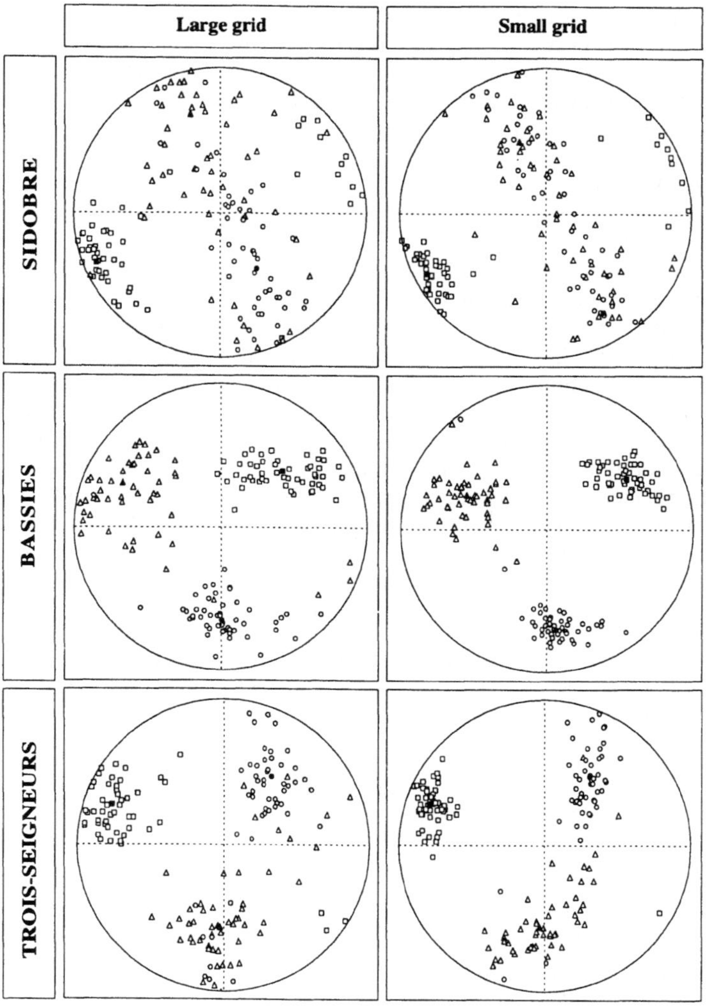

Figure 4. Principal axes of the ellipsoid of magnetic susceptibility plotted on stereonets (equal area, lower hemisphere). Squares: K1, magnetic lineation; triangles: K2; circles: K3, normal to the magnetic foliation. Open symbols: values of the 50 cores (100 specimens) per grid; black symbols: grid mean.

- For SI, the lineations fluctuate between NE-SW and ENE-WSW, with shallow plunges. This is particularly apparent on the small grid which displays a sigmoidal pattern, but harder to characterize precisely on the large grid;
- For BA, the lineations plunge moderately to the NE, displaying gradual fluctuations which outline a sigmoidal pattern, between NNE and ENE in trends on the large grid, compared with NE to ENE on the small grid;
- For TS, the lineations vary on both grids between NW and W in trends, with shallow plunges generally to the NW.

K3 axes, the normal to magnetic foliations, are generally more variable than K1 axes, as attested by σ (Table 2) ranging from 15° (BA-m) to 52° (SI-m). For SI, and to a lesser extent for TS, this variability is partly due to the zonal distribution of the foliations around a mean direction close to the mean lineation (Fig. 4), a phenomenon frequently observed in granitoids. For SI, this results in a significant difference in plunge between the mean K3 axes of both grids (30°: 146° / 52° against 149° / 22°), while, for TS both mean K3 axes are very close (34° / 41° against 32° / 42°). For BA, which displays no zonal distribution of K3, mean K3 of both grids are also similar (179° / 35° against 173° / 30°).

Within-core variability (Table 2) is the lowest for BA (σsp = 15°-7°; 86%-88% of the cores have an angular departure between individual k3 from 0° to 20°) and TS (σsp = 12°-18°; 86%-88% of the cores have well-grouped individual specimens), more important for SI (σsp = 23°-25°; 32%-28% of the cores have well-grouped individual specimens). The latter variability may account for the poor organization of the foliations in map view for SI, for which however, a NE-SW trend is visible, particularly on the small grid (Fig. 6). For the other two sites, the clustering of the foliations around a mean value is obvious. For BA, the dominantly E-W striking, northward steeply dipping foliations vary rather regularly between ENE-WSW and ESE-WNW for the large grid, and between ENE-WSW and E-W for the small grid. For TS, the dominantly SE-NW striking, southwestward moderately dipping foliations fluctuate between E-W and SE-NW, with rather sharp transitions from one direction to the other on the large grid, changing more gradually between ESE-WNW and SE-NW on the small grid.

TABLE 3. Km, MKm, mKm and σ: mean magnetic susceptibility, maximum, minimum and standard deviation of 50 cores (100 specimens); cv1 = (σ / Km) x 100; mdsp, σsp: mean and standard deviation of the departures between both specimens of a same core; cv2 = (σsp / Km) x 100.

	Km (10^{-5} SI)	Mkm (10^{-5} SI)	mKm (10^{-5} SI)	σ	cv1 (%)	mdsp	σsp	cv2 (%)
SI-dam	11.5	14.0	8.1	1.2	10	1.0	0.8	7
SI-m	11.9	13.1	10.2	0.7	6	1.0	0.7	6
BA-dam	17.3	18.8	15.5	0.8	5	0.9	0.7	4
BA-m	17.5	19.4	15.1	1.1	6	0.9	0.8	5
TS-dam	29.0	41.8	21.8	3.2	11	2.0	1.5	5
TS-m	29.2	34.9	22.5	2.4	8	2.0	1.9	7

MAGNETIC SUSCEPTIBILITY

Magnetic susceptibility magnitudes are very homogeneous for the three sites, as shown by the low values of σ and the coefficient of variation cv1 (σ/Km) (Table 3). On the large grid, the variability is about the same for SI as for TS (cv1 = 10% and 11% respectively), while it is much lower for BA (cv1 = 5%). On the small grids SI, BA and TS display similar variabilities (respectively 6%, 6% and 8%), lower than those of the large grid, except for BA.

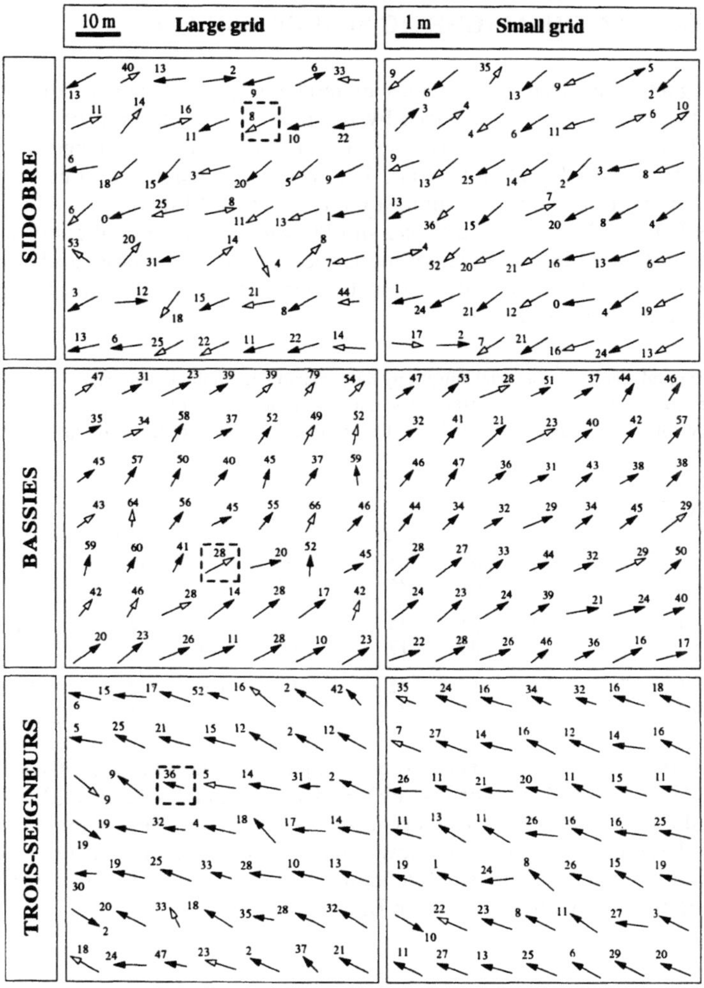

Figure 5. Magnetic lineations maps. The values at the tip of the arrows are the plunges. Black arrows correspond to the cores with an angular departure between individual k1 from 0° to 20°; higher than 20° for the white arrows. The square in the large grid represents the position of the small grid.

Within-core variability (Table 3) is rather high, with respect to the within-grid variability, for the three sites as shown by cv2 (= σsp/Km) spanning from 4% for BA-dam to 7% for SI-dam and TS-m. Despite these variabilities, the susceptibility magnitudes generally constitute several rather extensive zones of similar values (Fig. 7), and isolated values very different from their neighbours are rare. The boundaries of the main isovalue zones display orientations that do not depend on the rock magnetic orientation, except for SI-dam.

ANISOTROPY OF THE MAGNETIC SUSCEPTIBILITY

The total magnetic anisotropy percentages are moderately homogeneous for the different grids of the three sites. This is reflected by the rather large variation of cv1 from 17% (BA-m) to 36% (SI-m) (Table 4). The mean total anisotropy percentages are equal on both the large and small grids for SI and BA, while TS displays a larger Pp% on the small grid than on the large grid. Amongst the large grids, SI displays the highest variability, while BA and TS have weaker and similar variabilities. For the metric grids, SI is again the most variable, with a rather higher value than on the large grid, followed by TS displaying about the same value as on the large grid, and BA displaying a value weaker than on the large grid.

Table 4. Pp, MPp, mPp and σ: mean total anisotropy ratio, maximum, minimum and standard deviation of 50 cores (100 specimens); cv1 = (σ / Pp) x 100; mdsp, σsp: mean and standard deviation between both specimens of a same core; cv2 = (σsp / Pp) x 100. Fp, Lp: mean planar and linear anisotropy ratio of 50 cores (100 specimens).

	Pp (%)	MPp	mPp	σ	cv1 (%)	mdsp	σsp	cv2 (%)	Fp (%)	Lp (%)
SI-dam	2.3	4.1	0.9	0.7	32	0.9	0.7	32	1.2	1.1
SI-m	2.3	5.4	0.8	0.8	36	1.0	1.4	60	1.1	1.1
BA-dam	3.4	5.9	1.5	0.8	23	0.6	0.5	14	2.1	1.3
BA-m	3.3	4.8	2.2	0.5	17	0.5	0.3	9	2.1	1.2
TS-dam	5.1	7.7	2.5	1.1	22	0.9	0.6	12	3.1	2.0
TS-m	6.1	8.7	2.8	1.3	21	0.7	0.5	8	3.9	2.1

Within-core variability is rather strong, compared to the within-grid variability, for SI (cv2 = 32%-60%), moderate for BA (cv2 = 14%-9%) and TS (cv2 = 12%-8%) (Table 4). Despite these variabilities a zoned organization of Pp% appears on the different grids, with several large isovalue zones, and few isolated values (Fig. 8). The boundaries of the main zones of isovalues display slight tendencies to be oriented, for example parallel to N-S for TS-dam, and parallel to E-W on BA-m.

DISCUSSION

Homogeneity of the directional data, particularly the lineations, and their stability between scales, constitute undoubtedly the most spectacular result of this study. As an example, for the Sidobre site, the most variable amongst the three sites, the mean lineation is 248° / 9° for the dekametric grid, and 243° / 9° for the metric grid. Comparison between the lineations patterns at the grid scale (Fig. 5) and at the pluton scale (Fig. 1) also illustrates this homogeneity. Spatial variations of the directional data constitute an other important aspect of this study which is discussed hereafter with regard to intrinsic and external parameters. Scalar data, particularly the susceptibility magnitudes, also display bulk homogeneity. Spatial variations of these data, which are less clearly organized than for the directional data, are then briefly discussed.

DIRECTIONAL DATA

In map view (Figs. 5 and 6) the directional data generally vary gradually, but abrupt variations appear locally. These abrupt variations with respect to both the general trend and the neighbouring values, may be induced by different "intrinsic" causes, i.e. sample scale heterogeneities:

- i) grain size heterogeneity may explain some large differences between directional data of the two specimens of a given core (Tables 1 and 2: σsp), particularly in SI, which has the largest and most heterogeneous grain-size (Fig. 3). In this porphyritic granite, K-feldspar megacrysts may perturb the biotite sub-fabric, and therefore the magnetic fabric around them;

- ii) the deformation regime of the magma may cause specific variations. For example, the tendency for the foliations to be distributed in a zone around a mean direction as in SI and to a lesser extent TS (Fig. 4: partial girdles of K3), may correspond to a strong linear component of the fabric. This is observed at the grid scale, but is also evidenced at the core scale, particularly for SI where about 15% of the cores display very different plunges for both k3 of the same core. Note that the shape of the magnetic ellipsoid being dependent on the mineral marker, here the biotite flakes, the Lp%/Fp% ratio (see Table 4) cannot be used directly to determine whether the fabric of the rock is linear or planar;

- iii) the anisotropy magnitude is certainly, in most cases, the controlling factor for orientation variations, i.e. the lower the anisotropy, the less well defined will be the directional data. Within a given sample, this is reflected by k1-k2 axes permutations for one or two specimen(s) of the same core when the linear contribution of the total anisotropy is especially weak: this occurs in 5% of the samples in SI, the least anisotropic site. Similarly k2-k3 axes permutations are observed for specimens having particularly weak planar anisotropy: this concerns 8% of the samples in SI, 2% in TS and 1% in BA.

These different factors of within-core variability may combine, as for SI, the least anisotropic site, making variations at the grid scale difficult to interpret. On the contrary, when within-core variability is low, as for BA and TS, the variations of the directional data probably represent grid scale variations of the magmatic fabric. These variations are generally gradual, and some of them describe sigmoidal features, particularly in BA. For this site, the lineations mainly trend between N40° and N70° in the small grid, and between N20° and N70° in the large grid. This accounts for the rather high difference (9°) between the lineation means of both grids and indicates that the small grid represents a subdomain of the larger sigmoid evidenced in the large grid. In other words, the lineation strikes and plunges in the small grid do not reproduce the pattern observed in the large grid. It is proposed therefore that the wavelength of the undulations are larger than the size of the small grid. Conversely, for TS, the important angular departures sometimes observed between neighbouring directional data, particularly foliations in the large grid (Fig. 6), could be due to undulations with wavelengths smaller than the mesh of the grid.

It is tempting to ascribe the sigmoidal structures that appear on both the lineations and foliations maps to C/S patterns like those encountered in solid-state deformed rocks (Berthé *et al.*, 1979). For such an interpretation, a C orientation close to a shear plane must be identified along which the rock should be more strained than along a S orientation parallel to the finite stretch. But our maps do not display correlation between the highest magnitudes of Pp and particular structural directions, and so we cannot certainly identify C and S in our magnetic orientation patterns. Moreover, there is no evidence on the stereonets for bimodal distribution of the structural data. Alternatively, the waviness of the mineral fabric could be related to mechanical interactions in the crystal mush during crystallization of the magma, resulting in a fluctuation around a mean direction of stretch. Hence, even if the non-random nature of these sigmoids appears to be established, their significance is not clearly understood.

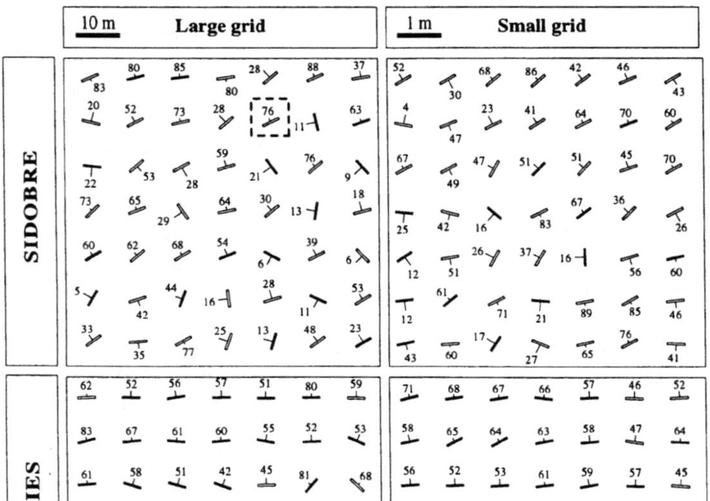

Figure 6. Magnetic foliations maps. The values are the foliation dips. Solid lines correspond to the cores having an angular departure between individual k3 ranging from 0° to 20°; larger than 20° represented by open lines.

SCALAR DATA

The magnetic susceptibility magnitudes are characterized by a rather strong clustering of the data around their means, while the anisotropy values are more variable (see cv1, Table 3 and 4). The scalar data have in common their rather heterogeneous spatial organizations. They are distributed, in map view, in different zones of isovalues, which might correspond to geologically organized domains having homogeneous biotite contents (K), and homogeneous strains (Pp).

Both for K and Pp, and for the three sites, several values of the small grids fall outside the range of values of the corresponding zones in the large grids, thus indicating heterogeneities in these zones. The most striking example is given by TS for which the average Pp of the small grid (6.1%) is clearly higher than for the large grid (5.1%). These heterogeneities may be attributed to the within-core variability, which is rather high compared with the grid scale variability, in particular for Pp for which cv2 ranges from 8% (TS-m) to 60% (SI-m). The within-core variability is probably mostly due to grain size and distribution heterogeneities, particularly for SI for which a single megacryst of K-feldspar can induce in a specimen important changes in biotite content and fabric. However, grain heterogeneity cannot account for all the variability of the scalar data. For instance, TS, the site having the most homogeneous grain size, displays a within-core variability similar to SI and BA for K values, and similar to BA for Pp values.

It may also be noted that it is difficult to correlate the isovalue lines, either between K and Pp, or between one of these parameters and the structural trends, or sigmoids that are observed on the lineation and foliation maps. For example, although in the large grid of SI the K-value zones display a NE-SW trend, more-or-less parallel to the lineation trend (Fig. 5), this correlation is not found at the metre scale, where the isovalue lines of K display an approximately N-S trend. Nevertheless, the presence of rather large zones of isovalue tends to demonstrate the robustness of zonation in K and Pp, on both scales.

The spatial organization of K is thus attributed to rock areas of several metres to dekametres, either enriched or depleted in biotite. This results either from an incomplete homogeneisation of the magma, or from a local partition between the solid fraction and the melt, by the end of crystallization. The spatial organization of Pp values probably indicates strain heterogeneities at the grid scales. As no clear correlation is evident between the orientation of the zones of anisotropy and the structural trends, no kinematic interpretation is proposed.

CONCLUSION

This study, which demonstrates the homogeneity of the directional data from a few metres to several tens of metres, complements fabric analyses previously performed in granites for which the directional data are often remarkably homogeneous in orientation on the scale of the whole pluton. This study also deals with the organization and random variations that appear in map views for both directional and scalar data. Transitions between neighbouring sampling points are generally gradual, and display consistent spatial organizations, such as sigmoids for directional data. Random variations are represented by few isolated and scattered values, at variance with the general trend. These are taken to be a function of heterogeneous grain size and local inhomogeneities of the magma.

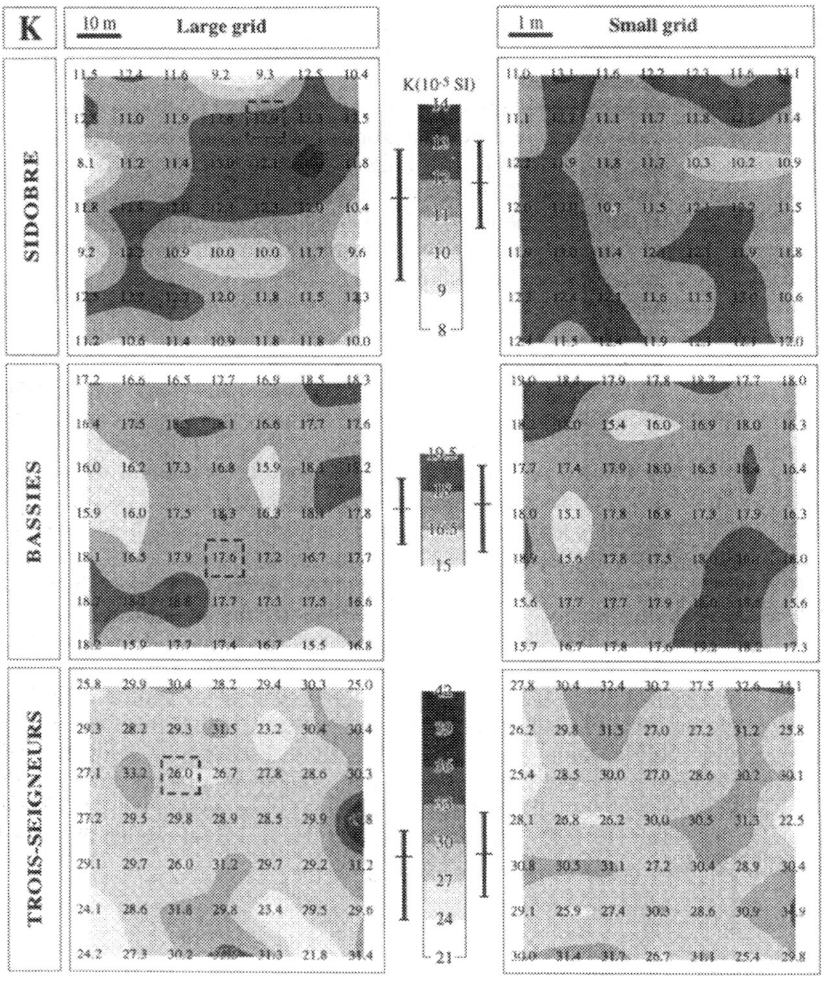

Figure 7. Magnetic susceptibility magnitude maps. Vertical bars on both sides of the scale include 90% of the values, and the position of the mean of the corresponding grid.

More extensive statistical analyses, based on these raw data, are now necessary in order to quantify better the spatially organized variations, like the geometry of sigmoidal features (K1, K3), the localisation of the strain (P%), or petrographic zoning (K).

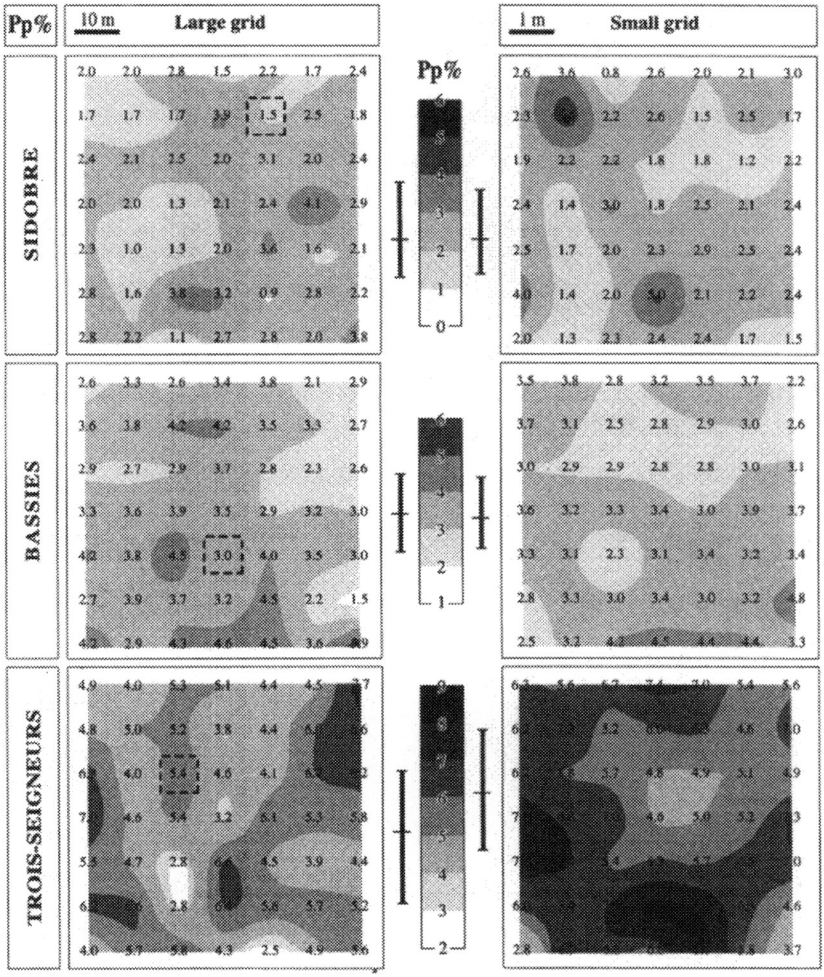

Figure 8. Anisotropy of magnetic susceptibility maps. Vertical bars on both sides of the scale include 90% of the values, and the position of the mean of the corresponding grid.

ACKNOWLEDGEMENTS

We are greatly indebted to Ed. Stephens and C. Teyssier for reviewing and improving this paper, to Ch. Cavaré and A.M. Roquet for technical assistance and to M. Lumpo for sampling assistance.

ANALOGUE 3D SIMPLE-SHEAR EXPERIMENTS OF MAGMATIC BIOTITE SUBFABRICS

Laurent ARBARET, Hervé DIOT
Université Blaise-Pascal
CNRS URA n°10, 5 rue Kessler
F-63038 Clermont-Ferrand, France
Jean Luc BOUCHEZ, Pierre LESPINASSE and
Michel de SAINT-BLANQUAT
Université Paul-Sabatier, CNRS UMR n° 5563
Laboratoire de Pétrophysique et Tectonique,
38 rue des Trente-Six-Ponts, F-31400 Toulouse, France

ABSTRACT

Experimental simple shear with $\gamma \leq 20$ was applied to three-dimensional shape fabrics of biotite grains embedded in a viscous matrix. This used a ring-shear apparatus deforming a medium made of silicone + plasticine + biotite grains at room pressure and 50°C temperature. The low-field anisotropy of magnetic susceptibility characterised the fabric of the samples. Two sets of experiments were performed: (1) with 5% and 10% in volume of biotite as the single solid fraction (Bi5% and Bi10%), and (2) with solid fractions augmented by wood-prisms (Bi10% + W6%) and spherical beads of Diakon (Bi15% + Di30%). Despite variations in biotite and total solid fractions, the resulting magnetic fabrics are all very similar. For $\gamma \leq 2$, the general fabric pattern and strength are greatly influenced by the initial fabric. For $2 < \gamma < 8$, the fabric orientation varies like in the model of Jeffery, but does not rotate as much and deviates 10° from that expected. For larger shear strains, the fabrics are consistently oriented, but a 10° off-shear-plane deviation remains. From these experiments we conclude that, even at low concentrations of crystals, magmatic biotite subfabrics are not expected to behave cyclically. Instead, with increasing shear strain, they should be constant both in orientation and magnitude. This behaviour is mainly attributed to the averaging effect of the various aspect ratios of the natural population of particles, and to the cumulative effects of their mechanical interactions.

J.L. Bouchez et al. (eds.) Granite: From Segregation of Melt to Emplacement Fabrics, 129–143.

INTRODUCTION

The relationships between grain shape preferred orientation, or shape fabric, and the direction of flow in rocks has been modelled previously in simple shear, pure shear and more general strain histories. Passive models consider markers rotating along with their matrix (March, 1932; Ramberg & Gosh, 1977), and the model of Jeffery (1922) applies to rigid markers (Bhattacharyya, 1966; Gay, 1966; Tullis, 1976). In the latter model the particles may behave cyclically, i.e. undergo multiple revolutions during bulk straining. In a magma, and at low concentrations of the solid fraction, the behaviour of the crystals is expected to be described numerically by the equations of Jeffery. They give the period of rotation of a revolution ellipsoid (the best approximation of an axisymmetrical particle) immersed in a fluid subjected to homogeneous simple shear flow. The azimuth and plunge of the particle long-axis with respect to the shear plane can be written as a function of the initial azimuth and plunge of the particle, its shape ratio (n) and the shear strain (γ) (see Fernandez & Laporte, 1991).

Several theoretical and experimental studies have tested the model of Jeffery and extended it to pure shear. Reed & Tryggvason (1974) modelled numerically the rotation of revolution ellipsoids and, using the equations of Gay (1968), compared the fabrics of populations of particles embedded in a matrix deforming in pure shear, with the case of simple shear. The experiments of Willis (1977) confirmed the validity of the model for axisymmetrical objects having various shapes, and Ferguson (1979) extended the domain of validity to particles immersed in a slow non-Newtonian flow, at low strains. On the same theme, Reed & Tryggvason (1974) and Freeman (1985) investigated the behaviour of isolated axi- and non axisymmetrical particles in simple shear, pure shear and simultaneous pure and simple shear. During such conditions of strain, Borradaile and Puumala (1989) investigate the domain of AMS fabric using ferromagnetic particles embedded in a plasticene medium.

More recently, several experimental simulations in two-dimensions, i.e. at the free surface of a deforming matrix, of the development of shape fabrics have been reported. The cyclic evolution of a whole population of mechanically non-interacting particles has been studied by Fernandez *et al.* (1983) and Ildefonse *et al.* (1992). Our experiments investigate the 3D fabric evolution of biotite grains embedded, at low concentrations, in a viscous fluid subjected to simple shear. These experimental fabrics may simulate those of biotite-bearing granites (Bouchez, this volume). The anisotropy of magnetic susceptibility (AMS) quantifies such fabrics and is commonly used in structural studies of rocks (Hrouda, 1982; Borradaile, 1988; Rochette *et al.*, 1992). The principal susceptibility axes of biotite closely approximate its principal dimensions, with the shortest susceptibility axis nearly normal to the cleavage plane, one axis precisely parallel to cleavage and the other nearly parallel to cleavage plane (see Borradaile & Werner, 1994). It is confirmed from field studies to the first order, and more precisely from laboratory studies, such as texture goniometry, that the magnetic fabric of populations of biotites, and more generally micas, closely reflect their shape fabrics, at least in orientations (Richter *et al.*, 1993; Bouchez, this volume).

First, we studied the fabric of biotite alone, at volume concentrations of 5% (Bi5%) and 10% (Bi10%). Then, we tested the effect of increasing the solid fraction, hence the frequency of mechanical interactions, by adding some « non magnetic » particles To test the influence of the presence of large, « non magnetic » crystals on the fabric, a small fraction (6%) of subrectangular wooden blocks about ten times larger than the individual biotite crystals was added to the suspension (Bi10%+W6%). This simulates the presence of large diamagnetic feldspar megacrysts. The case of a microcrystalline magma was simulated with spherical beads of Diakon, an acrylic material with a particle size comparable to the average biotite grain, where the fraction of Diakon largely exceeded that of biotite (Bi15%+Di30%). The resulting fabrics are presented as a function of the imposed shear strain, and discussed in terms of the effects of shape ratios and mechanical interactions between particles.

EXPERIMENTAL PROCEDURE

APPARATUS

Simple shear strains of up to $\gamma = 20$ were obtained using a ring-shear apparatus (Fig. 1 and caption), similar to that used by Passchier & Sokoutis (1993).

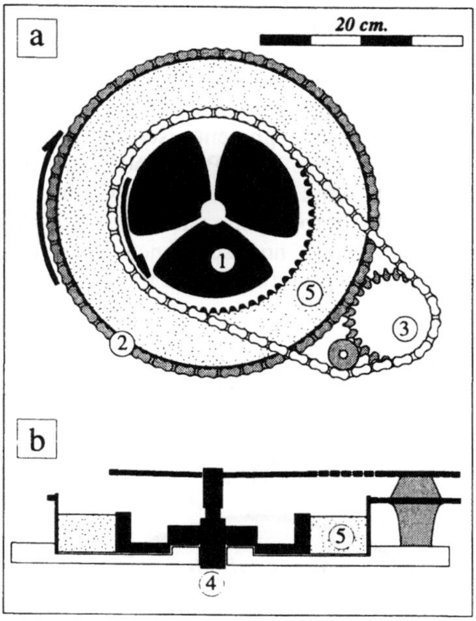

Figure 1. The shear apparatus: *a*: plan view, and *b*: cross-section. The inner cylinder (1) is driven by an electric motor with a reduction gear (4); the outer cylinder (2) is driven by a sprocket-wheel system (3). Interface between the matrix (5) and the immobile base of the apparatus is lubricated by a film of liquid soap. Shear deformation is vertical, shear direction is at a tangent.

A modification optimises strain homogeneity in the fluid (see Masuda *et al.*, 1995) by using two counter-rotating cylinders (Arbaret, 1995). The available ring section between the rotors is 107 mm-wide and 40 mm-thick, and provides a volume of up to 1650

cm^3 of material to be deformed by torsion. Equal and opposite angular velocities of the cylinders yield a constant shear strain rate of $d\gamma / dt = 8 \times 10^{-4}\ s^{-1}$, at T = 55°C, the working temperature which is maintained inside a hood surrounding the shear box. To avoid ferromagnetic contamination, the base of the box is lubricated with a 3 mm-thick layer of liquid soap; not with mercury as previously used in similar experiments (Passchier & Sokoutis, 1993).

ANALOGUE MATERIAL.

The analogue material derives from pink silicone putty, Silbione 70009 (Rhône-Poulenc Company), that was purified of its additives (barite, titanite and paramagnetic colour components) by washing in acetone. The substance remaining after this treatment is pure boropolysiloxane, a diamagnetic medium rheologically similar to the non-treated putty (Weijermars, 1986). This silicone putty is better suited to sample extraction than the classic, non-magnetic PDMS silicone, which remains fluid at low temperature, and is not easy to mix with other components. The fluid finally chosen for the experiments is thus a mixture of 56% boropolysiloxane, and 44% plasticine, a modelling material. The plasticine ("La Pierre Humide" Company, France) is itself a mixture of wax and colouring components of organic origin since no crystalline matter has been detected by X-ray diffraction.

The resulting fluid is diamagnetic (susceptibility magnitude: $-7.5 \pm 0.7 \times 10^{-6}$ SI), highly viscous at the experimental temperature of 55°C, and rigid at the temperature of sampling (-18°C). More precisely, the apparent viscosity of the matrix, measured on a Couette viscometer apparatus, is $5.6 \times 10^3 \times$ Pa.s at 50°C, the temperature of the experiments, and increases to about $4.6 \times 10^4 \times$ Pa.s at 20°C, the temperature of magnetic fabric measurements. Since the boropolysiloxane is a shear-thickening material, the Newtonian rheology of the mixture is not ascertained. Therefore, shearing tests at the experimental strain-rate have been performed to verify strain homogeneity in the matrix (Fig. 2).

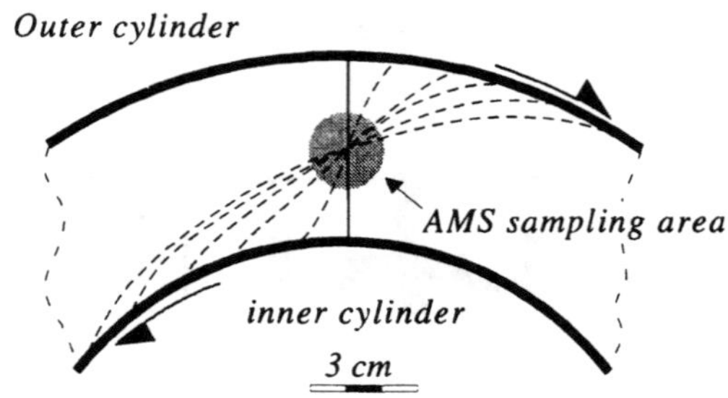

Figure 2. Experiment using a passive ink line perpendicular to the shear axis (full line), in order to test the homogeneity of the shear in the box with increasing strain (dashed lines). Shear strain is homogeneous in the sampling area (circle), and becomes inhomogeneous close to the cylinders.

The inhomogeneous shearing located close to the cylinders is explained by both the strain rate gradients in proximity to the cylinders and the non-Newtonian behaviour of the putty. However, in the central area of the apparatus (Fig. 2: circle), the shear strain is observed to be homogeneous at all shear strains. This is where we sampled our medium for AMS determination.

The biotite grains in the matrix were separated from various magnetite-free grains using high density liquids and the Frantz magnetic separator. For all experiments, we used the same 400-500 μm sieve fraction of biotite. Impurities, < 3% in volume, consisted principally of quartz and feldspar. We estimated biotite grain-morphology from a random collection of 100 grains (Fig. 3).

They are flat-shaped (mean c/a = 0.16) and, probably due to crushing, they appear rectangular (mean b/a= 0.74). The wooden particles used for the Bi10%+W6% experiments are much larger than the biotite grains (a=4.0±0.3 mm, b=2.8±0.4 mm; c=0.8±0.3 mm) but have similar aspect ratios (mean b/a= 0.7; mean c/a= 0.2). In the Bi5%+Di30% experiments, we added 30% of almost perfectly spherical Diakon pellets, which are diamagnetic and are from the same sieve fraction as biotite (400-500 μm).

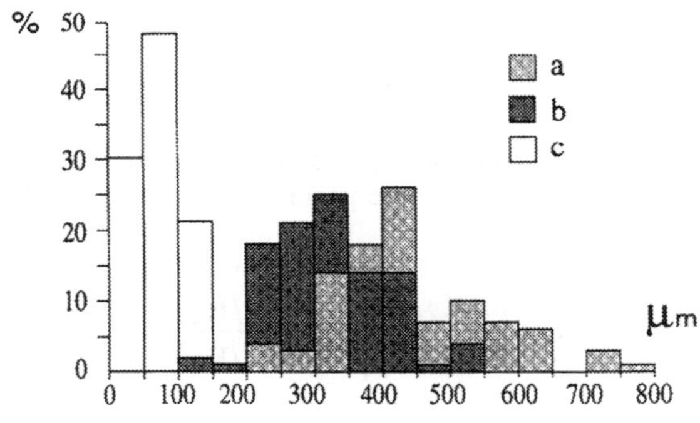

Figure 3. Morphology of the biotite used in the experiments. a, b and c are, respectively, the long, intermediate and short morphological axes of the grains (sieving fraction: 400-500 μm; 100 grains).

EXPERIMENTS

For each strain increment that we sampled for fabric measurement, the fabric of the medium was "randomized" before it was placed in the shear apparatus, being careful to minimise any development of an initial fabric. After deformation, the whole apparatus was cooled to +20°C, the cylinders and the sandwiched ring of medium was extracted and placed in a freezer at -18°C. After 3 hours, six 21mm-diameter and 40mm-long samples were cored with a non magnetic drill, from the homogeneous strain region between the cylinders (Fig. 2). These were strategically spaced and oriented with respect to the shear plane and sense of shear. From each core, a 18.5 mm-long specimen, 5.53 cm^3 in volume, was placed in a cylindrical plastic holder to facilitate transport, measurement and preservation. The remaining material was re-used for the next experiment, with supplementary material to maintain the volume.

FABRIC MEASUREMENTS

Measurements of the specimens were performed using the Kappabridge KLY2 apparatus (Geofyzika Brno), working at ± 3.8 x 10^{-4} Tesla and 920 Hz. It is a high-precision technique for quantitative determination of the magnetic fabric, and thus interpretation of the shape fabric of our paramagnetic particles (Bouchez, this volume). The magnetic fabric is defined by the principal axes $K_1 \geq K_2 \geq K_3$ of the susceptibility ellipsoid which has a mean susceptibility $K_m = [(K_1+K_2+K_3)/3]$.

The parameters used to characterise the shape of the magnetic ellipsoids are P or P_J, and T. $P= K_1/K_3$ gives a simple representation of the strength of the magnetic fabric. However, in order to reflect the role of the intermediate K_2 axis (Jelinek, 1981; see also Borradaile & Werner, 1994), we preferred the use of P_J:

$$Pj = \exp\left[2\left[\left(\ln\frac{K_1}{K_m}\right)^2 + \left(\ln\frac{K_2}{K_m}\right)^2 + \left(\ln\frac{K_3}{K_m}\right)^2\right]\right]^{1/2}$$

The shape of the fabric ellipsoid is represented by the shape parameter of Jelinek (1981), which varies from T=-1 for a pure prolate ellipsoid to T=+1 for a pure oblate ellipsoid, and is defined as:

$$T = \frac{\ln(K_2/K_3) - \ln(K_1/K_2)}{\ln(K_2/K_3) + \ln(K_1/K_2)}$$

MAGNETIC PROPERTY OF THE ANALOGUE MATERIAL.

The experimental medium and its cylindrical plastic container yield a diamagnetic susceptibility of -10. ± 0.7 x 10^{-6} SI. Their AMS measurements (Fig. 4) give erratic values for both the directions and the relative magnitudes of the principal susceptibility axes, in relation with confidence angles of Jelinek (1977) as large as ±36°, ±44° and ±36° for respectively γ=0, γ=1 and γ=3, suggesting near random orientations. At γ=3, however, K_3 clusters vertically, i.e. normal to the surface of the experiment. This may be attributed to increasing anisotropy of the matrix. However, such weak fabrics in the susceptibility domain of diamagnetism cannot interfere with the strong paramagnetic fabrics of the experiments.

Initial isotropy of the experimental medium is an ideal prerequisite in a study of fabric evolution; particularly for comparison between experiments and theory. Anisotropy may be inherited both from the stage when the particles were mixed with the matrix, and from its insertion in the apparatus. The experimental medium, once hardened by cooling and cut into small bricks is inserted randomly in-between the driving cylinders. Unfortunately, this procedure did not eradicate all initial fabric as shown in figure 5.

The initial fabric formed mostly by spreading of the material after its emplacement in the apparatus, and warming to the temperature of the experiments. The tendency of

K_3 to be vertical (i.e. biotite cleavage horizontal), and K_1 to be at a tangent to the ring (i.e. biotite cleavage in a zone around the extension direction during spreading) corroborates this. Thus we infer that very small strains suffice to imprint a fabric here related to coaxial strain during setting. This is also valid for the simple shear experiments.

BIOTITE-ALONE EXPERIMENTS

The two first sets of experiments were performed with 5% and 10% volume biotite. Specimens were collected at γ=0.5, 1, 2, 3, 4, 8, 12 and 20. Supplementary shear strain steps were sampled at γ=6, 10 and 16 for the Bi5% experiments, and at γ=0.25 and γ=0.75 for the Bi10% experiments. Shear strain steps at γ=0.5, 1, 2, 3, 4, 8, 12 and 20 are represented in the orientation diagrams (Figs. 6a and 6b), and all steps are represented in the α versus γ plot (Fig. 6c). We observe that, at the lowest γ values, γ=0.25 and γ=0.5, K_1 clusters parallel to the shear direction, and K_3 is vertical. Since the latter fabric is observed at γ=0 (Fig. 5), the fabrics at very low strains are interpreted as due to strain of the medium during the preparation of the experiment, i.e. during spreading parallel to the ring.

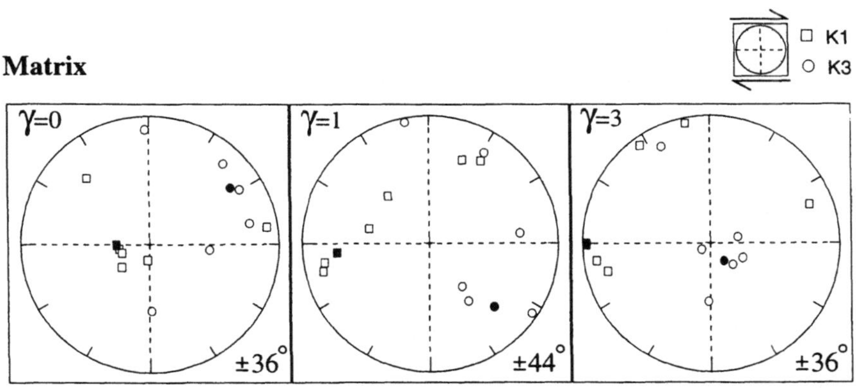

Figure 4. Diamagnetic fabric of the matrix at γ=0, γ=1 and γ=3 shear strains. Six specimens per diagram were measured. Note the very high values of the "confidence angle" between K_1 and K_3 at the 0.95 confidence level (after Jelinek, 1977), but noneless the tendency for K_3 to be vertical at γ=3. Full symbols: tensor averages. Lower hemisphere, equal-area projections.

The $\alpha(\gamma)$ plot of figure 6c shows that the orientation of K_1, characterized by its angle α with the shear direction, approaches its theoretical value at shear strains γ>2 . At still larger shear strains, the orientation of K_1 remains almost constant at an angle "above" the shear plane (Fig. 6c), i.e. counter-clockwise from the shear plane for the dextral senses of shear represented everywhere in this paper. Similarly, we observe that the magnetic foliation plane becomes vertical at γ>2, i.e. K_3 migrates to the periphery of the orientation diagram, then remains almost constant in orientation, at an angle close to α with the trace of the shear plane. Theses low simple shear AMS results are very similar to those obtained by Borradaile & Puumala (1989) using magnetite sand. In our case however, higher shear strain (Figs. 6 and 7) reveals that the foliation plane dips steeply toward the outer cylinder of the experimental device. This asymmetry may arise from the local distribution of the friction forces against the ring.

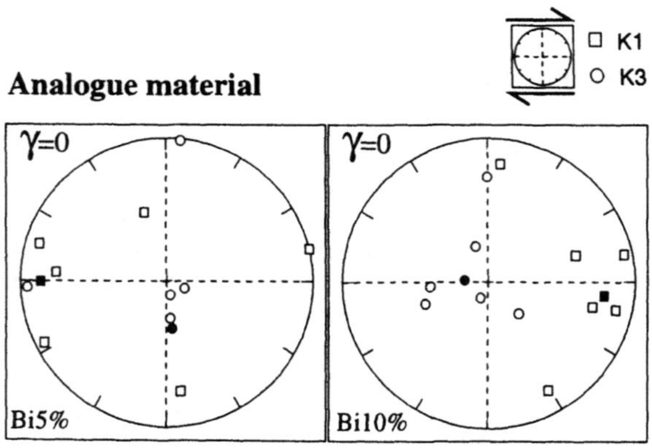

Figure 5. Magnetic fabric of the analogue material (Bi5% and Bi10%) at zero shear strain. Note the tendency for K_3 to be vertical and K_1 tangential in orientation, suggesting gravity spreading of the material during preparation of the experiment. Full symbols: tensor averages. Lower hemisphere, equal-area projections.

BIOTITE+SOLID FRACTION EXPERIMENTS

Figure 7 represents the orientation diagrams of the shear strain steps sampled at $\gamma = 2$, 3, 4, 8, 12 and 20. In figure 7a, the diagrams of Bi10%+W6% resemble those of figure 6, except for the K_3 axes tending to rotate around K_1, as indicated by the partial girdles of K_3 at $\gamma = 2$, 8 and 20, and by a more prolate magnetic ellipsoid attested by a lower mean T value (Fig. 8a). The α versus γ plot also reveals larger α values at shear strains $\gamma > 2$ for these experiments (Fig. 7c). In figure 7b, the Bi15%+Di30% experiments give similar results except that the biotite fabric tends to be more planar, as shown by tight clustering of K_3, at least in some experiments, and by the shape parameter T being more oblate (Fig. 8a). Another minor difference seen is the overall lower fabric strength for both kinds of solid fractions (Fig. 8b). This is due to the greater tendency for biotite to vary in its orientation, hence lowering the overall fabric strength, in the presence of an added solid fraction which increases the local flow heterogeneity.

DISCUSSION

The primary result of this study is the striking similarity between the orientation diagrams obtained from varying strains, particle concentrations and particle types. In particular the biotite and the biotite+solid fraction experiments differ only slightly. These are now considered and discussed using P_J and T as a function of shear strain, and departures of the observed $\alpha(\gamma)$ results from theoretical behaviour issued from shape fabric models. Geological applications will be considered subsequently.

P_J AND T VERSUS γ BEHAVIOUR

Although rather large variations in P_J and T may occur from one specimen to an other, and from one strain step to the next, the narrow range in the average values of P_J and T illustrates the little real variation between the different experiments.

All fabrics in all experiments are dominantly oblate (mean T=0.15) and rather strongly anisotropic (mean P_J=1.20). For comparison with real rocks for which the anisotropy degree P is commonly used to evaluate the fabric strength, the latter value of P_J corresponds to P=1.13 (= anisotropy percentage of 13%). This value is amongst the strongest anisotropy magnitudes observed in biotite-alone bearing granites (see Boûchez, this volume).

The highest positive T values are for Bi5% experiments, and the negative values of T come from some of the Bi10% and the Bi10%+W6% runs (Fig. 8a).

This suggests that mechanical interaction effects related to the increasing content in anisometric particles (biotite, wooden particles) enhance rotation of the biotite around the local stretch axis of the experiment. By contrast, the T values of the Bi15%+Di30% are positive despite the high biotite concentration. This is due to the sphericity of the Diakon beads, enhancing grain-scale rotation around a vertical axis, i.e. around their own vorticity axes within the shear plane and perpendicular to the plane of deformation, and thus opposing the tendency for grain rotation around the stretch axis.

The P_J values vary surprisingly little, from P_J=1.15 to P_J=1.30 (Fig. 8b). Except at low strains which are influenced by the initial fabrics, we observe a net inverse correlation between the solid fraction and the fabric strength. In particular, the Bi15%+Di30% runs have the lowest P_J at an almost constant value close to 1.16.

α VERSUS γ BEHAVIOUR

α represents the angle made by the long axis of the fabric ellipsoid of a population of particles with the direction of shear. The α versus γ curve is determined by calculating the tensor average, as a function of γ, of the orientation distribution of the different particles forming the population. Assuming simple shear, in 2D-flow with axisymmetrical particles having their long axes in the plane of deformation, the orientation θ of the long axis of an elliptical particle with respect to the shear direction, as a function of the particle aspect ratio (n=a/c) and shear strain (γ) may be derived from the following equation of Fernandez *et al.* (1983), based on Jeffery (1922) and Willis (1977):

$$Arc\tan(n\tan(\theta + d\theta) = Arc\tan(n\tan\theta) - \frac{\sqrt{1-k^2}}{2}d\gamma\,;$$

with n= a/c and k= $(n^2-1)/(n^2+1)$. This equation may be applied to a whole population of particles to determine its α fabric angle, by calculating the tensor average of the population as α function of γ. If we assume that all particles are identical with a/c=6.1

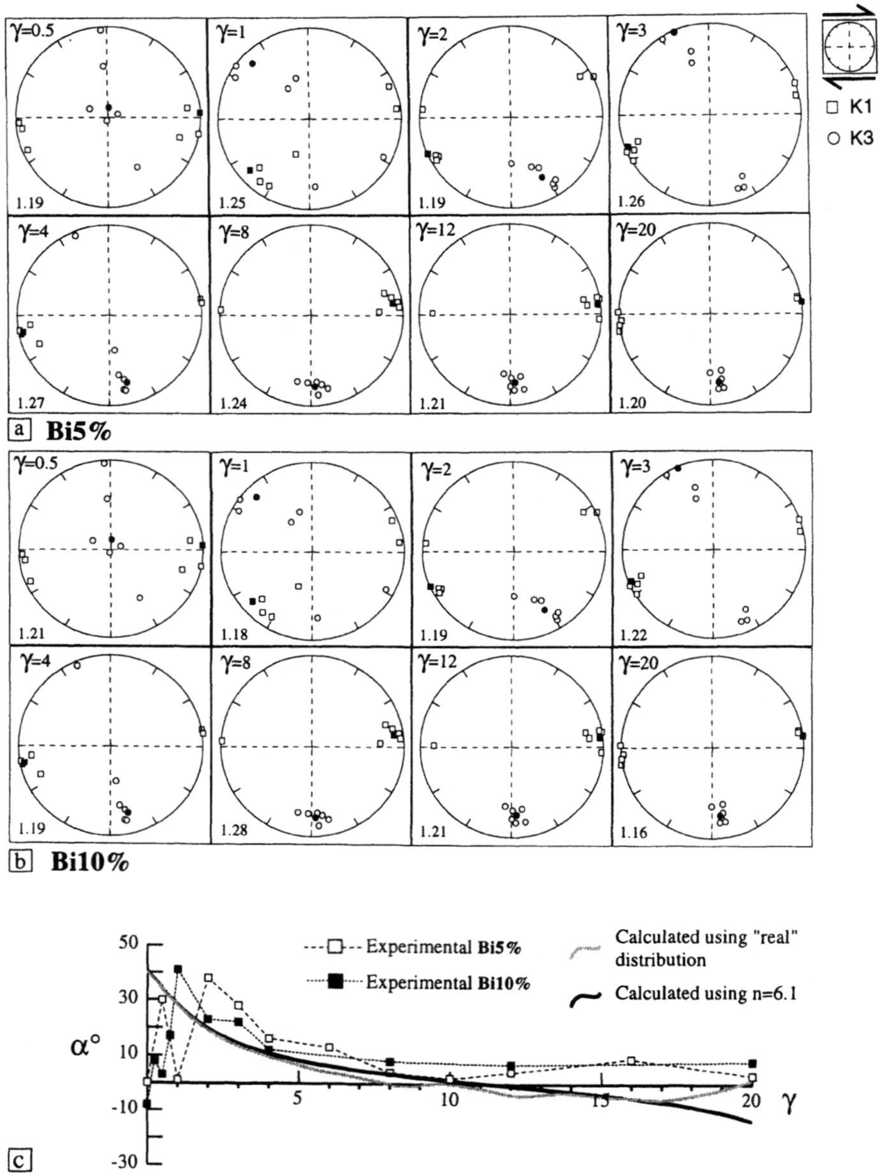

Figure 6. Biotite-alone simple shear experiments at 5% and 10% volume concentrations. *a* and *b*: orientation diagrams (lower hemisphere, equal area); *values in left corners*: P_J values, the fabric strength. *c*: angle of the fabric axis with the shear direction, α, as a function of the shear strain γ, the experimental data (symbols) are compared with the calculated ones (equations of Fernandez *et al.*, 1983), using the "real" grain shape ratio distribution (grey stippled), and the average grain shape ratio, n = 6.1, of Fig. 3 (solid curve).

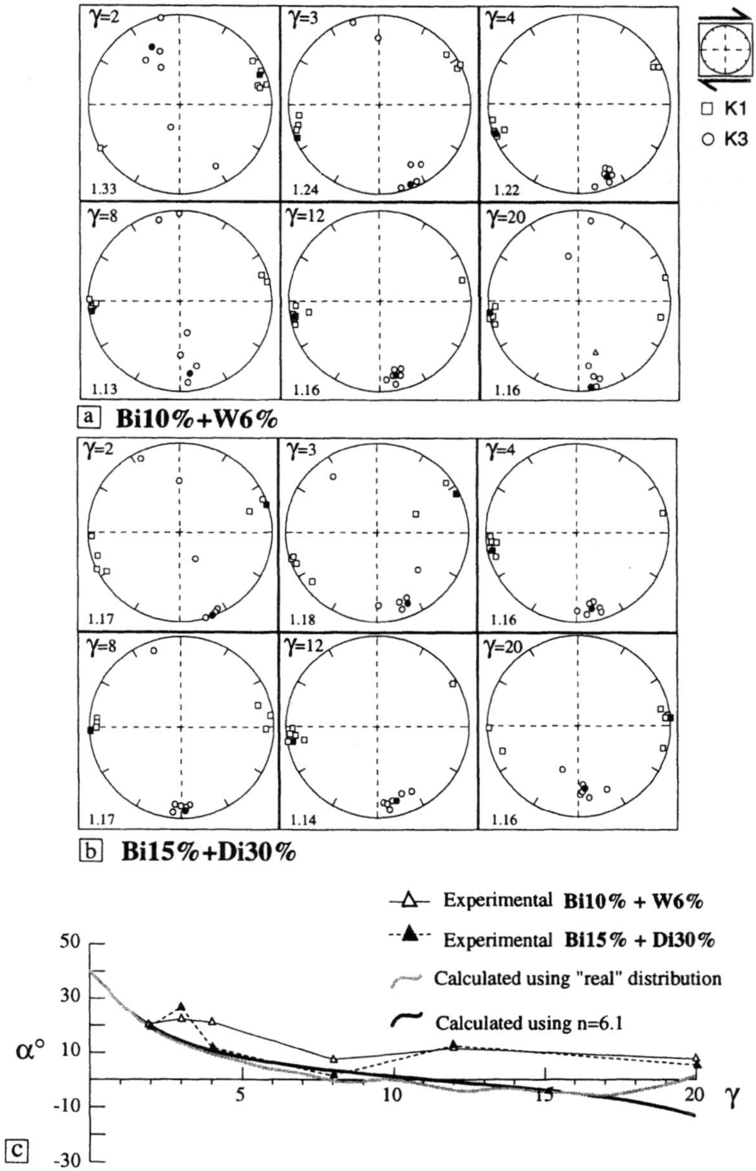

Figure 7. Biotite + solid fraction simple shear experiments. a: orientation diagrams (lower hemisphere, equal area) for 16% solid fraction (in volume) including 6% of wooden parallepipedic particles and b: 45% solid fraction including 30% of Diakon beads; values in left corners: P_J values, the fabric strength. c: same legend as in figure 6.

(the average aspect ratio of the biotites used in the experiments), this gives the solid curves of figures 6c and 7c. In fact, the particles are not identical, and the same calculation can be performed using the a/c values of the real population of 100 biotite grains (Fig. 3). This calculation gives the grey curve of figures 6c and 7c.

Departures between the two model-behaviours are naturally attributed to the varying aspect ratios of the real population. The fact that α becomes negative at $\gamma \approx 8$ in the model-behaviour using the real distribution (grey curve) is attributed to the rapid rotation of the shortest particles. At higher strains, α remains slightly negative, and this is likely the effect of the longest grains, before becoming positive again at $\gamma = 20$. This simple model shows that a real population of grains will hardly develop a cyclic fabric, and this is the basis for the steady state magmatic foliation concept of Ildefonse et al. (this volume).

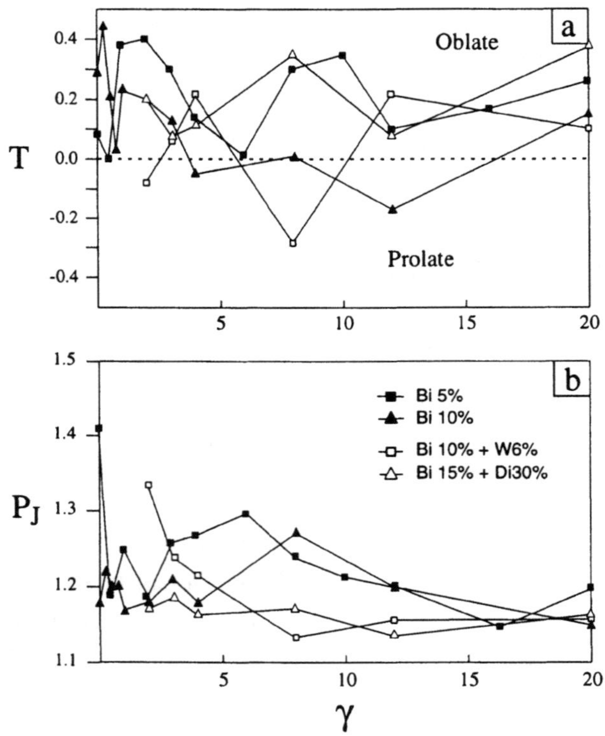

Figure 8. Evolution of P_J (a) and T (b) with the shear strain γ for all experiments. Full symbols: biotite-alone experiments; open symbols: biotite + solid fraction experiments.

In fact, our experimental data depart quite strongly from the model-behaviours with an evolution of the fabric which may be subdivided into three domains:

(i) For $\gamma \leq 2$, the scattering of α, T and P_J individual values, is naturally attributed to the influence of the initial fabric;

(ii) For shear strains up to $\gamma = 8$, α steadily decreases as in theory, but remains at more than +10°, i.e. counter-clockwise, with respect to the theoretical values and to the shear plane, and

(iii) Finally, for $\gamma \geq 10$, α becomes constant at a value around +10° with respect to the shear plane.

Although we do not understand all possible micromechanical effects in the matrix, the discrepancy, at $\gamma > 2$, between the measured and the theoretical α values are to be looked for in the assumptions for the model behaviours, namely 2D-flow, axisymmetrical grains, and no interaction between grains. Departure from 2D behaviour, along with (i), triaxial-shaped particles and (ii) interactions between particles are now considered sequentially.

(i) The 2D equation of Fernandez *et al.* (1983) assumes that the particles are two-dimensional. In three dimensions, as shown by Freeman (1985) and Ferguson (1979), the axisymmetrical particles may swing away from the plane of deformation, and the real triaxial particles may follow more complex trajectories. In fact, for shape fabrics with triaxial particles, both the period of revolution and fabric intensity drastically decrease with increasing strain (Jezek *et al.*, 1994).

(ii) Fabric "stabilisation" due to mechanical interactions in 2D concentrated suspensions, and very similar to our experimental $\alpha(\gamma)$ curves, has been reported by Ildefonse *et al.* (1992). If mechanical interaction readily explains the behaviour of the experiments with a high solid fraction (Bi15% + Di30%), it is proposed that it also stabilises the fabrics at low concentrations. Arguments for this are found in the 2D analogue experiments at low concentrations of Arbaret *et al.* (1996), in which the particles imbricate, or form tiling features, with a fabric orientation (α) remaining at an angle "above" the shear axis, contrary to fabric models which predict that fabric axis becomes parallel to shear axis.

Our biotite grains possess different aspect ratios a/c and grain sizes from those of natural rocks due to crushing and separation from the host-rock. However, our grain populations are more realistic than the identical particles used in the 2D experiments of Ildefonse *et al.* (1992) and Arbaret *et al.* (1996).

The concentration of particles also differs from nature. In the experiments, a large range of solid fractions, between 5% and 45%, has been considered. This could represent the concentrations at various crystallization stages in a magma, during which crystal shape fabrics develop. Unfortunately, our analogue experiments do not explore the case of syn-shearing concentration changes as in magmas. However, our experiments with varying solid fractions show that the orientation α of the fabric appears to be rather independent from the concentration of particles whatever the strain, as also noted by Arbaret *et al.* (1996). The only important deviations that we observed were those of T and P_J, at the lowest shear strains.

APPLICATION TO GRANITOIDS

We can predict three principal fabric patterns if the present experiments are realistic simulations of fabric acquisition in magmas with biotite as the main magnetic carrier. These are based on the three domains of the $\alpha(\gamma)$ experimental behaviour, and hence only on the fabric orientations since neither the fabric strength nor shape parameters are particularly distinctive of the fabric evolution with increasing strain.

The first type corresponds to a fabric pattern in a pluton, or part of a pluton, which is poorly organised as in low strain experiments ($\gamma < 2$). The pattern may be near random due effectively to low strain. However it is difficult to invoke low strain not knowing the complex history of the magma during its ascent and emplacement in the pluton. It is more realistic to consider, just like in our experiments, that the present ill-defined fabric results from the superimposition of previous fabrics that correspond to deformation episodes different in their geometries (in shear planes and directions), and that the latest homogenous strain episode ($\gamma < 2$) was insufficient to produce a consistent fabric. This interpretation of ill-defined fabrics is illustrated in the western part of the Monte Cappanne pluton (Elba Island, Italy), a zone where the vertical feeding of magma interferes with E-W lateral spreading under the pluton's roof, parallel to the local and regional extension direction (Bouillin *et al.*, 1993; Daniel and Jolivet, 1995).

In the two other cases, the fabric pattern is well-organised, a situation which is now well-established in plutons as attested by the many structural maps produced during the last decade, e.g. Bouchez and Gleizes (1995), Ferré *et al.* (1995) and Roman-Berdiel *et al.* (1995) among the most recent studies. The two cases differ by the intensity of the strain undergone by the magma. If the strain magnitude remains moderate ($\gamma < 5$) the long axis of the fabric ellipsoid (i.e. the magmatic lineation) will remain at an angle of $\alpha > 10°$ with the regional direction of shear. This case is illustrated by the NW-SE trending fabric of the Tesnou pluton (Hoggar, Algeria) which is spectacularly oblique with respect to the N-S trending, dextral (i.e. compatible with the obliqueness), late Pan-African shear zones that characterize this region (Djouadi and Bouchez, 1992). If the obliqueness between the fabric orientation in the pluton and the regional shear cannot be proven because geological information is lacking, one should be able to demonstrate that the strain magnitude is indeed moderate. This may be difficult since the fabric strength parameter seems insensitive to the strain magnitude undergone by the magma.

The third case is represented by plutons which are naturally qualified as syntectonic since they are involved in the regional deformational pattern. The corresponding magmas have undergone quite large strain magnitudes, either related to transcurrent shear zones, as in north-east Brazil (Archanjo *et al*, 1992), or related to low angle shear planes, as in the French Massif Central (Jover and Bouchez, 1986; Faure, 1995) or in the stratoid granite massifs of Madagascar (Nédélec *et al.*, 1994), among many other examples. In all these cases the magmatic lineations are considered as parallel to the direction of the regional shear over areas which may cover thousands of square kilometres. In comparison with these natural examples, it may remain, however an angle of a few degrees, as in our experiments at large strains, between the shape fabric long-axis and the shear direction. In fact, we do not know if this angle remains at large strains, not knowing precisely the shear direction. In real granites, a slight coaxial strain component, related to volume change during the latest stages of magma deformation by melt extraction to yield aplites (Bouillin *et al.*, 1993), may superimposed on the simple shear fabric, hence slightly modify the faint angular differences that may exist between shape fabric axis and plane and direction of shear.

CONCLUSION

Our experiments demonstrate that the magnetic fabrics, and therefore the shape fabrics of deformed suspensions of biotite crystals at various concentrations, are controlled by the rotation of the particles. However, the rotation of the long axis of the fabric ellipse toward the shear direction is slower than predicted by 2D models considering axisymmetrical and non-impinging particles. At large strains, the fabrics are constant in orientation and magnitude, regardless of concentration. This is due to the averaging effect of the various shape ratios of the triaxial ellipsoids, and to the role of their mechanical interactions in lowering their rotation rates. Whatever the strain magnitude, the fabric orientations are easier to analyse, and provide more information than the fabric strength and shape parameters.

Three stages of fabric evolution $\alpha(\gamma)$ are distinguished and compared with the three main types of fabric patterns often encountered in plutons. For $\gamma < 2$, the ill-defined fabric pattern is greatly influenced by the initial fabric; in plutons this stage corresponds to domains subjected to fabric superimpositions. For $2 < \gamma < 5$, the fabric orientation makes a significant angle with the shear direction. Although it is slightly higher than in theory, it reflects the obliqueness between the stretching and shear directions. Moderately deformed plutons may record obliqueness with the regional (or local) direction of shear, but this must be checked independently. At still higher shear strains, and for all sets of experiments, the fabric is in steady-state with respect to its orientation and magnitude, with an orientation remaining at several degrees off the shear plane, counted counter-clockwise for dextral shear. This illustrates the case of syntectonic plutons stricto sensu which have lineations *more-or-less* parallel to the regional (or local) shear direction.

ACKNOWLEDGEMENTS

This paper is a part of the Doctorate Thesis of L.A. at the University of Clermont-Ferrand. The apparatus was built by L.A., and details concerning its construction and the materials employed can be obtained on request. Graham Borradaile, Neil Mancktelow and Jean-Louis Vigneresse are warmly acknowledged for their punctilious reviews of the manuscript. This is a contribution of the "Unités de Recherche Associées au Centre National de la Recherche Scientifique", UMR CNRS n°10 (Clermont-Ferrand) and n°5563 (Toulouse).

3D BIOTITE SHAPE FABRIC EXPERIMENTS UNDER SIMPLE SHEAR STRAIN

Angel FERNANDEZ
Faculté des Sciences - Université de Limoges L.A.S.E.H.
123 Avenue Albert Thomas, 87060 Limoges Cedex, France
and Javier FERNÁNDEZ-CATUXO
Departamento de Geologia - Universidad de Oviedo Arias de Velasco s/n., 33740 - Oviedo, Spain

ABSTRACT

Twelve runs were performed with an analogue material consisting of a mixture of a synthetic resin mixed with kaolin, biotite and quartz grains. Mineral fabrics of biotite were obtained with the help of a 3D-reconstruction method using data obtained from two oriented sections, and confirmed by means of anisotropy of magnetic susceptibility measurements. In low concentration suspensions, the fabric develops in close agreement with theory. In concentrated suspensions, the interactions between grains predominate, giving fabrics with flattening symmetry and low intensity parameters. In some experiments, and in agreement with mineral fabric data and magnetic measurements, a biotite zone-lineation parallel to the Y-axis of finite strain develops. This effect is attributed to the rotation of the biotite grains around the Y-axis, induced by their close interactions with the more spherical grains of quartz. Comparison of experimental results with data from real samples provides an explanation for the differences commonly observed between feldspar megacryst subfabrics and biotite subfabrics.

INTRODUCTION

Among the various approaches devoted to study the mechanisms of magmatic intrusions, the patterns of foliation (F) and lineation (L), the orientations and shapes of enclaves, orientation of co-magmatic dikes and veins, as well as relationships between intrusive rocks and country rocks, are the most frequently used criteria. In the last ten years, anisotropy of magnetic susceptibility data (AMS) have been used as an alternative to the direct measurement of magmatic F and L (Bouchez, this volume). The AMS technique has the advantages of being sensitive and easy to apply.

J.L. Bouchez et al. (eds.) Granite: From Segregation of Melt to Emplacement Fabrics, 145–157.

The use of mineral fabric criteria other than the orientation of F and L is not common. This is due mainly to the fact that fabric analysis is time consuming, and that fabric interpretation presents many difficulties and is still matter of controversy. However, since magmatic rock fabrics record a crucial part of the intrusion and crystallisation histories, they might contain valuable information about the emplacement mechanisms of the magma and thus deserves a careful study.

Fabric development is complex since the growth and differential rotation of crystals with different shapes and sizes, their increasing interactions with progressive crystallisation and therefore change in magma rheology, play significant roles. Moreover, owing to the common late- to post-magmatic plastic deformation overprint, pre-full crystallisation textures of the magma may be significantly modified, although microstructural observations may detect these modifications (Paterson et al., 1989; Bouchez et al., 1992). Hence, a theoretical analysis cannot take into account all these factors and may be successful only in particular cases.

MODELLING OF MINERAL FABRIC

Three approaches have been followed to improve our understanding of magmatic fabrics, namely fabric measurements of rocks, theoretical modelling, and experimental modelling.

Natural mineral fabrics have been analysed by means of direct field and Universal Stage measurements, or using other techniques such as X-ray or neutron texture goniometry. At present, many structural studies on granitic rocks are based on AMS measurements. It is generally assumed that the magnetic fabric reflects the pattern of mineral fabric, and thus, that the AMS technique gives an indirect description of mineral fabrics. Most methods of fabric analysis aim at describing precise foliation and lineation patterns. Note however that even if relevant information is obtained from the F and L maps of magmatic rocks, mineral fabric studies *sensu* Sander (1948) stress the importance of understanding the significance of the rock internal structure at the grain-scale. Our present appreciation of fabric significance remains however limited. Fabric experiments appear therefore appropriate to contribute to the interpretation of magmatic fabrics. Up to now most experiments of magmatic fabric development have been performed in 2D. They have shown that, at low particle concentrations, needle-like markers develop preferred dimensional orientations (PDO) in agreement with theory (Fernandez, 1987). In systems containing markers with lower aspect ratios, i.e. closer to the real shape of magmatic crystals, the particles overpass the shear plane as shear strain increases, conformably with the prediction of cyclic fabric evolution (Fernandez et al., 1983). At high concentrations, interactions impede particle rotations (Ildefonse and Fernandez, 1988; Ildefonse et al., 1992). Studies in 3D are less common. After the early work of Tullis (1976) concerning uniaxial strain, it is only recently that experiments have been undertaken in simple shear (Fernández-Catuxo, 1994; Fernandez et al., 1995; Arbaret, 1995; Arbaret et al., this volume). These works have highlighted the importance of particle interactions in fabric development.

In this paper we present and discuss the results of experiments of 3D fabric development at low and moderate shear strains ($\gamma < 4$). The effects of interactions are analysed and compared with the subfabrics of natural specimens.

FEATURES OF MINERAL FABRICS

A 3D PDO may be entirely described by a fabric ellipsoid. March (1932) has given an analytical relationship between the orientation density and the radius of the finite strain ellipsoid in the corresponding direction in 3D, provided that markers are passive and perfectly linear or planar. In the March's model, the fabric ellipsoid and strain ellipsoid are alike for linear markers and inverse to each other for planar markers. Fernandez (1981, 1988) has given a generalisation of March's model for any population of axial rigid markers. Fabric constraints inferred from this generalised model are discussed below.

FABRIC PARAMETERS

Let X, Y and Z be the axes of the finite strain ellipsoid, its symmetry may be characterised by K (Flinn, 1962), where $K= [(X/Y)-1]/[(Y/Z)-1]$. Using the natural principal strains ε_1, ε_2 and ε_3 (Ramsay, 1967), the Lode's symmetry parameter ν can be defined as $\nu = (2\varepsilon_2-\varepsilon_1-\varepsilon_3)/(\varepsilon_1-\varepsilon_3)$. To define entirely the shape of the ellipsoid, an additional intensity parameter is also necessary, like for instance the octahedral natural strain (Nadai, 1963), $\varepsilon_s= 0.577 [(\varepsilon_1-\varepsilon_2)^2 + (\varepsilon_2-\varepsilon_3)^2 + (\varepsilon_3-\varepsilon_1)^2]^{1/2}$. Similar parameters may be used to describe fabric ellipsoids.

Let us deal with a population of axial markers with a shape parameter κ, where $\kappa = (n^2-1)/(n^2+1)$ and n the axial shape ratio of the marker (Willis, 1977), or with a real population modelled by axial ellipsoids with the same shape parameter. Generalisation of March's model yields the following relationships between the finite strain and the fabric ellipsoids for prolate markers under coaxial strain regimes:

$$\Lambda 1 = \lambda_1{}^{\kappa} \qquad \Lambda 2 = \lambda_2{}^{\kappa} \qquad \Lambda 3 = \lambda_3{}^{\kappa} \qquad (1)$$

where Λi are the principal quadratic semi-axes of the fabric ellipsoid, and λ_i, the principal quadratic extensions of the finite strain ellipsoid; for oblate markers, λ_1 (respectively λ_3) becomes λ_3 (respectively λ_1). Another important parameter is the maximum density of the conventional pole figures, expressed in multiples of the uniform distribution, and given by:

$$D_M = \lambda_1{}^{3\kappa/2}, \text{ or } D_M = \lambda_3{}^{3\kappa/2}, \text{ for prolate and oblate markers.} \qquad (2)$$

Note that, for a given total strain characterised by principal quadratic extensions λ_1, λ_2 and λ_3, the highest anisotropy magnitude possible of the mineral PDO for coaxial strain is given by the fabric ellipsoid defined in (1), and whatever the strain intensity, the orientation density has a maximum value given by (2). Values given by (1) and (2) correspond to fabrics developed under slow flow and quasi "free-rotation" conditions under coaxial strain. For the same total strain, the intensities of fabrics developed under simple shear strain are lower than, or at most equal to the values given by (1) and (2), owing to the periodicity effect and/or interactions between particles. Thus, whatever the strain regime, the fabric intensity developed at a given total strain by a given population of particles in a concentrated suspension is lower than in dilute suspensions (Ildefonse et al., 1992).

For a given fabric of markers of shape κ, expressions (1) define a "model strain ellipsoid" which is very useful for comparing different subfabrics corresponding to different mineral populations in a rock. This model strain ellipsoid also represents the minimum strain ellipsoid responsible for the observed fabric. Note again that the real strain may be much larger than the strain of the model strain ellipsoid.

3D BIOTITE FABRICS EXPERIMENTS

EXPERIMENTAL DEVICE AND ANALOGUE MATERIAL

The shear box used in the experiments (Fig. 1) is similar to that of Willis (1977). It consists of a 40 mm-wide endless belt driven by two vertical drums, 98 mm high and 40 mm in diameter, and placed inside a box made of plexiglas. One of the drums is driven by a variable speed electric motor ensuring that different strain rates can be applied. To perform the experiments, about 500 cm^3 of the analogue material is introduced between the belts. At the base of the box, a thin layer of mercury (5 mm) prevents any significant mechanical coupling between the analogue material and the box. The main drawback of this apparatus is that the rotation of the drums creates a marked flow heterogeneity in their vicinities. In the central part of the box, however, the strain is quite homogeneous provided that the total strain remains moderate ($\gamma < 5$). Even if data about the strain accumulated in a granitic rock during the intrusion history are still lacking, the latter value is considered to be large enough to cover most non-coaxial strain related to granitic intrusion processes.

Our analogue material is a mixture of natural crystals, forming the solid fraction, embedded in a liquid matrix. The matrix is of a synthetic resin polyester (trade mark: Stratyls Chloe, supplied as "Inclusal" by the Pierron Company, France) to which 25% in volume of kaolin powder has been added. This mixture presents a slight shear-thinning behaviour and, at the shear rate of the experiments, $d\varepsilon/dt = 1.5 \times 10^{-2}\ s^{-1}$, the apparent viscosity is 25 Pa.s. This viscosity prevents the solid fraction from sinking during experiments. The solid fraction itself is made of different proportions of biotite, 0.2 to 1.0 mm in diameter, and quartz grains, 0.2 to 0.63 mm in diameter.

The kaolin fraction, from Beauvoir (France), is very fined grained (48% by weight <2 μm; 96% <20 μm) and has an iron oxide content of less than 0.3%. The quartz fraction was separated from fluvial sands that was magnetically purified at high field. The biotite grains were separated from biotite-rich layers of the Confurco granite pluton (Spain), and purified magnetically to ensure elimination of the ferromagnetic fraction. Therefore, biotite is the only magnetic species in the analogue material, allowing AMS to be used to determine the magnetic fabric related to the PDO of biotite. In all experiments, the analogue material has the following properties: (i) constant resin/kaolin ratio in order to maintain a high enough viscosity, hence avoid grain sedimentation, and to keep the fluid rheology constant from one run to another; (ii) biotite volume fraction larger than 5% in order to maintain a good signal for the AMS measurement; (iii) total solid fraction (biotite + quartz) less than 40% to ensure that the magma analogue does not reach the solid-like behaviour of Arzi (1978).

EXPERIMENTAL PROCEDURE

After deformation, the analogue material is heated up to 40°C in order to accelerate its polymerisation hence avoid grain sinking. After perfect solidification a first section was cut, then polished parallel to the XZ plane of the strain ellipsoid. In this section, the best axis **x** (close to X) of the apparent 2D preferred orientation was determined, as explained in the next paragraph. Then, at least three sections were cut perpendicularly to **x** (i.e. **yz** sections). Independently, three cores were drilled vertically (i.e. parallel to **y**) and oriented with respect to the shear direction. The cylindrical specimens were cut to the standard length for AMS measurement (25 mm in diameter; 22 mm in height), taking care that their bases were not contaminated by mercury. With this procedure, a two-fold control on the reference fabric frame is then possible: (i) within the YZ section plane, parallelism of the best axis of the 2D biotite preferred orientation with the Y axis (vertical in the experiments), and (ii) in 3D, coaxiality of the principal magnetic fabric axes with the reference frame.

FABRIC ANALYSIS AND 3D RECONSTRUCTION

Sections parallel to **xz** and **yz** were studied by optical goniometry and image analysis, following methods developed at Limoges University. The goniometer used

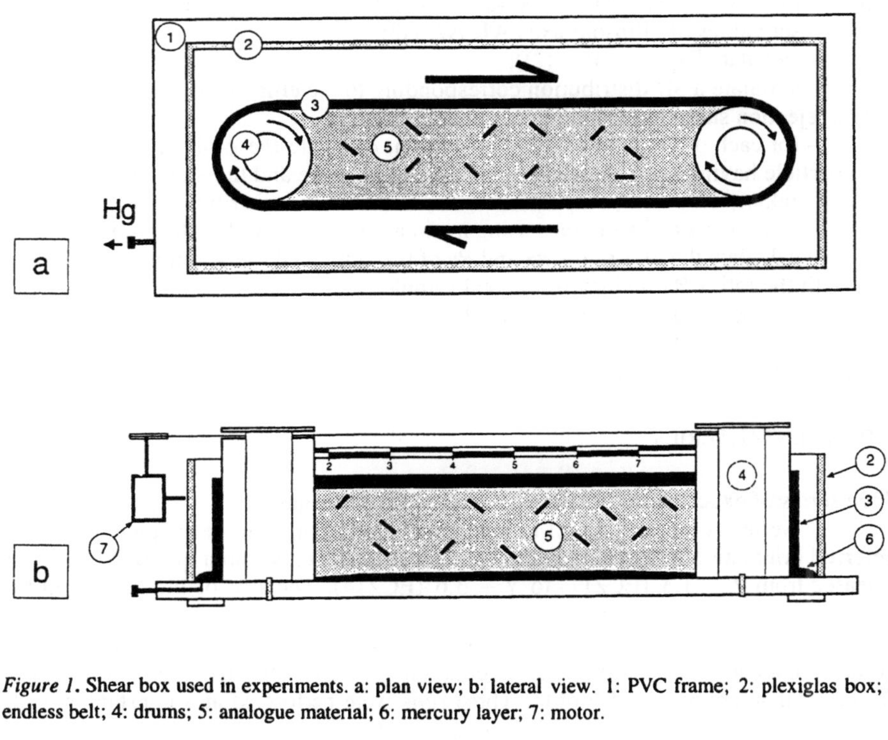

Figure 1. Shear box used in experiments. a: plan view; b: lateral view. 1: PVC frame; 2: plexiglas box; 3: endless belt; 4: drums; 5: analogue material; 6: mercury layer; 7: motor.

consists of a circular stage whose rotation axis is linked to an optical encoder. With the help of a X,Y moving specimen guide fixed on the rotating stage, each mineral is set in the centre of the stage and oriented parallel to the same reference line. The optical encoder sends the signal to a computer, which performs the treatment of the angular data (program GONIO). Our image processing unit is equipped with a PC-Eye card designed and manufactured by the company CPE (Lyon, France). The software used is MIPS, developed by the company Misis Image (Saint-Etienne, France). The images are treated and converted to binary following the usual techniques of image processing. The orientation of the markers is determined by means of a program written in Turbo Pascal. The first operation is the labelling of the markers over the image. The orientation of each marker is then determined by: (i) calculating the position of the centre of gravity of the marker, and (ii) search, by an iterative procedure, of the angle of rotation of the image that minimises the sum of the squares of the distances between the pixels of the image and the horizontal line passing through the centre of gravity of the marker. In each sample, the orientations of about 600 grains of biotite were determined within ± 0.5° with respect to the shear direction (in **xz** sections) and to the Y axis (in **yz** sections). The results are (Fig. 2): (i) the axial ratios of the fabric ellipses Rx (**yz** section) and Ry (**xz** section), and (ii) the corresponding histograms.

The axial ratios obtained for a sample may then be used for 3D fabric reconstruction using the program FAB2D3D. This program aims at finding a fabric ellipsoid such that if the corresponding orientation distribution was cut parallel to **yz** and **xz**, the distribution of the traces of the markers in both sections would yield the fabric ellipses of axial ratios Rx and Ry. This ellipsoid is calculated by iterating the following steps:

- set values of ellipsoid axes A1, A2, A3;
- calculate D_M (equation (2));
- simulate a 3D distribution corresponding to the ellipsoid of axes A1, A2, A3, using a rejection sampling technique;
- for each orientation of this distribution, calculate the probabilities Px and Py that a particle having such an orientation cuts the **yz** plane and the **xz** plane;
- taking Px and Py for each marker, simulate the distribution of the traces of the markers in the **yz** and **xz** planes using again a rejection sampling technique;
- calculate the axis ratios *Rx* and *Ry* of the simulated fabric ellipses and compare them with the ratios Rx and Ry of the real sample;
- if |Rx-*Rx*| and |Ry-*Ry*| are smaller than a given precision, A1, A2, A3 are considered to fit the 3D orientation distribution of the sample; otherwise, the values of the axes are modified and the loop is repeated.

EXPERIMENTAL SERIES

Three series of experiments, totaling 12 runs, were performed:

- Series A (EC211-EC214) was aimed at studying the development of biotite preferred orientation in low concentration suspensions, i.e., without quartz. Shear values increased from $\gamma \sim 1.8$ (EC211) to $\gamma \sim 3.6$ (EC214). The size of biotite ranged from 0.4 to 1 mm;
- Series B (EC206-EC210) was designed to test the effect of crystal content in the development of biotite preferred orientation. In these runs the grain-size of quartz was larger than grain-size of biotite. Three runs were performed at a constant $\gamma \sim 2.5$, but with increasing volume fractions of quartz (Q), from 10% to 30%, added to the

mixture (EC206, EC207, EC208). Complementary runs EC209 and EC210 were performed at Q = 20%, with shear values $\gamma \sim 1.7$ and $\gamma \sim 3.3$.

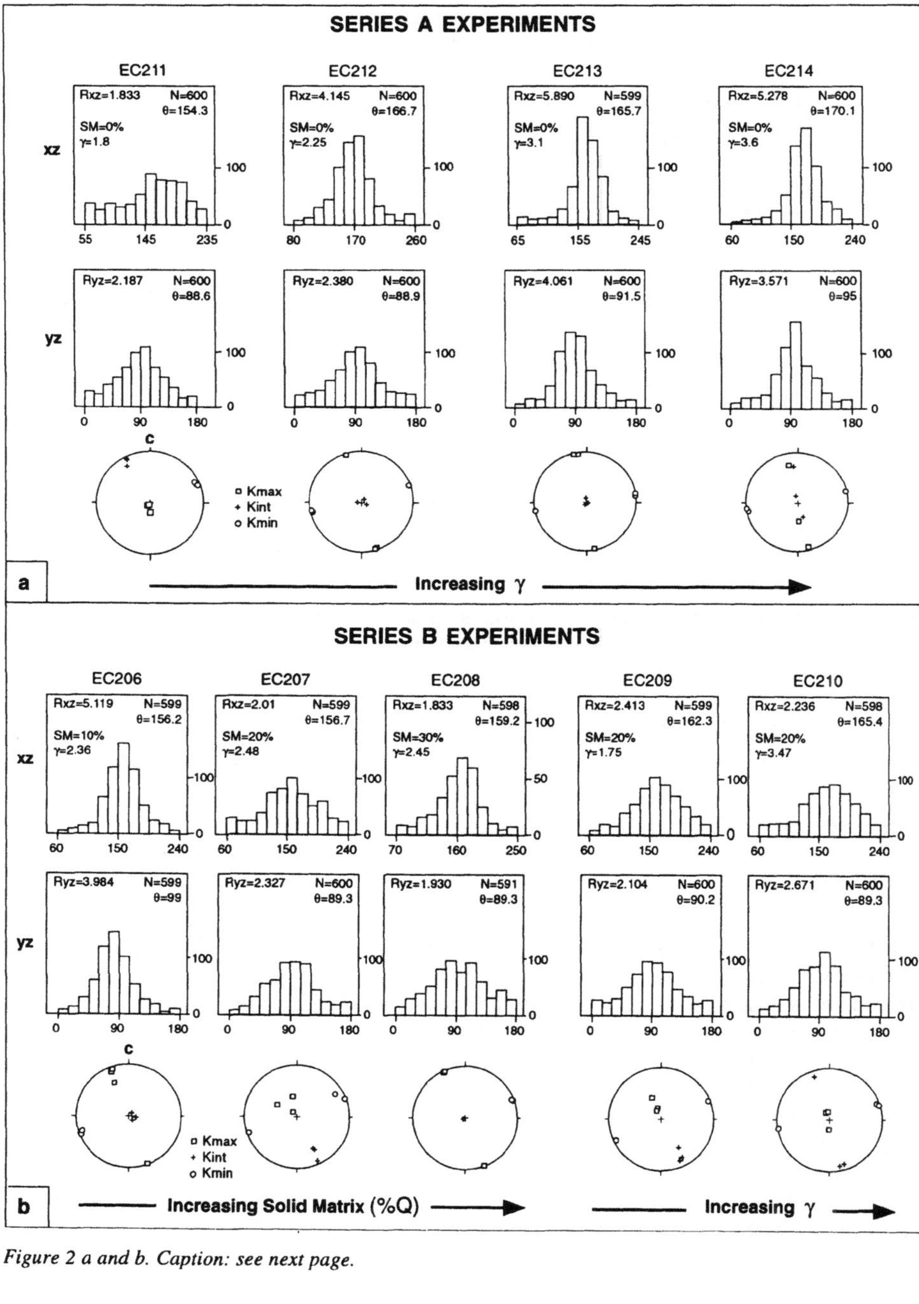

Figure 2 a and b. Caption: see next page.

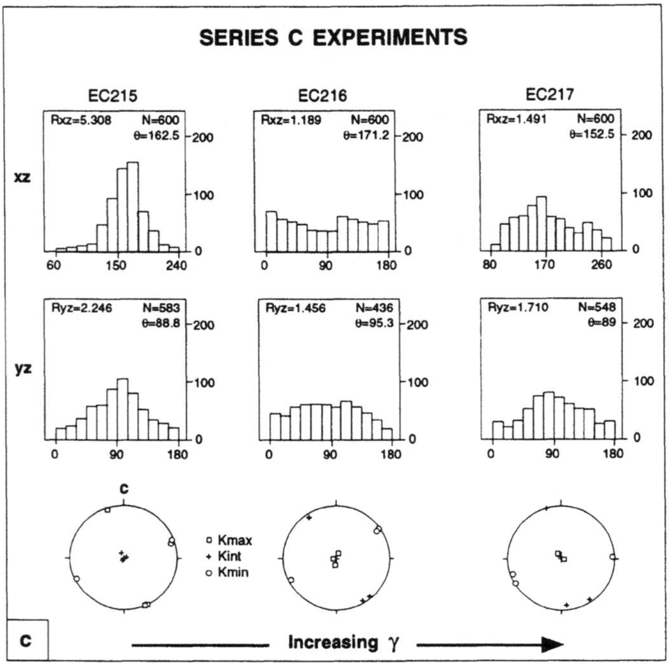

Figure 2. Orientation distribution of biotite crystals in **yz** and **xz** sections (600 grains per histogram) and AMS measurements (3 cores; stereoplots parallel to shear plane and shear direction **c**) of the corresponding runs. a: Series A experiments; b: Series B experiments; c: Series C experiments.

- Series C (EC215-EC217) had the same objective as series B but with the grain size of quartz smaller than that of biotite. The three runs were performed at a constant proportion of quartz (Q = 20%) but with increasing shear strains from $\gamma \sim 1.7$ (EC215) to $\gamma \sim 3.3$ (EC217).

RESULTS

The resulting fabrics are represented as orientation histograms for each structural section (**xz** and **yz**) together with the AMS ellipsoid in every run (Fig. 2). In general, good agreement is observed between mineral fabric data and AMS results.

In the Series A experiments, here taken as a reference series, the orientation histograms for **xz** and **yz** become sharper with increasing shear strain, indicating an increase in PDO intensity (Fig. 2a). The high symmetry of the orientation distributions in both sections enables the use of FAB2D3D to reconstruct the 3D fabrics (Fig. 3). The resulting pole figures correspond in all cases to fabrics with low K values, i.e., belonging to the flattening symmetry field. The fabric of sample EC211 has a low magnitude (D_M=4%) and presents an inversion of **x** and **y** fabric axes (**x** axis being subparallel to the Y strain axis). This inversion is probably related to the marked flattened shape of the fabric ellipsoid, and to the fact that a low total strain does not entirely erase the anisotropy of the initial orientation distribution. The decrease of fabric intensity observed in run EC214 is attributed to the increasing interactions between the biotite crystals with increasing shear strain. AMS measurements correlate well with the mineral fabric data: good fits are observed between the magnetic data (K1, K2, K3: stereoplots) and the PDO data (histograms). Moreover, the magnetic data from this series reveal that K1 varies similarly to the mineral fabric pattern, retaining parallelism with **x** in ten runs out of twelve. Figure 4 presents three micrographs of these runs from series A. Note that in **yz** sections, the vertical preferred orientation (i.e., parallel to the short sides of the photographs of Figs. 4a, and 4b) of biotite supports the choice of the reference frame. Figure 4c shows the excellent PDO of biotite in the **xz** plane.

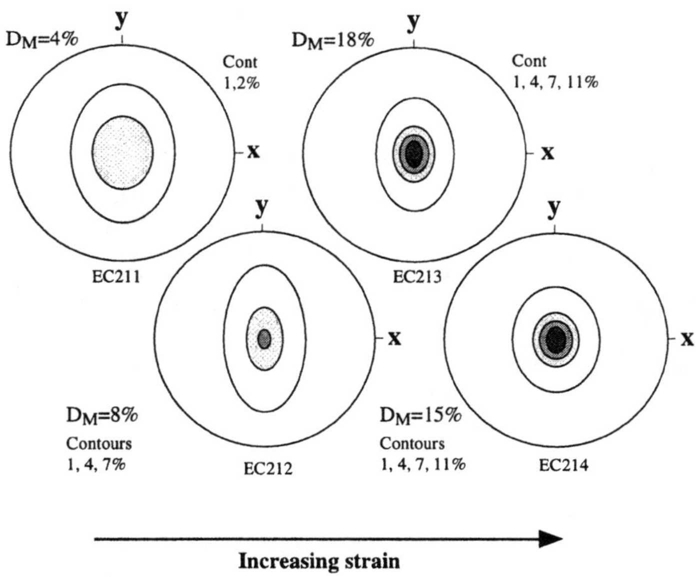

Figure 3. *Reconstructed pole figures corresponding to the fabrics of Series A experiments, obtained from the FAB2D3D program.*

Series B and series C experiments show the marked influence of the concentration of solids in the fabric evolution. In general, for a given strain, the axial ratios of the fabric ellipses in both the **xz** and **yz** sections diminish with increasing solid fraction (EC206, EC207, EC208: Fig. 2b). At the same time, the difference between Rx and Ry diminishes, indicating that the fabric becomes closer to the axial flattening symmetry. In concentrated suspensions, increasing shear strain produces a similar fabric symmetry evolution (series C experiments: EC215-EC217; Fig. 2c). The corresponding micrographs in **xz** sections illustrate the increasing disorder of the fabric with increasing shear (Figs. 4d, 4e and 4f). A surprising effect due to the interactions between the grains of biotite and quartz is observed in some cases. It is the inversion of the fabric axes **x** and **y**, revealed by the ellipse axis ratio in **yz** section, which is unexpectedly greater than the ratio in **xz** section (Fig. 2: runs EC207, EC210, EC216, and EC217).

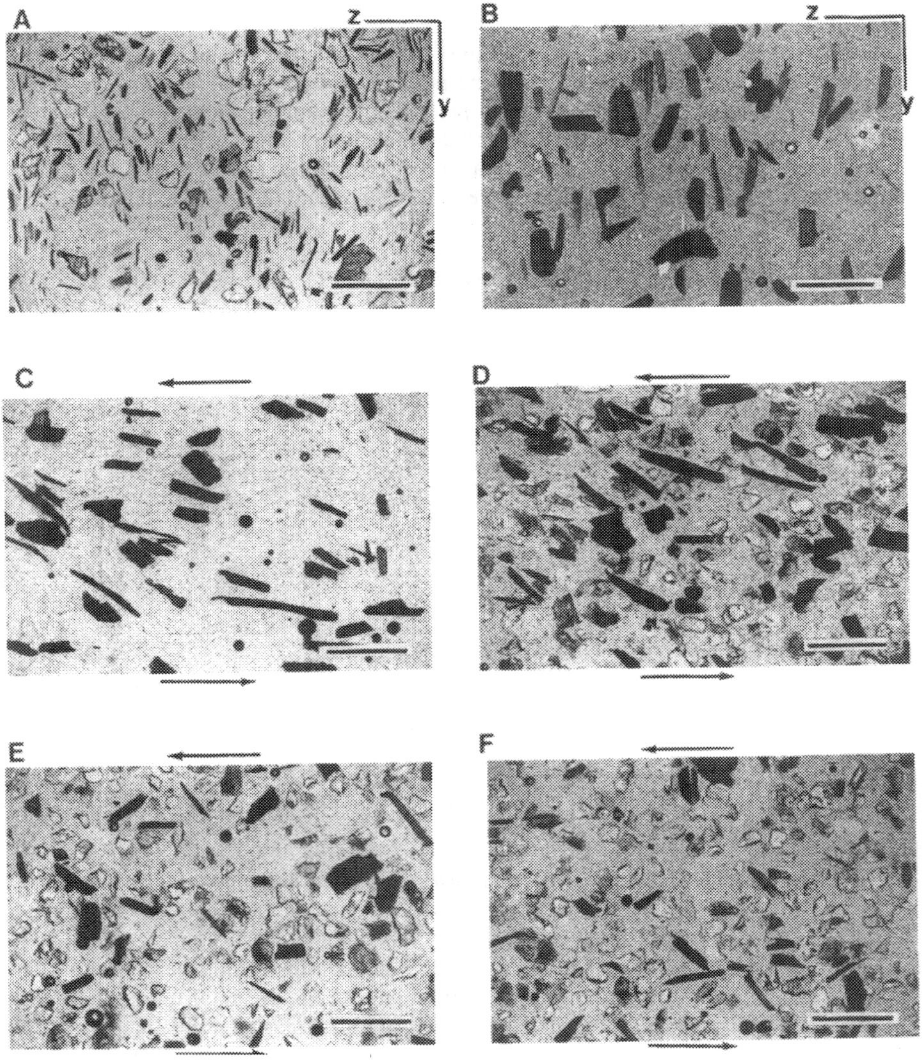

Figure 4. Micrographs from six experiments (scale bar: 1 mm). A and B: **yz** sections; note the preferred orientation of biotite parallel to **xy**. C: **xz** section: low quartz fraction (Q = 0%) and strong preferred orientation. D: **xz** section: high quartz fraction (Q = 20%), but low strain and rather high preferred orientation. E and F: **xz** sections: high quartz fractions (Q = 20%) and weaker preferred orientations. A: EC206 ($\gamma = 2.36$; Q = 10%); B: EC214 ($\gamma = 3.6$; Q = 0%); C: EC213 ($\gamma = 3.1$; Q = 0%); D: EC215 ($\gamma = 1.9$; Q = 20%); E: EC 216 ($\gamma = 2.5$; Q = 20%); F: EC217 ($\gamma = 3.28$; Q = 20%).

This change in the orientation of the biotite lineation is, in all these cases, accompanied by a change in the orientation of K1. The change in AMS pattern concurrent with the fabric axes inversion, strengthens our belief in the reality of this phenomenon. However, in runs EC208 and 209 (Fig. 2b), K1 (AMS) and **x** (PDO) do not coincide, but $|Rx-Ry| \leq 0.3$ suggests that this discrepancy is related to an almost axial fabric symmetry.

The evolution of α, the angle between **x** (PDO) or K1 (AMS) and the shear direction **c**, as a function of shear strain in the low concentration suspensions of series A experiments, is close to that of the theoretical models (Fig. 5). For comparison, α of previous runs using plaster as the matrix (Fernandez et al., 1995) are also given in figure 5. From these, α of run EC212 ($\gamma = 2.25$) differs by no more than 7° from the theoretical value that can be read on the $\kappa = -0.85$ curve. Finally, the most conspicuous result of the experiments in concentrated suspensions is the development of flattening symmetries ($\nu > 0$) with increasing shear strain, and the low intensity parameters, whatever the solid fraction.

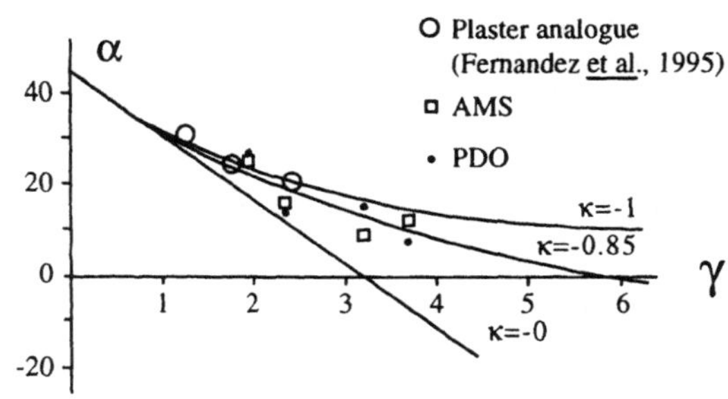

Figure 5. Experimental fabrics (PDO and AMS measurements), and theoretical behaviour (solid lines; see Fernandez, 1984) for particles ranging from equant ($\kappa = 0$) to perfectly flat ($\kappa = -1$), on a $\alpha(\gamma)$ plot. For comparison, the experiments of Fernandez et al. (1995), using plaster as an analogue material, are also given.

DISCUSSION AND CONCLUSIONS

Analogue experiments simulating 3D fabric development in simple shear, low concentration suspensions (~ 6% vol.) and moderate strains ($\gamma < 5$), are in good agreement with theoretical models, but the tendency to a pronounced flattening symmetry with increasing shear is observed. Concentrated suspensions under similar strain conditions clearly emphasize the role of the interactions between particles on fabric development. The principal effects of interactions are:

- the development of fabrics with fabric intensity lower than predicted by theory, or than in fabrics that developed in low concentration suspensions. Lower fabric intensity is attributed to the increasing disorder introduced in the structure of the suspension. This phenomenon, observed in both series B and C experiments, seems to develop independently of the relative sizes of biotite and quartz, at least in the range of the size ratios of the experiments (Bi/Q = 0.6-2.3);

- the general tendency to the development of flattening symmetries. In almost all the experiments the symmetry evolves from close to plane strain ($\nu = 0$), as in theory, to flattening types ($\nu > 0$), up to almost axial symmetry ($\nu = 1$). This evolution is attributed to the rotation of the flat biotite grains around the Y strain axis, induced by their contacts with the sub-spherical but irregular quartz grains, which themselves may have a strong tendency to rotate around Y;

- the frequent inversion of **x** and **y** fabric axes, with the development of a parallel-to-Y lineation. This surprising effect was observed, and confirmed by AMS measurements, in samples from the porphyritic facies of the Confurco granite pluton, Spain (Fernández-Catuxo, 1994), where the biotite lineations were almost perpendicular to conspicuous K-feldspar lineations in many parts of the massif.

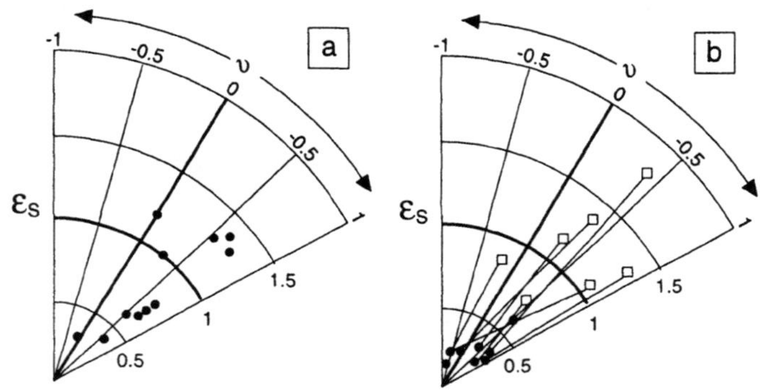

Figure 6. Hossak's or (ε_S, ν) diagram for a: the experiments of this study; note the strong development of fabric symmetry in the flattening field; and b: biotite (full dots) and feldspar (squares) subfabrics from the pluton of Confurco (data from Fernández-Catuxo, 1994); tie lines identify the different samples; note the significantly lower strength for biotite than for feldspar.

In the present experiments, a Y-lineation also develops in several runs (EC216, EC217, EC207, EC208, EC210) under conditions of the two main contributing factors, $\gamma > 2$ and quartz% >10. Therefore, in experiments as well as in nature, some conditions related to solid fraction or strain intensity may favour the development of a Y-axis zone-lineation. Both flattening symmetry and occasional Y-axis lineation could thus be related to the tendency of the biotite crystals to rotate around the Y-axis of finite strain, as a consequence of their interactions with the other grains. The PDO of biotite having both the well-known zone dispersion around X and a strong tendency to rotate around Y, may be at the origin of the observed flattening symmetry as well as the inversion of **x** and **y**. This effect is considered to be responsible for the structures observed in some samples from the Confurco granite pluton (Fernández-Catuxo, 1994).

Comparison of theoretical and experimental results with the subfabrics observed in natural cases permits a more robust interpretation of the fabrics of magmatic rocks. Differences in subfabric patterns may be better interpreted when using the concept of "model strain ellipsoid", which represents minimum strain values, allowing an almost quantitative comparison between subfabrics. Figure 6b presents subfabric intensity data for biotite and K-feldspar megacrysts from seven different outcrops of the Confurco massif. In all the cases, the intensity of the biotite subfabric is significantly lower than that of the feldspars. The low fabric intensity displayed by biotite cannot be explained by fabric stabilisation, e.g., steady-state fabrics with weak PDO's, as suggested by the 3D fabric experiments at high shear strains (Arbaret, 1995; Arbaret et al., this volume). Even if the strain regime responsible for these fabrics is not perfectly defined, since both biotite and feldspar have undergone the same strain history, subfabric strengths should be similar in both cases.

In the range of the biotite/quartz size ratios explored in these experiments (0.6 to 2.3) no significant difference in fabric development was observed as a function of the size ratios. However, the marked difference in the strength of subfabrics of feldspar megacrysts and biotites measured in seven outcrops of the Confurco granite (Fernández-Catuxo, 1994), and also observed in other porphyritic granites, where size of feldspar / size of biotite ≥10, suggests that under identical strain conditions, interference phenomena have a stronger dispersive effect on biotite than on the larger feldspar crystals.

We conclude, on the basis of our experiments and the observed differences in feldspar and biotite subfabrics in the Confurco pluton, that different subfabrics may develop as a function of the relative sizes of the markers responsible for the subfabrics and of the other crystals which constitute the solid matrix. When the markers are much larger than matrix crystals the subfabric in the markers develops in agreement with theoretical models, at least in low concentration suspensions. For markers similar in size to the matrix crystals, the subfabric in the markers develops like in our high concentration suspensions, and have low intensity parameters.

ACKNOWLEDGEMENTS

We thank J.L. Bouchez, B. Ildefonse, J. Jezek, K. Schulmann and W.E. Stephens for critical reading of the manuscript and for their comments that considerably improved this paper. We also thank A. Aranguren and J.M. Tubia (Univ. Pais Vasco) who kindly provided the AMS measurements.

PERPENDICULAR LINEAR FABRICS IN GRANITE: MARKERS OF COMBINED SIMPLE SHEAR AND PURE SHEAR FLOWS?

Karel SCHULMANN, Josef JEZEK and Zdenek VENERA
Faculty of Science, Charles University, Albertov 6,
128 43 Praha 2, Czech Republic

ABSTRACT

Recent results of numerical modelling of crystallographic fabrics of biotite and feldspar found in the porphyritic granite of the borehole EPS-1 at Soultz-sous-Forêt, France, are described and interpreted. Preferred orientations of biotite (001) planes and feldspar (010) and (100) cleavages, were measured by means of an optical goniometer. The poles of (001) biotites show stable orientations throughout the core with maximum eigenvector v1 oriented vertically, and minimum eigenvector v3, or biotite-lineation, constantly oriented in the NE-SW direction. Within the subhorizontal magmatic foliation plane, also defined by feldspar (010) planes, two mutually orthogonal linear fabrics were found to be NE and NW trending, defined by the feldspar long axes. Variations in biotite fabric intensity and symmetry in different sections of the vertical profile were used to determine a plausible flow geometry in these sections. Numerical modelling involving a component of coaxial flattening, with various amounts of extension parallel and perpendicular to stretch, in addition to simple shearing, agrees qualitatively with the observed data. The models demonstrate that, under combined coaxial and simple shear flow regimes, mutually orthogonal linear fabrics may result from progressive rotation of the shortest crystals.

INTRODUCTION

The problem of fabric development in magmatic rocks is often solved using a model of behaviour of rigid particles in slowly moving viscous fluids with Newtonian properties. Numerous papers have been devoted to numerical modelling based on the theories of Jeffery (1922) and March (1932). Most studies dealt with the motion of a spheroidal

J.L. Bouchez et al. (eds.) Granite: From Segregation of Melt to Emplacement Fabrics, 159–176.

particle in relatively simple flow geometries like simple shear (Jeffery, 1922), pure shear (Gay, 1968), axial compression (Debat et al., 1975). Particles of more general shapes were treated only exceptionally (Hinch & Leal, 1979; Freeman, 1985).

Systems composed of a large number of particles were first modelled by Reed & Trygvasson (1977) giving an insight into the possible fabric patterns of natural rocks. An important result of this model is the concept of cyclicity of fabric development and orientation in simple shear flows.

Progress in numerical methods and software has permitted the simulation of the development of fabric in any type of homogeneous flow, e.g. Willis (1977) for kinematic models of axial grains and Jezek et al. (1994) for general ellipsoidal particles. Willis (1977), Fernandez & Laporte (1991) and Jezek et al. (1994; 1996) studied more complicated flow systems involving combinations of coaxial and simple shear components and heterogeneous particle populations. Willis (1977) showed that unstable mutually orthogonal linear fabrics could develop as a result of a combined simple shear and axial flattening regime. These fabrics were produced by instantaneous superimposition of the different subfabrics characterised by shorter and longer particles, all tending towards the rotation axis of simple shear, although with different velocities. The development of stable mutually orthogonal linear fabrics was demonstrated by Jezek et al. (1996) for combined coaxial and simple shear flows with unequal elongation components of coaxial flow: the shorter particles rotated towards the rotation axis of simple shear while the longer ones found their stable orientations parallel to the flow direction. Occurrences of orthogonal feldspar lineations have also been reported from magmatic bodies, e.g. Bouchez et al. (1981).

Results of numerical modelling in more complicated flows cannot easily be tested by experimental analogue modelling nor compared with natural rocks. Analogue modelling (e.g. Ghosh & Ramberg, 1976; Fernandez, 1988; Ildefonse et al., 1992a,b) remains for technical reasons constrained to the simulation of simple flow regimes (mostly 2D and simple shear) and has more recently concentrated on the study of the effect of particle interactions. Moreover, simple shear analogue experiments permit the simulation of only small finite shear strains (Fernandez, 1988) except for the most recent experiments (Arbaret et al., 1995).

One of the most common techniques of fabric determination in natural magmatic rocks is the anisotropy of magnetic susceptibility (AMS) method which provides the bulk preferred orientation of magnetic minerals (Hrouda 1982, 1994; Bouchez, this volume). However, this technique provides integral information, i.e. it does not distinguish the shapes and axial ratios of individual particles. The Universal Stage mounted on a microscope provides more complete information since the size, axial ratio and orientation of crystals can be measured individually. However, this technique allows us only to measure very small minerals, and this is a serious restriction.

To avoid the above limitation, a biaxial reflection goniometer was constructed at the Institute of Petrology and Structural Geology, Prague (Venera et al., in press). It allows orientation measurements of large crystals like feldspars and biotite in rocks and was successfully used to compare natural with the results of numerical modelling of fabric development in complex types of flow.

This paper describes natural fabrics defined by the preferred orientations of biotite (001) planes and feldspar (010) and (100) planes which were obtained from the porphyritic granite recovered from the borehole at Soultz-sous-Forêt (France). An attempt is made to explain the two orthogonal linear fabrics observed in the rock using the numerical model of Jezek et al. (1994) which simulates combined pure shear and simple shear.

MEASUREMENT TECHNIQUE: OPTICAL REFLECTION GONIOMETRY

Magmatic planar fabrics or foliations are generally marked by flattened mafic enclaves, schlierens and shape preferred orientations of different platy minerals, principally biotite and feldspar (review *in* Paterson et al., 1989). Measurement of their orientations is possible when the enclaves are sufficiently planar and/or when the mineral shape preferred orientation (SPO) is sufficiently anisotropic. In the case of general triaxial-shaped enclaves and/or slightly anisotropic fabric, the XY plane of the finite strain ellipsoid is not simple to determine in the field. The magmatic lineation is even more difficult to define reliably due to the scarcity of elongate particles. The preferred orientation of the long axes of triaxial-shaped crystals, like feldspars, may definitely point to the direction of bulk stretch but only in the case of constrictional flow. Any other type of fabric symmetry may exhibit complex orientations of feldspar faces rendering the determination of the stretch direction ambiguous.

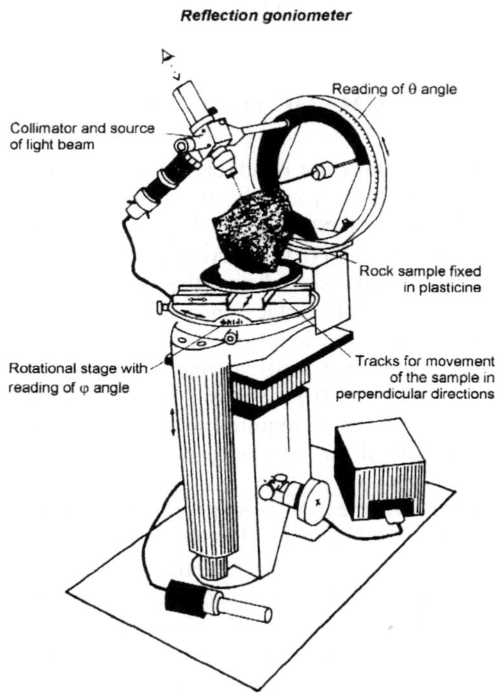

Figure 1. The reflection goniometer used for orientation measurements of cleavage planes (010) and (001) of large feldspar phenocrysts, and (001) cleavage of biotite

The technique of optical goniometry enables the determination of the SPO of feldspar phenocrysts, biotite crystals, and therefore precise determination of magmatic foliation and lineation. A detailed description of the technique is given in Venera et al. (1996) but the principles are summarised below. The two-axes optical goniometer (Fig. 1) provides orientation measurements of well-formed cleavage planes of crystals which are visible in the rock sample. Two angles, φ and τ, define the position of the measured reflecting cleavage plane. One essential component of the goniometer is the stage carrying the sample, permitting 360° rotation around the vertical axis. On the stage the sample can move along two perpendicular tracks. The sample is fixed on the tracks with plasticine in such a position that its oriented surface is parallel to the stage and the strike and dip marks are parallel to the tracks. Once fixed, the sample may be moved together with the stage in the vertical direction. The other rotating part of the apparatus is a vertical circle, bearing a collimator combined with a source of light beam, and which can rotate around the horizontal axis. The measurement technique consists of finding an orientation of maximum reflection of cleavage which is located when the light beam is exactly perpendicular to the measured plane. The selected cleavage is set to this position by simultaneous rotations of the sample and collimator around the vertical and horizontal axes, respectively. Azimuth angle φ of the cleavage plane is then read on the rotating stage, and its inclination τ on the vertical circle (Fig. 1).

DETERMINATION OF MAGMATIC FOLIATION AND LINEATION

The goniometer is designed to measure relatively large crystals, up to several centimetres in size, that have cleavage planes, mainly micas and feldspars in this study. In the case of micas, the orientations of their basal (001) plane may be determined. If applied to feldspar phenocrysts, the (001) and (010) cleavages may be measured. In most cases, these two cleavages may be distinguished during the process of measurement, according to crystal habit. The pole to the (001) surface, measured by reflection in the goniometer, is at an angle of about 25° with the long axis [c] of the crystal. Therefore the (001) poles form clusters disposed symmetrically around the main lineation defined by elongate shape of crystals. The pole to (010) is perpendicular to the largest face of the crystal. A minimum of 50 measurements is required to obtain a reliable picture of the phenocryst orientation distribution. If this number is not available in a given section, several sample sections are combined.

Orientation of foliation and lineation is conveniently determined using orientation tensor statistics of Scheidegger (1965) which comprise evaluation of the eigenvectors and eigenvalues of the tensor. The eigenvectors define the X, Y and Z axes and the principal planes of the fabric ellipsoid. The corresponding maximum, intermediate and minimum eigenvalues reflect the density of SPO around individual eigenvectors (Wallbrecher, 1986). The orientation of the magmatic foliation is determined precisely from the SPO of poles to biotite flakes or poles to feldspar (010) planes. The magmatic lineation is obtained from the SPO of feldspar (001).

The results are presented in the form of pole figures of biotite (001) planes, and feldspar (001) and (010) planes (Fig. 2). These data have been contoured using multiples of random distributions and treated by orientation tensor statistics (Fara and Scheidegger, 1963). Calculated eigenvectors are presented both in contour and orientation rose diagrams. Since the rocks under study come from a core-drill, variations of mean orientations with depth are displayed as orientation histograms. Ratios of eigenvalues of biotite subfabrics were plotted in Woodcock's (1977) diagram.

GEOLOGY AND SAMPLING STRATEGY

The granite chosen for this study comes from the deep borehole EPS-1 in Soultz-sous-Foret, France (Fig. 3), undertaken within the framework of the International Project 'Géothermie profonde', sponsored by the B.R.G.M. (the French Geological Survey). Variscan granitoids form a large part of the Vosges and Schwarzwald crystalline complexes (Boutin et al., 1995). Large NE-SW trending granitic to gabbro-dioritic bodies occur in both Moldanubian and Saxothuringian parts of Vosges, Schwarzwald and Odenwald (Krohe, 1991; Boutin et al., 1995). Granitoid massifs covered by Mesozoic sediments in the Rhine graben appear in magnetic and gravimetric maps where they form NE-SW trending structures (Edel, 1982). The Soultz granite belongs to the granitoid suite of the Rhine graben. It is a porphyritic biotite-bearing, medium- to coarse-grained containing numerous phenocrysts of K-feldspar ranging from 1 to 5 cm in size. The size of the biotite crystals varies between 0.2 and 0.5 cm.

Twenty four oriented samples from the EPS-1 borehole have been selected in the 1420-2180 m depth range, separated by approximately 50m each. The sample selection for the purpose of this study depended mainly on the quality of core orientation determi-

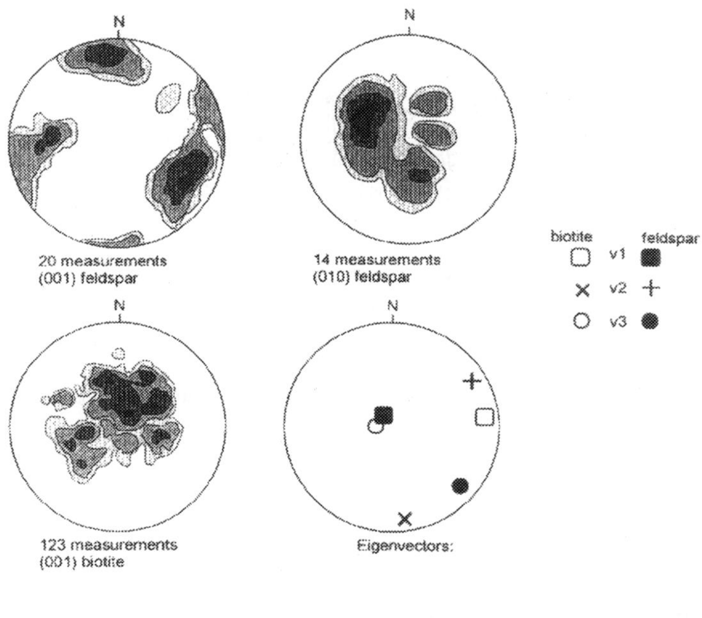

Figure 2. SPO optical goniometry measurements of feldspar and biotite. a: poles to feldspar (001) cleavage; b: poles to feldspar (010) plane, c: poles to biotite (001) cleavage. These diagrams point to a horizontal foliation. Lower hemisphere, equal area projection; contours: multiples of uniform distribution. d: corresponding orientations of the eigenvectors calculated from these SPO measurements, for biotite (open symbols) and feldspar (full symbols).

ned by a diagraphy method; a moderate content of phenocrysts and a low degree of alteration of the rock were other important factors.

For the purpose of reflection goniometry the cores were cut into 20 cm long sections (Fig. 4). The preferred orientations of biotite basal planes and feldspar prismatic and basal planes were then measured on broken uneven surfaces of the core. The process was reiterated to obtain sufficient measurements.

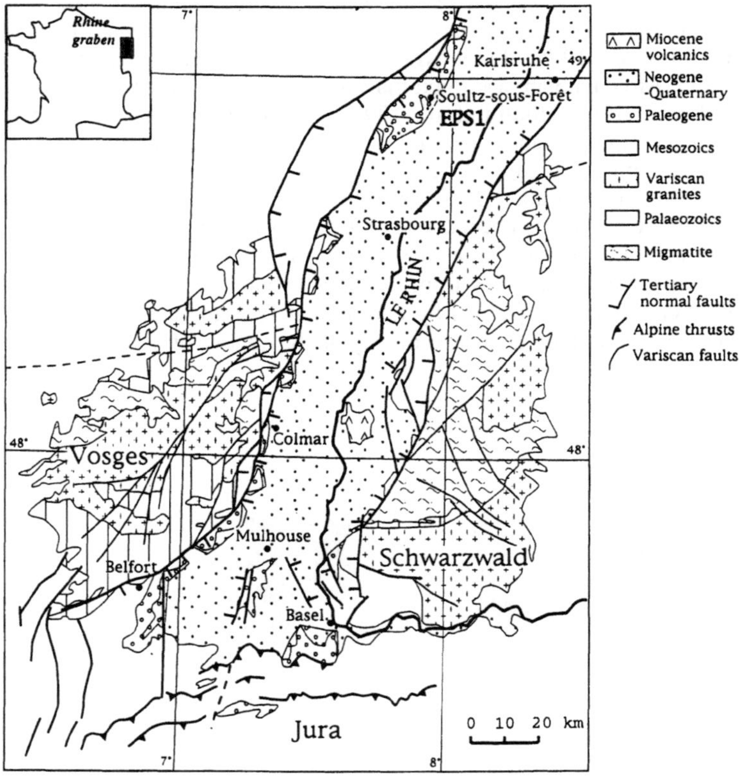

Figure 3. Simplified geological map of the Vosges and Schwarzwald areas, and location of the EPS-1 drill hole in Soultz-sous-Foret.

BIOTITE PREFERRED ORIENTATIONS

Each sample provided about 100 poles to (001) of biotite. The eigenvector v1 giving the direction of maximum density of these poles is more or less subvertical throughout the length of the bore hole (Fig. 5a). This indicates a stable subhorizontal orientation of the foliation defined by orientation of the biotite flakes. The minimum eigenvectors v3 is oriented mainly in the NW-SE quadrant; however, at some depth levels, v3 trends parallel to NW-SE, inverting its orientation with v2 (Figs. 5b and c). The rose diagrams of Figure 5 underline the constant NE-directed fabric of v3, contrasting with the NW trend of v2.

Fabric symmetry and intensity were determined using the following eigenvalues ratios, i.e. modified Flinn's parameter:

$$K = \ln(e1/e2)/\ln(e2/e3)$$

Figure 4. Photograph of a core-drill from the Soultz granite containing feldspar phenocrysts with well visible cleavages.

The corresponding plot (Fig. 6a) shows that almost all of the fabric ellipsoids are situated in the field of apparent flattening. In addition, their overall negative slope shows their increasing fabric intensity with increasing degree of oblateness. In figure 6b, K is reported with respect to the fabric intensity I of Lisle (1985), where e_i are the eigenvalues:

$$I = \frac{15}{3}\sum_{i=1}^{3}(e_i - 1/3)^2$$

In this diagram, the low fabric intensities (I < 1) exhibit K-values ranging mainly from 0.4 to 0.8 whereas at higher fabric intensities (I = 1 to 1.6) K ranges from 0.35 to 0.1, pointing to almost perfect oblate fabrics. The analysis of individual samples showed that the direction of v1 does not change with increasing fabric intensity nor with changing fabric symmetry.

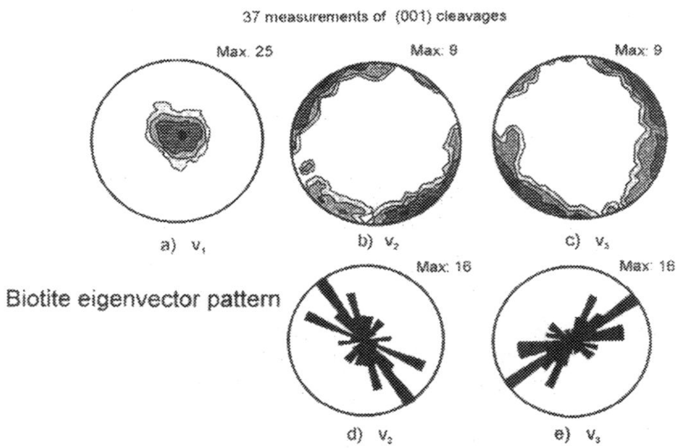

Figure 5. Orientation diagrams of the biotite fabric from the whole drill hole EPS-1. a,b and c: contoured orientation diagrams of eigenvectors v1, v2, v3 calculated from individual biotite fabrics (measurements of biotite (001) cleavage from 37 samples). d and e: rose diagrams showing the orientation of the intermediate v2 and minimum v3 eigenvectors, within the overall subhorizontal foliation defined by the biotite fabrics. v3 gives the direction of maximum stretch.

FELDSPAR PHENOCRYST PREFERRED ORIENTATION

On average, orientation measurements of fifteen basal (001) and twenty prismatic (010) cleavages were performed per sample. The study of preferred orientation of feldspars was more difficult because of the limited number of phenocrysts. Carlsbad twinning is a common feature and, where possible, two (001) cleavages of twinned crystals were measured. Moreover, it was not possible to distinguish in all cases between the basal and the prismatic cleavages. The maxima of poles to the prismatic planes were oriented systematically vertically clearly defining the subhorizontal foliation. The poles to the basal planes define great circles containing the v2 and v1 eigenvectors lying in the subhorizontal foliation plane. Note that v1 , the direction of maximum density of poles to basal planes, shows two nearly orthogonal maxima in the NE-SW and NW-SE directions, well defined in figures 7b and 7c.

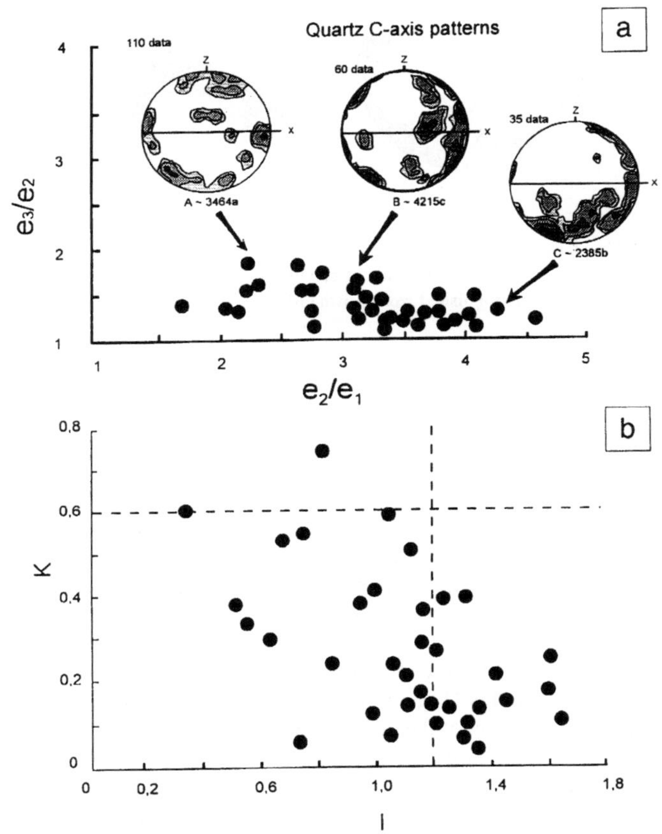

Figure 6. Eigenvalue statistics of biotite fabrics from the EPS-1 borehole. a: Woodcock's plot of eigenvalue ratios for planar markers. Quartz [c] axes patterns from different depths are also given, measured in samples having different fabric symmetries and intensities, as defined by biotite SPO. In samples 2385b (1528 m), 3464a (1855 m) and 4215c (2045 m), NE-SW trending lineation was determined by both biotite and feldspar SPOs. Lower hemisphere, equal area projection; contours: multiples of uniform distribution. b: relationship between biotite fabric symmetry, K, value and fabric intensity, I.

The feldspar, measured on both cut sections and individual crystals, are triaxial tabular in shape and elongated parallel to their [c] axes. Their long/short axial ratio varies between 1.5 to 4., with an average close to 2.7; between intermediate and short axes the ratio ranges from 1. to 2.3, with a maximum about 1.7.

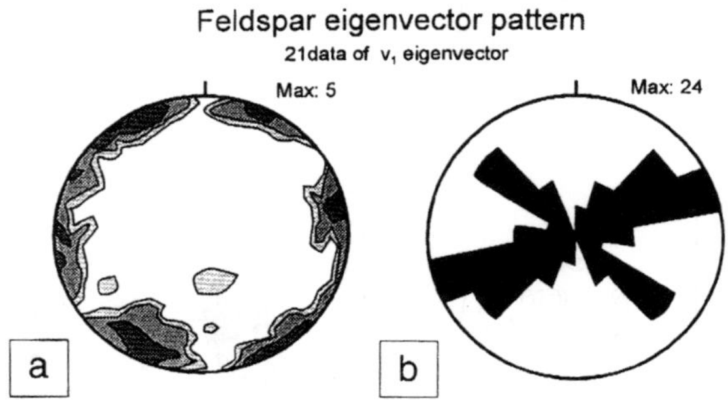

Figure 7: Feldspar (001) planes. a: orientation diagram of eigenvectors v1 from SPO of samples with well defined (001) cleavages; b: rose diagram of trends of calculated feldspar lineations using v1 from all samples where basal planes are not distinguished from prismatic ones.

FABRIC VARIATION WITH DEPTH

The subfabrics of biotites exhibit changes in symmetry and intensity throughout the borehole length (Fig. 8). Oblate fabrics, with K = 0.1 - 0.4, predominate in the whole profile except between 1700 and 1950 metres where prolate fabrics occur (K > 0.5, Fig. 8). The fabric intensities associated with fabric oblateness are restricted to a narrow range, I = (1. to 1.4), while between 1700 and 1950 metres the fabric strength varies between I = 0.2 and 1.4.

The eigenvector v3 of the biotite subfabrics, or X-axis of the shape fabric ellipsoid, shows dominant NE-SW trends (Fig. 5) along the whole profile with the exception of the deepest and shallowest parts of the borehole where NW-SE directions occur (Fig. 8). Since in the zones of high K values the NE-SW directed lineations do not fluctuate, the deviation from NE-SW is attributed to the oblateness of the fabrics. It is therefore proposed that the bulk flow direction effectively trends NE-SW along the whole profile, but that in zones of high flattening, the long axis of the shape fabric may scatter and depart from this mean direction, and may even be perpendicular to it.

The NE-SW lineation trend exists in the whole profile, but its direction, as marked by the feldspars, is slightly inclined with respect to that marked by the biotites (compare Figs. 5 and 7). We note, in addition, that weaker and more prolate biotite fabrics are associated with the NE-SW trending feldspar lineation. NW-SE and NNE-SSW oriented feldspar lineations are developed in zones of intense and oblate fabrics defined by the SPO of biotite. Finally, perpendicular feldspar lineations occur principally in samples having strongly oblate biotite fabrics (Figs. 6 and 8).

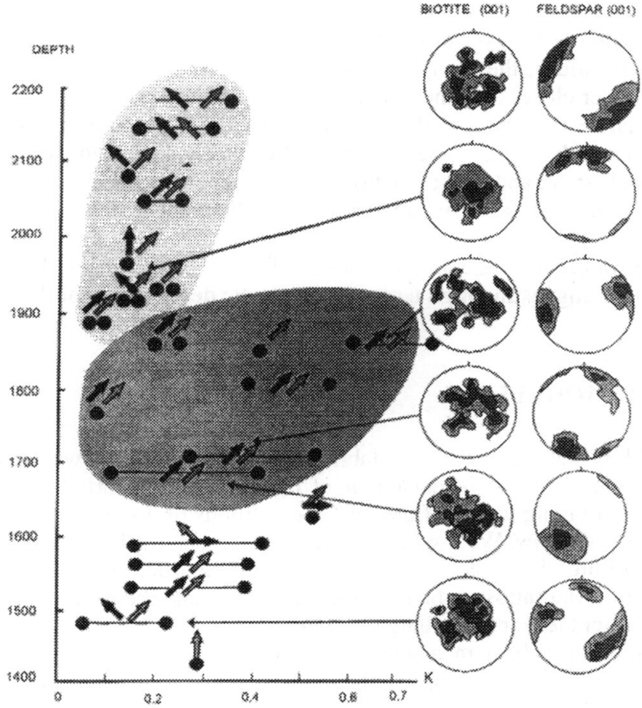

Figure 8: EPS-1 borehole: depth profile giving the K-values and the representative fabrics of poles to biotite (001) (i.e., foliation) and poles to feldspar (001) (i.e., feldspar lineation). Grey arrows indicate biotite lineations defined by v3 eigenvectors. Black arrows represent feldspar lineation orientations defined by v1 eigenvectors, calculated from SPO of (001) cleavages. The dark-shaded area has stable NE-SW lineations of both feldspar and biotite. The light-shaded area is a region of strongly oblate fabric and orthogonal biotite and feldspar lineations.

QUARTZ C-AXIS FABRIC

Quartz [c] axis measurement was employed as a complementary method on three test samples in which both biotites and basal planes of K-feldspars define the same zone-axis lineation trending NE-SW. Thin sections were cut parallel to this lineation and perpendicular to the foliation defined by biotite. In spite of a relatively small number of grains measured, between 35 and 110 (Fig. 6) due to the coarse grain size of quartz (0.5-1.0 cm), the three samples show similar fabric patterns. The quartz grains have lobate boundaries and commonly exhibit cross-hatched undulatory extinction with subgrain boundaries parallel and perpendicular to the foliation trace. The [c] axis fabrics show concentrations within, or close to the XZ plane, with sub-maxima lying close the lineation direction, and fall either close to Z or are distributed between X and Z (Fig. 6). We observe some correlation between the positions of the maxima of [c] axes and the fabric of biotite. Strong maxima of [c] axes close to the NE-SW lineation tend to occur in samples having higher K values; they may be interpreted in terms of the dominant activity of prism [c] slip under closer to plane strain, possibly non-coaxial flow (Blumenfeld et al. 1986). Small girdle distribution of [c] axes around the pole to foliation (Z) tends to occur in samples with intense oblate fabric and could be explained as due to flattening related to coaxial strain, and under conditions where [c] and basal slips may be combined (Lister 1981). Evidence from natural tectonites (Blumenfeld et al. 1986, Gapais and Barbarin 1986, Melka et al. 1992) indicate activity of prism [c] slip as characteristic of very high temperature, near the granite solidus. This is consistent with our microstructural observations suggesting high temperature plastic deformation of the Soultz granite.

NUMERICAL MODELLING

To understand how the observed shape fabrics may have developed, we used the model of Jeffery (1922) which studies the behaviour of rigid ellipsoidal particles (i.e. active markers) in a slowly moving viscous fluid, under conditions of non interacting particles and homogeneous steady flow. We used the software developed by Jezek (1994) that enables numerical modelling of the fabric evolution of a multiparticle system. The input parameters are (i) the axial ratios of the individual particles, and their initial orientations; and (ii) velocity gradient tensor describing the flow.

(i) The biotite flakes have measured axial ratios close to 4:4:1 on average, although up to 8:8:1 for some crystals. Thus, in the modelling biotite was represented by a population of spheroidal particles of axial ratios 4:4:1. Feldspars, with their measured long/short axial ratios ranging from 1.5 to 4, have been represented by three subpopulations of particles having long/short axial ratios of 2, 2.7 and 3.5. Therefore, four subpopulations have been considered. Initial isotropic distributions of the short axes of the biotites and of the long and intermediate axes of the feldspars were chosen.

(ii) The input flow parameters deduced from the observed fabrics, are summarised as follows. Mean vector v1, the normal to biotite (001), is almost vertical, does not change in orientation with increasing fabric intensity and symmetry variation, and its e1 associated eigenvalue is much greater than e2 and e3; mean v3 is mostly NE-SW and subhorizontal.

The corresponding flow is therefore modelled as a vertical flattening combined with simple shear (Fig. 9). The shear direction is close to NE-SW, the plane of simple shear is close to horizontal, and the principal elongation components of the coaxial compression are parallel to NE and NW in the horizontal plane. In such a model, the biotite fabric increases in strength with increasing degree of oblateness. Feldspar fabric confirms the subhorizontal biotite foliation by the subvertical orientation of its poles to (010) cleavages. It shows two weak and perpendicular maxima of long axes in NE-SW and NW-SE developed either in the same samples, and therefore coeval, or in separate samples.

Figure 10 shows some results of the modelling for three particular cases of the flow. In figures 10A and 10A', the following velocity gradient tensor was chosen and rotated in order to have the shear direction in NE-SW, and the shear plane slightly inclined with respect to horizontal:

$$\begin{pmatrix} 0.016 & 0 & 0 \\ 0.05 & -0.024 & 0 \\ 0 & 0 & 0.008 \end{pmatrix}$$

Therefore the compression axis of the coaxial component was almost vertical, and the elongation rate along NE was greater than along NW.

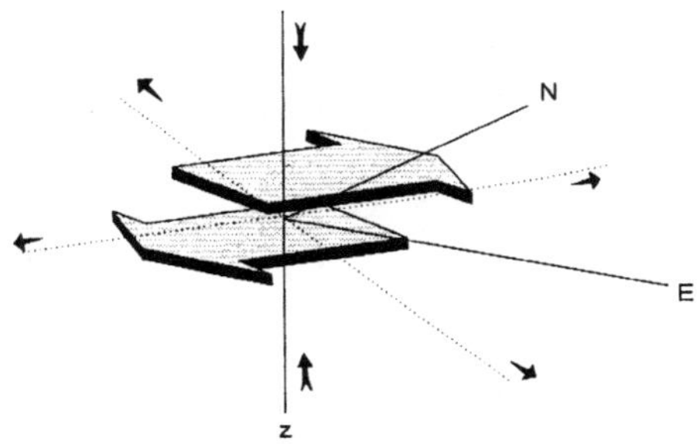

Figure 9. Geometry of the flow used in the numerical models. The large arrow indicates the simple shear component (horizontal shear plane; NE-directed shear direction). The vertical small arrows give the component of shortening of the superimposed coaxial strain, and the horizontal small arrows indicate the principal elongation directions of the coaxial strain component.

The multiparticle system is composed of four subpopulations, forming an initially isotropic distribution (Fig. 10A: t = 0). Three subpopulations represent feldspars with short (32 particles of axial ratios 1:1.5:2), intermediate (80 particles; 1:1.6:2.7) and long grains (32 particles; 1:1.7:3.5). The fourth subpopulation represents biotites (112 particles; 4:4:1). During flow, the poles of biotite form strong vertical maxima that are weakly elongate parallel to NW-SE, and the biotite-zone axis becomes oriented parallel to NE-SW. Feldspar initially forms a strong concentration parallel to the NE-SW lineation (Fig. 10A: t = 50 and t = 100); then the short and intermediate feldspars start changing their orientation towards NW-SE, the direction parallel to the rotation axis of simple shear (Fig. 10A: t = 150, t = 200), while the long feldspars remain mostly in NE-SW. Figure 10A' shows the orientations of the feldspar short axes during modelling. The modelled subhorizontal foliation of feldspar is in good agreement with biotite foliation except that the zone axis of the feldspar fabric pattern switches progressively from parallel to the shear direction (Fig. 10A': t = 50) to perpendicular to the shear direction (Fig. 10A': t = 200).

This example shows that, under a given velocity gradient, stable mutually orthogonal lineations may be created using both short and sufficiently long crystals. Based on the theory of Jezek et al. (1996), figure 11 illustrates how long a spheroidal particle must be to remain parallel to the simple shear direction (NE-SW), under the conditions of our model. For a given ratio of coaxial/simple shear rate of flow, a spheroidal particle having an axial ratio above the boundary line of figure 11 tends to the direction of the simple shear (NE-SW) direction, otherwise it will move towards the vorticity axis. i.e. perpendicular to the simple shear direction (NW-SE). Points S, I, L of figure 11 correspond to our short, intermediate, and long feldspars, if considered as spheroidal. In fact, the long feldspars (L) are slightly above to the boundary line, explaining why, in the model, they are mostly oriented parallel to NE-SW, with some parallel to NW-SE. The short (S) and intermediate (I) feldspars are below the boundary line and, therefore, rotate towards their stable NW-SE orientations.

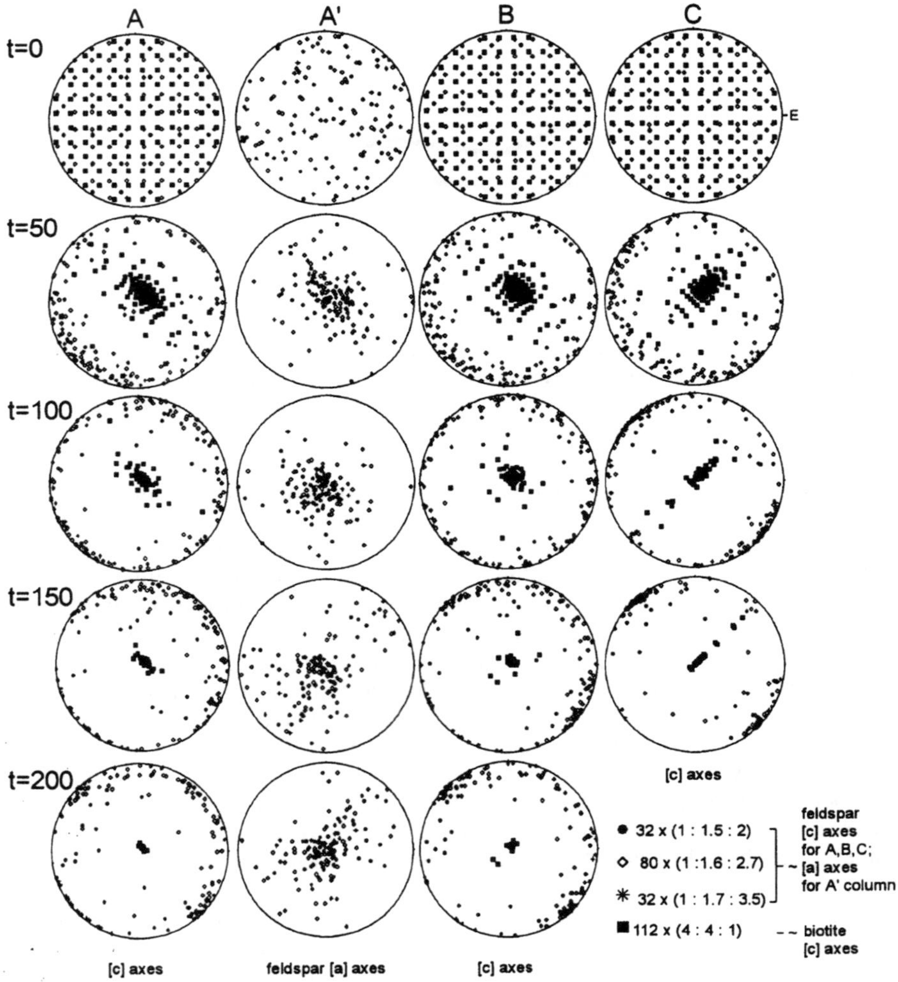

Figure 10. Numerical modelling of multiparticle systems simulating the development of biotite and feldspar fabrics during combined coaxial and simple shear flows. Equal-area projections of biotite and feldspar crystallographic [c] axes in A, B and C columns, and feldspar [a] axes in A'. A and A': elongation parallel to NE greater than parallel to NW. B: elongations parallel to NE equal to elongation parallel to NW. C: no elongation along NE.

Figure 10B simulates another flow reproducing the observed perpendicular lineations. The velocity gradient tensor was modified such that the elongation rate toward NE was equal to that toward NW:

$$\begin{pmatrix} 0.012 & 0 & 0 \\ 0.05 & -0.024 & 0 \\ 0 & 0 & 0.012 \end{pmatrix}$$

This simulates an axial flattening flow combined with simple shear. Jezek et al. (1996) showed that such a model cannot produce a stable orthogonal fabric since all the particles move towards the rotation axis of simple shear, i.e. NW-SE. Orthogonal fabrics may however appear temporarily in this model as already shown by Willis (1977). Depending on axial ratios and the number of particles in subpopulations, the orthogonal fabrics may be observed simultaneously or sequentially.

Finally, in the numerical modelling of figure 10C, the elongation parallel to NE-SW (the stretch direction of simple shear) was set to zero, and the elongation parallel to NW-SE (the rotation axis of simple shear) was set equal to the vertical compression:

$$\begin{pmatrix} 0 & 0 & 0 \\ 0.05 & -0.024 & 0 \\ 0 & 0 & -0.024 \end{pmatrix}$$

In this transpression model, as in the previous model, no stable orthogonal fabrics is obtained. However, in contradiction to the two previous models, the biotite-zone axis rapidly becomes oriented parallel to NW-SE.

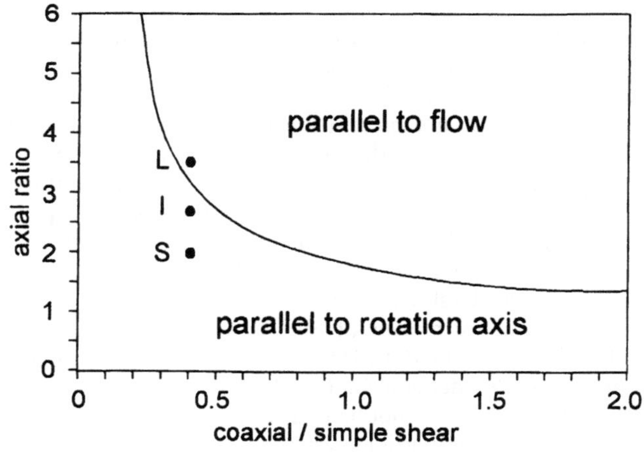

Figure 11. Stable subfabric orientations in transpressive flow as a function of the coaxial (press) / simple shear (trans) ratio and the particles axial ratio. The curve separates the area where the particles rotate parallel to the flow direction from the area where the particles rotate parallel to the axis of simple shear.

DISCUSSION

COMPARING NATURE WITH MODELS

The orientation data obtained in our study of the Soultz granite agree with structural studies in Schwarzwald-Odenwald and the Vosges crystallines, where the NE-elongation of plutonic bodies is associated with dominant NE-directed magma emplacement as suggested by several workers (e.g. Krohe 1991, Flottman and Oncken 1992). Our fabric data also show that the EPS-1 borehole is located somewhere in the central part of a possible laccolith where vertical shortening and NE-elongation is superimposed on simple shear with a NE-directed stretch, parallel to the elongation of the pluton.

However, the model of Jeffery, used to simulate observed fabrics, is based on a number of assumptions:

1) Rigid particles in a viscous fluid are considered. This assumes that the observed phenocrysts crystallised early in the magma, and that they behaved rigidly with respect to the surrounding liquid. This also assumes that the concentration of phenocrysts is sufficiently low to allow free rotation of crystals;

2) In the numerical modelling, we used a homogeneous steady flow which obviously cannot be valid for the whole profile. However, in view of the relatively consistent biotite data, this approximation is considered as valid;

3) The shapes of biotites and feldspars are approximated by ellipsoids. This widely used and simplifying assumption seems to work quite well, as shown in some papers (e.g. Willis 1977).

In the framework of the Jeffery's model and the above assumptions our numerical modelling shows that orthogonal lineations may, indeed, develop. Although having much higher fabric intensities, the modelled patterns agree well with observations. Several factors may explain the differences in intensities: i) the effect of interaction of particles may be responsible for a strong fabric intensity decrease (Ildefonse et al. 1992 a, b); ii) the shapes of particles may change during magma crystallisation, are far from ideal ellipsoids, and therefore may greatly influence the fabric intensity; iii) the biotites and feldspars may nucleate and grow at different times during a long time span during the cooling, hence these crystals are not present together in the flowing magma during its whole strain history; and iv) the initial distribution of crystals in the melt was possibly not random. We conclude that with respect to all the above assumptions of modelling and to the erratic character of our data, our demonstration is primarily qualitative.

The model of Jeffery is purely kinematic and the fabric intensity is controlled by time increments. The geologist may be interested in knowing the amount of finite strain corresponding to the modelled fabrics. This can be done by integrating the velocity gradient tensor. Finite strains for models of Fig. 10A and C, for times up to $t = 100$, fall in the range of X/Z ratios from 10 to 150. This corresponds to commonly measured strains in crustal materials (Pfiffner and Ramsay 1982). Iterations up to $t = 200$ correspond to very high strains which may not be realistic, but the calculation was performed to show the evolution of the multiparticle systems.

FABRIC CHANGES IN THE VERTICAL SECTION OF A MAGMATIC BODY

Our study highlights variations in crystal fabrics with depth along a 1 km-long profile of a granite body. It suggests that the viscous flow regime may vary with depth. It is observed that the magmatic foliation defined by biotite and feldspar SPOs remains stable throughout the profile. The lineation direction, defined as the zone axis of the biotites (eigenvalue e1 of biotite SPO), is also observed to be stable with the exception of two samples. However, the linear fabric determined by feldspar basal planes exhibits noticeable variations. These variations tend to correlate with the intensity and symmetry of the biotite fabric: for a weak biotite SPO and rather prolate biotite fabric ellipsoids, both the biotite and feldspar lineations are subparallel and trend NE-S. In contrast, for a stronger and more oblate biotite fabric, the feldspars show both NE-SW and NW-SE trending lineations. The latter planar and oblate fabrics predominate in the upper and lowermost parts of the profile, while in its central part (1700m-1850m), plane strain fabric dominates with NE-SW trending lineations defined by both minerals.

These fabric changes with depth call for two contrasting types of explanation:

1) In the central part of the pluton magma flow exhibits combined pure shear and simple shear with the principal elongation parallel to the simple shear direction, while in the upper and lower parts it is governed by non-coaxial flattening. Such a flow distribution could be interpreted in terms of partitioning into laminae having different flow regimes. This would imply the existence of mechanical incompatibilities between adjacent laminae. Arguments supporting different flow regimes could be looked for in the quartz [c]-axes fabrics which should reflect the symmetry of the imposed strain (Lister 1981), at least during the very latest stages of magma flow. For example, if the [c]-axis fabric of sample 2385b (Fig. 6) is accepted to belong to the small circle distribution type around the vertical (Z), it may attest to non-coaxial flattening during the late increments of deformation of the magma. In contrast, the fabrics of samples 3464a and 4215c could be seen as governed by dominant simple shear. Our poor quartz fabric data, however, based on a few measurements, should be interpreted with caution. The possible changes in the flow geometry could be related to successive supplies of magma during feeding, at different levels of the pluton having different degrees of crystallisation, hence different flow regimes.

2) Another explanation of the observed fabrics throughout the profile can be considered as a reflection of different stages of fabric development under a constant flow regime. This interpretation is also consistent with our numerical modelling in that, with increasing increments, changes in symmetry of particle orientation fabrics are observed. This would imply that at the different levels, the magma would undergo different amounts of total strain. In fact, at the scale of a pluton it is conceivable that magma flow may be highly heterogeneous.

Finally, to apply this approach correctly to a pluton and to discuss further the significance of linear fabric variations, a one-dimensional section, as in the present study is insufficient since the displacement of particles in a heterogeneously flowing magma is basically three-dimensional.

DEDICATION

This paper is dedicated to Radek Melka, our colleague and friend, who died two years ago at the age of thirty years. This work as many others started in close cooperation with this brilliant structural geologist. His bright ideas, systematic work and friendship are missed by all of us.

ACKNOWLEDGEMENTS

We are indebted to Albert Genter for his help in sampling the Soultz granite. This research was partly financed by French Bureau de Recherches Géologiques et Minières, and we are grateful to Philipe Elsasse and Zdenek Johan (BRGM), for their support. We thank Christophe Maurin (University of Strasbourg) for providing the geological map of the Rhine graben, and Vladimir Tolar for help in figure preparation. Ztìpánka Táborská (University Praha) helped in measuring some samples. The theoretical part of research was financed by the Czech Grant Agency, grant No. 30689 to Karel Schulmann.

RIGID PARTICLES IN SIMPLE SHEAR FLOW : IS THEIR PREFERRED ORIENTATION PERIODIC OR STEADY-STATE ?

Benoît ILDEFONSE
Laboratoire de Tectonophysique, CNRS URA 1764, ISTEEM-Université Montpellier II, 34095 Montpellier cedex 05, France
Laurent ARBARET and Hervé DIOT
Magmas et Volcans, CNRS URA 10, 5 rue Kessler, 63038 Clermont-Ferrand cedex, France

ABSTRACT

The theory of the rotation of isolated rigid particles within a linearly viscous fluid deforming in progressive simple shear is often invoked in models of Shape Preferred Orientations (SPO) of crystals in igneous rocks. A classical result of the theoretical model is that the SPO should rotate and pulsate with increasing strain, with a periodicity equal to that of the rotation of an individual particle. However, the initial theoretical model makes a large number of assumptions, many of which are unlikely to be satisfied by actual crystalline suspensions in magmatic melts. The purpose of this note is to review three of the reasons why periodicity of rigid particle SPO may not really be expected in igneous rocks: (i) SPO in igneous rocks are generally defined by suspensions of crystals that are concentrated enough to allow mechanical interactions between the crystals; (ii) the porphyroblast/matrix interface may not always be coherent; and (iii) the aspect ratios of the crystals defining the SPO are unlikely to be unique and constant, as assumed in the model. The last point is discussed on the basis of some 2D simple calculations of the development of SPO defined by different types of heterogeneous populations of particles. The combined effects of these deviations from the standard model point to two fundamental conclusions: (i) there is no simple relationship between fabric and finite strain, and (ii) the SPO in magmatic rocks may be considered as good markers of the flow, whatever the significance of the inferred flow pattern in terms of geodynamics and/or emplacement processes.

J.L. Bouchez et al. (eds.) Granite: From Segregation of Melt to Emplacement Fabrics, 177–185.

INTRODUCTION

The use of fabrics as strain and/or kinematic markers in igneous rocks has now become a popular technique for the structural analysis of magmatic intrusions. Shape preferred orientations (SPO) are obtained in rocks either from direct field measurements (e.g. Fernandez, 1988; Hippertt, 1994; Nicolas and Boudier, 1995) or from measurements and image analysis of sections/thin sections (e.g. Allard and Benn, 1989; Launeau et al., 1990; Philpotts and Asher, 1994). On another hand, the Anisotropy of Magnetic Susceptibility (AMS) technique is widely used for fabric measurements and has proved to be quite powerful when mapping fabrics otherwise difficult to recognize in the field (e.g. Guillet et al., 1983; Cogné and Perroud, 1988; Bouchez and Diot, 1990; Gleizes et al., 1993; Cruden and Launeau, 1994; Bouchez and Gleizes, 1995; Raposo and Ernesto, 1995).

It is widely recognized that fabrics in igneous rocks, when related to magmatic flow, are shape preferred orientations of rigid particles suspended in a viscous fluid (e.g. Blumenfeld and Bouchez, 1988; Paterson et al., 1989; Fernandez and Laporte, 1991; Nicolas, 1992). The theory of rotation of isolated rigid particles within a Newtonian fluid (Jeffery, 1922; Fernandez, 1987; Passchier, 1987; Jezek, 1994; Masuda et al., 1995) is invoked in models of SPO of crystals in rocks. A classical result of this model is that, in simple shear flow, the SPO should rotate and pulsate with increasing strain (Fernandez, 1987; Fernandez and Laporte, 1991; Ildefonse et al., 1992a; Jezek et al., 1994). This periodic behaviour contrasts with the steady-state orientation and progressively increasing intensity of the fabric obtained in a coaxial flow (Ildefonse et al., 1992a; Jezek et al., 1994; Masuda et al., 1995). In intermediate non-coaxial flows, the fabric develops towards a stable orientation or otherwise, depending on the ratio between the particle aspect ratio and the degree of non-coaxiality of the flow (Passchier, 1987; Jezek et al., 1994; Masuda et al., 1995).

Following this simple model, flow kinematics inferred from rigid particle SPO in igneous rocks may be highly questionable when the latter is periodic, i.e. when the flow is non-coaxial. However, the initial theoretical model makes a large number of assumptions, many of which are unlikely to be satisfied by actual crystalline suspensions in magmatic melts. The purpose of the present paper is to review some of the natural causes for a different rotational behaviour of rigid particle SPO in igneous rocks.

MECHANICAL INTERACTIONS

The most obvious modification of the rotational behaviour of the SPO is probably that due to mechanical interactions between the crystals of a magmatic suspension. The effect of interactions between rigid particles has been described experimentally (Ildefonse and Fernandez, 1988; Ildefonse et al., 1992a; Ildefonse et al., 1992b). It was shown that in concentrated suspensions, interactions are responsible for slowing down or blocking the rotation of neighbour particles, resulting in 1) the decrease in intensity of the SPO in both coaxial and non-coaxial flows, and 2) the break-down of the periodicity of the SPO development in simple shear flow. The latter is particularly important as the fabric then tends to align with the shear plane, resembling a passive fabric behaviour. This is best illustrated by figure 1, showing the difference in sub-fabric orientation between the disturbed and undisturbed particles of the same population (aspect ratio 2.5) in a given experiment (Ildefonse et al., 1992a).

More recent 2D experiments (Arbaret et al., 1996) have shown that the stabilising effect of interactions on the orientation of the fabric is still observed at lower concentrations of particles (as low as 13%).

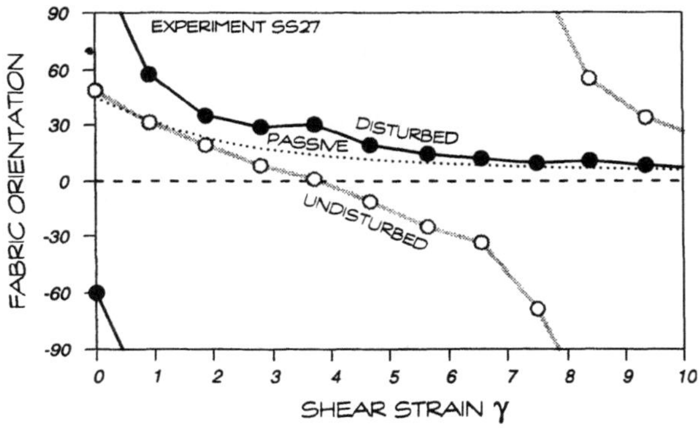

Figure 1. Experimental test of the effect of mechanical interactions between rigid particles on the development of shape preferred orientations (after Ildefonse et al., 1992a). Particle aspect ratio is 2.5; the fabric orientation is given with respect to the shear plane. Particles are considered as undisturbed if they were able to rotate more than 90 degrees during the experiment (if no interaction occurs, a particle with aspect ratio 2.5 should have rotated 180 degrees after a shear strain of 9.11). The results are compared with the evolution of a passive fabric.

SLIP AT THE PARTICLE/MATRIX INTERFACE

Another important parameter when considering the rotation of rigid particles in a viscous medium is the cohesiveness of the particle/matrix interface. Following Jeffery's model, the rigid particle suspensions in rocks have always been considered as a continuous medium, with a perfectly cohesive particle/matrix interface. However, it was argued in a previous paper (Ildefonse and Mancktelow, 1993) that this interface may be discontinuous and slipping in several geological situations, including the melt-lubricated porphyroblast boundaries in a partially molten rock or a crystallising magmatic mush. The effect of slip at the rigid particle/matrix interface has been studied experimentally using analogue paraffin wax models (Ildefonse and Mancktelow, 1993). In spite of the low maximum shear strain ($\gamma \approx 3$) that could be obtained in the simple shear apparatus, the result showed a marked tendency for the particles to rotate toward a stable orientation close to the shear plane, similar to passive markers. Compared to the case with a cohesive interface, the particle rotation becomes slower when approaching the shear plane; rotation may even be reverse (Fig. 2) for particles initially oriented between 0° and -30° (Ildefonse and Mancktelow, 1993). The corresponding SPO will thus be stronger at a given shear strain, and it will tend to acquire a stable orientation, approximately parallel to the shear plane.

VARIABLE CRYSTAL ASPECT RATIOS

So far, most theoretical and experimental studies of the development of rigid particle SPO considered the case of populations of particles with a unique aspect ratio. However, this obviously does not apply to natural situations, where crystals in a suspension will usually display variable aspect ratios. Heterogeneous populations have been considered by Fernandez et al. (1983) and Fernandez and Laporte (1991), but they only discuss the resulting SPO in non-coaxial flow in terms of symmetry and 3D shape of the fabric. The effect of having variable particle aspect ratios on the orientation of the global fabric as a function of strain was not discussed. We address this problem, on the basis of simple 2D calculations of SPO developed in simple shear flow with various types of heterogeneous populations (theoretical or real) of non-interacting particles.

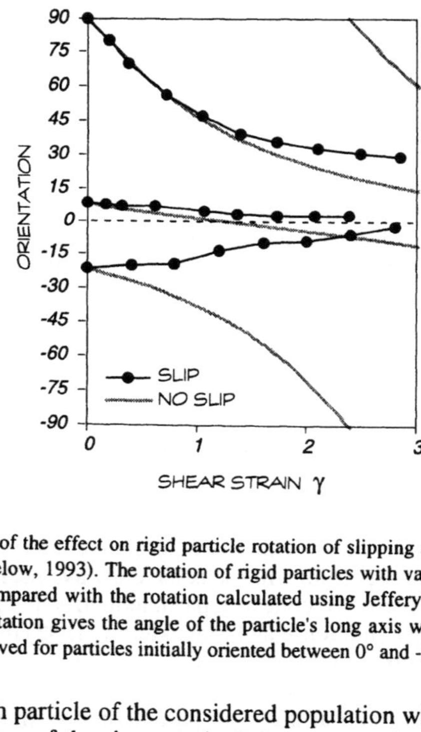

Figure 2. Experimental test of the effect on rigid particle rotation of slipping at the particle/matrix interface (after Ildefonse and Mancktelow, 1993). The rotation of rigid particles with various orientations and slipping interface (black lines) is compared with the rotation calculated using Jeffery's model (grey lines). Particle aspect ratio is 2.5; the orientation gives the angle of the particle's long axis with respect to the shear plane. Reverse rotations were observed for particles initially oriented between 0° and -30°.

The angle θ' of each particle of the considered population with vertical is calculated, at any time, as a function of the shear strain γ, its aspect ratio n and its initial angle θ, using the equation given in Fernandez et al. (1983) :

$$\frac{\sqrt{1-\chi^2}}{2}\gamma + arctg\left(\frac{\sqrt{1-\chi}}{\sqrt{1+\chi}} tg\theta\right) = arctg\left(\frac{\sqrt{1-\chi}}{\sqrt{1+\chi}} tg\theta'\right)$$

$$\text{with } \chi = \frac{n^2-1}{n^2+1} \text{ (shape parameter).}$$

At each calculated increment of deformation, the fabric intensity D and its orientation α are given respectively by the eigenvalues ratio and the first eigenvector of the orientation tensor (Harvey and Laxton, 1980).

BIMODAL POPULATIONS

We first calculated the evolution of a fabric developed within a mixture of two populations of particles with aspect ratios 6/1 and 6/5. The low aspect ratio particles have a period of rotation (which is a function of the particle aspect ratio; Fernandez et al., 1983) approximately three times shorter than the high aspect ratio particles. Calculations were performed for various proportions of the two sub-populations (Fig. 3).

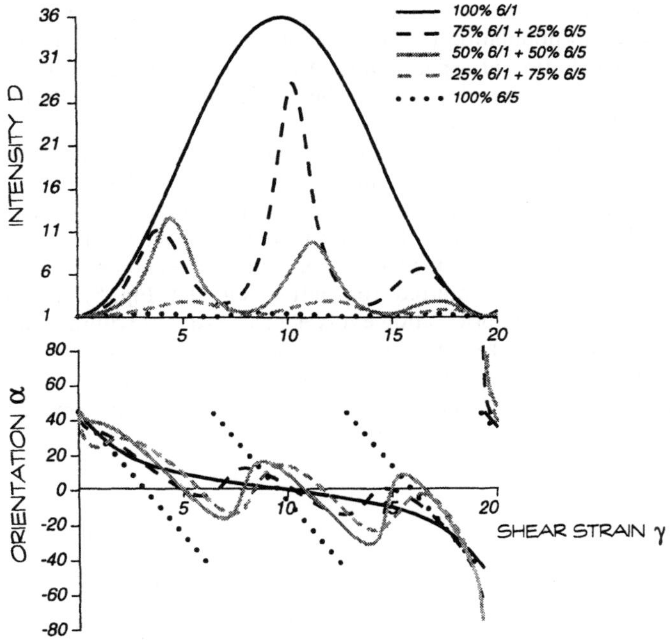

Figure 3. Evolution of SPO with various mixtures of two homogeneous populations of non-interacting particles in a simple shear flow (aspect ratios 6/5 and 6/1). The fabric intensity D and its orientation α are obtained respectively from the eigenvalues ratio and the first eigenvector of the orientation tensor (Harvey and Laxton, 1980).

The fabrics developed within the mixtures are significantly different from those calculated for the two end-member homogeneous situations. Within the γ range corresponding to the period of rotation of the high aspect ratio particle ($\gamma = 19.37$), the fabrics evolve following an intermediate and more complicated pattern than the single cyclic rotation calculated for homogeneous populations; the fabric intensity presents 3 different maxima. The most remarkable point is the reverse rotation of the fabric, which stays within a range of $\alpha=\pm20°$ around the shear plane approximately from $\gamma = 4$ to $\gamma = 12$. It is also note worthy that the mixed fabric momentarily disappears (D=1) when passing through the orientation 90° (perpendicular to the shear plane), instead of ±45° in end-member situations, when reaching the critical shear strain for the high aspect ratio particles.

GAUSSIAN POPULATIONS

In the preceding example, the calculation was made with a mixture of two distinct populations with a given aspect ratio. Practically, such a mixture may easily be divided in two sub-populations that can be analysed separately. Fernandez et al. (1983) have shown how one can use this situation to find the sense of shear. In nature, however, each determined sub-population (for instance each mineral phase) usually presents a heterogeneous distribution of aspect ratios. As a first attempt to estimate the effect of this heterogeneity on the development of the fabric, the calculation was made with a Gaussian distribution of aspect ratios.

Populations of 1080 particles have been chosen, in order to maintain an isotropic distribution of the orientations in the initial stage (6 particles/degree). The mean aspect ratio was 2.5 in every population, the standard deviation ranged from 0 to 0.5 (Fig. 4).

Results of the calculation show that with increasing standard deviation (i.e. increasing departure from the ideal homogeneous population) the curves slightly deviate from the classical cyclic behaviour inferred from Jeffery's model (Fig. 5); the fabric intensity D varies less and the fabric orientation evolves progressively within a shorter range of orientations around the shear plane. Again, one can note that when defined by a heterogeneous population of particles, the fabric undergoes a reverse rotation instead of disappearing at -45° and instantaneously reappearing at 45°. However, the deviation from the homogeneous case remains weak.

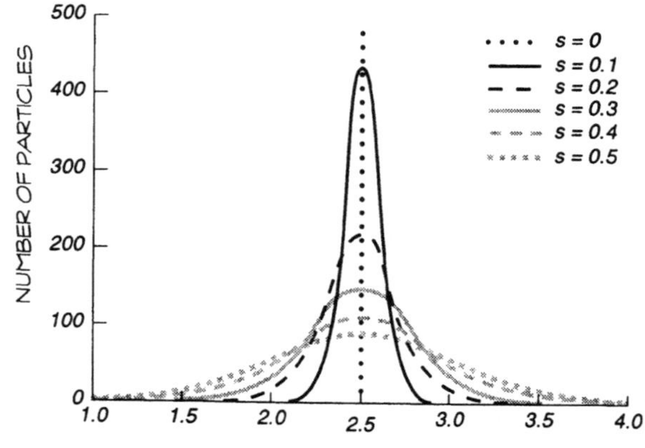

Figure 4. Normalized Gauss distribution representing the variation of particle aspect ratio in the chosen populations. s = standard deviation; the mean value is 2.5.

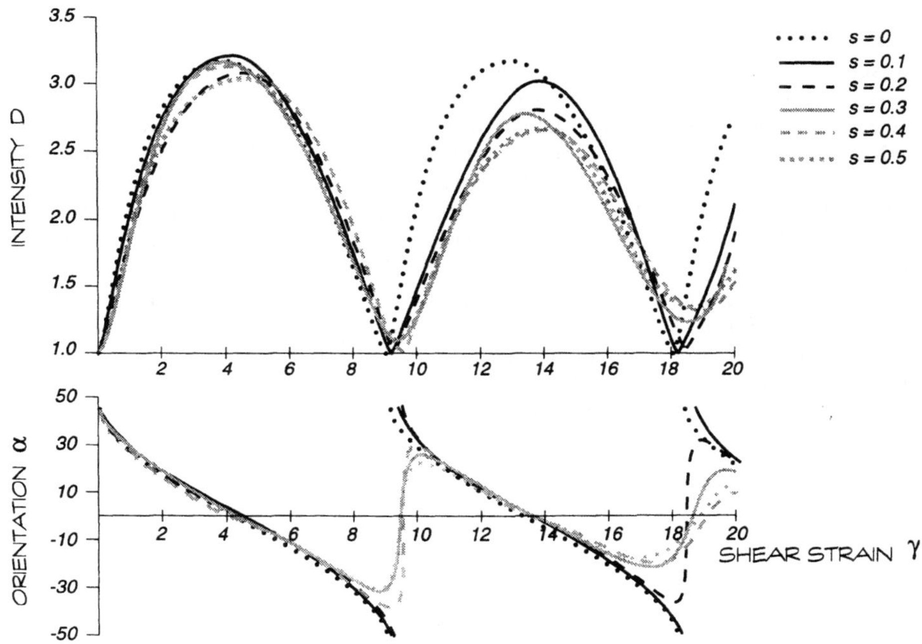

Figure 5. Evolution of SPO with heterogeneous populations of non-interacting particles in a simple shear flow. The particle aspect ratios vary following a Gaussian distribution (Fig. 4). The curve s = 0 corresponds to the end-member homogeneous distribution (aspect ratio 2.5). D and α are the same as in Figure 3.

NATURAL POPULATIONS

In order to have a better idea of the effect of the heterogeneity of particles aspect ratio on the development of the shape preferred orientation, the same calculation as in the previous examples has been performed with distributions of aspect ratios measured on natural populations of biotites and amphiboles (Fig. 6).

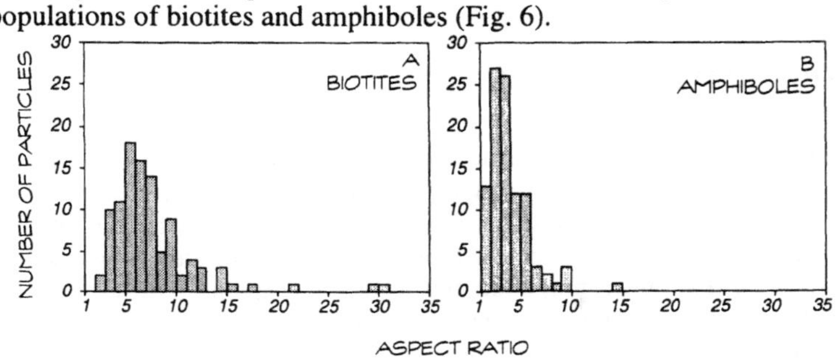

Figure 6. Distributions of the measured aspect ratios of 100 biotites (A) and 100 amphiboles (B) used as models of heterogeneous populations in the calculations.

These populations have been taken from 3D analogue experiments (Arbaret, this volume). The main characteristic of such natural populations is the heterogeneity of the measured shape, as shown by the histograms of the aspect ratios of the minerals (Fig. 6). Note that the distributions grossly display a Gaussian shape, with high standard deviations, but also show an important asymmetry with respect to the mean value.

Results are shown in comparison with the curves obtained for homogeneous populations with an aspect ratio equal to the mean value of the natural populations (Figs. 7 and 8). The discrepancy between the two situations is spectacular, in both cases (biotites and amphiboles). As soon as the fabric approaches the shear plane, the cyclic rotation inferred from Jeffery's theory totally disappears; the fabric remains sub-parallel to the shear plane, with a shorter range of orientations for the elongated particles (biotites, ± 5°; Fig. 7) than for the shorter particles (amphiboles, ± 10°; Fig. 8). The intensity of the fabric rapidly decreases and remains around the same value as strain increases. Again, the best stability is achieved with the particles having the higher aspect ratio (biotites, Fig. 7); fluctuations of the fabric intensity are more pronounced in the amphibole case (Fig. 8).

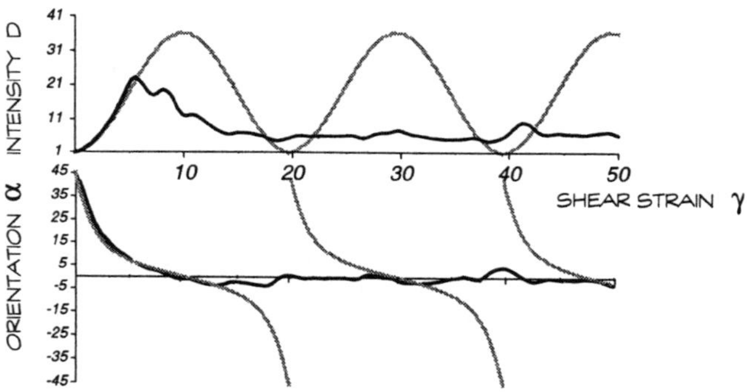

Figure 7. Calculated evolution of SPO with a heterogeneous population of non-interacting biotites (see Fig. 6a) in simple shear flow. The grey curve shows the evolution of a homogeneous population of particles with the mean aspect ratio (6.14), for comparison. The evolution is displayed for shear strain up to 50, in order to illustrate the quasi-steady-state nature of the SPO. D and α are the same as in Figure 3.

DISCUSSION AND CONCLUSION

The calculations presented herein demonstrate the influence of the variation in particle aspect ratio on the development of shape preferred orientation in a simple shear flow. The fabric tends to be more-or-less parallel to the shear plane while its intensity is decreased. This is due to the fact that any elongated particle rotates slower when parallel to the shear plane than when perpendicular to the shear plane. Furthermore the longer the particle the longer the time of residence around the shear plane. Consequently, when combining a number of different periods of rotation, corresponding to the different particles, there is always a larger number of particles more-or-less aligned with the shear plane. The only case where this effect is not very important is that of the Gaussian distribution, because of the symmetry of the aspect ratio distribution around the mean value 2.5, the slower rotation of the longer particles being then balanced by the faster rotation of an equal amount of the shorter particles.

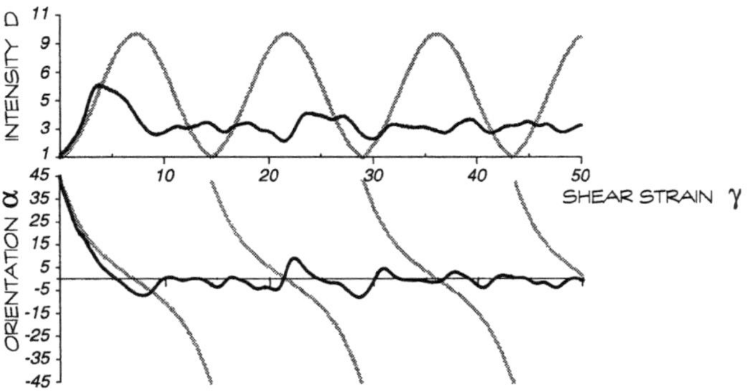

Figure 8. Calculated evolution of SPO with a heterogeneous population of non-interacting amphiboles (Fig. 6b) in simple shear flow. The grey curve shows the evolution of a homogeneous population of particles with the mean aspect ratio (3.1) for comparison. The evolution is displayed for shear strain up to 50, in order to illustrate the quasi-steady-state nature of the SPO. D and α are the same as in Figure 3.

The heterogeneous distribution of the particles aspect ratio has an effect on the development of SPO similar to the mechanical interactions and to a slipping particle/matrix interface, as reviewed above. Many deviations from the Jeffery's model may occur in rocks, and the present review is far from exhaustive. For example, the continuous input of new particles into the system, which is probably the normal situation in a crystallising magma, will have exactly the same effect as the variability of the particle aspect ratios (P.Y. Robin and P. Launeau, personal communication; Launeau et al., 1995). The triaxial shape of crystals, instead of the axisymmetrical shape assumed in Jeffery's model, should also be taken into account (Freeman, 1985; Jezek et al., 1994).

The various processes, suggested by the observation of rocks, and that modify the rotational behaviour of rigid particles and/or of corresponding fabrics, point to the inadequacy of the Jeffery's model to account for natural SPO. The major observed effect, clearly illustrated by Figures 7 and 8, is a break-down of the assumed pulsating fabric evolution: instead of undergoing its predicted cyclic development, the fabric tends to be a nearly steady-state feature. This implies two fundamental consequences :

1- there is no simple relationship between the fabric (both intensity and orientation) and finite strain,

2- the SPO in magmatic rocks may be considered as indicating the local flow direction, without any preconceived idea of the significance of the inferred flow pattern in terms of geodynamics and/or emplacement processes (see other contributions in this volume).

ACKNOWLEDGEMENTS

This paper has benefited from discussions with numerous persons, especially Patrick Launeau and Pierre-Yves Robin. Many thanks to Win Means and Jean Luc Bouchez for their careful reviews.

PART III

EMPLACEMENT OF GRANITE PLUTONS:

CASE STUDIES

SYNTECTONIC GRANITES AND THE PRINCIPLE OF EFFECTIVE STRESS: A GENERAL SOLUTION TO THE SPACE PROBLEM?

Donald H.W. HUTTON
Department of Geological Sciences
University of Durham
Durham DH1 3LE
United Kingdom

ABSTRACT

Syntectonic granite emplacement in dip slip and strike slip contractional shear zones is now well documented by a number of case histories including the spectacular 1200 km long, 20 km wide Great Tonalite Sill of North America. These examples show fundamentally that magma driving forces can overcome compressional tectonic stresses and suggest that in a contractional orogen it is not in general a necessity to make space for plutons by localised dilation along faults and shear zones. There are a number of other magma driving forces that are available, in addition to buoyancy, which can combine to exceed the tectonic compression, including one which is derived from the compression itself. These extra forces are optimised when ascent and emplacement is achieved along major transcrustal faults and lineaments. The principle of effective stress is applied to the general case of granitic magmas in crust undergoing tectonic horizontal compressive stress and it is argued that the magma pressure is an indistinguishable part of the regional (effective) stress field. This allows a general solution to the space problem in granite emplacement since in the lower crust, where the host rocks are weak, the emplacement strain will also become an indistinguishable part of the regional deformation field. This principle is also likely to underlie space creation mechanisms in the often highly heterogeneous middle crust, where it may be obscured by the observed opportunistic exploitation of dilation along tectonic structures and other weaknesses.

INTRODUCTION

During the last 15 years individual solutions to the problem of the accomodation of large granitic bodies in the Earth's crust ("the space problem") have been greatly extended by the realisation that emplacement space can be created in association with major tectonic faults and shear zones. This approach, which arose from the application of structural, tectonic and magnetic methodologies to deformed plutons, has now yielded a large number of studies and case histories for plutons of varying age from all over the world. A common feature of many of these studies is the demonstration,

J.L. Bouchez et al. (eds.) Granite: From Segregation of Melt to Emplacement Fabrics, 189–197.

through field, microscopic and laboratory measurements, of a clear connection between emplacing (i.e., non-fully uncrystallised) magma and active major tectonic structures, this being usually achieved through the examination of: a) the relative timing of deformation and crystallisation stages in the granites; b) the structural kinematics, strain types and strain gradients present and; c) their relationship to pluton geometry and external faults/shear zones. Despite the reservations of some authors (Paterson and Fowler, 1993; Tobisch et al., 1995), the ease with which such solutions fit the observed data is reflected in the most recent literature by the increasingly numerous publications on the emplacement of plutons in such settings (e.g., Guineberteau et al., 1987; Lagarde et al., 1990; McCaffrey, 1992; Archanjo et al., 1994; Grocott et al., 1994; Hutton and Miller, 1994; Jacques and Reavy, 1994; Speer et al., 1994; Fritz, 1995; Grocott and Taylor, 1995; Kohut and Janak, 1995; Musumeci and Pertusati, 1995; Rosenberg et al., 1995; Royse and Tarney, 1995; Schulmann and Venera, 1995; Verhaeren et al., 1995; Wilson, 1995; Aranguren et al., 1996, Ferré et al., 1996; Koukoukevlas et al., 1996; Mattern, 1996; Saint Blanquat and Tikoff, this volume; Jacques et al., 1997)

The vast majority of the models published on the emplacement mechanisms of these "fault-zone" granites envisage space being created in tensional or potentially dilational sites in, or adjacent to faults and shear zones e.g. at releasing bends, at jogs, between step-overs and between bifrucations in transcurrent structures, and in ramp-flat bends in extensional structures. Granitic sheeting which may be the key to pluton construction in such settings (Hutton 1992), offers a fast enough intrusion rate to allow magmatism to keep pace with the rates of movement on major tectonic structures (Paterson and Tobish, 1992; Hanson and Glazner, 1995), as well as providing a viable and efficient alternative to magmatic ascent by diapirism (Hutton, 1992; Clemens and Mawer, 1992; Petford et al., 1993). This has led to a general enthusiasm that magmas will always be able to find such dilating zones in the continental orogenic crust and that, when combined with the other, more well established emplacement mechanisms (cauldrons, stoping, doming and diapirism), the space problem, if not generally solved, is at least very solvable (see Hutton, 1988; Karlstrom, 1991; Brown, 1994). However, the extension of this space solving mechanism from the particular to the general may be flawed since it avoids the fact that the vast majority of the world's granites are not generated in transtensional or extensional tectonic regimes (although some, clearly are), but rather are formed, ascend and emplace in convergent margins and convergent zones where the stresses are fundamentally horizontally compressive, the strains contractional, and where the crust is shortened and thickened. Therefore it is argued here, that to *generally* solve the space problem we must understand how magmas can mobilise, ascend and emplace in orogenic crust undergoing contraction. This does not mean that emplacement in extensional/transtensional orogens does not occur, nor does it mean that emplacement in extensional/transtensional zones in otherwise contractional orogens does not happen either, but rather that knowledge of this type of emplacement may not directly address the main problem.

For many workers (e.g., Glazner, 1991) it would appear that emplacement of granite into active contractional structures such as thrusts and transpression zones is an unlikely scenario since it would seem intuitive that the tectonic compressive stress should keep the crust in contraction, that its magnitude would generally exceed the magmatic driving force and would thus prevent emplacement occurring. The essential objection is that compressional stress or contractional strain is *space denying*. Therefore a good place to begin this discussion is to draw attention to a number of documented field examples of granite emplacement in active contractional structures. If this is accepted then there can be no *a priori* reason to doubt that granitic ascent can occur in contractional tectonic regimes. The discussion then moves on to the likely physical mechanisms involved in driving magmas, particularly in crust under active horizontal compression. Finally the principle of effective stress is applied to syntectonic magmatism leading to what may be a general solution to the space problem.

EMPLACEMENT IN CONTRACTIONAL STRUCTURES

Until just a few years ago descriptions of plutons emplaced along synchronous dip-slip thrusts and high-angle reverse structures and strike-slip transpressional faults and shear zones were very rare. Blumenfeld and Bouchez (1988) showed that magmatic state fabrics in a granite sheet emplaced along a thrust were concordant with the contacts, intensified towards them, were coeval and continuous with similar fabrics in the immediately adjacent migmatitic country rocks and, perhaps most importantly, contained abundant evidence (magmatic tiling structures etc...) for thrust sense movements as did the coeval and concordant fabrics in the adjacent migmatites. The Great Tonalite Sill of Southeast Alaska and British Columbia (Brew and Ford, 1981), remains the most spectacular world example for this proposed emplacement setting (Hutton and Ingram, 1992; Ingram and Hutton, 1994; Davidson et al., 1992; Erdmer and Mortensen, 1993). This steeply inclined concordant series of sheets is generally less than 25 km wide but has now been traced out along the Coastal Cordillera, between the southwest Yukon and British Columbia for over 1200 km. It is emplaced within a dip-slip high angle reverse shear zone of similar dimensions and shares the steeply inclined fabrics of this zone, with steeply plunging stretching lineations. Although intensely deformed in the high tempreature solid state, particularly around its margins, moderately developed magmatic state fabrics are preserved in its central portions, and these show kinematic evidence, throughout the length of the sill, of horizontal contraction, and inferred horizontal maximum compressional stress, at this early stage in the crystallisation history. Ingram and Hutton (1994) argued that the sill was emplaced along the tectonic boundary of two colliding superterranes (i.e., geological units of considerable, sub-plate size). So, that if one accepts the field evidence then we must believe that magmatic forces can overcome the orogenic contractional forces, i.e., they can overcome the high normal stress resolved at the steeply inclined collisional interface of two large-scale colliding masses.

More recent papers in a variety of plutons of different age have reported similar field relationships between thrust zones, concordant axially emplaced granites, and dip-slip thrust sense kinematics from their magmatic state fabrics (McCaffrey et al., 1991; Druguet and Hutton, 1995; Howard et al., 1995; Kohut and Janak, 1995; Rosenberg et al., 1995; Koukoukevlas et al., 1996; Spanner and Krohl, 1996) and there seems little doubt that magmatic emplacement along synchronous contractional dip-slip structures does occur, however counter-intuitive this may be.

In addition to contractional dip-slip structures emplacement can also occur in contactional strike slip faults and shear zones (D'Lemos et al., 1992, Tribe et al., 1996). It may also be that many of the earlier generations of models which invoked emplacement in tensional zones of strike slip faults could stand re-interpretation as emplacement into transpressional segments.

In conclusion emplacement of syntectonic granites into active contractional dip slip and strike slip faults and shear zones where there is no tendency for tectonic space to be created and where the tectonic stresses would appear to directly oppose the emplacement, does seem to occur. This implies that granitic magmas can have an internal magma pressure which is large enough to overcome the tectonic compressional stress, as well as the strength of the crustal rocks. In the following section a brief review is presented of the likely mechanisms, in contractional orogenic settings, which might combine to create this magma pressure. The implications of these enhanced magma pressures to the general space problem are then explored in the final section.

MAGMA DRIVING FORCES IN CONTRACTIONAL OROGENIC SETTINGS

DENSITY-CONTRAST DRIVEN BUOYANCY

Much has been written about this driving force, which appears to be applicable to all ascent and emplacement mechanisms. In its simplest form the force per unit area produced is the difference between the density of the melt and the average density of the vertical crustal section it is attempting to traverse multiplied by the crustal depth and the acceleration due to gravity ($\Delta\rho hg$).

Petford et al. (1993) and Petford (1996) have modelled granitic ascent in dykes and shown for normal granitic viscosities that rapid ascent rates are possible as long as the dykes achieve a minimum critical width below which heat lost from the magma by advection into the surrounding country rocks will cause the dyke to freeze. The models, which do take into account friction between magma and country rocks, but otherwise assume a smooth open fracture from source to surface, rely only on density contrast buoyancy as a driving force and yet produce extremely high ascent rates (cm/sec). If correct, the models would presumably provide ample overpressures to exceed even horizontal compressive differential stresses in the wall rocks. Notwithstanding this there may be other sources of magma pressure.

CONFINING PRESSURE

Any melt forming at depth in the Earth will acquire a hydrostatic pressure appropriate to that depth, being equivalent to the weight of a column of that melt reaching from the source up the Earth's surface. This pressure can however only be utilised if the magma in the melt region is connected to a lower pressure area into which it may then move. For the situation of the source connected directly to the earth's surface, the force per unit area acting upwards on the melt will be the melt density multiplied by the crustal depth and the acceleration due to gravity (ρhg): potentially significantly greater than the density-contrast buoyancy ($\Delta\rho hg$). In general terms this force will only become important for substantial ascent lengths if considerable vertical crustal connectivity exists, as might be expected for example along already existing diapir trails, or where deep reaching faults, intersections between steeply inclined major faults (see Jacques and Reavy, 1994), or other conduits can allow magmatic accession to occur. If such connectivity does exist then this will drive the melt/magma down ambient pressure gradients, either locally, and more generally upwards towards the Earth's surface.

VOLUME CHANGE ON MELTING

Under conditions of dry melting there will be a volume expansion. As currently understood such volume increases may be fairly small (between 2% and 20%), generating small to moderate overpressures (10-100 MPa) (Clemens and Mawer, 1992). This addition to the magma driving pressure will be a non-renewable component acquired in the source which, as with the confining pressure, will cause the melt/magma to move down pressure gradients. The volume expansion may be important in initiating fractures around the source (Petford, 1996) and reactivating pre-existing crustal structures to enhance connectivity.

TECTONIC DIFFERENTIAL STRESS

If a saturated porous material is loaded and the solid parts deform around the liquid in the pores, the pore fluids will acquire a proportion of the loading pressure. This is a well known effect in saturated soils when the pore fluids are prevented from draining during loading experiments, or when loading occurs rapidly relative to the drainage rate. Fluid pressure increases of this type can be caused by the imposition of both hydrostatic confining stresses, *and directed differential stresses,* the later producing the "Shear Enhanced Compaction" of soil mechanics. This has been discussed by Rutter and Neuman (1994) in the context of granite production in the source region when the solid restite deforms and compacts around the granitic melt pockets giving the melt an overpressure (in addition to buoyancy etc.) with which to separate and ascend from the source. Because restitic material, and crustal rocks generally at these crustal depths, are weak and unable to accumulate large stresses the overpressures acquired by the melts may be relatively small (about 10-20 MPa for combined hydrostatic and differential loading, E.Rutter pers. comm., 1996).

Overpressures acquired in this way may, however, be very important. Firstly, they provide an actual mechanism which allows a proportion of the tectonic compressional differential stress to be acquired by the granitic melt as an overpressure which can be used for ascent and emplacement. Secondly, because this overpressure is partly derived from the tectonic compressional stress, it reduces the effectiveness of this stress as a space denying factor. Thirdly, the compaction effect occurs because of low drainage rates or high loading rates and because the solid material is relatively compressible compared to the granitic melt. It follows that as long as these conditions continue to be met this type of overpressure can be acquired anywhere in the crust during ascent. In the source the acquired overpressure comes from the deformation of the compressible restite against the melt pockets. At higher levels in the crust such overpressures could be acquired because of melt zones around granitic bodies. However they could also, in principle, be acquired from ordinary crustal rocks undergoing normal plastic creep deformation mechanisms in a contractional orogen, should they be more compressible than the granitic magmas. Finally, while melt overpressures acquired in the weak lower crust may be correspondingly low, magmas present in the stronger, load bearing mid-crust can acquire proportionally larger overpressures to oppose the higher compressional stress found there. The more general implications of magma in crust undergoing active orogenic contraction are explored in the final section of this contribution.

THE IMPORTANCE OF CRUSTAL FAULTS AND LINEAMENTS IN MOBILISING ASCENT FORCES

The above review of magma driving forces shows that there are many other driving forces potentially available than just density-driven buoyancy alone. This, together with the fact that one of these is derived from the space denying tectonic compression itself, may account for the fact that magma pressure can exceed the tectonic compressional force. To fully utilise these additional driving forces requires vertical pressure gradients to exist in the crust. The simplest way to link melts to lower pressures in the mid or upper crust is by deep reaching faults and shear zones. The common occurence of granitic plutons along such faults and shear zones may thus be due not only to the space creating potential of the structures, and to the exploitable weakness that their fault rocks and anisotropy brings to the crust, but also to the fact that ascent along such structures optimises the available driving forces. We know that steep deep reaching faults exist in the lower crust because of their lineamental nature in higher crustal levels. In many of these lineaments mantle derived basic magmas are preferentially located: ample evidence of the deep reaching nature of these faults and the fact that magmas do ascend from the source regions along major fractures (see, most recently, Jacques and Reavy, 1994, and Hutton and Alsop, 1996). In conclusion trans-crustal fractures linking source to emplacement are: a) vital in optimising the available ascent pressures so that even compressional, "space denying " stress can be overcome, and b) actually inferrable in the field.

SYNTECTONIC MAGMATISM AND THE PRINCIPLE OF EFFECTIVE STRESS

In recent years a paradigm has developed within the granite community (see for example Karlstrom, 1991; Hutton, 1988; 1992) that "all granites are syntectonic". This phrase expresses a belief that not only does granitic magmatism invariably occur synchronously with deformation and tectonics, but that there is an actual detailed causal relationship between the two with structures and deformation intimately linked to granite ascent, granite emplacement and even to melt extraction and petrogenesis itself. The essential idea here is that granitic melts/magmas occupy the crust in a non-fully-crystallised state in the presence of a stress field related to orogenesis which includes a major component of the far-field plate tectonics. As argued in the introduction most of the world's granitic magmatism occurs in convergent orogenic belts and so the plate tectonic stress component will be fundamentally horizontally compressive. Therefore, in a simple model of this we are dealing with a deformable solid (the crust) containing bodies of fluid (uncrystallised magma/melt) all subjected to a general stress in which the maximum principal stress is horizontal. This type of problem is well known in soil mechanics. Porous soils with fluid in the voids, subjected to a compressional differential stress ("end load") may be a small scale analogue for syntectonic magmatism in a convergent orogen. For a saturated soil, end loading will, under certain circumstances, allow the fluid parts to acquire a proportion of the loading pressure. The application of this aspect of soil mechanics to granitic melts in the presence of a solid framework has been discussed already and illustrates that magmatic fluids can dynamically interact with tectonic stresses on a small scale. If such dynamic interaction occurs in the larger scale situation then it allows the principle of effective stress to be applied to the problem of syntectonic magmatism.

The most common way of expressing the principle of effective stress is:

$$\sigma' = \sigma - \mathbf{u}$$

where σ' is the effective principal stress, σ the total principal stress, and $\mathbf{u}$ the pore fluid pressure.

This law was developed empirically from soil experiments, and an actual physical, phenomenological, explanation of it (Terzaghi, 1936; Terzaghi and Peck, 1967) has been discussed by a number of authors (Caper and Cassie, 1976; Bell, 1983; Milligan and Houlsby, 1984; Britto and Gunn, 1987; Craig, 1992; Mitchell 1993, among others). It is generally held to mean that the total stress within a body of soil is carried partly by the fluid in the pores (the fluid pressure) and partly by the contact forces between the particles (the effective stress), and that the fluid pressure acts to reduce the total imposed stress. Terzaghi (1936) also stated "all measurable affects of a change in stress, such as compression, a distortion and a change of shearing resistance, are exclusively due to changes in effective stress", i.e., the deformation and failure of soil are related only to the interparticle forces (the effective stress) (see Milligan and Houlsby, 1984, p.14). Thus the deformation of the soil as a whole is produced by the effective stress only, the strain making no distinction between the stresses producing them, it "seeing" only their combination (the effective stress).

If this is applicable to syntectonic regional magmatism in orogenic belts undergoing horizontal compression and convergence, then it has profound implications for the granite space problem. Thus (1) *within a continental crust undergoing horizontal orogenic compression, the presence of the granitic magmas (the magma pressure) will be experienced as an indistinguishable part of the regional (effective) stress field.* Therefore, if the country rocks are weak, as for example is generally held to be the case in the lower crust (2) *the strains (of ascent and emplacement) associated with these magmatic volumes will be distributed regionally as well.* Thus the space problem will be regionally solved. In effect because there is only one stress then there is only one strain.

This principle may be applicable, with modifications, at higher levels in the crust (Fig.1). In the mid crust, still out of the direct space-creating influence of the Earth's free surface, we enter a complex zone where heterogeneities seem to be the rule. In this zone stress amplification may occur (Kusznir and Park, 1984), and this may be exacerbated by lithological and rheological variations to generate pronounced layer dependent weaknesses. This is also the part of the crust where vertical shear zones are common and, if old crust with inherited faults exists below, lineamental reactivation will occur (Hutton and Alsop, 1996). It seems likely that the magma pressure will also be distributed regionally in this zone as part of the regional (effective) stress field, although there will be much more heterogeneity in this field than before. The space problem, as shown by the many studies that have been undertaken on granites intruded at these depths, is solved in a wide variety of ways with, particularly, the opportunistic use of "spaces and cavities" (volumes of dilation, see Hutton, 1988) in and around the fault and shear zone systems. If, as proposed, the magma pressure is still distibuted regionally in this part of the crust then, despite the crustal heterogeneity and the resultant strongly localised nature of the stress pattern and accomodation strain, there should be a more widespread regional component to this accomodation. This would correspond to the "near field and far field mass transfer processes" of Paterson and Fowler (1993). Accomodation of a small but significant proportion of the emplacement strain of a pluton by regionally distributed deformation in the heterogenous mid crust is likely to be a difficult thing to demonstrate, and there are probably few areas in the world where there is sufficient geological and structural detail available to attempt this. One such area may be in Donegal, Ireland where, for example, the emplacement of the Ardara pluton is considered to be synchronous with a regionally developed deformation phase (Meneilly, 1982). Such synchroneity implies regional accomodation of at least a part of the accommodation space.

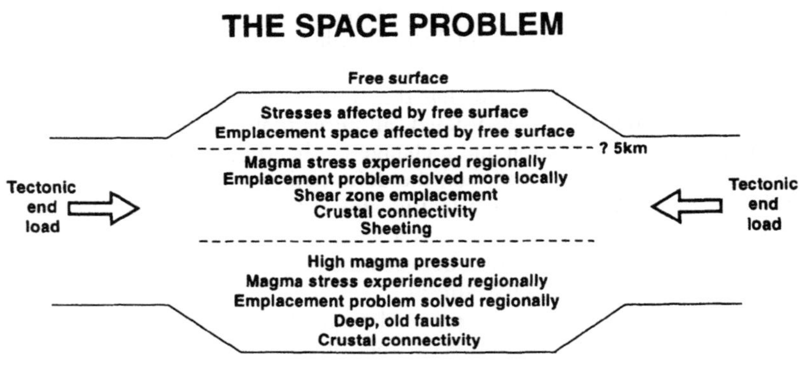

Figure 1. Simplified summary of the three realms of granite emplacement in a typical continental crust undergoing horizontal tectonic compressive stress. See text for details.

At still higher levels in the crust, in a sense the third realm of emplacement (Fig. 1), the dominant mode of space creation is by uplift and deformation of the free surface due to low overburden loading. This is allied with high viscosity and temperature gradients between magmas and country rocks leading to the familiar emplacement mechanisms of these levels (stoping, cauldrons, domes/balloons). Elevation of the free surface can also be easily achieved by the vertical inflation of sills (laccoliths) which have exploited weak sub-horizontal stratigraphic horizons. The spreading of magmas in this way may often obscure any fault which controlled ascent (Butler and Hutton, 1995; Roman-Berdiel et al., 1995), and indeed repeated laccolithic sill inflation at high crustal levels may offer a more realistic alternative to the traditional view that batholiths are huge, steep sided, trans-crustal bodies with all the inherent space problems implied (McCaffrey et al., 1997).

A key question is to what depth does the space creating affect of the free surface extend? This may vary depending on lithologies, structure and geological history. In the SW Highlands of Scotland the country rocks are mid to high greenschist facies metasediments of variable lithology and variable pre-granite structure, with the plutons controlled in their siting and much of their emplacement by steep transcurrent shear zones and steep cross cutting lineaments. Here, evidence of ring dykes is seen at emplacement depths down to 4 km, pluton inflation (ballooning) dominates down to 8 km, and below this (down to 10-12 km) granite sheeting along the major structures occurs (Jacques et al., 1996). This particular data set implies that the influence of the free surface is obvious down to approximately 4 km and perhaps more cryptic below this to depths of about 8 km.

CONCLUSIONS

The essence of the idea of syntectonic magmatism, given that most of the world's granites are created, ascend and are emplaced in convergent orogens, is that of uncrystallised granite magmas being present in orogens undergoing tectonic horizontal compressional differential stress. A small but growing number of world examples shows that granites can indeed easily overcome such compressional tectonic stresses. These relationships allow the principle of effective stress to be applied to contractional orogens containing syntectonic granitic magmas. The pressure of these magmas will interact with the applied tectonic compressional stress and become an indistinguishable part of the regional (effective) stress field. In the general case, and especially in the weak rocks of the lower crust, this will create only one general strain field, which will combine the regional shortening in the orogen and the emplacement/ascent space requirements of the granites. Thus the granite space problem, in the general case, is regionally solved.

ACKNOWLEDGEMENTS.

I am very grateful to Ernie Rutter and Michel de Saint Blanquat whose patient, thoughtful and careful reviews were instrumental in a considerable revision of the original manuscript of this paper. I would also like to thank Professor Peter Attewell of the University of Durham for encouraging me with some simple soil mechanics. Amicalement, also, to the senior editor, and his patience.

GRANITE PLUTON GEOMETRY AND EMPLACEMENT MODE INFERRED FROM COMBINED FABRIC AND GRAVITY DATA

Laurent AMÉGLIO [1,2], Jean Louis VIGNERESSE [2]
and Jean Luc BOUCHEZ [1]
[1] *LMTG, Equipe de Pétrophysique, UMR CNRS n° 5563, Université Paul-Sabatier, 38 rue des Trente-Six-Ponts, 31400 Toulouse, France.*
[2] *CREGU, B.P. 23, 54501 Vandœuvre-lès-Nancy, and ENS Géologie Nancy, France.*

ABSTRACT

This review is based on several years of joint structural and geophysical surveys on granites. Fabric measurements include lineations and foliations, the former being the more useful in understanding the magma strain pattern in a pluton. Through the inversion of gravity data, which is particularly sensitive to density contrasts, the shape at depth of the pluton, and depth of its floor, may be derived with good confidence. Vertical lineations, when associated with a deepening of the pluton's floor, are interpreted as feeder zones. Correlation between the inner-pluton lineation pattern and the regional deformation field helps in inferring the emplacement mode of the pluton. Combined geophysical and structural data therefore allow us to describe two main types of plutons. Flat-floored plutons are rather thin (a few kilometres), extend largely in every horizontal direction, and have several feeder zones. They are emplaced as sills in the upper crust, or within rather ductile environments in extensionnal tectonic contexts. These contrast with the thick (more than 10 kilometres) wedge-shaped plutons, extending largely in one direction along which a few root zones appear, which are V-shaped in transverse section, and correspond to the infilling of dilatant volumes of the brittle crust during transcurrent tectonics.

INTRODUCTION

Granitic intrusions constitute a major mode of continental crustal growth. They have been studied for centuries (e.g. Hutton, 1795) and a wealth of papers in petrology and geochemistry is devoted to them. They also serve as a reference, e.g. the Westerly granite, for the physical properties (such as seismic velocity and rheological properties) of the upper crust (Clark, 1966). As a major constituent of the basement, granites have also been widely investigated for their structure. However, their emplacement mode and

J.L. Bouchez et al. (eds.) Granite: From Segregation of Melt to Emplacement Fabrics, 199–214.

internal dynamics are still non fully ascertained. With time, ideas and paradigms have changed (Pitcher, 1993), evolving from in-situ assimilation, forceful intrusion, diapirism, in-situ ballooning, or ascent through dykes and pluton growth by imbricated sheets or growing cavities in reponse to crust dilatancy during regional deformation. However, underlying most of these concepts is a dearth of observations of the deep processes to support these ideas.

Field data on granites are essentially restricted to surface observations. In the case of a deeply eroded massif, a vertical section on the order of one kilometer may be accessible from the surface, but this may constitute a small fraction of its total vertical extension. To infer deeper geological features, structural and geophysical data are required. Until recently, the inner-intrusion structures have been neglected. They are now better taken into account, since the realisation that "granite is never isotropic" (Bouchez, this volume) and that the internal structures may provide key information on pluton emplacement. Technical advances, such as high sensitivity magnetic fabric measurement and texture studies using numerical image analysis, also played an important role in fabric studies of plutons.

After ten years of observations and measurements following the first study on the Plouaret granite massif (Guillet et al., 1985), this paper aims to review and discuss the advances of such studies which integrate internal structures determination and gravity measurements to infer the morphology at depth, hence (hopefully) the emplacement mode of granites.

METHODS

Amongst the geophysical tools applied to granitic bodies, seismic profiles at usual frequencies (<10 Hz) fail to show the rock-type variations and *a fortiori* their internal fabrics, because of their low impedance contrasts (Matthews, 1986). Conversely, high-resolution seismic reflection profiles reveal fault structures in granites (Mair and Green, 1981) but plutons remain transparent, inhibiting the revelation of deep boundaries such as the floor of the pluton. Magnetic surveys are well suited to provide information on contacts between pluton and country rock, since magnetic halos are appropriate to delineate surface contours (Schwarz, 1992), but the technique lacks the resolution at depth to reveal the pluton's floor. Fabric measurements, mainly by anisotropy of magnetic susceptibility (AMS), are particularly well-adapted to determine the internal structures of plutons (Bouchez et al., 1990), and gravity measurements are best suited to investigate the shape of plutons at depth (Vigneresse, 1990). Both techniques, used in the same way for all the examples given below, in order to permit comparison, are now examined in some detail.

FABRIC MEASUREMENTS AND INTERPRETATION

Information on the magma flow is obtained from crystal shape preferred orientation (SPO) measurements (Oertel, 1955; Bouchez, this volume). In a magma, the early crystallized minerals, if anisometric in shape, induce a planar fabric, or foliation, marked by the average orientation of their most developed face. Within the foliation plane, a lineation may be defined as parallel either to the average elongation of the crystals, or to the zone axis of the orientation of their most developed face. These crystals are mostly Fe (± Mg) bearing silicates (biotite, amphibole), but plagioclase and megacrysts of K-feldspar are also used to determine the internal fabrics of a granite pluton. In the field, a compass is used to measure the SPO of the crystals by estimating their average planar and linear orientations. The results are approximate, especially on equigranular and fine-grained rocks. Grain by grain studies are also performed, using a U-stage mounted on a microscope (Pons, 1970; Bouchez et al., 1981), but these are time consuming, hence not applicable routinely. Owing to the magnetic properties of biotite and amphibole and/or magnetite grains, magnetic fabrics (Borradaile, 1988) are now routinely performed in granites (Van der Voo and Klootwijk, 1972; Guillet et al., 1983; Archanjo et al., 1995) and favorably compare with the classically measured SPO (Bouchez et al., 1981; Amice et al., 1991).

The foliation, or flattening plane of the shape fabric ellipsoid, may be seen as the plane of finite flattening undergone by the magma during its emplacement. Similarly, the lineation points to the finite stretching during this event. Note that the finite strain axes can be equated with the shear plane and direction provided that a large enough homogeneous strain, say a few γ, has been recorded by the magma under study (see Arbaret et al., this volume). It is possible to argue about the relative importance of determination and use, in map interpretation, of lineation rather than foliation. Returning to basic considerations, the path followed by particles in a deforming body serves as an indicator of flattening or stretching (Nicolas, 1989). In this framework, flattening-dominated rocks such as sedimentary rocks where compaction dominates, are hardly lineated, and lineation is much less important in their strain/kinematic reconstruction than foliation. Conversely, rocks undergoing large shear strains, or magmas travelling through the crust, have a linear component of their strain path which dominates the planar one.

Hence the lineation, which records the distortional part of this path (i.e. ignores its translational part), is richer in kinematic information than the foliation. In addition, the foliation plane is more sensitive to the local flow geometries than the lineation. For instance, vertical lineations will reflect a vertical flow, or upwell, of the magma, such as in figure 1c in the Cabeza de Araya pluton (Amice et al., 1991), and this flow will be necessarily carried along vertical foliation planes. In contrast, a vertical foliation may reflect the same flow provided the lineation is vertical, but may also reflect a horizontal flow, controlled by a wrench deformation along a vertical plane, and shown by a horizontal lineation. Indeed, but only in particular instances (local geometries, magma compaction), it could also indicate a flattening flow regime, for example against the pluton's wall if the lineations are randomly oriented within the vertical foliation plane. Hence, because of the inherent ambiguities carried by the foliation plane, we claim that the lineation is of greater use in magma flow studies than the foliation plane.

GRAVITY DATA.

Since aspects of gravity data acquisition and processing are much less widely used than the structural approach, the different stages of the gravimetric technique will be described step-by-step.

Field experiments

The quality of a gravity survey depends on the accuracy of the gravity measurements, but also and chiefly, on the precision of elevation measurements. The latter should be better than 0.5 metre if a 0.1 mGal anomaly detection is expected. This is obtained using topographic maps with accurate elevation benchmarks, and a precision baro-altimeter checked several times per day at benchmarks.

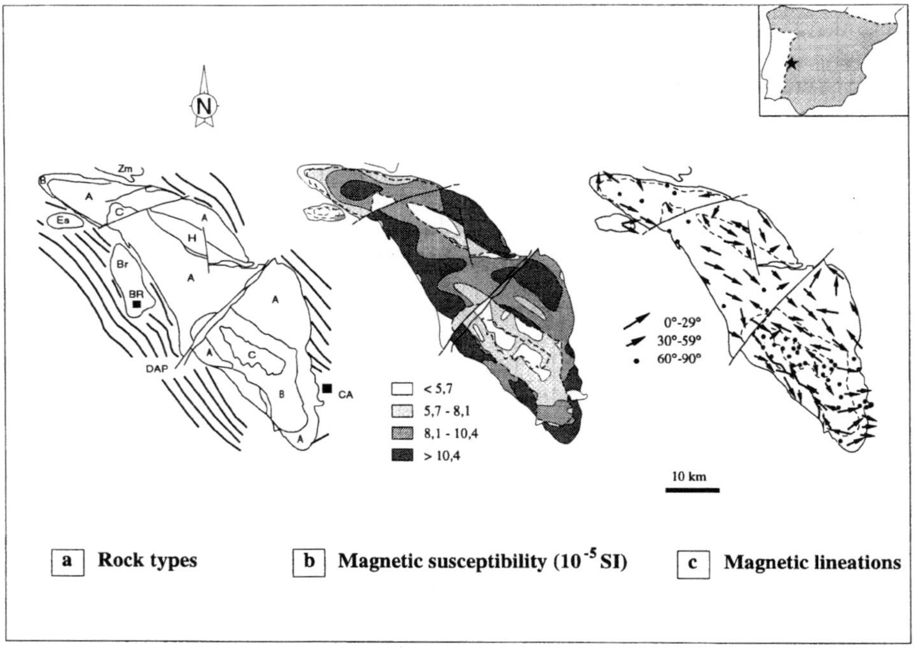

Figure 1. The Cabeza de Araya pluton (Extremadura, Spain). (a) Major granitic facies (A, B, C, H) modified from Corretge (1971): (A) porphyritic; (B) two-mica bearing (biotite dominant) and coarse-grained; (C) two-mica and fine-grained, and (H) heterogeneous with C and B subtypes. DAP: Alentejo-Plasencia dyke; Br (Brozas), Es (Estorninos) and Zm (Zarza la Mayor) are three satellite massifs. Full squares: cities of Caceres (CA) and Brozas (BR). (b) Magnetic susceptibility map (Amice, 1990). Four domains of susceptibility magnitudes (in 10^{-5} SI), separated by broken lines, have been selected, stressing their good correlation with the petrographic types (see a). (c) Magnetic lineations. Synthetic map obtained from 230 anisotropy of magnetic susceptibility measurements (Amice et al., 1991). Two main orientations are evidenced: NW-SE with low plunges, and sub-vertical mainly related to petrographic subtype C (see a).

The grid mesh of gravity measurements is an important choice for the operator since it determines the frequency content of the row-data maps in 2D, or profiles in 1D, hence the minimum size of the objects which may be detected. In practice, through the high frequencies of the data analysis, the mesh of the sampling grid will determine the smallest size of the body that one can attempt to interpret. Conversely, the depth of the source causing the anomaly will condition the map extension, or the profile length. To use correctly the Fourier transform data processing (Regan and Hinze, 1976), the side of the area to be covered by gravity measurements should be six times the maximum depth of the body. Finally, to adequately represent the gravity effects of density variations in plutons, as well as the roughness of the floor, a regularly spaced coverage of one station per square kilometer is recommended over the pluton and its nearby surroundings. After the usual correction of the raw gravity data (Bouguer, latitude and elevation corrections), the Bouguer anomaly map is established (density correction: d = 2.7).

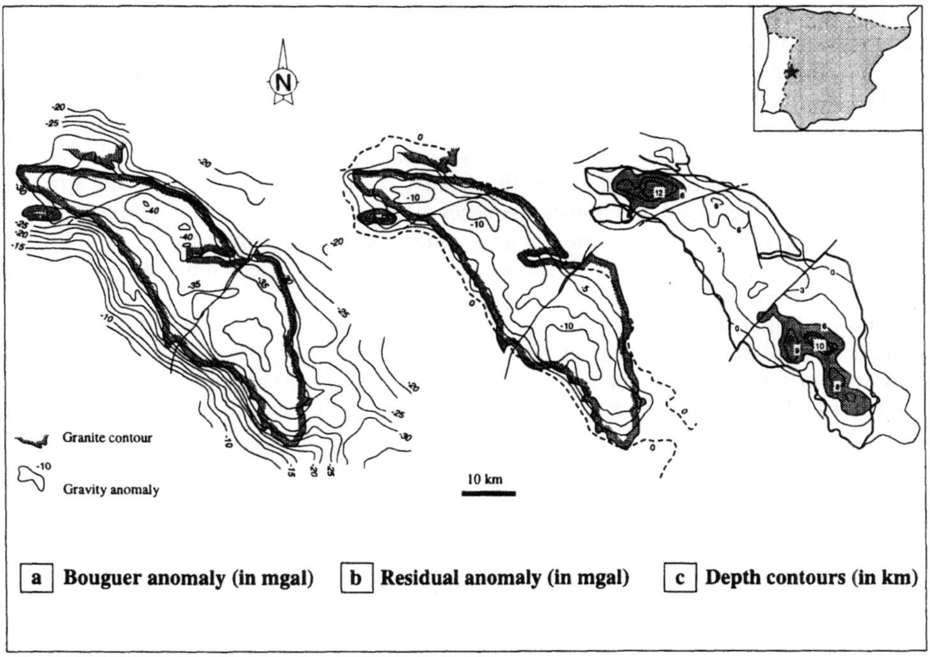

Figure 2. Gravity anomalies over the Cabeza de Araya massif and its surrounding rocks, and depth of the granitic floor (modified from Audrain et al., 1989a). (a) Bouguer anomaly map in milligals (density reduction: 2.7). The granitic body appears as a depression with anomalies ranging from -30 to -25 mGal (equidistance: 2.5 mGal). Grey shaded line: granite border. (b) Residual anomaly (mGal) obtained after substraction of the regional anomaly. Except in the south-eastern part, the correlation between the outcropping limit of the granite (shaded) and the zero contour of the residual anomaly (dashed line) validates the regional anomaly input in the model. (c) Depth contours (in km) calculated from gravity data inversion. Contour lines, at each 3 km, point to a pronounced deepening along elongate zones in both the north and south of pluton. The strong correlation between the vertical plunges of the lineations (Fig. 1c) and zones of greater depth (below 6 km: shaded) is interpreted as the root zones from which magma upwelled.

Extracting the anomaly caused by the granite

The Bouguer anomaly reflects the heterogeneous distribution of masses at depth. Since gravity potential is a direct function of the mass, both the total volume of the granitic intrusion and its density contrast with respect to its surrounding, have a linear effect on the anomaly. In addition, the deeper the source, the lower will be the amplitude, but this variation is non linear with depth since the gravity potential varies as the inverse of the square of the distance to the source. In practice, for a given density contrast, the larger the volume of the pluton, the larger the amplitude of the anomaly, and for a given volume, the larger the density contrast, the larger the amplitude of the anomaly. In the Cabeza de Araya example, the Bouguer anomaly map (Fig. 2a) evidences a high-amplitude gravity anomaly (≈ 10 mGal) with a kilometric wavelength and strong gradients at the border zones of the pluton. A thick (≈ 10 km) and deeply rooted (13 km) body, larger than 5000 km^3 in volume, has been proposed (Audrain et al., 1989a). Conversely, the Sidobre granite massif (Fig. 3a) is thin (≈ 2 km) and shallowly rooted (3 km), with a volume 30 times smaller, and is characterized by low amplitudes (< 3 mGal) of the gravity anomaly and low gradients at its borders (Améglio et al., 1994).

The amplitude of the anomaly varies with both depth of the source and density contrasts, and carries two contrasting types of information. On the one hand, large and deep-seated heterogeneities induce smooth gravity anomalies. These anomalies, called regional because of their extension, form the lowest frequency component of the gravity signal. On the other hand, the higher frequency component of the signal corresponds to the more superficial heterogeneities, among which the granite body itself.

Before local interpretations are undertaken, the regional effect must be substracted from the gravity measurements to determine the residual anomaly. Various techniques, manual or digital, using frequency filters, averaging gridded values, substracting polynomial or hand-drawn surfaces obtained after profile filtering, have been progressively improved to define the regional field (Telford et al., 1990). Whatever method is used, it is suggested here that a convenient result is obtained when the zero contour level of the residual anomaly map outlines the border line of the granite. In the example of Cabeza de Araya, its outcropping limits correspond roughly to -30 mGal on its northern boundary, and is close to the -25 mGal iso-value contour of the Bouguer map on its southern border (Fig. 2a). When the regional anomaly, modelled by a polynomial adjustment all over Extremadura, was removed, the residual anomaly map (Fig. 2b) satisfactorily isolated the effect of the pluton, except at its south-eastern tip where a concealed granite body is suspected, probably connected to other plutons more to the southeast (Castro, 1986). At the northwest extremity of the pluton, remaining discrepancies are, again, attributed to concealed granitic satellites, Estorninos (Es) and Zarza la Mayor (Zm) (Fig. 1a).

Determining the shape of the granite body

The next step is the processing of the residual anomaly in terms of mass distribution, hence, in terms of the pluton's shape and/or depth of the pluton's floor. Three main techniques have been evaluated by Vigneresse (1990). (i) The direct modelling, used for instance to characterize peculiar bell-shaped plutons, assumes an *a-priori* mass distribution (i.e. shape of the source body and densities) and computes the corresponding anomaly which in turn is compared to the measured anomaly. Shapes and densities are modified step-by-step until the computed and measured anomalies equate. (ii) Indirect processing of the measured anomaly, using some filtering operations (high- and low-pass filters) and field transformation encompassing first or second vertical derivative, or downward continuity, is applied to measurements to extract one parameter carrying information on the source body (e.g. Bott, 1962; Stanley and Green, 1976). (iii) The inverse technique allows the removal of (most of) the *a-priori* information on the source body. Model parameters are computed directly from measurements. Among the several methods of data inversion, we recommend the 3D iterative methods (Cordell and Henderson, 1968). In these methods the domain under study is sliced into elementary and regularly juxtaposed vertical prisms, each having a constant density. Provided gravity measurements are distributed on a regular grid in map, and this is obtained by interpolation techniques, then the thickness of the prism centered at each mesh point is calculated. The resulting anomaly is then calculated from the 3D structure defined by all the prisms, and the differences between the computed anomaly and the measured anomaly are used to slightly modify the depth at each point of measurement. This procedure is repeated until the computed and observed anomalies become alike. It has been improved (Vigneresse, 1990) by assigning a specific density contrast to each prism, and by integrating the densities at depth of the different granitic rock-types and country rocks. However, due to the variety of input data (choice of densities, tracing of surface outlines and shape of the regional anomaly), the result is obviously not unique. Its stability must be tested, mainly with respect to the density contrasts.

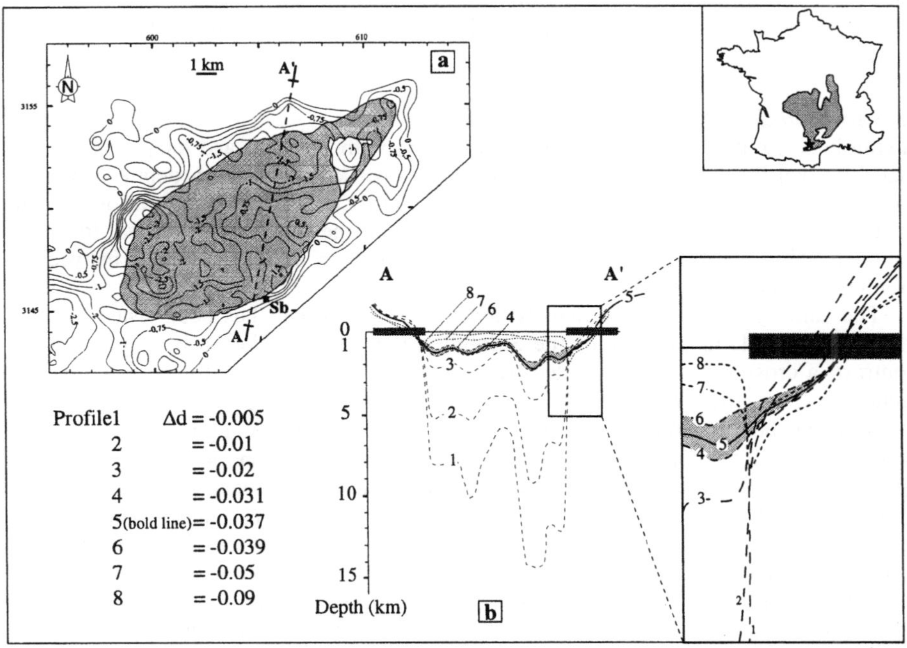

Figure 3. The Sidobre granite massif (after Améglio et al., 1994). (a) Depth (in km) of the granite floor (density contrast: Δd = -0.037). The broken line refers to the section given in b. Sb: St Salvy de la Balme. (b) Depth profiles (in km) of the floor from 3D gravity inversion using various density contrasts Δd between the granite and its country rocks. No vertical exaggeration. Grey shading: sections with Δd = [-0.031; -0.039], in agreement with the density measurements (profile 5, Δd = -0.037).

Densities determination

As data inversion requires knowledge of the density contrasts between the several subtypes in a granite body and surrounding rocks, it is recommended that systematic density measurements are made. This may be done using the cylindrical specimens collected for the AMS study, by means of the classical picnometer technique. Since the cores are obtained from surface drilling, the porosity volume must be removed from the bulk volume to yield the true density. This is established by the procedure described by Vigneresse and Cannat (1987) in which the cores are set under vacuum, then impregnated by water and weighed, then slowly dried to avoid thermal cracking, and weighed again. Such measurements, performed on Hercynian granite plutons from Spain (Cabeza de Araya; Amice, 1990) and France (Mont-Louis-Andorra: Gleizes, 1992; and Sidobre: Améglio et al., 1994) yield densities ranging from 2.63 (leucogranites) to 2.83 (tonalites) depending on the rock iron content (biotite, amphibole). Because these granites are paramagnetic (i.e. magnetite-free), the first-order linear correlation between iron content and magnetic susceptibility (Rochette et al., 1992; Gleizes et al., 1993), explains the excellent linear correlation between granite density and magnetic susceptibility (Fig. 4). In country rocks, densities are more variable, sometimes lower than in granite (e.g. 2.53: limestones; 2.60: quartzites) but usually much larger, in the range 2.65 to 3.00, for example 2.65 in low-grade schists, 2.80 in gneiss (our measurements; see Clark, 1966).

Sensitivity to density contrast

A simple calculation using rock volume variation with pressure and temperature shows that rock density increases by about 1.0-1.5 10^{-3} per kilometer with depth. Since this variation more-or-less similarly affects the different rock types, i.e. the various granitic facies and the surrounding rocks, it has no bearing on the model with respect to density contrasts. Thus the model is very sensitive to changes in density contrasts between the granite and its country rocks.

The validity of assigning a density contrast to each prism has been tested by Améglio et al. (1994) in the case of the Sidobre granite massif. For each assigned density contrast, a cross section is extracted from the depths calculated with the 3D inversion technique (Fig. 3b). When the density contrast is too low, the routine is faced with a mass deficit and accumulates mass at depth: the calculated thickness of the massif reaches 14 km for a density contrast $|\Delta d| = 0.005$ (profile 1), and 4 km for $|\Delta d| = 0.02$ (profile 3). An increase of the thickness increases the high frequency component of the anomaly, a feature not observed in the measurements. As a consequence, fictive masses are added laterally to compensate the resulting mass deficit. Conversely, when the density contrast is too large ($|\Delta d| \geq 0.04$; profiles 7 and 8), the routine reacts by generating a very thin granite massif, but there is a need to compensate the mass deficit by adding extra masses at the border of the pluton (Fig. 3b: cartoon). The acceptable solution (Fig. 3b: grey shaded), which does not present any numerical oscillation, is in a very restricted range of density contrasts ($0.031 \leq |\Delta d| \leq 0.039$). Ideally, it closely corresponds to the average contrast ($|\Delta d| = 0.037$) calculated from densities measured from surface samples (profile 5).

Therefore, a map of surface densities, and ideally of integrated densities at depth, for all granite types and surrounding rocks of a pluton should be incorporated in the computation, in order to best constrain the nominal densities of the corresponding prisms. This has been realized in the pluton of Cabeza de Araya, by using the geological map of Corretge (1971) (Fig. 1a), the magnetic susceptibility map of Amice et al. (1991) (Fig. 1b), and density measurements of surface rocks (Amice, 1990). However, in fact we do not know the mass distribution with depth, hence, there is no certainty that the chosen densities for the prisms are correct. Any geological information concerning both the organization of the different rock types in the pluton, such as the petrographic zoning or layering, and the nature and organization of the country rocks, are potential inputs in the model which will strengthen confidence in the results.

In model computation, minimization of numerical oscillations, as quoted above, may be a good criterion optimizing the solution. The computation procedure could also be improved by checking the density maps at various depths after each iteration to reinforce the validity of the assumed parameters (density contrasts) and help to select the appropriate densities between facies, and to understand the spatial distribution of facies.

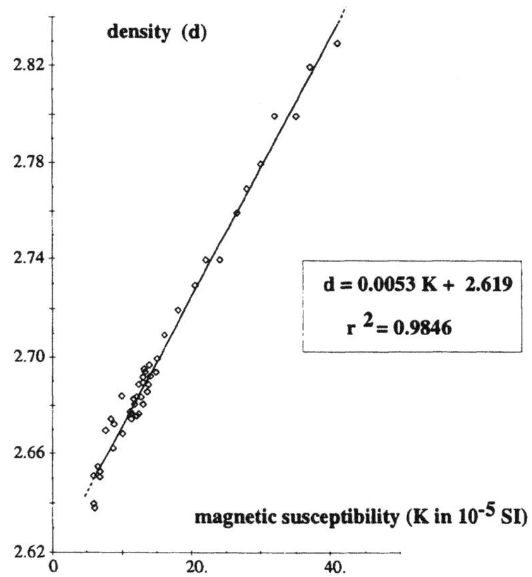

Figure 4. Linear correlation between the measured magnetic susceptibility (K) and measured rock densities (d) of corresponding specimens. r^2 = correlation coefficient; 45 samples from the Sidobre (Massif Central, France) and the Mont-Louis-Andorra (Pyrenees, France) paramagnetic granites. Error on measured densities is ± 0.008.

Estimating errors

We are confident that the computed depth values of the floor, although inexact, give a good first-order picture of the pluton. Precision of the raw data, attributable to the apparatus (better than 0.1 mGal) and the elevation data including uncertainties in topographic corrections (≈ 0.1 mGal), and errors in the far-field gravity anomaly estimate (≈ 0.5 mGal), contribute to less than half of the total precision on the depth values. The main factor of uncertainty is obviously related to the choice of the densities attributed to the prisms, not to the measurement in the laboratory of the surface rocks, which has a precision better than 1%. In turn, this reflects the uncertainty on the density contrasts at depth. Considering all these sources of uncertainties, the calculated depth values are considered to be better than ±15%. However, concerning the overall shape of the pluton, for a given choice of densities, the depth values are affected by the same error everywhere in the pluton, the dips of the pluton's walls will also vary in the same manner everywhere, hence the shape will not vary significantly.

INFERENCE ON PLUTON EMPLACEMENT

Plutons may reach some thousands of square kilometres in area, and petrographically may be subdivided into several facies that may reflect different magma pulses. Thus, it is reasonable to enquire whether their morphology at depth reflect their mode of emplacement. Elements for correlation between shape and emplacement mode are now discussed in detail, based on granite intrusions observed in several regions of the Hercynian belt.

All surveyed massifs, mostly peraluminous, are syntectonic. Their regional deformation takes several forms depending on the considered area (Vigneresse, 1995) and the overall context seems to indicate that the same crustal level of erosion was reached for all the massifs. Finally, isotope chronology studies give ages around 300 Ma.

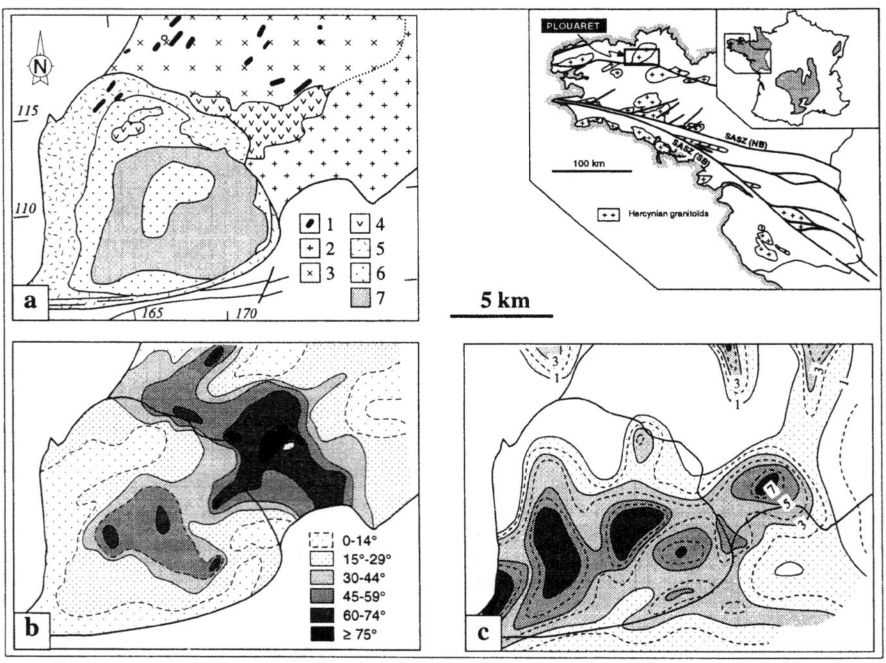

Figure 5. The Plouaret granite complex (Brittany, France), after Guillet (1983) and Guillet et al. (1985). (a) Petrographic types (modified from Chauris and Garreau, 1983). 1- Lanvellec diorite; 2- Bégard granodiorite; 3- Tonquédec granite; 4- St Carré and Bruillac leucogranites; 5- Ponthou granite; 6- Croaz-Illiès and Guerlesquin leucogranites; 7- Loguivy-Plougras leucogranite. (b) Contour map of domains of equal plunge (in degrees) of the magmatic lineations. (c) Depth (in kilometres) of the pluton's floor, derived from the gravimetric study.

ROOT ZONE DETERMINATION

Among the several correlations between shape at depth and internal structure of a pluton, the easiest to interpret is the root-zone, or feeder zone. As a matter of fact, zones where magmatic lineations are steeply plunging (≥ 60°) often correlate with strongly deepening, funnel-shaped zones evidenced by gravity data. Root zones have been discovered in the Plouaret pluton (Guillet et al., 1985) in Brittany (Fig. 5: compare 5b and 5c), and also illustrated in the Cabeza de Araya pluton (compare Fig. 1c and 2c). These zones, a few 10 to 100 km^2 in area and related to a marked deepening of the floor, were probably active as feeders all through the duration of pluton emplacement. This is attested by the rock types at the surface, i.e. vertically above the feeder, which are characterized by chemically more evolved magmas, or late intrusive facies, such as the two-mica leucocratic subtypes B (coarse-grained) and C (fine-grained) in Cabeza de Araya (Fig. 1a), and leucogranites in Plouaret (types 4, 6 and 7 of Fig. 5a).

The case of a deepening of the pluton's floor that does not correlate at the surface with steeply plunging lineations, can be explained by the rotation of the flow lines of the upwelling magma, for instance parallel to the pluton's roof if the erosion level is close to the roof. Conversely, if areas with clusters of steeply plunging lineations do not correlate with a deepening of the pluton's floor, unrooting and bulk translation of that volume during pluton feeding may be invoked, among other explanations. In conclusion, the vertical lineations and the accompanying rock-types reflect the frozen-in latest emplacement stages of the pluton. Except in the case of the Zaër pluton (Bouchez and Diot, 1990) where subvertical lineations dominate over all the granodiorite, the scarcity of highly plunging lineations in structural maps is attributed therefore to the change to horizontal magma flow after the feeding stage, during its deformation when spreading away from the root-zone.

FLAT-FLOORED PLUTONS VERSUS WEDGE-SHAPED PLUTONS

Cross-sections with respect to the present level of erosion, established at the same scale and with no vertical exageration, are drawn parallel (||) and perpendicular (⊥) to the longest axis of a number of plutons (Fig. 6). Two families of plutons readily appear. The flat-floored plutons (H/l < 0.5), Saint-Sylvestre and Sidobre (Fig. 6), but also Guitiriz (Spain; Aranguren et al., 1996), Fichtelgebirge OIC (Bavaria, Germany; Hecht, 1994) and La Marche (Guineberteau et al., 1989) appear to be quite thin (a few kilometres), have several feeder zones protruding downward from an overall subhorizontal floor, and extend in each horizontal direction. In contrast, the wedge-shaped plutons (H/l ≥ 0.5), Cabeza de Araya and Mortagne (Fig. 6), but also Pontivy and Guehenno (Britany, France; Vigneresse, 1983), may be thicker than 10 km, and have a preferred horizontal extension along which a few feeder zones appear, and perpendicularly, have typical V-shaped sections with steep walls, steepening with depth. These examples, whatever the type, correspond to the flattened cupola-shape of granite plutons, emphasized by Vigneresse (1995), where horizontal extension is always much larger than vertical.

In these plutons, whatever the shape, the lineations are mostly subhorizontal or shallowly plunging and record the horizontal stretch undergone by the magma during its spreading at the site of emplacement. The original data for the Saint-Sylvestre pluton are given in Mollier (1984) (see also Bouchez, this volume, Fig. 3); many other data which are similar and come from the same area (French Massif Central) are given in Jover (1986), and data for the Mortagne pluton may be found in Guineberteau et al. (1987).

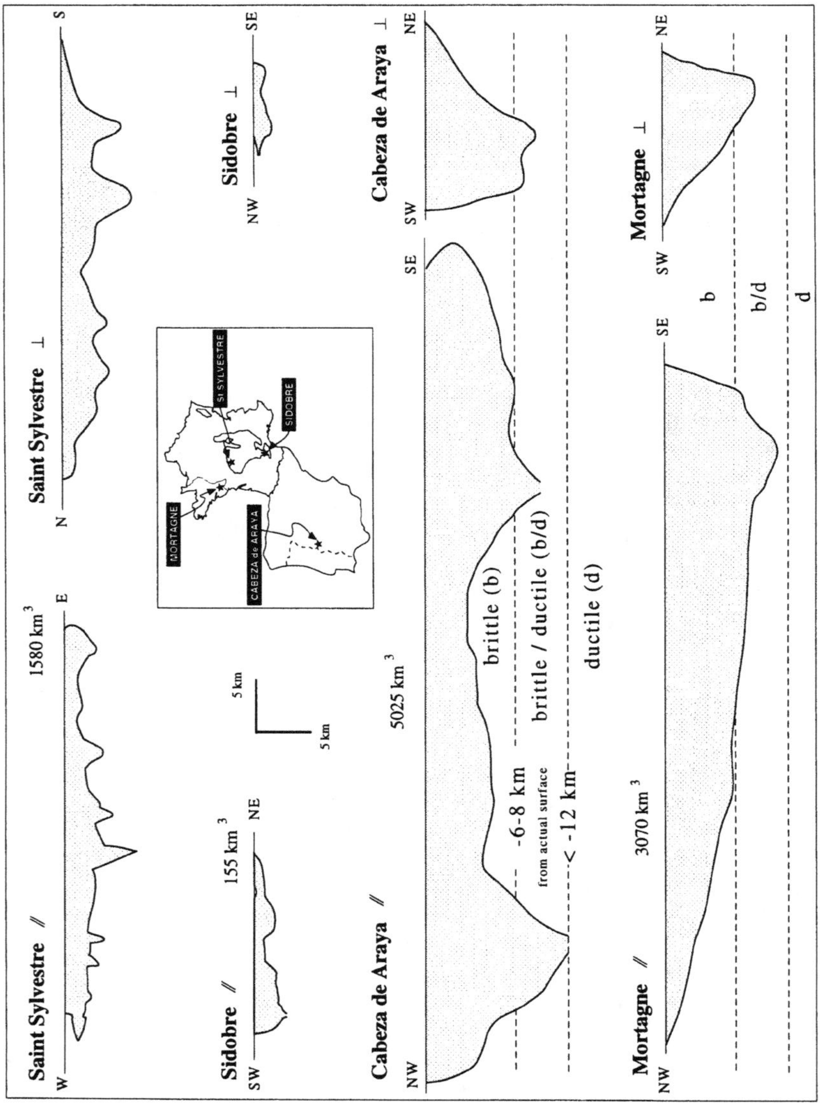

Figure 6. Cross-sections through four granite massifs discussed in the text, showing their morphology with depth, as inferred from 3D inversion of gravity data. All sections are shown at the same scale with no vertical exaggeration. Approximate volumes in km^3 are given. Left-hand cross sections are parallel (//) to the longest axis of the massifs, and right-hand sections are perpendicular (⊥). Roots zones are evidenced by abrupt changes in dips of the walls.

The shapes and internal structures of these plutons suggest that their emplacement is largely controled by the anisotropy and rheology of the crust particularly around the brittle-ductile transition: (i) flat-shaped plutons would spread parallel to a tectonically layered crust, in an eventually still plastic crustal domain, whereas (ii) wedge-shaped plutons would infill more-or-less vertical fractures in the brittle crust.

(i) In the examples cited of flat-shaped plutons, a wealth of geological considerations, synthesized by Faure and Pons (1991), argue in favor of their emplacement in late-orogenic, gravity spreading or extensional tectonic environments. In the Saint-Sylvestre pluton (Fig. 7), the dominantly subhorizontal and NW-SE trending lineations reflect the overall direction of spreading in this part of the French Massif Central, at least during magma emplacement. In the Sidobre pluton (southeast Massif Central) the direction of crustal extension is reflected by perfectly oriented NE-SW trending and subhorizontal lineations (see Olivier et al., this volume). Extensional tectonic environments are the more favourable for the emplacement of sheet like plutons and the « room problem » (Bowen, 1948) is readily solved. The several roots or pathways through which the magma infills the pluton may reflect the extensional state of stress, and allow large areas to be fed by reducing the distance that a given batch of magma must travel. In such a context where the geotherms may rise to very high levels, the crust may be ductile at relatively low levels and sheared horizons such as listric faults may bound the pluton at its roof. Some granites of this type may also intrude the very upper crust and form thin sills that developed into local, mainly subhorizontal, roughly bedding-parallel dilatant volumes. The reduced thickness of these sills correlates with their small volumes (< 300 km^3) compared with the large massifs of flat-shaped pluton type (≈ 1500 km^3). An example of such a small massif is given by the Flamanville granite body (Brun et al., 1990).

In higher metamorphic conditions, the « stratoid » granites of Madagascar belong to this type. They form subhorizontal layers of granite, a few hundred metres thick, with remarkably constant WNW-ESE lineation directions having shallow plunges. They are considered to represent intrusives into the lower to intermediate crust during the late Pan-African extension event (Nédélec et al., 1994).

(ii) As for wedge-shaped plutons, it is now rather well accepted that they represent the infilling of dilatant volumes of the brittle crust undergoing deformation. This process also elegantly solves the room problem. Local dilatancy of the brittle crust may be achieved by several means, under both conditions of compression and extension, but dominated by transcurrent movements. The wedge-shaped plutons exemplified in figure 6 were emplaced during regional contractional and shear regimes. In Cabeza de Araya (Amice et al., 1991), the progressive bending of a dextral shear zone globally parallel to the pluton's elongation has been presented, similar to the retreating bend mechanism of Hutton (1982). In Mortagne (Guineberteau et al., 1987), the progressive opening of a sinistral jog during the propagation of a shear zone parallel to pluton has been proposed as the principal control on emplacement.

The rheological level of emplacement of these plutons can be inferred from the following considerations, when observed in perpendicular-to-elongation sections (figure 6: ⊥ sections). The shear zone above which the pluton develops behaved plastically, hence was probably at T > 400°C, i.e.10-13 km deep for gradients 30°- 40°C/km. Rather high thermal gradients are justified by the fact that these Hercynian S-type granites mostly derive from the partial melting of a fertile middle crust in the context of post-collisional thermal rebound after crustal thickening (Cassard et al., 1993). The pluton's roots may be therefore connected to a zone where no substantial extension is visible since strain is accomodated by ductile mechanisms. The roots have noticeably steep walls that may represent the brittle-ductile transition zone, and may be up to 5 kilometres in thickness. The remaining upper part of the pluton is within the brittle crust, the base of which may be 6-8 km deep from the present surface. In contrast to the ductile crust, the brittle crust may provide large magma reservoirs delimited by clear cut walls which are less steep than along the roots, but may have different dips depending on the side of the pluton (see perpendicular sections in figure 6). In Cabeza de Araya, asymmetry of the walls has been correlated with the sense of strike-slip and rotation along the fault plane (Amice et al., 1991).

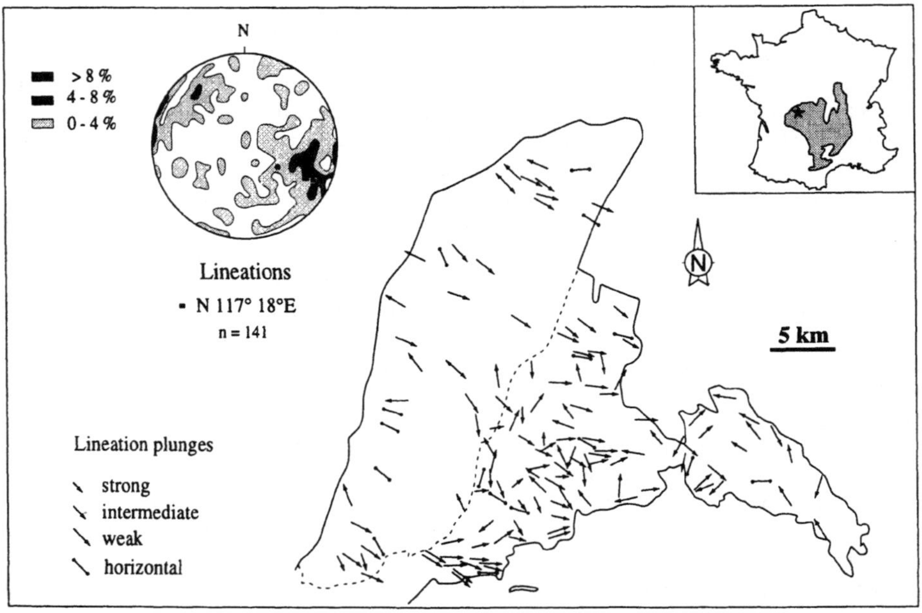

Figure 7. Conventional magmatic lineation map (141 measurements) of the Saint-Sylvestre massif (Massif Central, France) and corresponding orientation diagram (Schmidt, lower hemisphere), after Mollier (1984).

INFERRING EMPLACEMENT MODE FROM STRUCTURAL AND GRAVITY DATA

Combining systematic gravity and structural measurements has demonstrated its potential in geological reconstruction of granite plutons. Note that this review, based on experience within the Variscan orogeny in France and Spain, does not necessarily represent all possible situations. However, it points to features that may be generalized to most environments of pluton emplacement.

(i) Both structural and geophysical information are essential to infer the emplacement mechanism. We claim that lineations provide more information than foliations since we are interested in determining the stretching directions during syn-emplacement magma flow.

(ii) A correlation between the lineation pattern and the bulk direction of the regional deformation field is worth studying. When shear deformation is recorded on a regional scale, the lineations in the pluton are often horizontal and parallel to the regional stretching direction (Archanjo et al., 1992; Pons et al., 1992). Conversely, when the pluton is bounded by a strike-slip fault that was active during pluton growth, its lineation pattern may be parallel to the fault, as in the Bassiès pluton in the Pyrenees (Gleizes et al., 1991). The lineation pattern may be oblique to the fault and subparallel, either to the latest increments of the opening, as proposed by Bouillin et al. (1993) for the Monte Capanne pluton (Elba Island, Italy), or to the local finite stretch undergone by the pluton during its emplacement, as in the late Pan-African Tesnou pluton (Hoggar, Algeria; Djouadi and Bouchez, 1992).

However, these patterns reflect only the surficial deformation conditions induced by faulting, and only those recorded just before solidification of the magma, i.e. corresponding to the final few gammas of strain (see Arbaret et al., this volume). They do not carry information on the mode of opening and emplacement a few kilometres deeper in the pluton.

(iii) Vertical lineations may indicate the presence of a magma feeder zone if they correlate, in map view, with a deepening of the pluton's floor, identified by the gravity data. The presence of differentiated, late-magmatic granite subtypes further attests to the presence of a feeder zone.

(iv) A relationship between the bulk orientation of the pluton and the orientation(s) of the root-zone(s) may also be observed. Two main dispositions appear: root-zone(s) may be elongated and/or may align parallel to the bulk elongation of the pluton, as in Cabeza de Araya (Fig. 2c), or perpendicularly to pluton elongation like in Guitiriz, (Aranguren et al., 1996). In the first case, the deformation field probably remains the same at different structural levels of the massif. In the second case, root zones oblique to pluton elongation, reflect the fact that the direction of crack opening at the brittle-ductile transition, where root zones initiate, is not parallel to that prevailing in the remaining brittle crust. This may be due to the role of magma pressure which controls fracture orientation (see Nicolas, 1989, p. 34). For a given regional principal stress direction, fracturing of the crust occurs closer to the stress direction than the magma pressure is higher, and the crust more brittle. Post-emplacement shear deformation may also modify the apparent shape of the pluton, thus obscuring the regional strain field acting during emplacement, as in the Mont-Louis-Andorra pluton (Bouchez and Gleizes, 1995).

(v) Finally, the number of roots may reflect the overall tectonic setting. Plutons emplaced during regional transcurrent shear may display a single root (e.g. Mortagne; Fig. 6) or a few main roots, as in Cabeza de Araya (Fig. 2c and Fig. 6) and most plutons of Brittany (Vigneresse, 1983), e.g. Plouaret (Fig. 5c). This contrasts with plutons emplaced during large-scale extension where many tiny roots are evidenced all over the granite complex. This is the case for the late-Variscan pluton of the french Massif Central (Faure and Pons, 1991), e.g. Saint Sylvestre (Audrain et al., 1989b) (Fig. 6), and Sidobre (Améglio et al., 1994) (Fig. 4).

CONCLUSIONS

An integrated methodology has been devised to describe and interpret the structures and shape at depth of granitic plutons. The methodology mostly consists of surveying the area with detailed gravity measurements which permit, by 3D inversion treatment of data, the computation of the depth and the shape of granites. It also benefits from structural measurements which record the last movements of the magma while still not fully crystallized. Both sets of data are required to infer with some confidence the mode of magma emplacement.

Observations obtained from gravity and structural studies enable root zones to be identified beneath granitic plutons. The root zones are determined where vertical lineations occur above zones of deeper floor with, in some cases, differentiated or late magmatic facies outcropping.

Deformation plays an important role in determining the crustal position of plutons. Bulk shape, internal structure (essentially lineations), and regional deformation, characterize two main families of plutons. In a flat shaped pluton, magma upwells from multiple roots and spreads parallel to a tectonically layered crust at the brittle/ductile transition in a still plastic crustal domain. In contrast, for wedge-shaped plutons, magma is permissively emplaced within an extensional zone in the brittle crust controlled by transcurrent movement.

The paleo brittle-ductile transition field has been estimated from the shape of the pluton floor as being the place where the floor changes in dip with noticeably steep walls. In several regions of the Hercynian belt, this transition ranges between 6 to 8 km with respect to actual surface.

We have determined several features that may be of more general application to plutons, in order to understand their emplacement mode. We suggest that such integrated studies should take place systematically before sophisticated geochemical studies are undertaken.

ACKNOWLEDGEMENTS

This paper is based on many gravity and structural data coming from surveys conducted with the help of students from the Universities of Nantes, Nancy and Toulouse. Christiane Cavaré-Hester is particularly thanked for the reworking of the drawings. Funding for the gravity and structural surveys come from grants from the CNRS, INSU, DBT program, CREGU, and from COGEMA and CISA uranium-mining companies. Drs J.P. Lefort (University of Rennes) and Dr. R. Trzebski (University of Göttingen) are warmly thanked for their constructive reviews.

THE LOS PEDROCHES BATHOLITH (SOUTHERN SPAIN): POLYPHASE INTERPLAY BETWEEN SHEAR ZONES IN TRANSTENSION AND SETTING OF GRANITES.

Aitor ARANGUREN
Departamento de Geodinámica. Facultad de Ciencias.
Universidad del País Vasco. 48080 Bilbao, Spain.
Francisco Jose LARREA, Manuel CARRACEDO
Departamento de Mineralogía y Petrología. Facultad de Ciencias.
Universidad del País Vasco. 48080 Bilbao, Spain.
Julia CUEVAS and Jose MariaTUBÍA
Departamento de Geodinámica. Facultad de Ciencias.
Universidad del País Vasco. 48080 Bilbao, Spain.

ABSTRACT

The Los Pedroches batholith (LPB) in the Variscan belt of southern Spain is a 200 kilometres long magmatic body which has intruded low-grade metamorphic rocks of the Central Iberian Zone. The LPB is composed of granodiorites, a dyke swarm, and granites emplaced during three successive magmatic events. Field structural data from both the country rocks and the LPB indicate that the northern boundary of the LPB is a transtensional shear zone, the Conquista shear zone, dipping 50° to the NE. This shear zone is sealed by the Cerro Mogábar granite which was emplaced during the last magmatic event. A detailed magnetic susceptibility study, using 2236 oriented samples from 559 stations, of the main part of the batholith yields flat magmatic foliations, and magmatic lineations with a mean east-west trend. The structural, kinematic and magnetic susceptibility data indicate that the emplacement of the LPB was controlled by a crustal scale and dextral transtensional shear zone, parallel to the batholith elongation. The eastern part of the granodiorite takes the shape of a large laccolith prolonged westwards by four NW-SE elongate domes. These contrasted geometries reflect the upwelling of magma through the transtensional zone that vanishes towards the west. Intrusion of dykes along Riedel shears related to the regional transtension took place in a second magmatic event. The main transtensional shear zone is interpreted as a lateral ramp that accommodated the crustal thinning developed in the Central Iberian zone during the collapse of the Iberian Variscan belt.

J.L. Bouchez et al. (eds.) Granite: From Segregation of Melt to Emplacement Fabrics, 215–229.

INTRODUCTION

Concepts about the transfer of granitic melts within the crust and the subsequent emplacement of granites have changed greatly in recent years (see Pitcher, 1993, for a review). The long-held diapiric model, with magma rise driven by the buoyant upwelling of large volumes of magma overlain by denser country rocks, has been shown to occur only at deep crustal levels, where the thermal and viscosity contrasts between the magma and its metamorphic cover are weaker than in shallow crustal levels (Cruden, 1990; Kukowski and Neugebauer, 1990). Instead, permitted ascent of magma through crustal scale shear zones or propagating fractures (Clemens and Mawer, 1992) is considered to be a much more efficient mechanism for magma transfer than diapirism, and tectonically-controlled models of emplacement of granites are now widely recognized in different orogens (Hutton, 1988).

Syn-tectonic plutons have been related to both compressional (Tobisch and Paterson, 1990) and extensional shear zones (Hutton et al., 1992). Nevertheless, many granitic plutons are associated in both space and time with strike-slip shear zones, but in structural contexts as different as: (i) tensile fractures (Castro, 1986), pull-apart sites (Guineberteau et al., 1987; Bouillin et al., 1993) or P-shear tensional bridges (Tikoff and Teyssier, 1992) related to a major strike-slip shear zone, (ii) cracks opening at tip areas of a moving shear-zone (Hutton, 1988) or, (iii) transpressional shear zones (Hutton and Reavy, 1992). In short, strike-slip shear zones may play a significant role in magma transfer within the crust. However, as the shape and structures of granitic plutons mainly reflect deformation mechanisms at the emplacement level, the evidence for previous stages of magma ascent is often scarce. Another consequence of this is that in many instances the related shear zone does not show up clearly at the present erosion level and its existence is only deduced from indirect structural or geophysical observations. This paper focuses on structural and anisotropy of magnetic susceptibility (AMS) data from the Los Pedroches batholith (LPB), one of the largest magmatic complexes of Spain. In the LPB, the occurrence of multiple intrusive events provides an excellent opportunity to analyse the interaction between granite emplacement and contemporaneous shear zones developed in transtensional conditions.

LOS PEDROCHES BATHOLITH: FIELD RELATIONSHIPS AND PETROGRAPHY

The LPB is a narrow and elongate magmatic body, 5-25 kilometres wide and *ca.* 200 kilometres long, located in the southern part of the Hercynian (also called Iberian) massif of Spain (Fig. 1). Following the original tectonic subdivision of the Iberian massif (Lotze, 1945), most authors consider that the LPB marks the limit between the Central Iberian Zone (CIZ) to the north and the Ossa-Morena zone to the south (Julivert et al., 1972; Ribeiro, 1990; Apalategui et al., 1990). The tectonic setting of the LPB is, however, still controversial since the batholith and its southern metamorphic envelope have been proposed to belong to the CIZ (Díez Balda et al., 1990). Relevant for any regional tectonic consideration is that linear batholiths are usually the expression of deep thermal anomalies that were closely related to blind crustal, or lithospheric, shear zones (Nicolas et al., 1977; Hutton and Reavy, 1992; Tomassi et al., 1994). The LPB is concordant with the 120° E trending Hercynian structures, folds and shear zones affecting the surrounding metasedimentary rocks. It also coincides with an elongate, negative gravity anomaly reaching -30 mgal over the eastern half of the batholith (Instituto Geográfico Nacional, 1976).

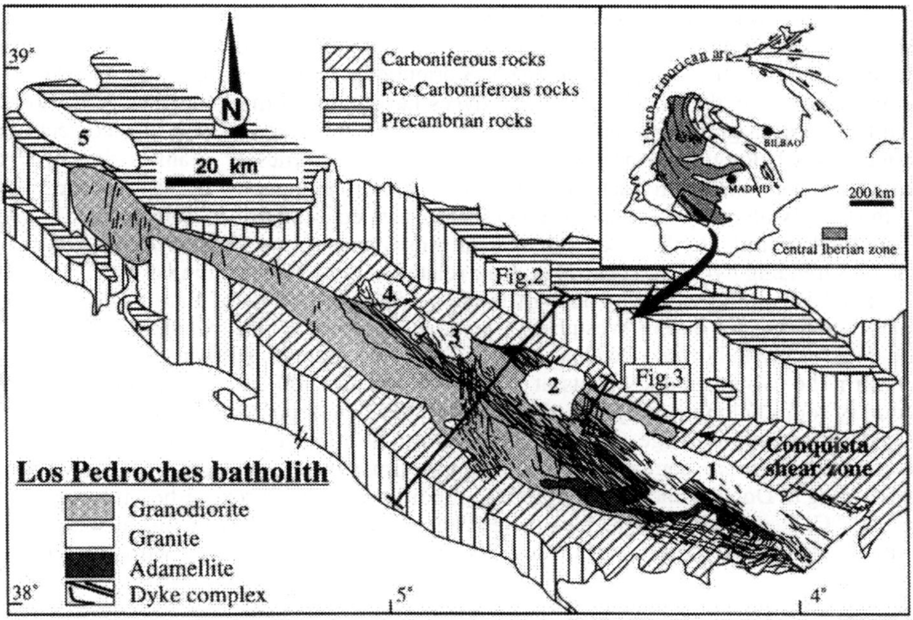

Figure 1.- Schematic map of the Los Pedroches batholith, showing the intrusive relationship between the main petrographic groups. 1) Cardeña granite, 2) Cerro Mogábar granite, 3) El Guijo granite, 4) Santa Eufemia granite, and 5) Campanario-La Haba granite. Insert map location of the Los Pedroches batholith within the Iberian belt.

The LPB results from three successive magmatic pulses (Carracedo, 1991). The oldest one encompasses the Pedroches granodiorite, the Cardeña granite and, between them, transitional adamellites forming a continuous band of more than 40 kilometres long and 3 kilometres wide (Fig. 1). The second magmatic pulse is represented by a well-developed dyke swarm intruding the Pedroches granodiorite and the Cardeña granite. Finally, several granite plutons intruded the former rocks (Fig. 1). The Pedroches granodiorite is fine- to medium-grained with dominant equigranular textures. It is composed of quartz, a large amount of feldspars (plagioclase An_{25-45} and perthitic K-feldspar), biotite and some hornblende; allanite, apatite, ilmenite, sphene and zircon appear as accessory minerals. It displays an impressive mineralogical and textural homogeneity, rarely disturbed by subtle layering defined by grain size contrasts or minor mineralogical variations. Microgranular enclaves of tonalitic to quartz-dioritic composition are widespread throughout the Pedroches granodiorite. The latter granodiorite also contains metasedimentary xenoliths which can be larger than 30 m in the vicinity of the country rocks but only reach a few centimetres away from the contacts. The late granitic plutons have abrupt contacts with the Pedroches granodiorite and are concentrated mainly within the eastern half of the batholith. They correspond to porphyritic rocks with K-feldspar megacrysts in a fine- to coarse-grained groundmass composed of quartz, plagioclase (An_{14-34}), biotite, K-feldspar, ± cordierite and secondary muscovite. This granitic group also includes minor amounts of two-mica leucogranites and aplitic leucogranites.

The dyke swarm displays two structural systems. The largest system is composed of 130°E trending dykes of granitic composition and the second system of granodiorite to monzonite dykes displaying an average N-S orientation (Fig. 1). Cross-cutting relationships demonstrate that the N-S trending dykes have intruded the Pedroches granodiorite and the Cardeña granite earlier than the 130°E trending dykes. Along the contacts of some dykes belonging to both systems the host granodiorites or granites developed a noticeable solid-state deformation. This deformation is represented by asymmetric folds with strongly plunging axes and discrete shear zones concentrated in narrow bands, 1 to 5 cm thick, parallel to the dyke walls. Shear senses are sinistral in shear zones related to the N-S trending dykes and dextral in those related to the 130°E trending dykes.

The LPB is intrusive into low-grade metasedimentary rocks of Late Proterozoic to Carboniferous ages during Westphalian-Stephanian times. Rb/Sr, whole-rock radiometric data for granodiorites (Défalque et al., 1992) and granites (Fernández et al., 1990) confirm the two magmatic events at 308 ± 6.5 Ma and 300 ± 6 Ma respectively. The country rocks belong mostly to a thick, about 3000 m, sequence of slates of Lower to Middle Carboniferous age (Fig. 1). By contrast, the underlying Palaeozoic rocks are dominated by more competent materials, like limestones, quartzites and graywackes. According to Donaire and Pascual (1991), conditions of the pyroxene-hornfels facies (T = 800 °C, P < 2 kbar) were locally reached within xenoliths from the metamorphic aureole, which extends some 1000 to 1500 m away from the contact with the igneous rocks.

STRUCTURAL ANALYSIS

The LPB shows sharp contacts with the country rocks. The late granitic plutons crosscut the country rocks at an angle whereas the Pedroches granodiorite is concordant with their structural elements. In the Pedroches granodiorite there is a strong contrast between the southern and northern contacts (Fig. 2). The southern contact dips at 30° to the southwest and does not show solid-state deformation structures. The northern contact partly corresponds to a ductile shear zone, the Conquista shear zone, dipping 50° to the northeast (Carracedo et al., 1994). The remaining part of the latter contact displays similar dips to the northeast.

STRUCTURAL ELEMENTS IN THE COUNTRY ROCKS

One phase of regional folding followed by subsequent thrusting and a phase of late-stage kink-banding are recognized in the metamorphic envelope on the LPB. The earliest Hercynian deformation is represented throughout the sector by the S_1-cleavage, well-developed only in the slates. S_1 is axial planar of tight F_1-folds (Díez Balda et al., 1990), strikes northwest-southeast and dips steeply to the northeast, except along the southern contact of the batholith where it has been rotated during the emplacement of the subjacent granitic rocks. Outcrops of host rocks are considerably more abundant on the northern side of the LPB. There, minor F_1-folds are asymmetric and S-shaped when viewed towards the northwest, and are consistent with south-westward vergence deduced from local S_0-S_1 relationships. This asymmetry indicates that the Carboniferous slates in this area form part of the normal limb of a major and northeast-dipping synform (Fig. 2). The Carboniferous slate sequence lies below older Palaeozoic rocks, as a result of a thrusting event. A southward motion of the upper block is deduced for this thrust from the widespread presence of 10°E-directed slickensides on minor thrust surfaces, and geometry from the asymmetry of numerous intrafolial folds observed in the contact zone.

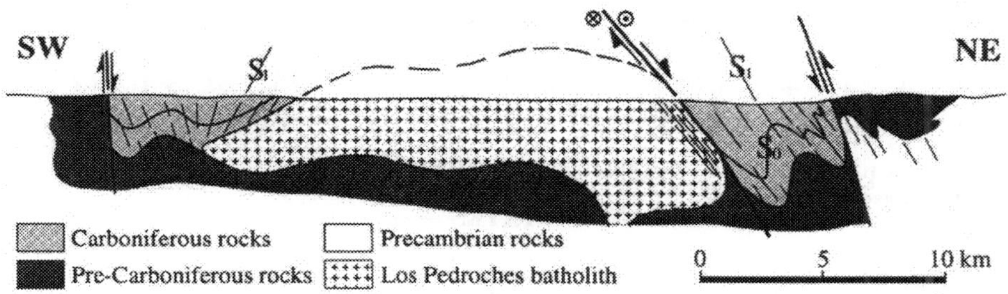

Figure 2.- Schematic cross-section through the Los Pedroches batholith, showing the S_0-S_1 relationships in the country rocks. Note that the sSouthern contact is concordant with S_0 whereas the northern contact coincides with a north-dipping shear zone (see Fig. 1 for the location).

THE CONQUISTA SHEAR ZONE

A large part of the northern border of the Los Pedroches granodiorite is a ductile shear zone, referred to as the Conquista shear zone after the village of the same name where we have established the type-section (Figs. 1 and 3). The shear zone, 1 kilometre-thick and more than 50 kilometres-long, is sealed by the Cerro Mogábar granite (Fig. 1).

Deformation associated with the shear zone increases towards the granodiorite/slate contact (Fig. 3). The subhorizontal, weak magmatic fabrics of the granodiorites are progressively replaced by pronounced planar fabrics defined by S-C structures when traced into the shear zone. The subspherical shape of microgranular enclaves also changes progressively into flattened ellipsoids. Close to the contact the mylonitic foliation becomes parallel to the shear zone which dips 50° to the northeast. The stretching lineation lying on the mylonitic foliation is oblique to the long axis of the batholith, and plunges at 30° to the east-northeast (Fig. 3a).

The granodiorite from the Conquista shear zone is dynamically recrystallized and, therefore, has mainly recorded the last increments of strain. Fine-grained, highly recrystallized mylonitic granodiorite provides quartz c-axis diagrams with a strong maximum concentration around the Y-axis of the finite strain ellipsoid (Fig. 4a). Similar c-axis patterns have been obtained in mylonitic quartzites from the metamorphic

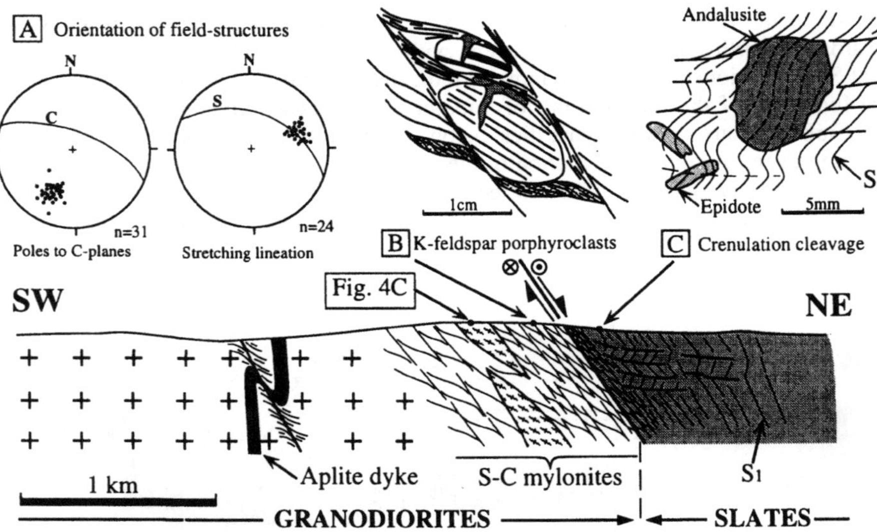

Figure 3.- Schematic section across the Conquista shear zone (location in Fig. 1) showing the main related structures and microstructures. Lens-shaped domains in the mylonitic zone represent granodiorite bodies where high-temperature deformation is partially preserved. a): orientation diagrams for C-planes and stretching lineation. Each point represents the mean orientation of more than 10 measurements in the selected outcrops.

aureole (Fig. 4b). Such diagrams are representative of deformation under moderate temperature, probably higher than 450°C. However, less severely deformed granodiorites, in relict aggregates of quartz display mosaic microstructures preserved from the late magmatic stage, defined by square- or rectangular-shaped grains with boundaries at right angles (Fig. 4c), which clearly demonstrate that solid-state deformation began close to the solidus temperatures (Blumenfeld et al., 1986). These quartz aggregates provide pronounced preferred crystallographic orientations with c-axes concentrated at less than 30° to the stretching lineation (Fig. 4c). Similar quartz fabrics have been described in the syntectonic Hermitage granite massif by Gapais and Barbarin (1986), and interpreted in terms of preferential growth and migration recrystallization of quartz grains along the c-axis during high-temperature solid-state flow. This interpretation is preferred here, since we have not found basal subboundaries indicative of intracrystalline deformation by c-slip as proposed for other syntectonic granites (Mainprice et al., 1986; Aranguren and Tubía, 1992). Our suggestion that the Conquista shear zone originated at sub-solidus temperature conditions is supported by additional microstructural and field evidence: (i) development of myrmekite rims along the long sides of K-feldspar porphyroclasts; and (ii) abundant wedges of undeformed quartz lodged in intracrystalline cracks of K-feldspar porphyroclasts (Fig. 3b). Micro-shear planes of mylonitic microstructures and crystallisation of red biotite are common; in contrast, those underlined by chlorite or other retrograde low-temperature minerals are scarce throughout the shear zone, and suggest that the Conquista shear zone was only active at fairly high temperatures. This interpretation is consistent with the fact that the mylonitic granodiorite is cross-cut by the Cerro Mogábar granite (Fig. 1), thus indicating that solid-state deformation in the shear zone predated total recrystallization of the batholith.

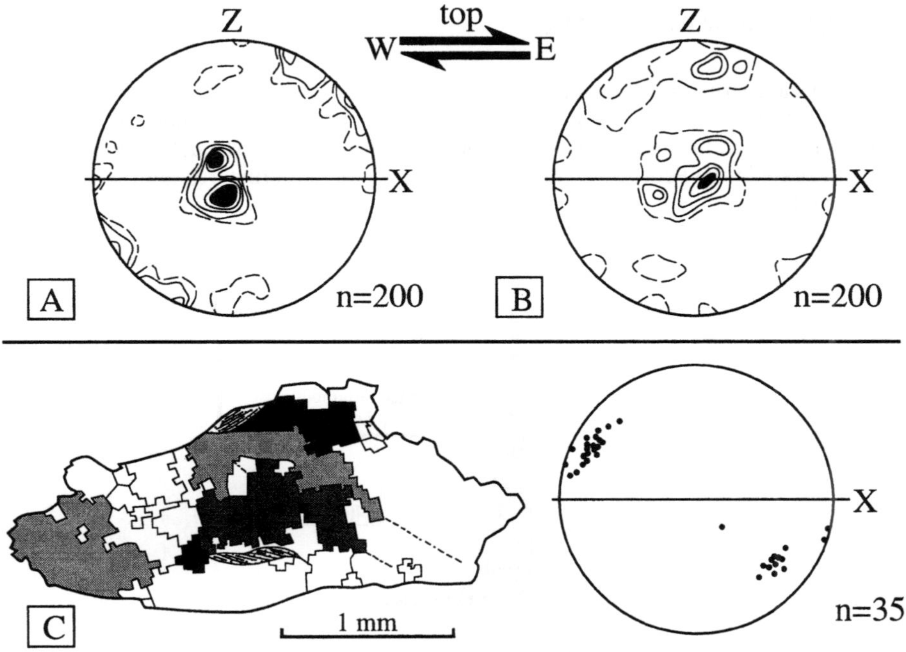

Figure 4.- Quartz c-axis patterns of mylonitic rocks deformed at intermediate temperatures. a: granodiorite; b: quartzite from the metamorphic aureole (XZ section; lower hemisphere, equal-area diagrams; contour intervals: 1, 2, 4, 6, ≥ 8%). c: relict high-temperature fabric of quartz showing the characteristic mosaic microstructure and the corresponding c-axis diagram (see the text for discussion).

Figure 3 shows that the shear zone is also imprinted in the country rocks and that isolated minor shear zones deform the dyke swarm. The structural imprint of the shear zone in the country rocks includes an asymmetric extensional crenulation cleavage, in the sense of Platt and Vissers (1980), parallel to the shear zone, and a subhorizontal crenulation cleavage already described in many epizonal areas of the CIZ and related to a late extensional event (Mira et al., 1987; Díez Balda et al., 1990). In our sector, the subhorizontal crenulation cleavage was clearly developed during the emplacement of the Los Pedroches granodiorite. This is evidenced by: (i) its increasing abundance towards the Conquista shear zone, and (ii) its nucleation at bulges along the grain-boundaries of the andalusite neoblasts, and its overprinting by randomly oriented neoblasts of epidote, as observed in oriented thin sections from the metamorphic aureole of the Los Pedroches granodiorite (Fig. 3c).

The structures described above and kinematic criteria consistently indicate that the Conquista shear zone is a dextral-transtensional shear zone. The Carboniferous slates are therefore inferred to have moved downwards above the Pedroches granodiorites. The dextral component of motion is also suggested by the obliquity between the stretching lineation and the batholith's long axis.

MAGNETIC SUSCEPTIBILITY

In order to gain structural information on the LPB, we have performed a detailed magnetic susceptibility study of the main part of the batholith. This technique allows us to measure planar and linear structures in apparently isotropic rocks (Bouchez et al., 1990). This is the case in the Pedroches granodiorite where internal structures are imperceptible to the naked eye.

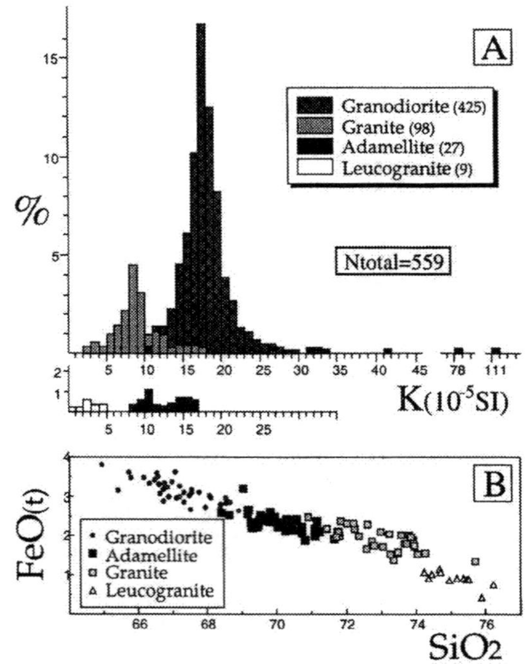

Figure 5.- (a) Histograms of the bulk magnetic susceptibility (K) values in the main petrographic facies of the LPB. (b) Variation in the Fe content (Harker diagram) for the main petrographic facies of the LPB. Note that K values increase with increasing Fe contents.

Structurally this technique is interesting because the magnetic susceptibility is a petrophysical property which can be expressed by a symmetric second rank tensor and geometrically represented as an ellipsoid (Jelinek, 1981) whose principal axes $K_{max} \geq K_{int} \geq K_{min}$ often correlate to the main directions X, Y and Z of finite strain (Borradaile, 1991; Hrouda, 1993). The bulk susceptibility magnitude corresponds to $K = (K_1 + K_2 + K_3) / 3$. The reader is referred to Tarling and Hrouda (1993) for an up-to-date review of theoretical fundamentals of this technique, and to Bouchez et al. (1990) and Bouchez (this volume) for specific applications to granitic rocks. Magnetic susceptibility and anisotropy measurements (AMS) were performed with a KLY-2 magnetic susceptibility bridge manufactured by Geofyzika (now Brno; Czech Republic). This instrument operates in a low magnetic field ($\pm$ 4 x 10^{-4} T and 920 Hz) with a resolution better than 5 x 10^{-8} SI units.

The study is based on a regular sampling of 559 stations: 439 from the granodiorites and 120 from the granitic plutons. Since the work concerns the overall structure and emplacement mechanism of the LPB, we have not included the specimens from the dykes. A minimum of four samples were taken per station using the standard palaeomagnetic field techniques for collecting oriented cores. Our study is therefore based on magnetic measurements of 2236 oriented cores.

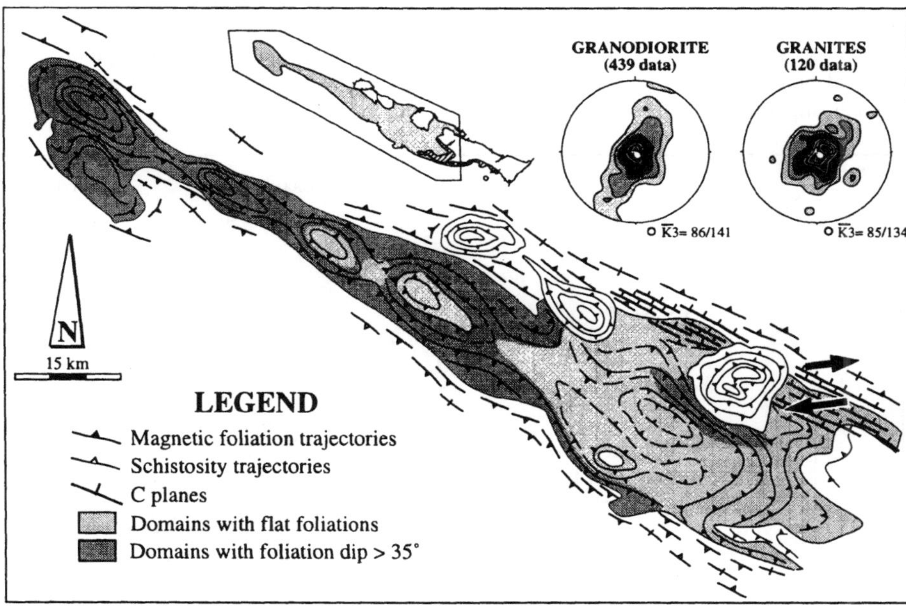

Figure 6.- Simplified map of foliation trajectories in the LPB. Note the presence of several granodiorite domes from the El Guijo granite to the West and the great extent of the eastern domain characterized by flat dipping foliations. Stereonets (Schmidt projection; lower hemisphere diagrams) for K3 (pole to magnetic foliation) magnetic axes are shown for granodiorite and granite samples. White dots indicate the best pole of the magnetic foliation. Contour intervals in multiples of uniform distribution.

All but three granodiorite samples give magnetic susceptibility values between K = 0.7 x 10^{-5} SI and 33.6 x 10^{-5} SI (Fig. 5). Rochette (1987) and Jover et al. (1989) have proposed that in granites such low K values mainly reflect the magnetic contribution of paramagnetic minerals, like biotite and hornblende. Figure 5 shows that the granodiorites have the highest K values and the leucogranites the lowest K values, the granites and adamellites being intermediate. The paramagnetic origin of the magnetic susceptibilities in the LPB is further supported by figure 5 where the susceptibility values linearly increase with increasing Fe-content of the sample, indicating the lack of substantial ferromagnetic contribution.

AMS DIRECTIONAL DATA

Despite the great extent of the Pedroches granodiorite, the corresponding orientation diagrams of the magnetic lineations (K_1) and the poles to magnetic foliations (K_3) reflect a surprising structural homogeneity. Poles to foliations define a girdle perpendicular to batholith's elongation, with a maximum pointing to the dominance of flat-dipping foliations (Fig. 6). Data from the granites reveal a still stronger concentration of K_3-magnetic axes around the vertical, and consequently point to sub-horizontal magnetic foliations (Fig. 6). Magnetic lineations are subhorizontal, and show a mean east-west orientation in the Pedroches granodiorite (Fig. 7) whereas, in the granitic plutons, they are scattered at the periphery of the diagram (Fig. 7). The shape of the AMS ellipsoid is slightly oblate, and similar in both the granodiorites and granites. Consequently, the scattered pattern of magnetic lineations in the granites can be interpreted as due to a radial arrangement.

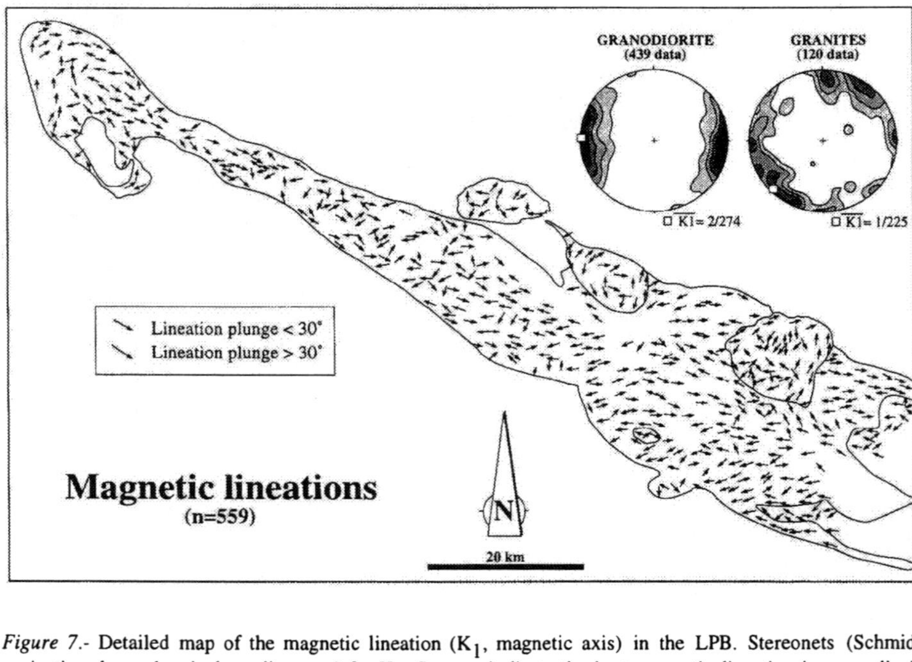

Figure 7.- Detailed map of the magnetic lineation (K_1, magnetic axis) in the LPB. Stereonets (Schmidt projection; lower hemisphere diagrams) for K_1. Squares indicate the best magnetic lineation in granodiorites and granites. Contour intervals in multiples of uniform distribution.

The previously described structural homogeneity of the Pedroches granodiorite is less obvious if we consider the maps of the LPB instead of the whole orientation data (Figs. 6 and 7). In fact, spatial variations in the trend of foliation trajectories help to recognize four domes at the west of the Pedroches granodiorite. In addition, the flat-dipping foliations are mainly found in the eastern part of the granodiorite outcrop, whereas higher dip values, > 35°, are concentrated in the domain of the domes (Fig. 6). The above mentionned variations in foliation orientations suggest the existence of a large, subhorizontal sheet of granodiorite in the easternmost part of the Los Pedroches granodiorite, in clear contrast with the polydomal nature of the western part (Fig. 6). A similar conclusion can be derived from the spatial variations of the lineation trends, since they keep a rather uniform east-west trend inside the domain of flat-dipping foliation, but diverge by more than 60° in trend, from 60°E to 120°E in the domain of the domes (Fig. 7). In the granitic plutons, the foliation trajectories give concentric patterns that are concordant with the granite contact, and magnetic lineations display radial patterns. Some disturbances of foliation trajectories in the granodiorites such as an isolated zone with high dip-values adjacent to the southern contact of the Cerro Mogábar granite, or the foliation triple point observed around its western border, could be due to reorientations in relation with the intrusion of the Cerro Mogábar granite.

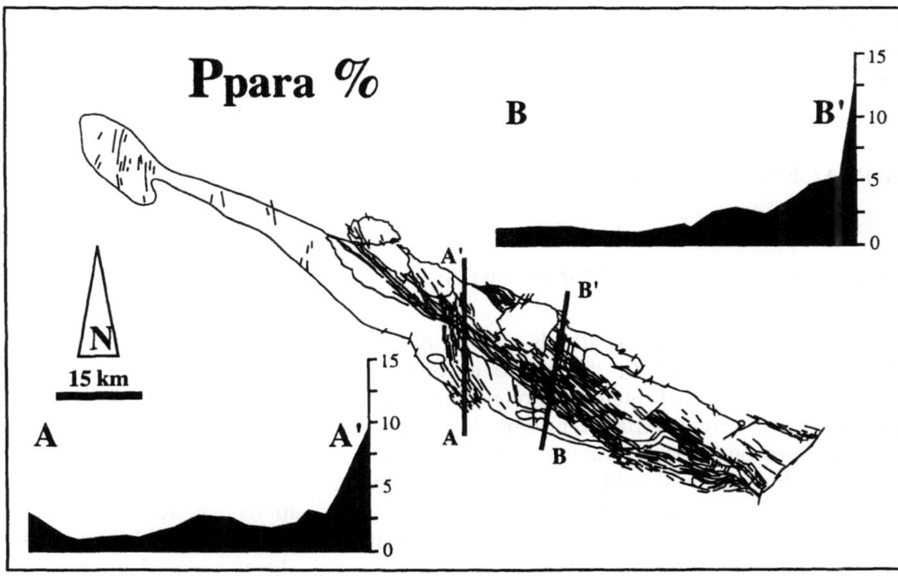

Figure 8.- Variation of the bulk anisotropy (P_{para} %) of the magnetic susceptibility in two traverses across the Los Pedroches granodiorite. High anisotropy values are systematically concentrated along the Conquista shear zone.

AMS SCALAR DATA

Given the low K-values of rocks having magnetic susceptibility dominated by paramagnetic minerals, the AMS parameters calculation requires that the constant contribution, $K_{dia} \sim -1.4 \times 10^{-5}$ SI, of diamagnetic minerals, mainly quartz and feldspar, be corrected (Hrouda, 1986; Rochette, 1987; Bouchez et al., 1987). The anisotropy percentage of a rock due to paramagnetic minerals is therefore defined as $P_{para}\% = [((K_1-K_{dia})/(K_3-K_{dia})) - 1] \times 100$. $P_{para}\%$ ranges from 0.5 to 12.5, with rather low mean values of 2.0 in granodiorites and 2.25 in granites. As is expected in granites with magmatic fabrics (Bouchez et al., 1987) the great majority (97%) of the studied samples provide values of $P_{para}\% \leq 5$. Most of the samples having larger anisotropy values display solid-state deformation features and were collected inside the Conquista shear zone. The variations of $P_{para}\%$ across the LPB, show a pronounced increase from the southern contact towards the Conquista shear zone (Fig. 8).

Parsons and Thompson (1993) proposed that intrusion of magma along vertical dykes is an efficient mechanism in increasing the horizontal stresses. This is because, for a sufficient magma pressure, inflation of dykes emplaced perpendicularly to the least principal stress direction can create locally compressional conditions. These considerations could explain the finding of relative highs in the $P_{para}\%$ values observed along the two profiles across the LPB, because they correspond to the areas where the dyke swarm is best developed (Fig. 8). This interpretation is strengthened by the presence of weak solid-state deformation in narrow granodiorite bands in contact with many dykes, reinforcing the hypothesis that compressional conditions were locally reached during dyke intrusion.

EMPLACEMENT MODEL

GEOMETRY AND TYPOLOGY OF INTRUSIONS

The Los Pedroches batholith affords a series of relationships between deformation and magmatism which must be taken into account when modelling its emplacement at shallow crustal levels. Principal amongst these, is the shape of the batholith and the links between different intrusive groups. Inferences about the geometry of the plutonic suite are based on the structural maps of the batholith.

The predominance of subhorizontal foliations within the granodiorites and granites (Fig. 6) can be interpreted in different ways, since such foliations may develop in several structural settings: (i) close to the roof of the intrusive body (Bouchez and Diot, 1990), (ii) in syn-tectonic granites intrusive along extensional shear zones (Hutton et al., 1990), or ductile thrust zones (Aranguren and Tubía, 1992; Karlstrom et al., 1993), or (iii) in subhorizontal sheeted granites or sills (Corry, 1988). Our reconstruction of the pre-erosional configuration of the LPB suggests that the subhorizontal foliations of the granodiorite and the granitic plutons reflect intrusive bodies of sheeted shapes with horizontal extensions much larger than vertical ones. This interpretation is supported by: (i) the widespread presence of xenoliths of country rocks near the contacts and their absence inwards, and (ii) except in the Conquista shear zone, solid-state deformation structures in the LPB are scarce, contrary to what would be expected in granites emplaced along extensional shear zones or ductile thrust zones. Furthermore, within the Pedroches granodiorite, the mineral homogeneity maintained across the outcropping area and the low dip of the southern contact also support a sheeted geometry.

Our reconstruction of the three dimensional shape of the Pedroches granodiorite, as represented in figure 9, bears many similarities with a laccolithic intrusion, particularly the concordant relationship with country rocks and the sheeted geometry (Corry, 1988). However, we suggest that the Pedroches granodiorite should not be considered as a single intrusive body, but rather as a composite laccolith composed by at least five coalescent plutons (Fig. 6). No simple explanation has yet been given for the feeding and growth of laccoliths. The problem has been addressed by Clemens and Mawer (1992) and Corry (1988). They have shown that fracture propagation is an efficient transport mechanism of granitic magmas, and that the change from vertical to horizontal migration of magma leads to progressively growing laccoliths by the elevation of the overburden. The horizontal spreading of magma requires a rheological horizontal boundary within the intruded materials which can act as a barrier for the rising magma (Corry, 1988; Román Berdiel et al., 1995). These conditions are met in the LPB, since the existence of 130°E trending shear zones acting as magma conduits is strongly suggested by the presence of the Conquista shear zone, the alignment of the granodiorite domes, and the orientation of the dykes within the swarm (Fig. 7).

Hence, we consider that magma intruded upwards along a regional-scale fracture affecting the mechanically strong pre-Carboniferous country rocks. But, as the magma was unable to migrate further upwards into the overlying Carboniferous shales because of their plastic nature, it spread laterally along the boundary between the brittle pre-Carboniferous and the weak Carboniferous rocks.

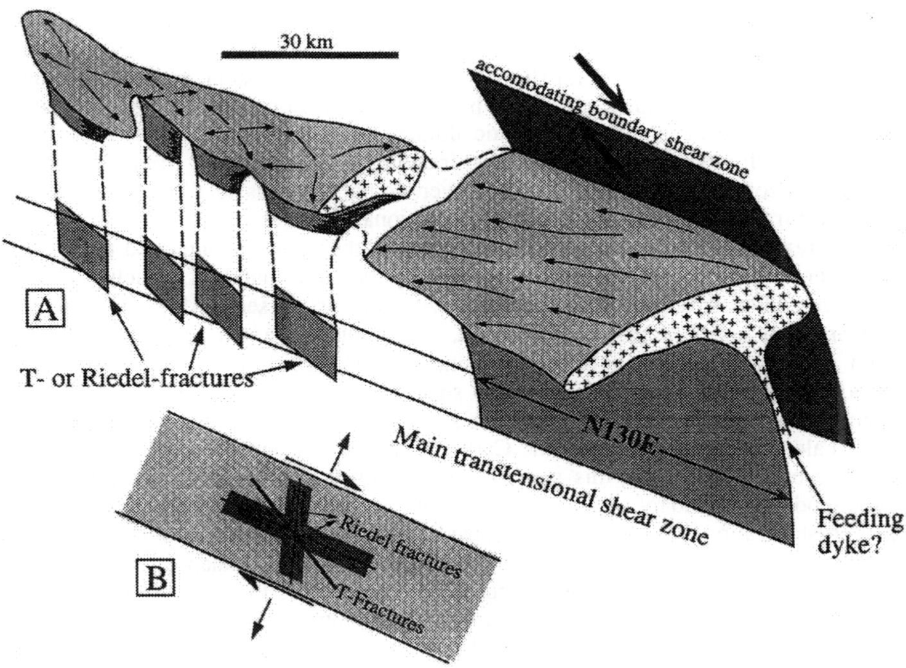

Figure 9.- Schematic three dimensional geometry proposed for the Pedroches granodiorite, showing its relationship with a blind main transtensional shear zone and with the Conquista shear zone, interpreted as an accommodating boundary shear zone (for discussion, see the text).

SEQUENCE OF MAGMATIC PULSES

The LPB was emplaced during three magmatic pulses characterized by different tectonic constraints. The oldest is represented by the Pedroches granodiorite and the Cardeña granite, and the second corresponds to the intrusion of dykes into the former rocks. These two magmatic events led to linear magmatic bodies that were probably related to the development of a single transtensional shear zone at crustal scale. According to Corry (1988) the thickest part of many laccoliths is located at the vertical position of the feeder dykes. In the LPB, it is tempting to consider that the thickest part and, consequently, the feeding zone should follow the Conquista shear zone, particularly if we take into account its transtensional nature. This is also suggested by the analogue models of laccoliths formed in extensional settings of Román Berdiel et al. (1995).

However, we think that the feeder zone is located close to the alignment defined by the *en échelon* granodioritic domes. We interpret this alignment as the surface expression of a major, blind transtensional shear zone (Fig. 9a) which provides an easy conduit for the upwelling of magma (Fig. 9b). Our choice for a transtension shear zone is supported by the parallelism of the dykes belonging to the principal family with the shear zone itself. This is based on the interpretation that the dykes were emplaced along Riedel shears, hence, as proposed by Sanderson & Marchini (1984), in transtension the Riedel shears and tension fractures can rotate until they reach a stable orientation enabling accommodation of the tensional component (Fig. 9b). It is worth noting that the weak solid-state deformation structures observed in the host rocks close to the dyke walls do provide constant criteria of dextral shear sense which are consistent with the motion of the predicted Riedel fractures of figure 9b. This interpretation also agrees with an *en échelon* arrangement of the granodioritic domes, since their small sizes suggest that they were fed through several parallel short fractures at an angle with the main transtensional shear zone. Our structural data, however, do not allow us to discriminate whether these short fractures correspond to Riedel or tensional fractures.

The transtensional shear zone vanishes towards the west, as suggested not only by the presence of the *en échelon* domes, but also by the more-than-half reduction in width of the granodioritic domes when compared with the eastern sector (Fig. 6). The shear zone probably also becomes less localized, then vanishes in the upper crustal levels, given the plastic behaviour of the Carboniferous slates found at such levels. These slates, located at the hanging-wall of the shear zone, apparently acted as a trap for the magma rising through the shear zone. The successive magma supplies which apparently ponded along the basal contact of the Carboniferous slates, led to the thickening of the intrusive body (Fig. 9). In this scenario the Conquista shear zone is interpreted as an accommodating boundary shear zone (Clemens and Mawer, 1992) that formed before final crystallization and cooling of the intruding magma. This interpretation implies that steady magma supply produced a dome in the overburden which balanced the thinning of the crust, and controlled subsequent crustal extension since the partially melted granodiorite was much weaker than the solid and cooler country rocks.

Numerous vertical dykes then intruded the quasi-isotropic and brittle materials of the Pedroches granodiorite and Cardeña granite, after their complete solidification. Some dykes clearly evolved into sills when reaching the overlying country rocks, thus showing a structural control of emplacement similar to that of the LPB.

Finally, the last magmatic pulse was responsible for the emplacement of the remaining granitic plutons (Fig. 1), which also define a 130°E trending alignment. However, the location of this alignment does not coincide with the dyke swarm or granodioritic dome alignments, suggesting that there was a northward migration of the youngest magmatic event.

REGIONAL IMPLICATIONS

The southern part of the Iberian Variscan belt presents numerous strike-slip shear zones with an orientation similar to that of the Conquista shear zone but, unlike the dextral nature of the latter, they show sinistral displacements (Fig. 1: insert). According to recently published structural syntheses of the Central Iberian Zone (Díez Balda et al., 1990, and references therein) the sinistral shear zones were produced during a collisional event leading to crustal thickening. Other dextral strike-slip shear zones have been recently recognized in the northern areas of the CIZ (Castro, 1985; Díez Balda et al., 1990). From the structural study of the Cabeza de Araya granite, Castro (1985) deduced the existence of an east-west trending dextral shear zone, the Central Extremadura shear zone, that was interpreted as the lateral boundary of a crustal wedge thickened by eastward verging thrust faults.

Extensional shear zones generated during the gravitational collapse of the thickened Variscan belt are now envisaged as a common structural element of the CIZ (Díez Balda et al., 1990; Hernández-Enrile, 1991; Escuder Viruete et al., 1994). Because the transtensional shear zones described for the first time in this paper are kinematically compatible with the extensional shear zones recognized within the CIZ (Hernández-Enrile, 1991; Escuder Viruete et al., 1994), and also because the beginning of the extensional event should have taken place between 325 and 310 Ma (Escuder Viruete et al., 1994) the emplacement of the LPB and its related transtensional shear zones are proposed to be related to late extensional evolution of the Iberian Variscan belt.

The extensional shear zones of the CIZ are subhorizontal whereas the Conquista shear zone and the main transtensional shear zone at the base of the Pedroches granodiorite strongly dip towards the northeast (Figs. 3 and 9). Since the subhorizontal and transtensional shear zones have an eastward directed motion of the hanging-wall plate, we consider that both structures can be integrated in a single tectonic scenario in which our transtensional shear zones should correspond to lateral boundary ramps and the subhorizontal shear zones to detachments of a crustal-scale extensional system.

Our reconstruction of the LPB emplacement suggests that the extention of the transtensional shear zones reaches deep levels in the crust. As a consequence, we expect that subhorizontal detachment contacts do exist at deep levels in the crust of the CIZ. This possibility should be taken into account when interpreting the structural history of this sector of the Iberian Variscan belt using seismic reflection profiles.

ACKNOWLEDGEMENTS

This study is supported by the research projects PB93-1149-C03-03 and UPV130.310-EA090/95. We thank J.L. Bouchez, J.L. Vigneresse, W.E. Stephens, M. de Saint Blanquat and R. Law for their careful revision, criticism and improvement of the manuscript.

DEVELOPMENT OF MAGMATIC TO SOLID-STATE FABRICS DURING SYNTECTONIC EMPLACEMENT OF THE MONO CREEK GRANITE, SIERRA NEVADA BATHOLITH

Michel de SAINT BLANQUAT
CNRS - UMR 5563, Laboratoire de Pétrophysique et Tectonique, Université Paul-Sabatier, 38 rue des Trente-Six Ponts, Toulouse, France
and Basil TIKOFF
Department of Geology and Geophysics, University of Minnesota, Minneapolis, USA

ABSTRACT

Interpretation of rock fabric is a key objective in obtaining kinematic information about crustal dynamics. In order to constrain the connection between tectonism and magma emplacement, we have conducted a structural, microstructural, and magnetic fabric study of the Mono Creek Granite (MCG), one of the youngest plutons of the Sierra Nevada batholith, California (USA). In addition to field measurements, we have used the Anisotropy of Magnetic Susceptibility (AMS) to investigate granite fabrics. The magnetic susceptibility (K) of the MCG varies between 0.4 and $4x10^{-2}$, and the total anisotropy (P) reaches 1.6. Thermomagnetic measurements (not given), and reflected and optical microscopic observations indicate that Ti-poor multi-domain magnetite is the primary AMS carrier, accouting for the high K values. The AMS intensive parameters, which are difficult to interpret in ferromagnetic rocks, provide semi-quantitative information: K correlates with rock mineralogy, and P is a good indicator of the rock texture. The AMS directional data of the MCG are characterized by good clustering at each sampling station, and by a sigmoidal pattern of foliation and lineation. This pattern results from deformation in the Rosy Finch shear zone during syn-magmatic dextral shearing deformation event. The AMS data, combined with the microstructural study, indicate that dextral shearing occurred under magmatic, high-temperature solid-state, and low-temperature solid-state conditions. The geometrical, structural and temporal continuity between the magmatic and solid-state deformations strongly support a model of continuous shearing during emplacement of the MCG. Our study demonstrates that the fabric of the MCG retains a record of all the pluton construction, which is syn-tectonic. Therefore, within the MCG, the magmatic and solid-state fabrics are both "intrusion-induced" and "tectonic-induced".

J.L. Bouchez et al. (eds.) Granite: From Segregation of Melt to Emplacement Fabrics, 231–252.

INTRODUCTION

The evolution of a magma can be summarized chronologically in four steps: segregation, ascent, emplacement, and cooling. As soon as a magma contains crystals, a fabric (foliation and lineation) forms, and evolves continuously, reflecting the strain condition at each step. Kinematic information about the flowing and cooling magma, particularly related to emplacement and subsequent deformation, can be recorded by mineral fabrics.

A fabric is defined by the shape, orientation, and spatial distribution of minerals, and by the mechanical condition of its formation. For magmatic rocks, the significance of the fabric is similar to that of other rocks: it is a deformation feature that describes both the orientation and the shape (e.g., flattening, constriction) of the strain ellipsoid. As in a deforming magma the early-formed crystals rotate and undergo relative translations, fabrics indirectly inform on intensity and orientation of strain. An example is the shape preferred orientation (SPO) of the early crystallizing minerals (mainly Fe-Mg silicates and feldpars). Information about strain magnitude is only qualitative, because the molten nature of the material does not permit an exact correlation between the strength of the fabric (degree of SPO) and the magnitude of total strain to be specified: we do not know when the strain-clock starts. Thus, fabrics in magmatic rocks could track either infinitesimal or finite strain fields. Like with non-magmatic rocks, the fabric records only the strain component (stretching plus internal rotation) and not the translation component of the deformation (see Means, 1990). The translation component might be inferred from the pattern of finite strain, such as in (simple-shear) laminar flow, but in general it must be determined independently.

Different parts of a plutonic body can acquire fabrics at different times in the crystallization history. Interpretation of fabric patterns in terms of strain-time evolution may be conducted, for example, with the help of a map of microstructures, which informs on the rheological condition of formation of the fabric. The distinction between intrusion-induced fabrics and tectonic-induced fabrics (Hutton, 1988) can be solved if we can determine the relative ratio between internal forces and tectonic forces (Brun et al., 1990). In fact, real cases are most likely intermediate between these end members, and there is no reason to assume that a pluton is characterized only by one type of mechanism (Saint Blanquat and Tikoff, 1995). Further, the distinction between intrusion-induced and tectonic-induced fabric is irrelevant if magma emplacement is controlled by tectonic processes. In that case, the fabric records the interactions between internal driving forces (buoyancy) and tectonic forces. An emplacement fabric *is* a tectonic fabric if we can prove that a regional deformation event is responsible of the pluton construction.

Mapping the microstructure, orientation, and intensity of fabrics is a very efficient tool to characterize the space/time/temperature/deformation relationships within a pluton. The interest in magmatic fabrics, particularly of granitic rocks, is in large part due to rapid pluton emplacement and cooling, making granites short lived geological events, often less than 10^5-10^6 years (Clemens and Mawer, 1992; Hanson and Glazner, 1995), and critical recorders of crustal deformation (Bouchez et al., 1981; Hutton, 1988; Karlstrom, 1989; Hutton and Reavy, 1992; Karlstrom et al., 1993; Bergantz and Dawes, 1994). This interest is often tempered by questions about the significance of fabrics in granite (see Bergantz, 1995, for a recent review). If fabric within plutons is always recorded very late in the crystallization process, it may not reflect magma emplacement. If this hypothesis is true, emplacement of magma may take place for exemple in tectonic «lulls », between deformation events (Paterson, 1989; Paterson and Vernon, 1995).

The aim of this paper is to show, 1) that the fabric of granitoids could have a non-negligible structural memory and provide valuable information about kinematics of magma emplacement, and 2) the connection between tectonism and magma emplacement in the specific example of the Mono Creek Granite. To document the first point, we need a rigorous documentation of fabric within granite. This goal is difficult because of the low anisotropy of granitic rock, especially if only magmatic-state deformation is recorded. In addition to field measurements, we have utilized the Anisotropy of the Magnetic Susceptibility, or AMS (Hrouda, 1982; Borradaille, 1988; Rochette et al., 1992), to quantify the fabric of granitic rocks (see Bouchez, this vol.). AMS provides a rapid insight into the bulk internal structure of a pluton and allows the definition of planar and especially linear fabrics which are commonly hard or impossible to recognize in the field. Numerous studies have demonstrated a close parallelism between the ellipsoid of magnetic susceptibility and grain shape fabrics (Launeau et al., 1990; Darrozes et al., 1994; Cruden and Launeau, 1994; Archanjo et al., 1995; Grégoire et al., 1995; Sigiesmund et al., 1995) and AMS is now almost routinely used in granitoids for the analysis of both solid-state fabrics and more subtle magmatic fabrics.

To document the connection between tectonism and magma emplacement, we need a pluton which is not overprinted by later intrusions and/or deformation events. The Mono Creek granite (MCG) is one of the youngest plutons of the Sierra Nevada batholith of California and structural studies suggest a relationship between magma emplacement and transpressional deformation for this pluton (Tikoff and Teyssier, 1992; Tikoff and Saint Blanquat, in review). Application of AMS techniques on this pluton provides critical data to constrain the significance of fabrics, and therefore the emplacement model. We first present results of the field and microstructural studies, followed by the AMS methodology and results. The discussion will focus on the interpretation of magnetic fabrics in ferromagnetic granites, and their interpretation in terms of regional tectonics and emplacement of magma.

GEOLOGICAL SETTING

The Mono Creek Granite (MCG) is located on the eastern flank of the central part of the Sierra Nevada Batholith (SNB) (Fig. 1). The SNB is a 600 km long belt of a continental magmatic arc stretching discontinuously from southeastern Alaska to Baja California, and was constructed from the Triassic to the late Cretaceous. The Cretaceous Sierra Nevada batholith was active from 120-83 Ma (Stern et al., 1981; Bateman, 1992), and was the result of the eastward subduction of the Farallon plate below North America (Engebretson et al., 1985). The latest magmatic event within the SNB is called the Cathedral Range Intrusive Epoch (Evernden and Kistler, 1970), and is Late Cretaceous in age (93-81 Ma). The plutons of this intrusive epoch constitute about 50% of the batholith surface area (Chen and Moore, 1982). Several intrusive series were emplaced along the crest of the Sierra Nevada during this event: from north to south, the Sonora pluton, the Tuolumne Intrusive Series (TIS) (Bateman and Chapell, 1979; Kistler et al., 1986), the Mono Pass Intrusive Series (MPIS) (part of the John Muir Intrusive Suite of Bateman, 1992), and the Mount Whitney Intrusive Series (MWIS) (Hirt, 1989). Each Intrusive Series outcrops over an area between 1000 and 2000 km^2, and the thickness of the plutons is estimated to be about 10 km (Oliver, 1977). The series are characterized

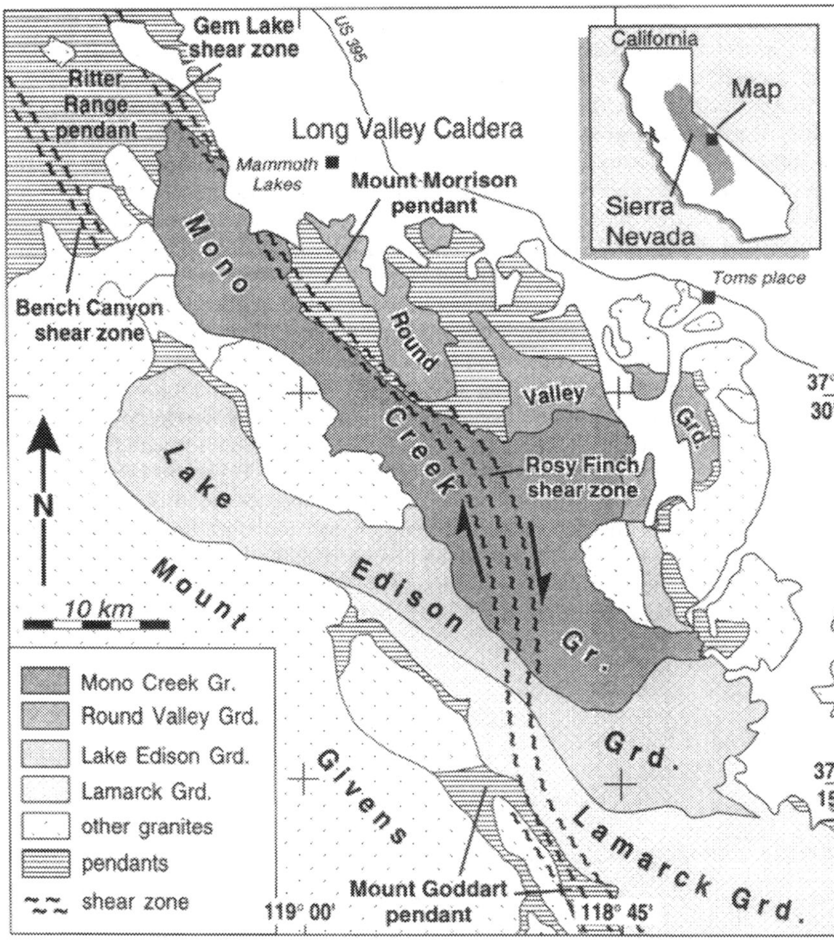

Figure 1. Generalized geological map of the east-central Sierra Nevada (inset), showing the Mono Creek granite and surrounding rocks of the John Muir Intrusive suite (modified from Bateman, 1992; Gem Lake shear zone after Greene and Schweickert, 1995; Bench Canyon shear zone after McNulty, 1995).

by NNW-elongated plutons, with a shape ratio of up to 5:1, ranging in composition from granodiorites for the older plutons to granite for the younger plutons. Plutons within the Intrusive Series are comagmatic and show crystallization from the outer pluton inward, from similar parent melts. Each series is composed of about 3-4 plutons, which are nested (TIS and MWIS) or arranged en échelon (MPIS). They were emplaced between 92 and 83 Ma (Stern et al., 1981; Chen and Moore, 1982; Kistler et al., 1986; Tobisch et al., 1995) at upper-crustal depths (2 - 6 km) (Ague and Brimhall, 1988).

Structural evidence for strike-slip deformation associated with emplacement of these late granitic rocks has been recently documented. Proposed large-scale transcurrent and/or transpressional faults/shear zones include the proto-Kern Canyon fault (Busby-Spera and Saleeby, 1990), Rosy Finch shear zone (Tikoff and Teyssier, 1991, 1992), and Gem Lake shear zone (Greene and Schweickhert, 1995). In the central Sierra Nevada, shear zones which were active before 90 Ma show dip-slip movement and associated plutons appear to be emplaced in an extensional context (Tobisch et al., 1993; McNulty, 1995; Tobisch et al., 1995), whereas shear-zones active after 90 Ma show strike-slip movement (Busby-Spera and Saleeby, 1990; Tikoff and Teyssier, 1992; Greene & Schweickert, 1995). Considering the evidence from regional studies that the Sierra Nevada magmatic arc underwent normal contraction coeval with pluton emplacement and strike-slip faulting, Tikoff and Teyssier (1991, 1992) proposed a transpressional setting during emplacement of the Late Cretaceous granitoids of the Sierra Nevada batholith. In particular, TIS, MPIS, and MWIS plutons may have been emplaced within tensional bridges between crustal-scale en échelon P shears during dextral shearing of the arc in response to oblique convergence at the North America plate boundary. This hypothesis offers a solution to the long-standing paradox that large volumes of granitoids apparently intrude a crust which is in overall compression.

The Mono Creek Granite (Lockwood and Lydon, 1975) is the youngest pluton of the MPIS (86 Ma, B. Carl, pers. com. 1996), and one of the youngest plutons of the Cretaceous SNB. Wall rocks of the MCG are primarily the slightly older plutons of the MPIS, the Lamarck (Frost and Mahood, 1987), Lake Edison, and Round Valley granodiorites, and Paleozoic and Mesozoic metasedimentary and metavolcanics rocks. The north-east side of the pluton is cut by the western part of the 0.7 Ma Long Valley Caldera.

FIELD DATA

The MCG is a 55 km long pluton elongated in the NW/SE direction, and is about 3 to 13 km wide (about 600 km^2). It is characterized by a northeast oriented and 10 km wide protrusion or "bulge" along its eastern margin (Fig. 1). Vertical relief, between 1900 and 4200 m, is about 2300 m. Exposûre is excellent, except in the northern part, where it is covered by recent lava flows. The MCG is a porphyritic granite, very light coloured in outcrop. Quartz, K-feldspar, and plagioclase make up 90% of the rock. Biotite and magnetite are the ubiquitous mafic phases, while titanite, hornblende, and allanite are accessories mafic phases. Average grain size is about 3-4 mm except for large K-feldspars which average 10-20 mm and are as large as 40 mm. Plagioclase crystals show oscillatory zoning. One of the most impressive features of the pluton is its homogeneity: no sheeting or internal contacts have been observed throughout the main body. Mafic enclaves are very rare but more present along the margins.

Our detailed field study and thin section analysis shows some consistent and gradual lateral mineralogical variations, independent of the elevation. We observe a transition between a mafic facies (biotite + titanite + hornblende), an intermediate facies (biotite + titanite), and a more felsic facies (biotite) (Fig. 2a). All of these facies contain quartz, feldspar, magnetite, and other accesory minerals. The mafic facies is located in the bulge

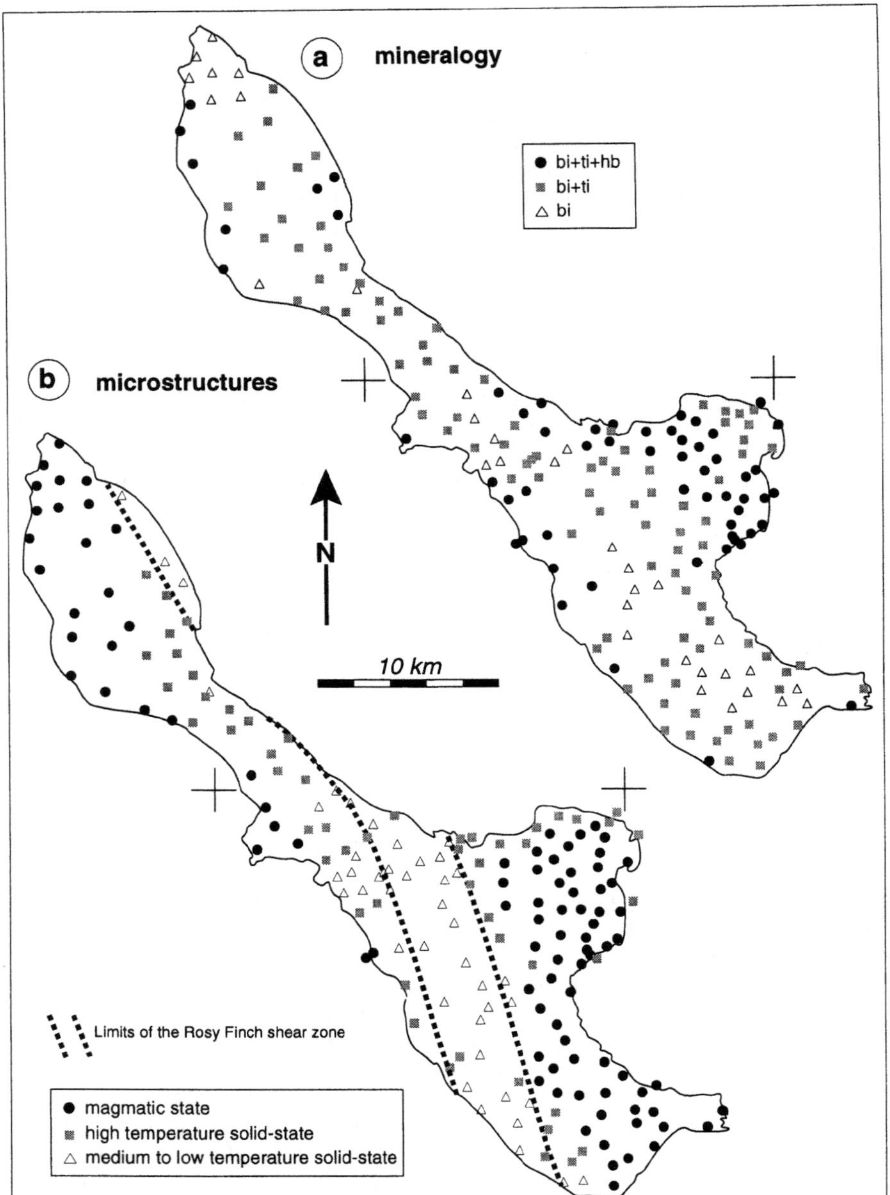

Figure 2. Map view of the main results of the thin section analysis for the 183 AMS sites within the MCG and 7 sites in adjacent plutons. (a) mineralogy; bi : biotite, ti : titanite, and hb : hornblende; the bi+ti+hb sites are located on the pluton's margin and in the bulge; (b) microstructures and field map of the RFSZ. Note the evolution of the deformation zone from magmatic, to medium to low temperature deformation.

and along the margins of the pluton. The felsic facies rocks are found in the center of the pluton, parallel to its long axis, and in the northernmost part of the pluton. These observations correlate well with the mapping of alkali feldspar megacryst abundance in the Mount Abbot quadrange by Lockwood (1975), who showed a systematic decrease toward the center of the pluton and no relation between megacryst abundance and elevation. There is also a good correlation between the high content of megacrysts (> 8%), mafic facies, and "high" (≈ 2.66) rock-density values found by Lockwood (1975). These data clearly demonstrate that the bulge is slightly different from the main body of the pluton.

The most striking structural feature of the pluton is the Rosy Finch shear zone (RFSZ) (Tikoff and Teyssier, 1992). The RFSZ is a N20°W subvertical shear zone which is at least 80 km long and 1-4 km wide. It crosscuts the MCG and can be followed toward the south through the entire MPIS. The RFSZ is ductile within the MCG and ductile, ductile-brittle, and/or brittle within the older plutons of the MPIS (Lake Edison and Lamarck granodiorites). Toward the north, the RFSZ is in continuity with the Gem Lake shear zone (Tikoff and Greene, 1994; Greene and Schweickert, 1995). The RFSZ in the Mono Creek granite is easily recognized in the field as an orthogneiss, in which solid-state foliation and lineation are developed. The style of deformation varies slightly along strike, but in general, foliation is defined by the elongation of quartz aggregates and by mica folia wrapping around well preserved feldspar porphyroclasts. The lineation is typically defined by the elongation of quartz grains and by a streaky lineation on mica folia, resembling a slickenside. Biotite, and/or hornblende, is generally retrogressed to chlorite, epidote, and/or white mica within these streaky lineations, which tend to occur in the middle of the shear zone where strain becomes localized. The shear zone parallels the Mono Creek granite/Mt. Morrison pendant contact on the northeastern side of the granite (160° or N20°W), is oriented N-S through the middle of the pluton (Fig. 1). At the south end of the pluton, the shear zone mostly continues into the Lake Edison granodiorite although dextral shearing also parallels the southwestern side of the Mono Creek granite (N20°W). Although the shear and foliation planes within the shear zone are consistently sub-vertical, the plunge of the lineation varies along strike. In the northeastern section of the granite, where the pluton parallels the Mt. Morrison roof pendant, the lineation plunges 0° to 50° to the south. In the middle section of the RFSZ the lineation is dominantly sub-horizontal and in the southwestern section, by the contact with the Lake Edison granodiorite, the lineation plunges shallowly northward. Outcrop scale-patterns indicate bulk dextral shear, including folded aplitic dikes, S-C relationships, asymetric porphyroclast tails, and porphyroclast imbrication (Tikoff and Teyssier, 1994). Dating of minerals (U-Pb on zircon, Ar-Ar on hornblende and biotite) from the RFSZ within the Lake Edison pluton suggests ductile movement on the RFSZ between 88 and 83 Ma (Tobisch et al., 1995), i.e. during emplacement of the MCG. Other Ar/Ar studies on white mica adjacent to the MCG indicate the possibility of shearing lasting until ~ 80 Ma (Sharp et al., 1993; Segall et al., 1990).

Brittle, brittle-ductile, and ductile sinistral shear zones are seen in all the plutons of the MPIS, and are oriented ENE-NE with subhorizontal lineations (Lockwood and Moore, 1979). These faults in the Lake Edison granodiorite displace dikes from the Mono Creek granite, and the faults become increasingly more ductile toward the contact with the Mono Creek granite, pointing to a close time relationship between magma emplacement in the MCG and deformation in the surrounding and already solidified plutons. Displacements on these faults are at least one order of magnitude smaller than displacement along the RFSZ.

The main evidence for magma emplacement during contraction are sub-vertical foliations throughout the MPIS and sub-horizontal dike swarms. Sub-vertical, NW foliations are found throughout the Mono Creek, Lake Edison, and Round Valley plutons and their orientation cross-cuts pluton contacts, arguing for trans-arc contraction during emplacement. Sub-horizontal dike complexes emanate from several plutons of the Cathedral Range Intrusive Epoch (Tikoff, 1994), but are particularly well-developed emanating from the MCG. These aplitic dike complexes are granitic in composition, locally constitute 50% of the adjacent rocks, dip between 0° to 30° W on the east side of the MCG, and extend for up to 2 km from the edge of the granite. The horizontal orientation of this dike swarm implies a maximum instantaneous strain - or least compressive stress - in the vertical direction, consistent with contractional deformation during magma emplacement. However, within the MCG, the dikes are more subvertical and strike N-S to NE-SW. This orientation could reflect a different stress /strain orientation within the magma chamber during its construction, such as an horizontal orientation for the least compressive stress axis, consistent with the tensional bridge of the P-shear model (Tikoff & Saint Blanquat, in review).

MICROSTRUCTURES

Direct examination in the field and microscope examination of thin sections from each of our 183 AMS sites were analyzed to determine magmatic versus solid state deformation, in order to compare with the AMS study. Magmatic fabrics form above the magma solidus and no solid-state strain occurs after full crystallization (Blumenfeld and Bouchez, 1988; Paterson et al., 1989). A submagmatic fabric forms above the solidus, but with a high content in crystals making a rigid framework allowing brittle or plastic behavior of the crystals despite the presence of melt (Bouchez et al., 1992). In our study, we have grouped both magmatic and sub-magmatic textures as magmatic (M). A solid-state fabric forms below the solidus with plastic and/or brittle deformation of minerals. Solid-state fabrics can be subdivided into different thermal steps if microstructural criteria are presented (Gapais and Barbarin, 1986; Blumenfeld and Bouchez, 1988; Gapais, 1989; Paterson et al., 1989; Miller and Paterson, 1994). In our study, we have distinguished between high-temperature solid-state (HT) and medium to low temperature solid state (MLT), described in more detail below. If the pluton has a simple thermal history, including no re-heating by later intrusions, a map of fabrics - magmatic (M), high-temperature solid-state (HT), and medium to low temperature solid state (MLT) - provides a relative chronology between the different textures.

MAP DISTRIBUTION

The microstructural types observed in thin section are reported in map view in figure 2b. The distribution of the texture is more or less symmetric with respect to the RFSZ, in good agreement with the field mapping of the shear zone. MLT textures are found everywhere within the RFSZ, and HT textures are found adjacent to the RFSZ. M textures are preserved farther from the shear zones, although there is a continuous transition between the shear zone and the magmatic texture. The deformation zone characterized by M textures is broader than the solid state zone, suggesting localization of strain with increasing deformation and decreasing temperature. This observation leads us to interpret the RFSZ as a syn- to post- magmatic (or syn- to post-emplacement) shear zone: solid-state deformation follows continuously magmatic deformation. This result was previously suggested by Tikoff and Teyssier (1994), who showed from 2D modelling of porphyroclast interaction that the strength of the K-feldspar megacryst fabric in the granite is higher than that predicted from just solid-state strain determination. It was concluded that the feldspar fabric was partly acquired during magmatic flow, and then served as an initial fabric for later solid-state deformation.

OUTSIDE SHEAR ZONE

The texture of the MCG, outside the RFSZ, is magmatic and characterized by large and equant quartz grains, poorly aligned mafic minerals, and phenocrysts of feldspars that lack solid-state deformation. In thin section, plagioclase and K-feldspars are undeformed. Some sites are characterized by a slight chess-board pattern of subgrains in quartz which denote some deformation under subsolidus condition, but at high temperature, with the overall magmatic texture remaining unmodified. The bulk fabric of the rock remains that of a dominant magmatic imprint. HT textures are found toward the RFSZ and are difficult to differentiate in the field from the magmatic textures: the former show a slight decrease in the quartz grain size and a more pronounced shape preferred orientation of K-feldspar megacrysts. In thin section, this texture is characterized by systematic chess-board patterns of sub-boundaries in quartz, evolving into new grains, ductilely deformed plagioclases, and kinked biotites. We observe a few basal sub-boundaries (c-slip) and perfectly polygonized prismatic sub-boundaries which are frequently associated with grain-boundary migration (bowed grain boundaries). The high temperature of deformation is attested by the stability of biotite, the rather large size of the new grains of quartz, and c-slip in quartz (Blumenfeld et al., 1986; Okudaira et al., 1995). This deformation probably occurred just after the full crystallization of the granite.

WITHIN SHEAR ZONE

Within the RFSZ we found orthogneissic to mylonitic textures, which are attributed to medium to low temperature solid-state deformation. They are characterized by a significant reduction in grain size, development of shear band and S-C structures, and pressure shadows adjacent to broken feldspar megacrysts. The quartz aggregates defining the foliation are elongate in the direction of finite extension, consistent with the elongation of the rare xenoliths, and show no evidence of large strain gradients on the outcrop scale. In thin section, the grains have irregular, lobate shapes suggesting grain boundary migration and subgrain rotation, indicating high temperatures during deformation (Gapais and Barbarin, 1986; Hirth and Tullis, 1992). Where deformation is localized, the quartz aggregates form elongate ribbons, and elongate new grains are often visible. These structures represent deformation at a slightly lower temperature (Hirth and Tullis, 1992), and are particularly well developed immediately adjacent to the wallrocks of the Mt. Morrison pendant. In thin section, mica fish, σ and δ porphyroclasts, and "bookshelf" displacement on feldspar cleavages (Simpson and Schmidt, 1983; Passchier and Simpson, 1986) all indicate dextral shear. Hornblende and biotite are commonly partially retrogressed to chlorite. Shear bands consisting primarily of quartz and micas are observed in thin section, and indicate dextral shear. They vary in orientation, and wrap around larger feldspar crystals. Microstructurally, K-feldspar megacrysts sometimes show evidence of recrystallization in the pressure shadow areas, although more commonly quartz or myrmekite are found in this location. These large K-feldspars commonly contain cracks infilled by quartz and/or plagioclase, which are interpreted as resulting from brittle deformation at the grain scale in the presence of residual melt (sub-magmatic deformation, Bouchez et al., 1992; John and Stünitz, this volume). These microstructures show the existence in a single outcrop of an early magmatic and/or high temperature deformation before the medium to low temperature episode. Thus, deformation seems to have occurred at a range of temperatures consistent with strain modeling and interpretation of the RFSZ as a syn-magmatic shear zone.

AMS ANALYSIS

THEORY

The magnetic susceptibility (K) is defined by M=KxH, where M is the induced magnetization of the material and H is the inducing magnetic field. As both M and H are expressed in Amperes per meter, the volumetric susceptibility is dimensionless. Susceptibility varies according to the applied magnetic field and temperature, and may also vary according to the measurement direction resulting in a non-parallelism between H and M vectors. The output of an AMS measurement of one rock sample is an ellipsoid of magnetic susceptibility defined by the length and orientation of its three principal axes, $k_{max} \geq k_{int} \geq k_{min}$, which are the three eigenvectors of the susceptibility tensor. Depending on their mineralogy, granitic rocks may have paramagnetic or ferromagnetic behavior. In paramagnetic granites, AMS is induced by the Fe-bearing silicates such as biotite or amphibole. In ferromagnetic granites, i.e. magnetite-bearing granites, the paramagnetic contribution of Fe-Mg silicates may become negligible with respect to the ferromagnetic contribution because of the high intrinsic magnetic susceptibility of magnetite. Hence, the magnetic fabric results primarily from the shape anisotropy of the grains of magnetite. This effect was postulated by Stacey (1960), examined by Khan (1962) and Uyeda et al. (1963) by microscopic observations on different rock types, and by Ellwood and Whitney (1980) and Grégoire et al. (1995) on granites. A component of the magnetic anisotropy may also result from the anisotropy of grain distribution, often referred to as distribution anisotropy (Hargraves et al., 1990; Stephenson, 1994; Grégoire et al., 1995).

The magmatic texture is defined by the shape preferred orientation (SPO) of ferromagnesian minerals and feldspar phenocrysts. Their SPO must be associated with the orientation and the characteristics of the magmatic flow: they are early-crystallized minerals which are oriented in the flow while the magma is still viscous. The shape and distribution of magnetite grains should reflect the geometrical characteristics of the SPO of these early-crystallizing minerals, regardless of when the magnetite crystallized during the emplacement history of the granitoid: early crystallized magnetite grains would become aligned with the earlier-formed minerals, while later crystallized magnetite would preferentially grow along the pre-existing grain-shape fabric of the ferromagnesian minerals (Knight & Walker, 1988) (see below). The magnetic anisotropy that results from the mean orientation of these minerals is thought to reflect the orientation and shape of the finite strain ellipsoid. The plane of maximum flattening (foliation) and the direction of maximum elongation (lineation) thus defined can be mapped over an entire pluton and, along with microstructural data, are used to interpret the structural history of the granitoid.

METHODOLOGY

Two accurately oriented hand-samples were collected in the field at each of the 183 stations distributed in the MCG as uniformly as possible. Rock cylinders, 25 mm in diameter, were cored and oriented in the lab. A minimum of two cores (maximum 4) were drilled in each sample and were cut into 22 mm-long specimens, and so 1040 samples were measured for the whole pluton, with a mean of 6 samples per station. Oriented thin sections were prepared from either chips, from the excess part of the cores, or from oriented hand samples. The magnetic mineralogy of the sample was characterized by Curie temperature determination (using the CS2 apparatus of Geofyzika Brno), and examination of polished thin sections using reflected light microscopy. Bulk susceptibility (magnitude of the susceptibility) and anisotropy (AMS) values were determined using a Kappabridge KLY2 (Geofyzika Brno) susceptometer working at 4×10^{-4} T and 920 Hz, with a detection limit of about 4×10^{-8} SI. The orientation and magnitude of the three principal axes of the AMS ellipsoid for each sample were then calculated from the bulk susceptibility measurements and the orientation of the specimen. The tensor average of the individual sample measurements and their variability were computed for each site with the program EXAMS (Saint Blanquat, 1993). The orientation data are the foliation, which is the plane (k_{max}, k_{int}) perpendicular to the smallest axis (k_{min}) of the ellipsoid, and the lineation, which is the line parallel to the longest axis of the ellipsoid (k_{max}). The average, or bulk, susceptibility is given as $K = (k_{max}+k_{int}+k_{min})/3$. The principal intensity parameters are the total anisotropy $P= k_{max}/k_{min}$, the linear anisotropy $L = k_{max}/k_{int}$, and the planar anisotropy $F = k_{int}/k_{min}$. To characterize the shape of the magnetic susceptibility ellipsoid, an other intensive parameter is used: $T = (2(\ln k_{int} - \ln k_{min})/(\ln k_{max} - \ln k_{min}))-1$, from Jelinek (1981). A value of T between -1 and 0 indicates a prolate shape, and T between 0 and 1 indicates an oblate shape.

MAGNETIC MINERALOGY

Magnetic mineralogy has been studied in order to determine the mineral species responsible for the induced magnetization leading to the magnetic fabric. Thin section analysis shows two principal habits for the opaque phase: a) euhedral to subhedral grains or elongate aggregates of grains associated with the main mafic mineral phases (biotite, hornblende), which acted as mafic clots aligned parallel to the preferred orientation of larger grains (K-feldspar, biotite and hornblende) and b) isolated euhedral to subhedral grains occuring as inclusions in other minerals (mainly K-feldspars and biotite). Susceptibility versus temperature measurements for five samples corresponding to the complete range of the susceptibilities measured in the pluton show a Curie temperature Tc (temperature above which the mineral loose its magnetization) around 580 °C. This indicates that the ferromagnetic phase in all the samples is made of nearly pure magnetite. This is confir-

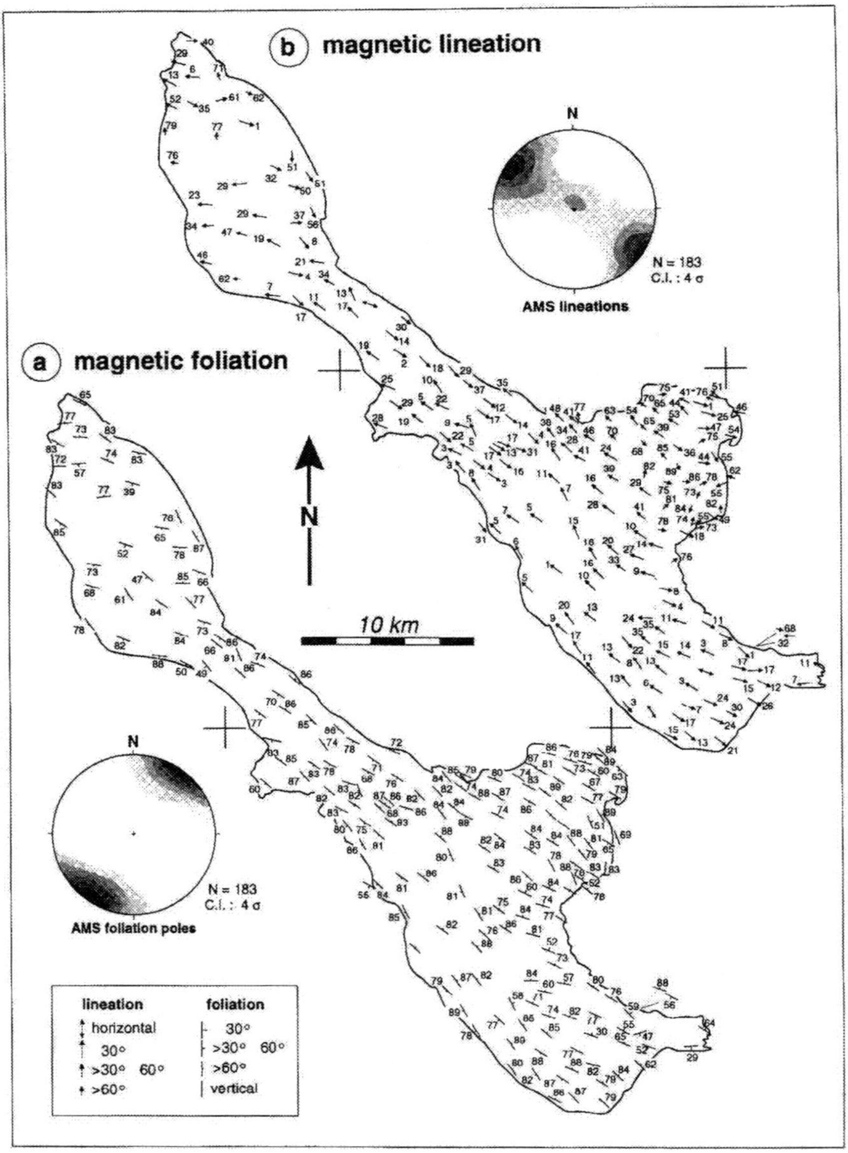

Figure 3. AMS orientation parameters in map views of the MCG. (a) Magnetic foliation (plane perpendicular to k_{min}), and (b) lineation (k_{max}). Kamb contoured equal-area, lower hemisphere projections of k_{max} and poles to k_{min} are also shown (contour intervals 2, 6, 10...σ).

med by reflected light microscopy which shows that the opaque phase is dominated by magnetite. The magnetite grains are euhedral to subhedral, with a size up to 0.5 mm (mean around 0.15 mm), and often form elongated clusters associated with the other mafic minerals (biotite, amphibole, and titanite). Grain boundaries of the magnetite are straight to arcuate, indicating early crystallization in equilibrium with the mafic phase of the granite. Small elongate grains of magnetite may be present as inclusions within the cleavage planes of biotite, due to a secondary chloritization of the biotite. Regardless of these, it is clear that large grains of multidomain magnetite, with low Ti-content, are responsible for the high susceptibility magnitude of our rocks. The susceptibility map is then an image of the amount of magnetite in the granite. Elongation (or shape anisotropy) of individual grains, or cluster of grains, are responsible for the anisotropy of the susceptibility.

AMS RESULTS

Directional data

A NW-SE striking and vertical foliation appears throughout the pluton (mean value: 125° N 88°) (Table 1). A prominent sigmoidal pattern, where the magnetic foliations are rotated to a NNW-SSE orientation, correlates to the trace of the RFSZ (Fig. 3a). The AMS foliation map is in very good agreement with field data, particularly within and adjacent to the RFSZ. Outside the shear zone, the magmatic foliation plane is defined by the flattening plane of (rare) xenoliths or orientation of feldspar phenocrysts. However, at some localities no field foliation could be measured, particularly in the northwestern part of the pluton. For most sites, the magnetic foliation agrees with the planar fabric measured in the field to within 10°. Thus, the trajectories of the magmatic and magnetic foliations, to a first approximation, are considered parallel. Within the RFSZ, the AMS foliation is sub-parallel to the macroscopic foliation (grain flattening) rather than to the N-S-striking shear planes, or is at an intermediate orientation between these two planes.

The magnetic lineations (mean value: 304°/8°) shows two groups of data (Fig. 3b): subhorizontal lineations which characterize the main body of the pluton, and steeply-plunging lineations which are localized in the bulge and along the northwest margin of the pluton. The lineations follow the same trend as the foliations, because of the manner in which they are defined (parallel to k_{max}) and the vertical orientation of the foliation. Magnetic lineations are progressively more subhorizontal toward the RFSZ. The northern part of the RFSZ within the pluton is characterized by steep (> 50°) magnetic lineations, in excellent agreement with field data. The magnetic foliation and lineation patterns suggest that initially, i.e. during the first stages of magma emplacement and before solid-state deformation, the lineations were steep and the foliations were WNW/ESE striking. The sigmoidal trajectories near the RFSZ demonstrate both the dextral movement and the syn- to late-magmatic style of deformation.

Variability of the directional data

Depending on the consistency of the orientation of the 3 principal axes, the sites could be classified as follows (Fig. 4a) : 1) all axes are strongly clustered (51% of the sites), 2) moderately clustered (28%), 3) k_{min} is well defined (7%), 4) k_{max} is well defined (11%), and 5) the axes are dispersed (3%). The degree of dispersion of the axes' orientations within one site is defined by the "average angular departure" (arithmetic mean of the angular departures between the principal axes of the different samples and the mean of the site). Figure 4b depicts the variation of this parameter: for k_{max} (k_{min}), the average angular departure is less than 20° for 72% (83%) of the sites. This distribution confers a high confidence to the magnetic foliation and lineation maps.

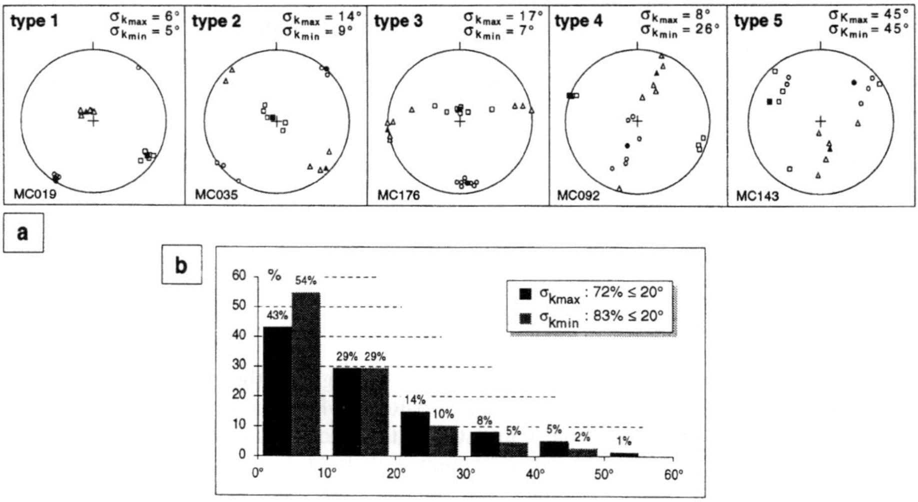

Figure 4. AMS variability. (a) the five types of AMS sites depending on the mutual arrangement of the 3 axes of the susceptibility ellipsoid: square = k_{max}, triangle = k_{int}, circle = k_{min}, open symbols = samples, black symbols = site mean values; (b) frequency histogram of the within-site angular variability for k_{max} and k_{min}. See text for details.

Susceptibility

Susceptibility values are high, ranging from 4 to 40 x 10^{-3} SI, with a mean value of 21 x 10^{-3} SI, indicative of a maximum of about 1 wt % of magnetite content; 52% of the samples fall between 15 and 25 x 10^{-3} SI (Fig. 5a). The map of the magnetic susceptibility (Fig. 5b) shows a NW-SE zonation in the eastern bulge and a general NW-SE trend of the isovalues lines. A correlation should be noted between susceptibility and mineralogy (Fig. 2a): the susceptibility is generally higher in the areas containing the more mafic assemblages. Since the susceptibility map is essentially an image of the magnetite content of the granite, and given the observation of the textural association between mafic minerals and magnetite (see above), this correlation K/petrographic facies, established for paramagnetic granites by Gleizes et al. (1993), is not surprising in ferromagnetic granites. The ratio between the maximum and the minimum mean susceptibility within one site could be up to 9 (with a mean of 1.7), without affecting the orientations of the principal axes. This within-site variability is attributed to the anisotropic distribution of magnetite at the grain scale, such as schlierens, clusters or inclusions in larger minerals (K-feldspar, biotite...).

Anisotropy

The total anisotropy P ranges from 1.03 to 1.54, with a mean value of 1.15, and 73% of the sites fall between 1.05 and 1.25 (Fig. 5c). The anisotropy map (Fig. 5d) is a qualitative image of the amount of strain undergone by the magma. The RFSZ is clearly outlined by P, and also evidenced by the straightforward correlation between P and microstructures (Fig. 2b). A value of P above 1.20 appears to be a reliable indicator of the presence of solid-state deformation in the MCG. Moreover, the mean value of P for the magmatic sites is 1.09, whereas it is 1.16 for the high temperature sites, and 1.28 for

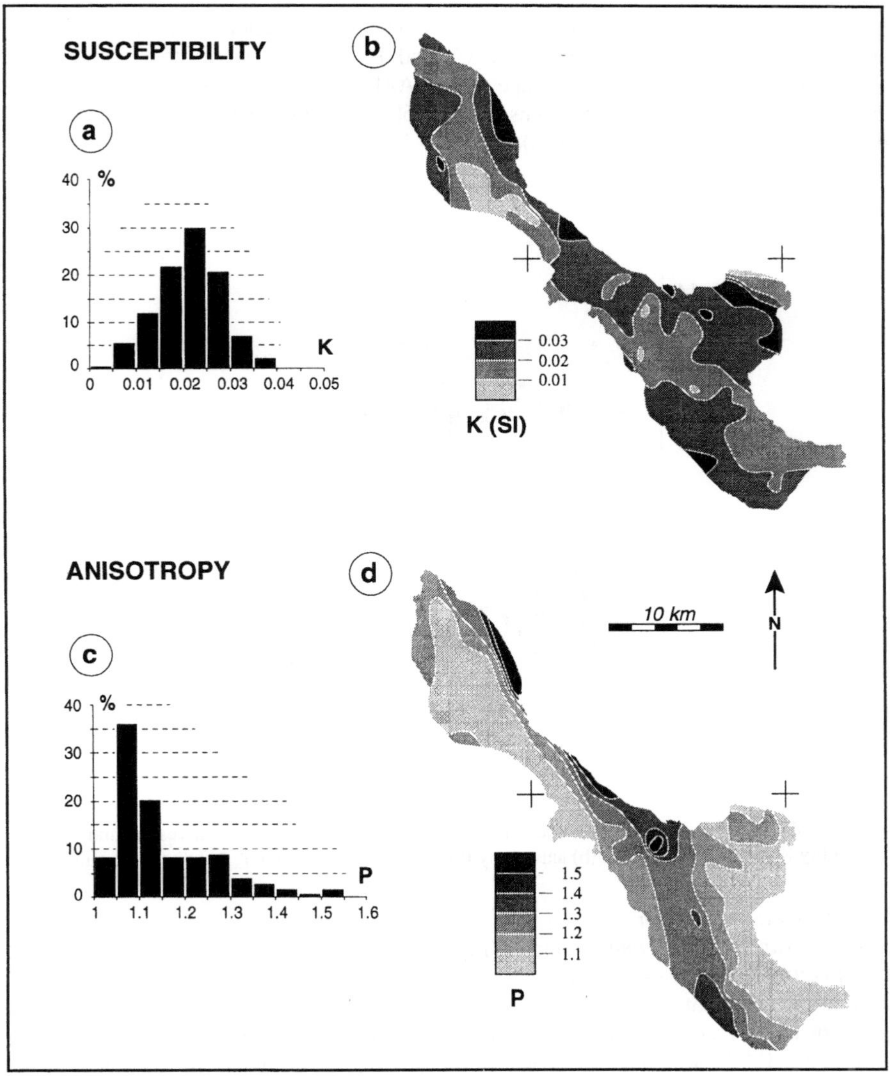

Figure 5. AMS intensive parameters. Susceptibility: (a) frequency histogram; (b) map obtained from manual interpolation between the mean values of each site. Anisotropy: (c) frequency histogram; (d) map (same method). Note the good correspondence between mineralogy and susceptibility (Figs. 2a and 5b), and between microstructures and anisotropy (Fig. 2b and 5d).

the medium to low temperature sites (Table 1), without substantial changes in the mean susceptibility values. Domainal distribution of the planar and linear anisotropies F and L also clearly delineate the RFSZ.

Shape parameter

The shape parameter T ranges from -0.47 to 0.85, with a mean of 0.27; 85 % of the sites have an oblate magnetic fabric (T > 0), and T shows no significant domainal distribution. As outlined below, the shape of the AMS ellipsoid seems to be independent of both susceptibility and strain intensity (Fig. 6b) (although we observe a more flattened shape in the more deformed sites).

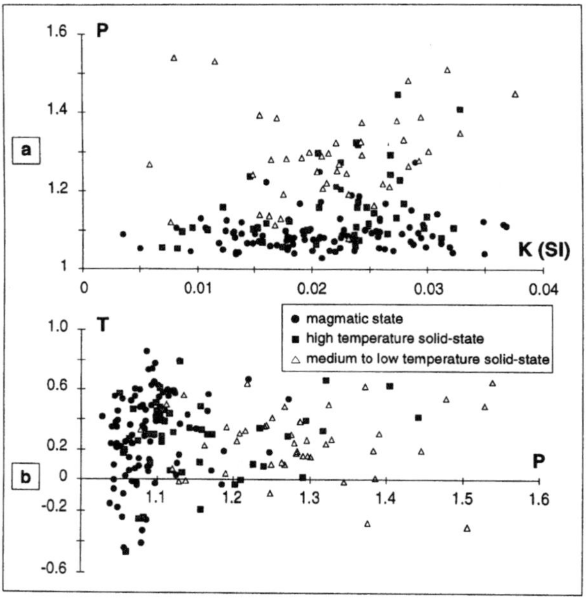

Figure 6. AMS intensive parameters variations, with distinction of the three microstructural types. (a) susceptibility K versus anisotropy P. (b) anisotropy P versus shape parameter T. See text for details.

AMS parameter variation

There is a correlation between P and K (Fig. 6a) as, except for a few high P-low K sites, total anisotropy increases with increasing susceptibility. Further, the more solid-state deformation, the greater the rate of increase in anisotropy with susceptibility. This kind of pattern is very common in magnetite-bearing granites (Archanjo et al., 1994; Rochette et al., 1992; Bouchez, this vol.). In the absence of a change in magnetic mineralogy, this pattern can be explained by an increase in magnetic interactions between magnetite grains. The simultaneous increase of strain and/or concentration of magnetite grains could initiate a distribution anisotropy (Hargraves et al., 1991), which, when added to the SPO of magnetite grains, will significantly increase the overall AMS. This explanation is valid for solid-state textures, because solid-state deformation generally yields strong geometrical organizations of grains and, therefore, strong distribution anisotropies. Another possibility is the change in shape of the magnetite grains due to plastic deformation (Housen et al., 1995), although this may require relatively high T conditions.

Plots of T versus K (not shown) demonstrate that there is no correlation between the shape of the susceptibility ellipsoid and the susceptibility magnitude. A slight correlation between the shape parameter T with the anisotropy ratio could be noted (Fig. 6b) : solid state deformation tend to flatten the ellipsoid of the susceptibility. As shown on the Flinn type diagram (Fig. 7a), most sites fall in the oblate field, with $T > 0$ and $F > L$. The plot of P versus the standard deviation on k_{max} orientation σ_{kmax} (Fig. 7b), demonstrates the correlation between the strength of the magnetic fabric and the clustering of the samples' k_{max} orientations around the site average orientation (i.e. small values of σ_{kmax}). The same result is obtained with k_{min} (not shown). We observe that magmatic sites are not very sensitive to this clustering effect, but HT and MLT sites are sensitive to it.

DISCUSSION

SIGNIFICANCE OF THE MAGNETIC FABRIC

The magnetic fabric of the MCG is characterized by a clustering of AMS axes at each station, and by a remarkable sigmoidal pattern of both the foliation and lineation (Saint Blanquat et al., 1993; Saint Blanquat & Tikoff, 1995). For the pluton as a whole, magnetic lineations show a girdle distribution around the foliations, which are very well clustered (Fig. 3). This configuration is likely attributable to a specific tectonic setting, like transpression, which allows a better defined foliation in comparison to lineation, than to an undefined magnetic mineralogy effect. The AMS intensive parameters are generally difficult to interpret in ferromagnetic rocks, because of the unknown relationship between the fabric of magnetite (low volume %) and the fabrics of the main phases (K-feldspars, biotite). However, in our case, the AMS parameters provide some semi-quantitative information: the susceptibility correlates with the rock mineralogy, and the anisotropy ratio is a good indicator of the rock texture. In addition, comparison of AMS, field data, and thin section observations permit us to establish a direct correspondence between the magnetic fabric and the rock fabric. For example, the usual observation of crystallization of magnetite in association with titanite, hornblende, and biotite suggests that the shape of the magnetite grains are controlled by the crystalline faces of those minerals. If the fabric of magnetite is a "template" of the Fe-silicates, then the magnetic fabric is a good image of the rock fabric. This "template" behavior has been demonstrated on some examples of ferromagnetic granites with quantitative image analysis of thin sections (Archanjo et al., 1994; Cruden and Launeau, 1994).

THE EFFECTS OF DEFORMATION ON THE MAGNETIC FABRIC

The magnetic fabric of the MCG is clearly affected by the RFSZ. The magnetic lineation tends toward a more horizontal orientation, together with a slight rotation from WNW to NW, within the shear zone, where the magnetic foliation is steeper and has a more NW orientation. As shown above, solid-state deformation yields tight within-site angular variations for both magnetic foliation and lineation. While the orientation of the magnetic fabrics is sensitive to magmatic and HT solid-state deformation, it seems less affected by the MLT solid-state deformation. For example, the N-S striking shear planes of the RFSZ in the middle of the MCG are not systematically recorded by AMS. Intermediate orientation of the magnetic foliation between the foliation plane and the shear plane have been documented within ductile shear zones (Aranguren et al., 1995). This observation implies that the fabric of the magnetite grains is not parallel to the shear plane. A reasonable explanation is that the amount of strain was not sufficient, hence the magnetite grains remained within the S plane. For our study, a possible factor is a sampling effect, the areas between shear bands, rather than the narrow shear bands themselves, comprising most of the analyzed samples. The latter factor could explain the observation that the magnetic lineation and foliation appear to record orientations intermediate between S and C planes. Regardless of this, the eventual lack of resolution in orthogneiss zones is not a problem, since these features are easy to measure in the field.

The intensive parameters of the magnetic fabric, especially the anisotropy degree P, are very sensitive to all types of deformation and are effected by the MLT deformation in the RFSZ. The susceptibility K shows a slight increase (Table 1) with the intensity of solid-state deformation, which normally would be attributed to an increase in the magne-

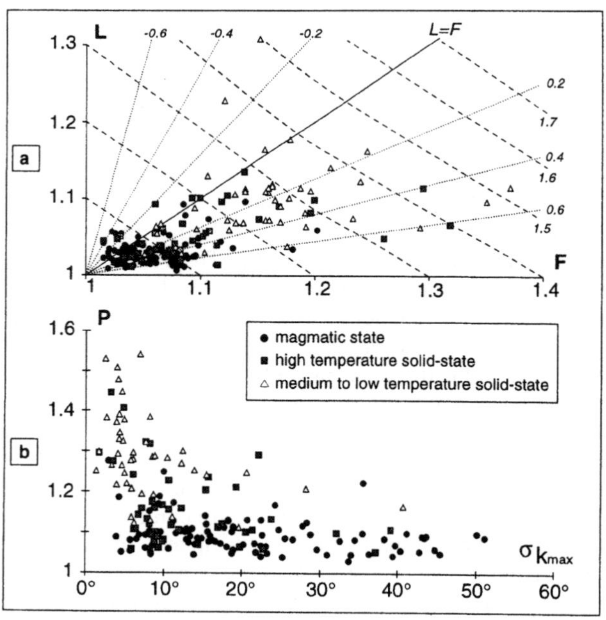

Figure 7. AMS intensive parameters variations, with distinction of the three microstructural types. (a) Flinn type diagram: linear anisotropy (= k_{max}/k_{int}) versus planar anisotropy (= k_{int}/k_{min}); dashed lines are the isovalue lines of the total anisotropy (=k_{max}/k_{min}); thin lines are the isovalue lines of the shape parameter T, most of the sites plot in the flattening field (F > L), and where the deformation is more intense, the shapes are more dispersed; (b) plot of the within-site angular standard deviation σk_{max} versus site total anisotropy P. See text for further explanation.

tite content. However, a map of the mineralogical facies shows the contrary, that the more deformed zones are more felsic. Thus, the reason for the increase of K is attributed to magnetic interactions, as shown theoretically by Stephenson (1994) and experimentally by Grégoire et al. (1995). The total anisotropy P shows a net correlation with strain intensity. The increase of P could be due to 1) an increase of the intensity of the SPO of the magnetite grains, 2) plastic deformation of magnetite grains leading to more elongated grains, and/or 3) a distribution anisotropy within the RFSZ. The shape parameter T shows a slight decrease in the flattening of the magnetic ellipsoid and a better within-site homogeneity in the shear zone. Thus, the magnetic fabric is a precise measure of the rock fabric, in orientation and intensity, both within and outside the RFSZ.

TABLE 1. Mean AMS values for the MCG as a whole, and for each type of microstructure. K is the mean susceptibility, P is the total anisotropy, F and L the planar and linear anisotropy respectively, T the shape parameter of Jelinek (1981), Fp is the Flinn parameter (= L/F) (after Flinn, 1962), k_{max} (k_{min}) is the longest (shortest) axis of the susceptibility ellipsoid, and Nb is the number of sites. σ is the standard deviation and Cv is the coefficient of variation (= 100 x standard deviation/mean value) for each parameter. HT for high temperature solid-state, and MLT for medium to low temperature solid-state.

mean values of : for :	K	σ_K	Cv_K	P	σ_P	Cv_P	F	L	T	Fp	σ_{Fp}	Cv_{Fp}	k_{max}	σ_{kmax}	k_{min}	σ_{kmin}	Nb
all sites	0.0214	0.0031	15%	1.15	0.02	2%	1.10	1.05	0.27	0.65	0.41	54%	304 / 8	16°	215 / 2	12°	183
all magmatic sites	0.0212	0.0032	16%	1.09	0.02	1%	1.06	1.03	0.29	0.64	0.58	76%	294 / 15	21°	211 / 5	15°	91
all HT sites	0.0215	0.0028	15%	1.16	0.02	2%	1.11	1.05	0.27	0.66	0.27	40%	308 / 9	13°	33 / 1	11°	48
all MLT sites	0.0217	0.0031	15%	1.28	0.03	3%	1.18	1.10	0.25	0.66	0.18	25%	314 / 2	8°	43 / 0	8°	44

SIGNIFICANCE OF THE ROCK FABRIC

The most important results of this study are given in Figure 8, which synthetizes the trajectories of the magnetic lineation superimposed on the map of microstructures. The transitions between the different textural zones - magmatic, HT solid-state and MLT solid-state - have been represented gradually, despite the discrete location of data, because no sharp contact between these zones was observed on the field. Figure 8 demonstrates the geometrical, structural, and temporal continuity between the almost E-W foliated/steeply lineated magmatic sites and the almost N-S foliated/horizontally lineated solid-state sites, as evidenced by field study, AMS patterns and microstructural data.

The fabric recorded with AMS is both syn-emplacement and tectonically-induced. Support for this interpretation is provided by the superimposition of solid-state fabrics onto magmatic fabrics, including: 1) our microstructural observations in the pluton; 2) a porphyroclast imbrication study (Tikoff and Teyssier, 1994) suggesting that magmatic shearing preceded solid-state shearing; 3) unstraining the "magmatic" RFSZ provides a "magmatic" displacement of about 10 km, and suggests that the fabrics in the northwest and southeast terminaisons of the pluton could have been close together and could represent the early stage of magma emplacement; and 4) the continuity between solid-state fabrics in the MCG and lower temperature regional deformation, for example the Gem Lake shear zone (Greene and Schweickert, 1995). Further evidence of the syntectonic emplacement of the MCG is also provided by estimates of the displacement along the RFSZ and the Gem Lake shear zone. Estimates for offset on the latter zones are greater than 20 km (Greene and Schweickert, 1995). However, displacement on the RFSZ is between 2 and 8 km for the solid-state shearing, as estimated from quartz analysis and porphyroclast imbrication (Tikoff and Teyssier, 1994), and reach about 10 km if the AMS foliation and lineation patterns are « unstrained » (Saint Blanquat and Tikoff, in prep). Because of the continuity between these two shear zones, this displacement must have occured during and slightly after (and possibly before !) the emplacement of the MCG. Strong evidence of syntectonic emplacement is also provided by the absence of any offset of the MCG contact despite the >20 km of displacement along the RFSZ. Since the AMS records magmatic fabrics, the lack of contact offset strongly suggests that displacement was at least partially accommodated by emplacement of new magma. Geochronological data also suggest syn-tectonic emplacement of the MCG. Tobisch et al. (1995) reported $^{40}Ar/^{39}Ar$ ages of 87 Ma on hornblende and 83 Ma on biotite from within the RFSZ in the Lake Edison pluton, south of the MCG (Fig. 1). The RFSZ in the Lake Edison pluton records ductile, dextral deformation (Tikoff, 1994). Since the Lake Edison (88 ± 1 Ma; U/Pb on zircon) is known to be older than the MCG based on field relations, ductile shearing must have occured during emplacement of the MCG (86 Ma). Geochronological constraints are also available in the metavolcanic rocks in the Ritter Range segment of the Gem Lake shear zone. Sharp et al. (1993) obtained $^{40}Ar/^{39}Ar$ ages of 85 Ma for syntectonic amphibole, and 80 Ma for syntectonic biotite and white mica. While these data may be cooling ages, they clearly show a close tie between deformation and emplacement.

THE STRUCTURAL MEMORY OF THE ROCK FABRIC

From the above arguments, synchronism between tectonism and magma emplacement during the construction of the MCG is established. Therefore, the fabrics recorded in the MCG are both syn-emplacement and tectonically-induced. These observations run counter to the inference that the fabric in a pluton is acquired late in the crystallization and may not reflect emplacement. A typical and very common field feature which is cited to support the concept that magmatic fabrics are acquired late in the cooling history of the pluton is the crosscutting relationships between magmatic foliation and petrographic zonation (see examples in Paterson and Vernon, 1995). However, fabrics which form during the early stages of emplacement do not reflect the complete history of pluton evolution, but rather a "snapshot" of magma movement at that time (if magmatic deformation is recorded): as new magma arrives in the magma reservoir, the « old » magma may be deformed, which could lead to cross-cutting relationships between petrological zonation and magmatic structures *during emplacement.*

By correlating the microstructures and the AMS in the MCG, we can document a chronology from the first set of magma emplacement until the late solid-state fabrics. In that sense, fabrics reflect both magmatic flow *and* emplacement-related strain. The fabrics in the MCG could not have been acquired late in the history of pluton, once the magma set « in place », because a later dextral shearing event with 20 km of displacement could not explain the pattern of internal strain and the geometry near the pluton contact. It is inconceiveable that a post-emplacement tectonic event would completely destroy and reorganize the fabric in presence of melt, with no relationship with the regional deformation which obviously controls the pluton construction. To destroy a fabric and reconstruct a new one, we must involve a dynamic event, such as magma emplacement and/or regional deformation. Therefore if magmatic microstructures are preserved and their kinematics can be determined, even if the fabric is acquired late in the crystallization process, it must reflect a magmatic and/or tectonic event related to pluton construction. The fabric of the MCG is an example of this behaviour, and demonstrates that fabrics within granitoids have a significant structural memory. Plutons are not completely amnesic after all...

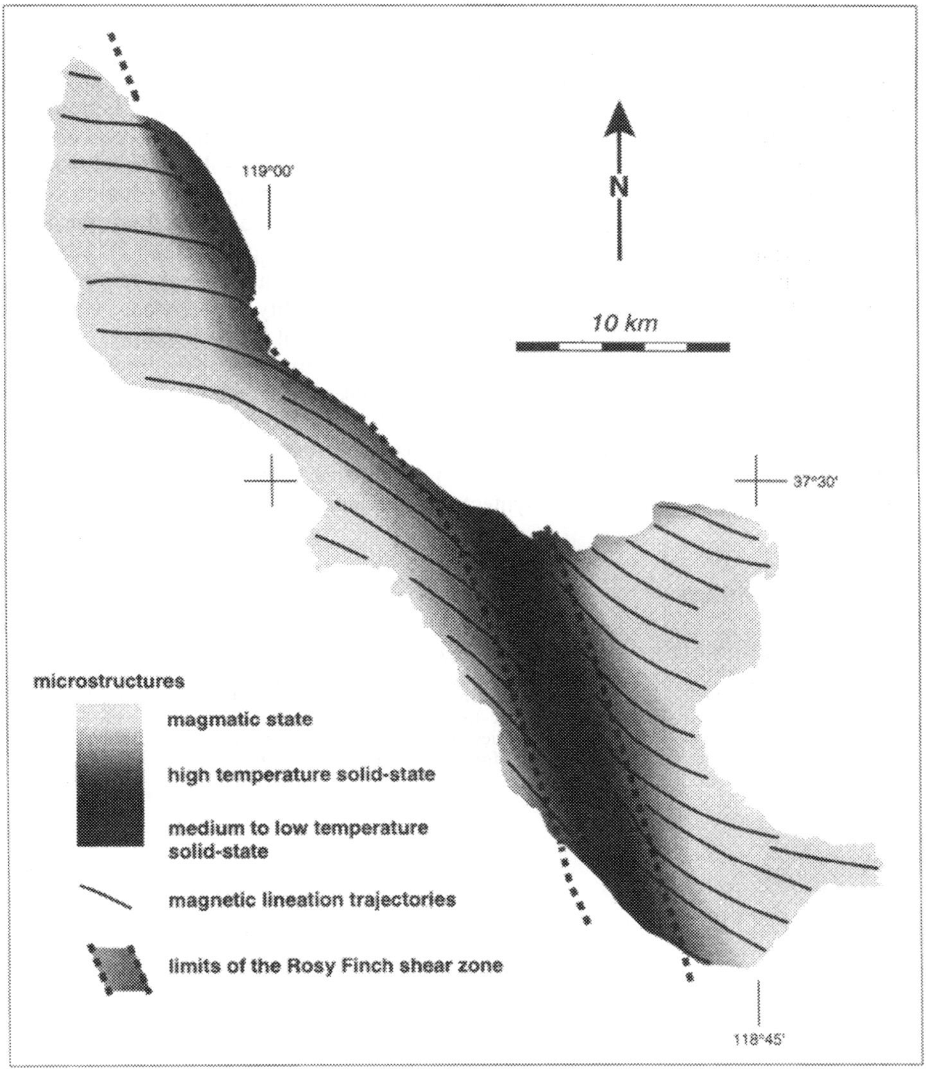

Figure 8. Synthetic map outlining the main results of this study. The magnetic lineation trajectories are superimposed on an interpolated map of the microstructures determined from thin section analyses. The geometrical, structural, and temporal continuity between the magmatic and solid-state deformations support a model of shearing during emplacement of the MCG.

CONCLUSIONS

Field and AMS investigations of the fabric of the Mono Creek Granite demonstrate the synchronism between magma emplacement and tectonism. The fabric of the Mono Creek Granite, as evidenced by AMS, has recorded the history of the pluton emplacement. Early-formed magmatic fabrics, which are characterized by WNW-ESE steeply-dipping foliations and steeply-plunging lineations, are in structural continuity with later-formed magmatic fabrics which show NW-striking, steeply-dipping foliations and shallowly-dipping lineations. As the pluton cooled, dextral deformation concentrated into high-temperature and, eventually medium-low temperatures solid-state fabrics, the latter being represented by the Rosy Finch shear zone. The progressive rotation of the AMS lineations within the shear zone allows us to estimate the offset along the Rosy Finch shear zone, in both the magmatic and solid-states, to be greater than 20 km. The AMS intensive parameters, which are generally difficult to interpret in ferromagnetic rocks provide semi-quantitative information: the susceptibility K correlates with rock mineralogy, and the anisotropy degree P is a good indicator of the rock texture.

The fabric of the MCG has apparently recorded all the history of the pluton construction. The continuous progression from steeply plunging lineations, representing an initially vertical magma flow, to shallowly plunging lineations indicating horizontal stretch, and accompanying the continuous textural evolution from magmatic to solid-state when approaching the dextral RFSZ, indicate that these fabrics have a certain amount of memory, and support the syntectonic emplacement of the MCG.

ACKNOWLEDGEMENTS

Supported by CNRS (INSU/DBT, MDRI and UMR 5563) and NSF grants (EAR 9305262). Philippe Olivier, Sven Morgan, Amy Miller, Carol Ormand, Bobby Bothman, Logan Colby, and A. Bradshaw Murray are gratefully acknowledged for sampling assistance. Technical assistance in the lab was provided by Christiane Cavaré-Hester, Anne Marie Roquet, Pierre Lespinasse, and Laurent Améglio, who are gratefully acknowledged. Discussions with Christian Teyssier and Jean Luc Bouchez were very helpful. We thank D.H. Hutton, K.E. Karlstrom, and B.A. McNulty for their constructive comments.

DRAINAGE AND EMPLACEMENT OF MAGMAS ALONG AN INCLINED TRANSCURRENT SHEAR ZONE: PETROPHYSICAL EVIDENCE FROM A GRANITE-CHARNOCKITE PLUTON (RAHAMA, NIGERIA)

Eric FERRÉ, Gérard GLEIZES, M.Toufik DJOUADI, Jean Luc BOUCHEZ
Pétrophysique et Tectonique, UMR CNRS n° 5563, Université Paul–Sabatier, 38 rue des 36-Ponts, 31400 Toulouse, France
and Francis X.O. UGODULUNWA
Department of Geology and Mining, University of Jos, PMB 2084, Jos, Nigeria

ABSTRACT

A Pan-African granite-charnockite pluton of northeast Nigeria has been core-sampled at an average grid spacing of 2 km. Detailed petrographic and microstructural observations are combined with an AMS study of 172 stations. The continuum between charnockites and granites is revealed by their constituent minerals and the magnetic susceptibility map. The contemporaneity of emplacement of all the petrographic types is demonstrated by their similar mineral- and magnetic-fabrics. Inclined strike-slip shear zones were active during and slightly after the emplacement of this pluton. Earlier geochemical data indicate that magmas originated from several vertically juxtaposed lower-crust sources. The proposed model of emplacement implies that the shear zones were responsible for the migration, drainage and tapping of magmas from the source region. The inclined attitude of the shear zones would have acted as a roof-guide for the up-rising magmas. Finally, the transcurrent movement most likely enhanced the upward migration of magmas.

INTRODUCTION

Granites and charnockites are often associated in plutons of the Pan-African belt of Nigeria (Fig. 1; Rahaman, 1988). Modal composition of such complexes range from Hbl-Bt granite to Opx-Cpx-Hbl-Bt ± Fa Qtz-monzonite. On the basis of the apparent lack of intermediate compositions, these rock units were previously considered to be unrelated to each other, but no structural data were available on them before this study. This hypothesis can be tested from an investigation of their internal structures. For this pur-

J.L. Bouchez et al. (eds.) Granite: From Segregation of Melt to Emplacement Fabrics, 253–273.

pose, the magnetic fabric technique (Bouchez, this volume) is powerful to unravel the internal fabrics of granitic rocks as exemplified elsewhere both in magnetite-free granites (Bouchez et al., 1990; Bouillin et al., 1993; Leblanc et al., 1994) and magnetite-bearing granites (Archanjo et al., 1994 and 1995; Ferré et al., 1995). Such studies have shown that plutonic bodies can be useful in deciphering the kinematic evolution of an orogenic belt. We present here a detailed study of the Rahama Complex (Fig. 1), the best exposed and a type example of granitic-charnockitic intrusions in Nigeria (Wright, 1971). Our study includes field, side-looking airborne radar (SLAR), microscopic and magnetic susceptibility data. These techniques aim to explain the relations between the two major rock types, their mode of emplacement, and ultimately constrain the kinematic evolution of the Pan-African Belt.

REGIONAL SETTING

The Transaharan-Braziliano Belt is a 4000 kilometres long, and several hundred of kilometres wide Pan-African orogen stretching north-south from Hoggar in West Africa to Brazil in South America (e.g. Caby, 1989). The width of the belt results from exotic terrane amalgamation during an oblique collision as it has been recently demonstrated, to the North of Nigeria, in the Tuareg Shield, by Latouche et al. (1993), Black et al. (1994) and Liégeois et al., (1994). Nigeria is made of two contrasted lithotectonic domains (Affaton et al., 1991; Castaing et al. 1993; Fig. 1). Metallogenic, lithostructural and geochronological evidences (respectively Woakes et al., 1987; Ajibade and Wright, 1989 and Ferré et al., in 1996) suggest that they may correspond to two terranes separated by a major lineament identified by Ananaba and Ajakaye (1987). Western Nigeria is characterised by Neoproterozoic metapelitic and quartzitic Schist Belts (Turner, 1983) whereas Eastern Nigeria is made of high-grade gneisses and granite-gneisses (Wright, 1971). A voluminous Pan-African calc-alkaline plutonism occurs in both domains between 750 Ma and 500 Ma (Tubosun et al., 1984; Egbuniwe et al., 1985; Fitches et al., 1985; Dada et al., 1995). From the structural point of view, the western Schist Belts have been studied in detail (Burke et al., 1976; Turner, 1983; Annor and Freeth, 1985) because strain markers are more prominent in schistosed rocks than in gneisses and granites. Hence, several Pan-African N-S to NNE-SSW trending, dextral strike-slip shear zones are identified in this western realm (Fig. 1). In Eastern Nigeria, numerous shear zones in granitic terrains have not been distinguished from surrounding orthogneisses and thus only a few of them are presently recognised (Turner, 1986).

The Rahama Complex (450 km^2; Fig. 2), north of Jos (Fig. 1), one of the largest Pan-African calc-alkaline granite-charnockite plutons of the eastern domain, has been previously mapped by Wright (1971) and Olarewaju (1978). This complex, roughly circular in shape, and dated at 668 ± 128 Ma by Rb-Sr method (van Breemen et al., 1977) was emplaced under a pressure of 540 ± 60 MPa as indicated by Al-in hornblende geobarometry on 10 samples. Hornblende forms corona textures around pyroxenes and therefore it has crystallised in situ. These pressure estimates are compatible with mineral assemblage of the country rock. The Rahama complex is intrusive in anatectic Bt ± Grt ± Opx granites, Sil-Grt-Bt gneisses and migmatites, whose regional foliations strike between N160° and N30° with dips constantly to the west. All of these rocks are crosscut by numerous Jurassic alkaline anorogenic granites (Fig. 2) that have been studied in detail by Bowden et al. (1987).

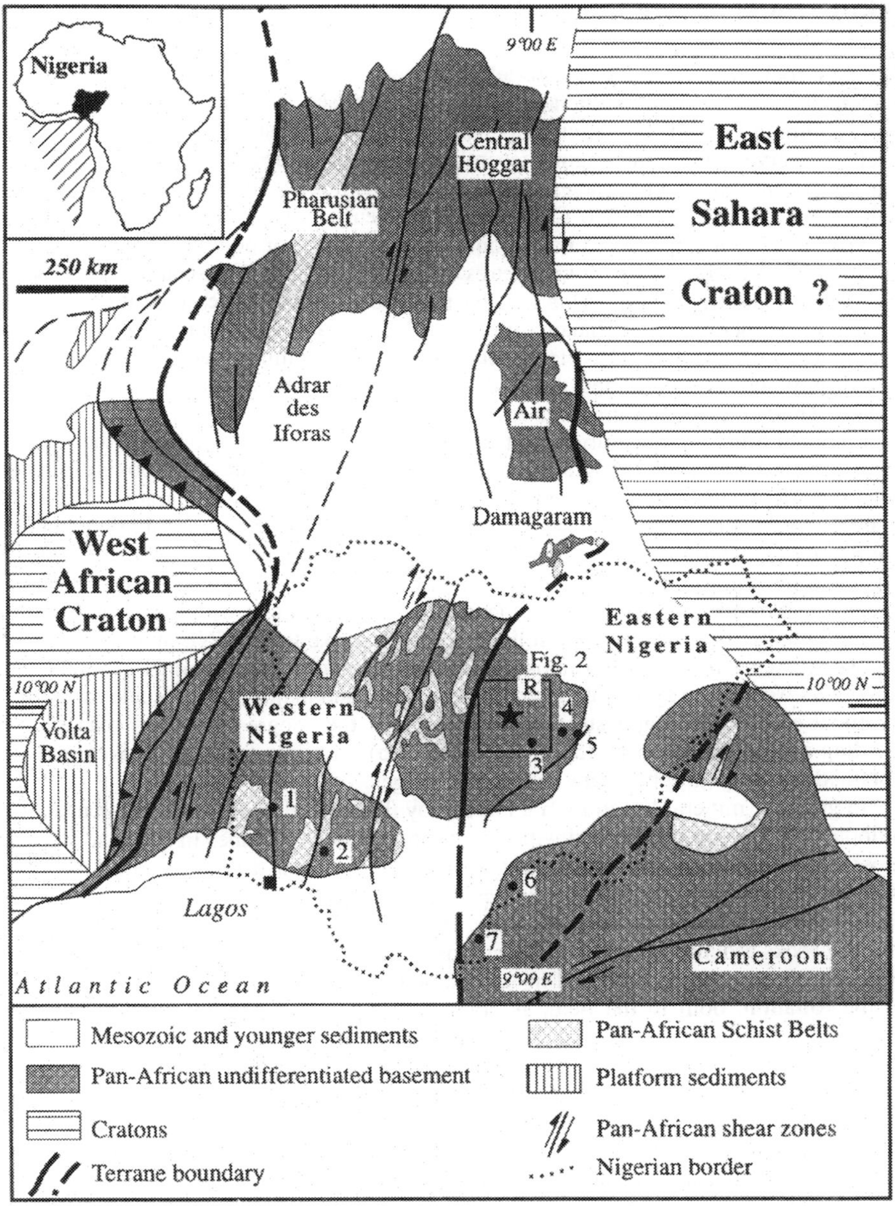

Figure 1. Geological setting of the Pan-African Belt of Nigeria in West Africa (modified after Caby 1989; Ajibade & Wright 1989; Affaton *et al.* 1991). Numbers refer to occurrences of granite - charnockite complexes: R. Rahama 1. Iwo, 2. Ikerre, 3. Toro, 4. Bauchi, 5. Dindima, 6. Obudu, 7. Oban. Jurassic alkaline granites omitted.

STRUCTURES IN THE COUNTRY ROCKS

The main rock types, banded quartzo-feldspathic gneisses, nebulitic migmatites, homogeneous leucocratic granites, and aplo-pegmatitic gneisses, show gradational boundaries. Their common Grt-Sil±Ms assemblage and the boudins of amphibolite or calc-silicate locally preserved within the migmatites indicates that the series derives from metapelites that underwent amphibolite facies metamorphic conditions. Muscovite-out and partial melting conditions at T° ≥ 650°C are attested by leucosomes containing Qtz-Kfs-Bt (< 5% vol.) and Pl (An15 to 25). In addition, several outcrops or boudins of kinzigites, i.e. Bt-Grt-Sil-Crd-Gr-Rt gneiss, 5 kilometres south of Lere, and Opx-bearing gneisses, 5 kilometres southeast of Dan Alhaji, have recorded granulite facies conditions. At the immediate periphery of the Rahama pluton, the rocks are mainly anatectic granites, or migmatites with abundant Grt-Bt±Sil leucosomes. The granulite facies evidence is preserved only within lithologies that did not undergo extensive partial melting. Thus, hydrous anatexis postdates granulite facies conditions.

Regional foliation trajectories, sketched in Figure 2, were obtained from the interpretation of 1:50,000 scale topographic maps and 1:250,000 scale side-looking airborne radar (SLAR) images and compared with field measurements. On topographic maps, the analysis of the fine-scale drainage pattern provides the direction of foliation, and the asymmetry of natural slopes, including inselbergs (Δh ≈ 10 m) and river banks (Δh ≈ 1 m), yield the dip direction. On SLAR images (MARS, 1977) visual interpretation of lineaments provided foliation strikes that are in good agreement with those inferred from the topography. In the study area, foliations mostly strike to the NNE and dip steeply to the west, except in the westernmost zone where the foliations trend N-S. Field measurements were taken systematically along the roads and major tracks. Rock exposure is excellent except near major streams. Results from the Jengre-Pambegua road section, yielding a mean foliation at 15° W 60° (Fig. 2b), confirm foliation strikes and dips obtained from topographic maps and SLAR images.

In granites, mineral lineations are marked by the parallel orientation of elongate amphibole or Kfs crystals observed within an exposed foliation plane. In gneissic material, most of the constituting minerals mark a stretching lineation within the foliation plane. Since these lineations in both rock types most likely mark the maximum principal finite stretch undergone by the rock, they can be used to map strain trajectories. In the country rocks around the complex, lineations trend N-S and plunge slightly to the north (Fig. 2b). Shear criteria observed in sections parallel to the lineation and perpendicular to the foliation, both in the field, as given by asymmetric boudins or S-C structures (Plate 1a), and under the microscope (Plate 1b), always show a dextral sense of shear. Textures, such as highly polygonized grain-boundaries, recovered internal structures and minute brown biotite recrystallised in C-planes indicate high- to medium-temperature ductile deformation. Two steeply dipping shear zones, oriented NNE to NE, where gneisses are interlayered with granitic dikes (Plate 1c), bound the pluton along its northwest and southeast margins. These shear zones, of a few kilometres in width, are characterised by a reduction of grain size and grain elongation ratios in XZ sections indicating clearly strain gradient. Other similar shear zones have been recognised further to the southeast (Ferré et al., 1995; Fig. 2). They all developed under high-temperature conditions, close to anatexis, in the stability field of biotite and sillimanite.

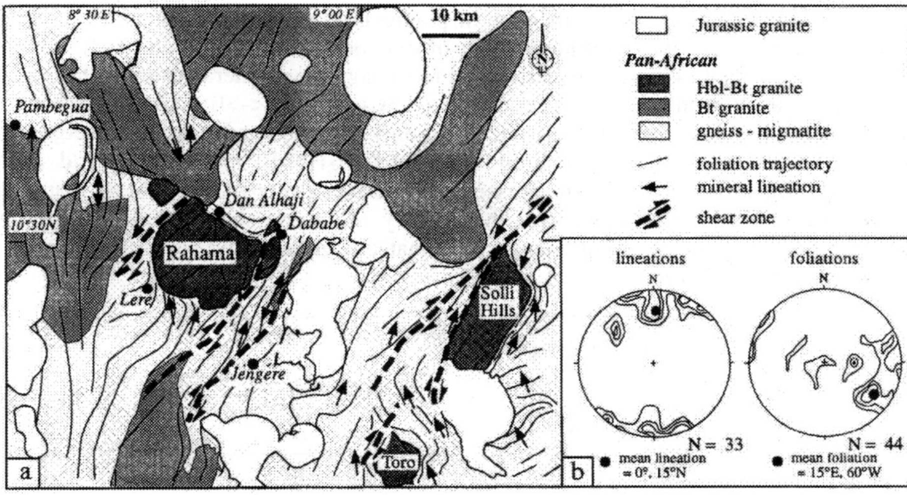

Figure 2. The Rahama Complex and its geological environment. (a) Schematic geological map modified after MacLeod et al. (1971) and Olarewaju (1978). Foliation trajectories are drawn from field data, morphological analysis and SLAR image interpretation. (b) Lower hemisphere stereonets of lineations and foliation poles of the metamorphic basement along a cross-section between Jengere and Pambegua. N = number of stations

PETROGRAPHY OF THE RAHAMA COMPLEX

Field and petrographic observations are combined with magnetic susceptibility data and published geochemical data to define the petrological zonation of the complex. Thin sections were systematically prepared from oriented cores that were collected for susceptibility and anisotropy of magnetic susceptibility (AMS) measurements. The distribution of sampling sites is shown in Figure 3a. Field and microscopic observations show that the Rahama Complex consists of two plutonic series (Fig. 3a): (i) a zoned granite-charnockite series forming a main pluton with two satellites, and (ii) a diorite-norite series forming mafic stocks. The two satellites, located at the northwest and at the east of the pluton, were not distinguished on previous maps (Wright, 1971; Olarewaju, 1978).

The granite-charnockite series (≈ 400 km^2) is porphyritic and displays a prominent zonation around an arcuate central zone (Fig. 3a). From margin to core, three main petrographic types are observed (Fig. 3): (1) Hbl-Bt monzogranites; (2) Cpx-Hbl-Bt monzogranites, and (3) Opx-Cpx-Hbl-Bt charnockites (Wright, 1971). Orthopyroxene is an eulite-orthoferrosilite (Fs80 - 95), clinopyroxene is an hedenbergite (Di ≈ 5) and amphibole is a ferrohastingsite. Note that fayalite has been observed in the charnockites by van Breemen et al. (1977). Some Hbl-Bt syenogranite also outcrops near the margins. The zonation is gradational and characterised, from rim to core, by a decrease in quartz and alkali feldspars, a lack of microcline and titanite in the Cpx-Hbl-Bt monzogranites and charnockites, and an increase in plagioclase, ferromagnesian minerals and magnetite. In addition, the anorthite content of plagioclase increases from An15-25 in syenogranites to An30-45 in charnockites, and ferromagnesian mineral assemblages vary according to the "Bowen" reactional series (Opx<->Cpx<->Hbl<->Bt). Quartz vermicules or bleached patches in the amphibole, commonly observed under the microscope in the Hbl-Bt monzogranites near the limit with the Cpx-Hbl-Bt monzogranites (dashed line of Fig. 3a), strongly suggest the presence of a former clinopyroxene. At its margin, the Complex has sharp contacts with the country rocks and shows numerous aplite dikes, metric in size inclusions of country rock anatectic granites and xenoliths of gneiss and of Hbl-Bt microdiorite.

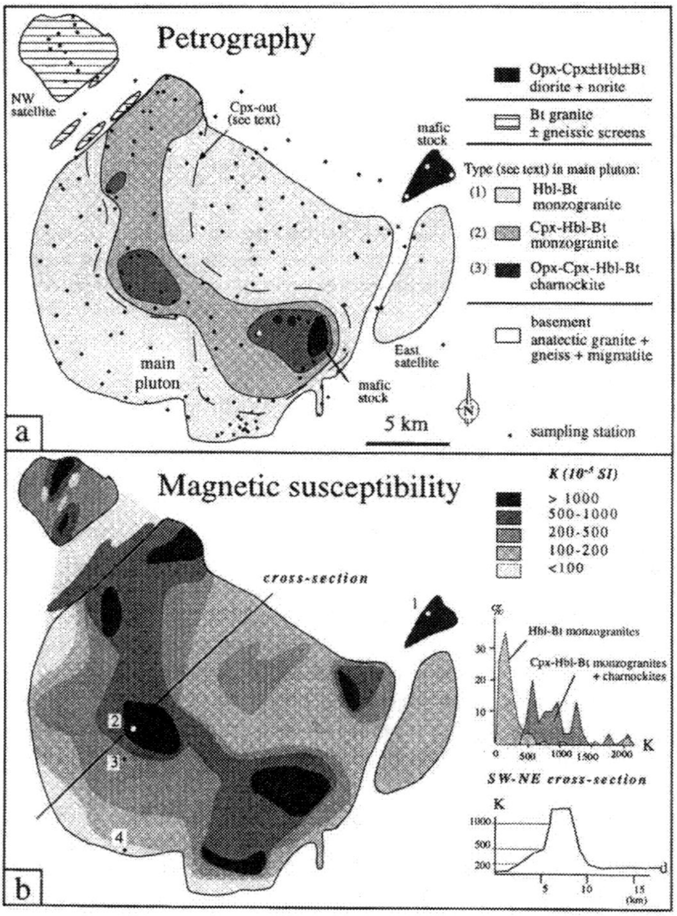

Figure 3. Zonation of the Rahama Complex. (a) Petrographic map: note the reverse zonation revealed by the systematic sampling. (b) Magnetic susceptibility map: the zoned pattern is confirmed and detailed. The frequency histogram shows a limited overlap between Kgranites and Kcharnockites. XX' cross-section: note the step of susceptibility magnitude across the charnockite body. Numbers 1 to 4 refer to samples analysed in figure 4.

The mafic stocks (≈ 4 km^2) outcrop both at the core of the charnockites within the pluton and in the country rocks (Fig. 3a). These stocks, mainly diorites and norites, display Opx-Cpx-Hbl-Bt assemblages and a calcic plagioclase (An_{45-65}). The largest isolated stock is a coarse-grained norite, at the northeast of the pluton, rimmed by a fine-grained diorite which is interpreted as a chilled margin. Mingling features between Opx-Grt bearing anatectic granite and diorite are observed in the shear zone linking this northeastern mafic stock to the main pluton. The mafic bodies within the pluton consist of fine-grained homogeneous diorites, similar in petrography, chemistry and magnetic mineralogy to those of the Toro Complex (Fig. 2; Déléris et al., 1996).

The northwestern satellite (≈15 km^2) is heterogeneous in composition and made of porphyritic Bt granite with little or no hornblende, and interlayered gneisses. It is separated from the main pluton by a 3 kilometre-wide shear zone in the country rock. The eastern satellite (≈ 20 km^2) is made up of homogeneous Cpx-Hbl-Bt monzogranite with few elongate mafic enclaves.

The petrographic zonation can be mapped in more detail with the help of the low-field magnetic susceptibility magnitude (K). In granitic rocks, K results mainly from two types of magnetic behaviours (Rochette, 1987): ferromagnetism, generated by magnetite or accessorily by pyrrhotite; paramagnetism controlled by ferromagnesian mineral. Diamagnetism originating from the quartzo-feldspathic matrix is quantitatively negligible. The K value of magnetite is two orders of magnitude higher than that of any ferromagnesian mineral and therefore magnetite-bearing granites are dominantly ferromagnetic. On the other hand, K in magnetite-free granites results mainly from the ferromagnesian minerals and these granites have paramagnetic properties. Thus, the classic magnetite-series and ilmenite series of granites (Ishihara, 1977) correspond respectively to the ferromagnetic and paramagnetic groups of granites. In magnetite-bearing granites, the magnitude of K increases in the more mafic units (Archanjo et al., 1995; Ferré et al., 1995; Déléris et al., 1996).

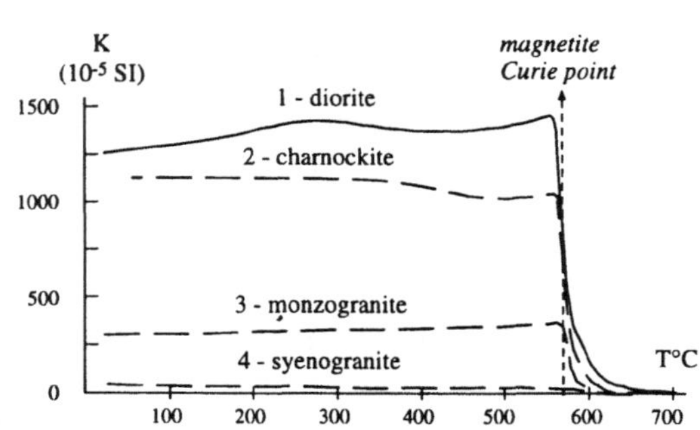

Figure 4. Magnetic mineralogy of rocks in the Rahama Complex: magnetic susceptibility magnitude versus temperature curves of four specimens representative of the Rahama Complex; note the abrupt drop at 580°C, stressing that magnetite is the main carrier of magnetisation, except for the syenogranite which is paramagnetic. Numbers correspond to samples localised in figure 3b: 1 = RA150, 2 = RA94, 3 = RA95 and 4 = RA99.

Nearly all the sampling stations within the Rahama pluton have K > 40 x 10^{-5} SI, a value above which the ferromagnetic behaviour becomes dominant (Rochette et al., 1992). Susceptibility magnitude versus temperature curves of representative samples

(Fig. 4) indicate unambiguously that magnetite, characterised by a Curie temperature of 580°C, is the dominant carrier of the low–field magnetic susceptibility. In this case, the contribution of monoclinic ferromagnesian minerals (e.g. clinopyroxenes) to the rock magnetic fabric is negligible. The susceptibility map of Figure 3b shows striking similarities with the petrographic map (Fig. 3a), and documents also a continuous and progressive zonation of the magnetic susceptibility magnitude on sides of an arcuate high-susceptibility NW-SE trending zone (Fig. 3b, cross section). The southern margin of the pluton is narrow and has a relatively low K (< 100 x 10^{-5} SI); it corresponds to a porphyritic syenogranite with inclusions of country rocks. From the latter rock-type, and progressing to the northeast, the susceptibility increases continuously up to more than 10^{-2} SI inside the Opx-Cpx-Hbl-Bt charnockites. The distribution of susceptibility magnitudes is bimodal (Fig. 3b: susceptibility frequency histogram), the values of K larger than 10^{-2} SI corresponding to the charnockites, diorites and norites. However, some "anomalies" are observed: (1) in the granites, two high-susceptibility zones (K> 10^{-2} SI) in the northeast and southeast quadrants correspond to magnetite-enriched granites probably due to higher oxygen fugacities during magma crystallisation since magnetite is primary; (2) outside of the main pluton, the northwest satellite displays highly variable susceptibility magnitudes, K ranging from 9 to 2156 x 10^{-5} SI, in agreement with its heterogeneous nature; the low-K values correspond to the anatectic granite layers of the country rock, and the high-K ones correspond to granites mingled with mafic rocks. The diorite forming the northeast stock has very strong susceptibility magnitudes, K> 6.10^{-2} SI. The country rocks generally have low K (< 20.10^{-5} SI) indicating a paramagnetic behaviour.

The reverse petrographic zonation of the main pluton, supported by the magnetic susceptibility study, is confirmed by the available geochemical data (Olarewaju, 1978; Batchelor and Bowden, 1986; Dada et al., 1995; Dickin et al., 1991). Dada (1989), using major, trace elements, Sr, Nd and Pb isotopes, has demonstrated that the monzogranite-charnockite series has a sub-alkaline geochemical signature distinct from the calc-alkaline one of the diorite-norite series. On the basis of trace and Nd-Sr isotope data, Dada (1989) considers that both series derive from partial melting and progressive extraction of contrasted and lower continental crust protoliths. The wide range of Nd model ages for the Nigerian charnockitic series, 1.23 to 1.96 Ga (Dada et al., 1995; Dickin et al., 1991), suggests a derivation from various protoliths.

STRUCTURES AND MICROSTRUCTURES IN THE RAHAMA COMPLEX

The measurements of foliation orientations were obtained by three different methods. (i) When the fabric could be observed in outcrop, it yielded a reference measurement of the magmatic planar fabric (Fig. 5a); (ii) geomorphological interpretation of topographical maps yielded valuable information in remote places that were not visited for magnetic fabric sampling (Fig. 5b), and (iii) anisotropy of magnetic susceptibility (AMS) measurements which gave orientation data that were, in most cases, in agreement with field measurements; these data are particularly useful in weakly foliated rocks (Fig. 6a). The lineation measurements were obtained only from the magnetic fabric study (Fig. 6b).

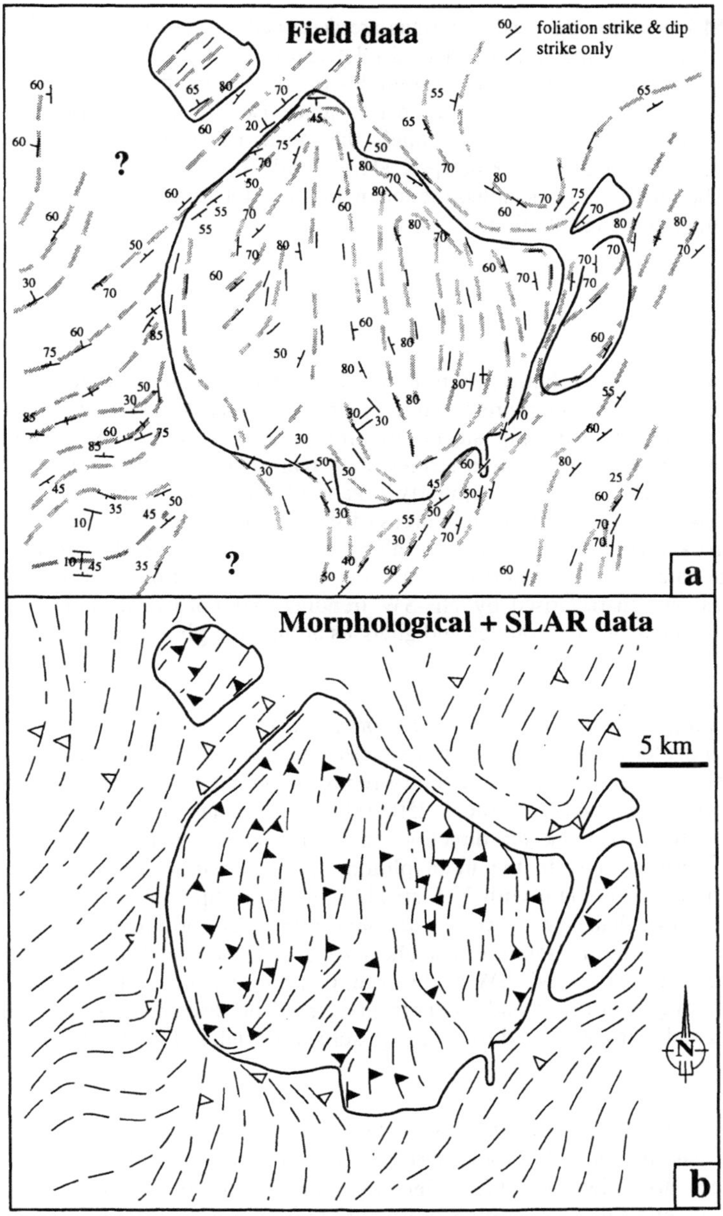

Figure 5. Rahama Complex foliations and trajectories. (a) Compilation of field data; in the country rocks: data are from this study and modified after Wright (1971) and Olarewaju (1978); in the pluton: data are from this study. (b) Interpretation of topographical maps at 1:50,000 scale and of SLAR radar image at 1:250,000 scale (MARS, 1977).

The AMS methodology and apparatus are described in Leblanc et al. (1994). Each core drill yielded two specimens, oriented relative to the geographical frame. The measurements provide orientations and magnitudes of the 3 principal axes of the specimen AMS ellipsoid, $k_1 \geq k_2 \geq k_3$. Each station is characterised by the average AMS parameters of a minimum of 4 specimens, namely the mean susceptibility axes $K_1 \geq K_2 \geq K_3$, the susceptibility magnitude $K=(K_1+K_2+K_3)/3$ and the total anisotropy magnitude $P=K_1/K_3$. The long axis K_1 is called the magnetic lineation, and the short axis K_3 is the normal to the magnetic foliation.

FOLIATIONS

Field data

Field measurements of foliations within the pluton and its country rocks and interpolated foliation trajectories are represented in Figure 5a. The eastern side of the main pluton is characterised by N-S striking, steeply dipping foliations, even at the northern margin of the domain where the foliation strikes abut against the country rocks. The western side shows less steep and less regularly organised foliations. It should be noted that foliation strikes are poorly defined in areas of shallow dips. The limit between these two zones is marked by a N-S median line along which foliations strike mostly N-S. The two granite satellites at the northwest and the east of the pluton, and their contiguous shear zones, are characterised by NE-SW trending and northwest dipping foliations. Foliations in the anatectic granites and gneisses around the pluton are roughly parallel to its contour.

Geomorphological and side-looking airborne radar (SLAR) data

Foliation trajectories (Fig. 5b) were extracted from the 1:50,000 scale topographical maps for both the pluton and its country rocks following the method described for the whole area (Fig. 2), and from the 1:250,000 SLAR image for the country rocks. In most areas, the strikes and dip directions inferred from the morphology are consistent with field observations. In the main pluton, the trajectories confirm the distinction between an eastern domain with N-S striking and west dipping foliations, and, on the other side of a median line, a western domain with a less prominent pattern. The northwest dips in the granite satellites and shear zones are confirmed. In the country rocks, the foliations that strike NE-SW at distance, come into conformity around the pluton especially along its northern and southern margins. The discrepancies observed between Figures 5 a and b are attributed to poor exposures especially along the NE margin of the complex.

Anisotropy of magnetic susceptibility (AMS) data

Precise measurements of the rock fabric are provided by the AMS technique (described by Bouchez, this volume) at 172 different stations located in Figure 3a. For each station, the magnetic foliation, or plane normal to the average orientation of the short axis of the AMS ellipsoid (K3), has been calculated based on measurements performed on four specimens per sampling site. In 88% of sampling stations, the angular dispersion of the four measurements was less than 25°; hence the vectorial mean AMS is considered as meaningful. Internal consistency of the magnetic data is attested by measurements that are comparable between neighbouring stations. The map of magnetic foliations (Fig. 6a) is more complete and gives more information than those derived from field and

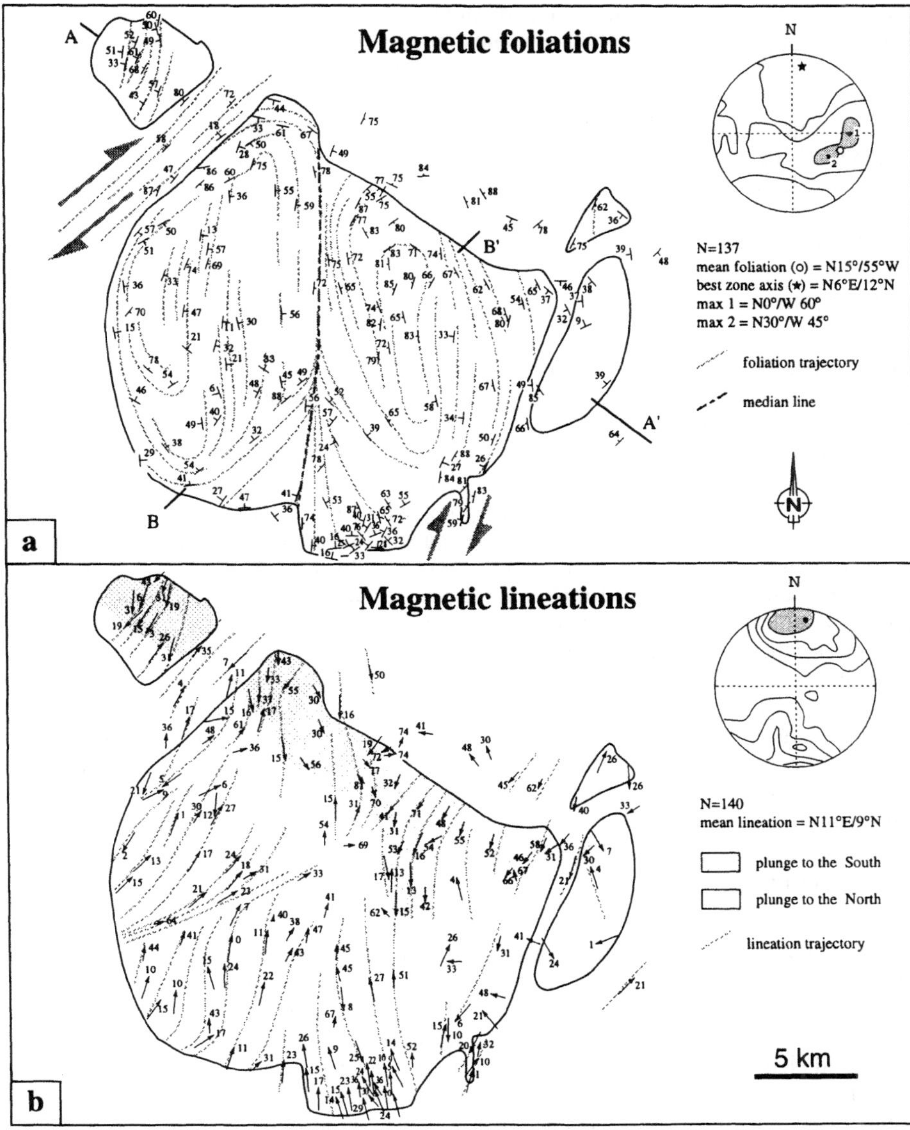

Figure 6. Magnetic fabrics of the Rahama Complex. Stereonets: Schmidt lower hemisphere; contours: Kamb; C.I. = 2σ; N = number of stations. (a) Foliations and trajectories; note the N-S trend in the main pluton and the NE trend in the NE shear zone. AA' and BB' refer to the cross sections of figure 8. (b) Lineations and trajectories.

geomorphological observations (Fig. 5). Foliation poles on the stereonet of Figure 6a are distributed in a partial girdle with a best zone axis oriented at 006°/12°, and display two maxima of west dipping foliations. Maximum 1 giving a mean foliation plane at

0°W 60° corresponds to the dominant foliations in the main pluton, and maximum 2 giving a mean plane at 30° W 45° represents the foliations in the northwest satellite and its contiguous shear zone. Analysis of the map shows that foliations tend to dip moderately (< 50°) in the western side of the main pluton and more steeply in the eastern side. Foliation trajectories possibly define two lobes. Along the median line separating these two sides, N-S and sub-vertical foliations are mostly observed. The southern margin of the pluton shows gentle dips mostly to the north. The northwest granite satellite has foliations variably dipping (30-60°) to the northwest, while the eastern granite satellite has foliations dipping gently to the northwest.

LINEATIONS

The long axis K1 of the magnetic ellipsoid is equated with the magmatic lineation (see Bouchez, this volume, for explanations). The magnetic lineations shown in figure 6 represent the vectorial mean of four specimens having a within-station angular variability less than 25° around the mean in 98% of the stations. Therefore, the average is used as the average mineral lineation at each station. The southernmost part of the pluton was the site of a detailed preliminary sampling where the high homogeneity of linear structures was demonstrated. Thus kilometre spacing for sampling was deemed sufficient for the remaining study. Although the lineations mostly trend N-S, several subdomains can be distinguished. In the main pluton, the plunges are gentle to moderate to the north and northeast, except along the northern margin (Fig. 6: shaded) which is characterised by rather steeply southwest plunging lineations. Lineations trajectories display asymmetrical patterns on sides of the shear zone separating the northwest extremity of the main pluton and the northwest satellite. This pattern fits with a dextral sense of shear. In the country rocks in the vicinity of the pluton, lineations, although rarely measured, are often consistent with those in the granite. The stereonet of Figure 6b shows that the average lineation of all the AMS measurements (011°/9°) is close to the foliation zone axis of figure 6a (006°/12°).

MICROSTRUCTURES

A systematic investigation of microstructures in the core samples used for the petrographic and AMS studies shows that the rocks preserve evidence of a continuum of deformation from magmatic- to high-temperature solid state (for a definition, see Paterson et al. (1989) and Bouchez et al. (1992)). The magmatic type is always found in the mafic stocks and charnockites (Fig. 7a, Plate 1d). High-temperature incipient plastic deformation microstructures are ubiquitous in the granitic types of the main pluton. They correspond to a few kinked biotite crystals and a few subgrain boundaries in quartz. Pervasive solid-state deformation features are observed mainly in the northwest and east satellites, and also in the southernmost part of the pluton (Fig. 7a). Orthogneissification (Plate 1f) appears along the southeast and northwestern margins of the pluton. In these zones, solid-state deformation and gneissification occurred under high-temperature conditions as demonstrated by the co-stability of hornblende and biotite within the foliation and C-planes. The shear sense criteria associated with these microstructures always indicate a dextral sense, consistent with that observed in neighbouring shear zones affecting the country rocks.

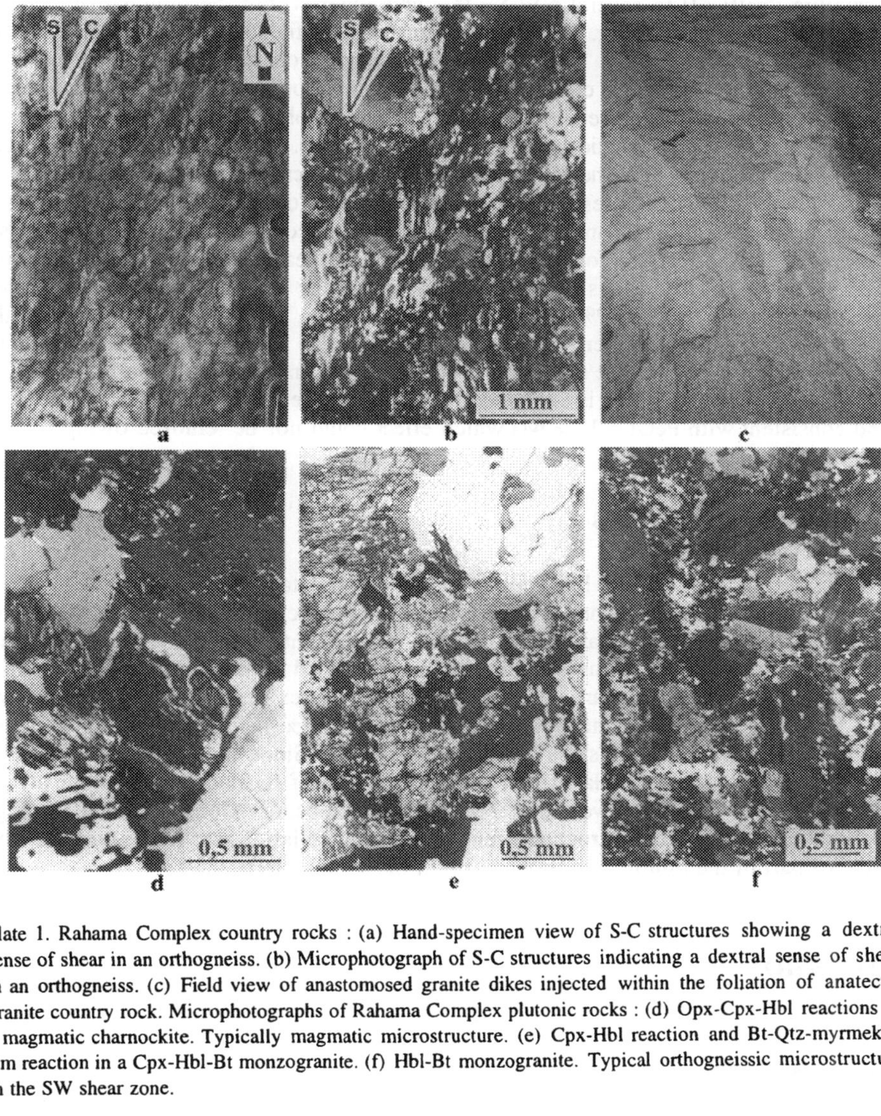

Plate 1. Rahama Complex country rocks : (a) Hand-specimen view of S-C structures showing a dextral sense of shear in an orthogneiss. (b) Microphotograph of S-C structures indicating a dextral sense of shear in an orthogneiss. (c) Field view of anastomosed granite dikes injected within the foliation of anatectic granite country rock. Microphotographs of Rahama Complex plutonic rocks : (d) Opx-Cpx-Hbl reactions in a magmatic charnockite. Typically magmatic microstructure. (e) Cpx-Hbl reaction and Bt-Qtz-myrmekite rim reaction in a Cpx-Hbl-Bt monzogranite. (f) Hbl-Bt monzogranite. Typical orthogneissic microstructure in the SW shear zone.

MAGNETIC ANISOTROPY

The total magnetic anisotropy (Fig. 7b), here expressed in percentage $P\% = 100 \times ((K_1/K_3) - 1)$ ranges from 3% to 40% and varies smoothly at the exception of local high gradients along shear zones. In 50% of stations, the magnetic anisotropy is between 5% and 10%. The lowest anisotropies are found at the southwest margin of the pluton and correspond to the lowest susceptibility magnitudes. The highest anisotropies correspond: (1) to the shear zone located between the pluton and the northwest satellite and within the satellite itself; (2) to several NW-SE trending zones within the pluton (parallel to the petrographic zonation, Fig. 3), (3) to the northeast mafic stock and (4) to the shear zone at the SE of the pluton. This shows (1) a correlation between anisotropies and petrographic types and (2) a correlation between anisotropies and strongly deformed areas. Indeed, the positive correlation between P% and K, obvious in Figure 7c, is also observed on other ferromagnetic granites (Archanjo, 1993). When the modal abundance of magnetite, which is here the main magnetic carrier, increases, the likelihood of aligned magnetite grains increases yielding a distribution anisotropy (Grégoire et al., 1995) added to the magnetite shape anisotropy. To reduce the effect of petrographic variations, an average P% is calculated for each station on the 15 nearest values from the list of increasing susceptibility values. The curve of these average P% in Figure 7c shows the positive correlation between P% and K due to petrography. The stations for which high P% are mainly due to a higher strain (empty circles) plot above this curve. These stations correspond mostly to the shear zones (shaded zones in Fig. 7a). Finally, the high P% of the diorite-norite stock is explained by a high magnetite content (attested by Fig. 4 and consistent with FeOt ≈ 12% wt) which effect could not be removed owing to the small number of samples.

SYNTHETIC CROSS-SECTIONS

Two cross sections have been built out of the structural data (Fig. 8). Section AA' (Fig. 8a) normal to the regional trend (see Fig. 6a), illustrates the northwest dips of the foliations in the country rocks, parallel to the two transcurrent zones shear where gneissic screens of country rocks in the granites, and anastomosed granite dikes interfoliated in the country rocks (Plate 1c) are observed. The charnockites have been given a concave shape in this section to take into account their arcuate shape in map view (Fig. 3a) and their inward foliation dips on sides of the "median line". Since the interior of the main pluton is free of the felsic enclaves that are characteristic of its margins, the roof of the pluton was likely much above the present level of exposure. Along section BB' (Fig. 8b), parallel to the regional structural trend, the dips of foliations are to the south in the northeast part of the pluton, and to the north in the southwestern part.

DISCUSSION

MAGMA SOURCES

The magmas of the Rahama complex, like other Nigerian monzogranite-charnockite series, derive from partial melting and progressive extraction from a lower continental crust source (Dada et al., 1995). The presence of hypersthene in, at least some of the granitic rocks of the Rahama complex, clearly points to an origin from water undersaturated magmas. All this is in agreement with igneous charnockites being derived from

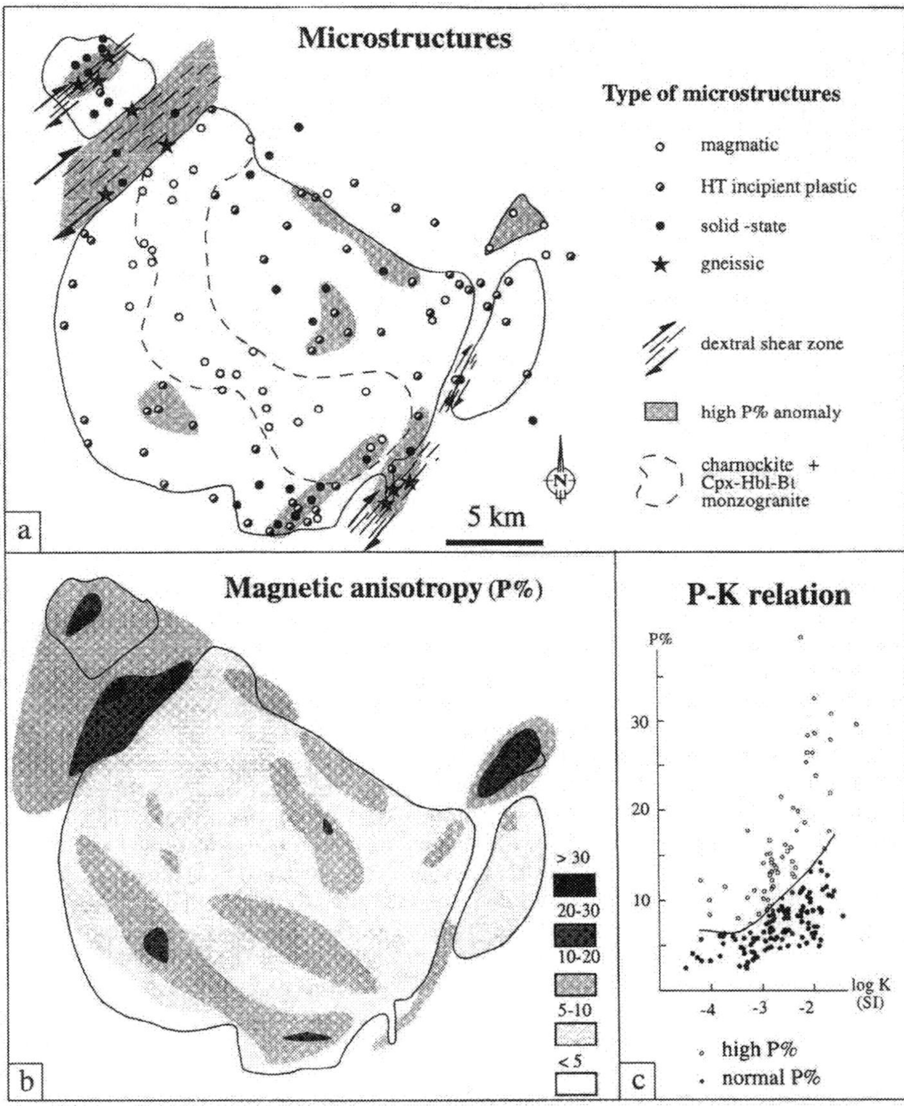

Figure 7. Microstructures and magnetic anisotropy in the Rahama Complex. (a) Microstructures; note the magmatic microstructures in the charnockites and the solid-states to gneissic microstructures along the NE-trending shear zones. (b) Raw magnetic anisotropy map; note the increase of P% related to the amount of strain (shear zones) and to petrographic variations (mafic stocks). (c) P%-log K diagram; note the positive correlation between P% and K, the solid curve discriminates stations where P% is consistent with petrographic type from stations exhibiting high P% due to the strain amount or to an exceptional composition (see text).

hot (≈900°C), dry (≤ 2 %) and deep (900 ± 100 MPa) crustal magmas as shown by Kirkpatrick and Ellis (1992). The depth of emplacement can be inferred from Al-in hornblende geobarometry on the specimens of the Rahama Complex. Pressures of 540 ± 60 MPa (Ferré et al., in progress) are recorded, which correspond to depths of about 15 kilometres. The present-day thickness of the continental crust in the Rahama area is about 35 kilometres (Fairhead and Okereke, 1987). It corresponds to the minimum portion of crust below the Rahama Complex at the time of emplacement since, after the intrusion, the crust did not undergo any thickening nor magma underplating events. Thus, the total crustal thickness at the time of emplacement was at least of 35 + 15 = 50 km.

ORIGIN OF THE GRANITE-CHARNOCKITE REVERSE ZONATION

Our petrographic and fabric study, based on a closely spaced sampling, has evidenced a progressive reverse zonation of the Rahama pluton, from Hbl-Bt monzogranites to Opx-Cpx-Hbl-Bt charnockites (Fig. 3a). Such a zonation might be common in the charnockitic plutons of Nigeria as most previous studies describe a progressive contact between charnockites and granites (Olarewaju, 1978; Tubosun et al., 1984). This progressive change in magma types supports the idea of a continuous feeding of magma through the same pathway during the emplacement of the complex, and strengthens the proposition of root-zones in the tentative sections of Figure 8.

Some mineralogical features of the charnockites and Cpx-Hbl-Bt monzogranites, namely the systematic presence of clinopyroxene and the weak zonation of plagioclase, are typical of an origin from water undersaturated magmas (Vance, 1962; Naney, 1983; Whitney, 1988). This undersaturation is compatible with the amphibole dehydration melting model of Tepper et al. (1993). The granite-charnockite series shows a range of decreasing whole rock water contents, from 0.65 wt% in the more felsic rocks to 0.35 % in the more mafic ones (Dada et al., 1995). We interpret these data as resulting from the increasingly water-depleted and deeper or hotter sources of these magmas, in agreement with the lower continental crust stratification model of Glazner (1994). This model predicts that mafic magmas tend to equilibrate within a high-temperature and low viscosity environment at their level of neutral buoyancy, i.e. the denser, the lower.

Following the latter considerations that suggest vertically juxtaposed sources for the magmas within a stratified lower crust becoming more mafic downwards (Fig. 9), the pluton zonation likely results from the successive emplacement of more mafic and more water depleted magmas, and tapped from increasingly deeper sources. Indeed, two main factors may work towards the emplacement of the granites prior to the charnockites: 1) production horizons of the former at higher levels, and thus easier access to the emplacement site and 2) their lower densities faster transits through the crust.

Interpretation of structures and mode of emplacement. The deep emplacement of the complex, > 15 km, inferred from geobarometry on the monzogranites, agrees with the high temperature (T°C ≥ 650°C) of the country rocks during emplacement. This is also attested by the lack of grain size reduction in the granite near its margins, and agrees with the absence of low-temperature deformation features in the neighbouring country rocks. Instead, the presence of in-situ melted anatectic granite, at both the pluton contact and in the eastern satellite where the anatectic melt is interfoliated with the Hbl-Bt monzogranite (Plate 1c), indicates that intrusion occurred within the ductile domain of the continental crust. This is also supported by the elongate and folded shape of anatectic granite xenoliths in the Hbl - Bt monzogranite (Plate 1c).

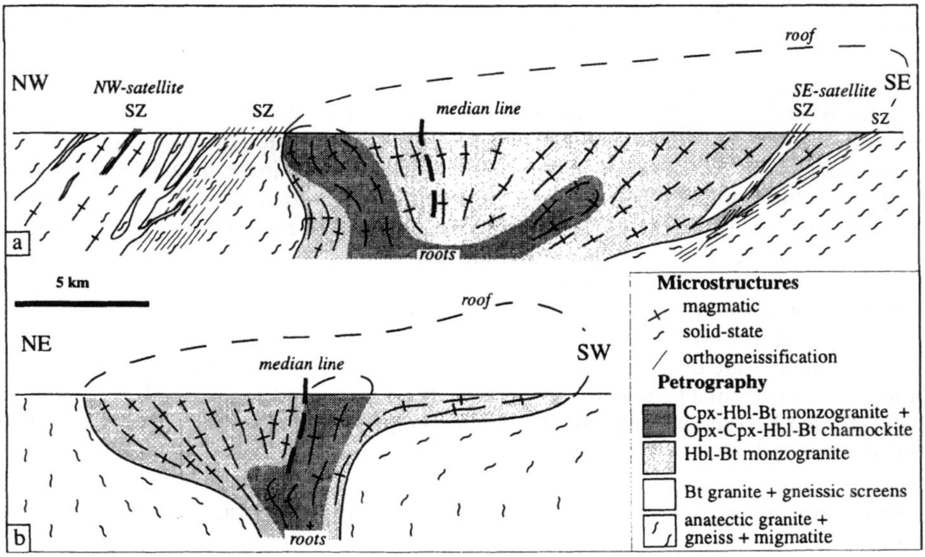

Figure 8. Anatomy of the Rahama Complex: Interpretative cross-sections. (a) NW-SE section, normal to the regional structural trends. Note: 1- foliations in the country rocks dip to the NW, 2- foliations in the pluton dip inward on both sides of the median line, 3- the main pluton is flanked by two high temperature shear zones, 4- the roof of the pluton is believed to be not very close to the present-day level. (b) SW-NE section, parallel to the regional stretching direction. Note: 1- conformity between structures in the pluton and in the country rock; 2- the steep NE margin and the shallow SW margin.

Pluton emplacement kinematics can be assessed by comparing within-pluton structures with regional structures away from the intrusion. Regional structures are characterised by NNE-SSW striking and steeply-dipping foliations (mean at 15° W 60°), and N-S trending lineations, slightly plunging to the north (mean at 000°/15°). Such regional strike-slip tectonics shows that the entire area has been penetratively sheared. However, shear strain was higher in the kilometre-wide strike-slip shear zones. These regional structures are the same as those in the pluton as stressed by the similarities of the stereonets of Figures 2b and 6: mean foliation at 15° W 55° and mean lineation at 010°/10°. All these observations and the systematical dextral shear senses point to a dextral strike-slip syntectonic emplacement of the pluton with possibly an extensional component. Furthermore, the magmatic microstructures observed near the pluton margins show the lack of substantial post-emplacement periplutonic deformation that may have accounted for the wrapping of regional foliations around the main pluton.

Within the pluton, the foliations and lineations are independent from the boundaries between rock types and, as already mentioned, conformable with the regional structures. These observations demonstrate that all the plutonic facies have undergone the same final deformation, at least to some extent. Therefore, they were emplaced during the same regional event. The absence of solid-state, or even incipient plastic deformation, in the central body of charnockites and Cpx-Hbl-Bt monzogranites (Fig. 7a) is compatible with their late emplacement. A detailed analysis of lineations in the pluton (Fig. 6b) reveals their tendency to follow asymmetrical trajectories depicting shear zones, among which the SW-NE trending and dextral shear zone affecting the western half of the pluton is the most conspicuous. Since the observed lineation pattern constitutes the latest strain undergone by the pluton, these shear zones trace the places where strain was localised by the end of magma emplacement. In addition, since the microstructures associated with these shear zones are almost ubiquitously of magmatic type (see Fig. 7a), the corresponding late-emplacement strain was definitely acquired in the magmatic state. The north-south median line (Figs. 6a and 8), separating the pluton in two halves, is suspected to be the latest dextral magmatic shear zone, probably responsible for the $\approx$ 3 km of dextral offset of the pluton outline. Alternatively, it may represent the contact between two coalescent lobes of the complex. These two intraplutonic shear zones likely post-date the shear zones of the northwest and southeast margins of the pluton which display high-temperature microstructures with a kinematic continuum of dextral shear strain from magmatic state to solid-state. At this late, post-emplacement stage, the pluton was probably less deformable than its country rock, hence remained undeformed in the solid-state. These observations suggest that magma emplacement and solidification occurred during a major dextral strike-slip regional event.

A MODEL OF DRAINAGE OF MAGMAS ALONG INCLINED TRANSCURRENT SHEAR ZONES

In orogenic domains, crustal structures such as foliations, thrust planes and shear zones and their directions of dip are probably inherited from earlier collision episodes (Allmendinger et al., 1987). Steep shear zones at upper-crustal levels are relayed at mid-crustal levels by inclined shear zones and at lower-crustal levels by low-angle shear zones (Davies, 1987). The lower crust is most likely horizontally compositionally stratified (Glazner, 1994) and therefore the fertile zones are expected to be horizontal. Partial melting in the lower crust is likely to occur in amphibole or biotite-rich protoliths by dehydration melting (Tepper et al., 1993), especially along rheological discontinuities where melts will gather (Tommasi et al., 1994). Thus, crustal-scale shear zones are suitable to tap magmas at different depths. Magma upward migration is achieved both by density contrast and by sucking due to the strike-slip movement along shear zones (D'Lemos et al., 1992). Magmas injected along shear zones can act as lubricants and thus enhance strain localisation (Hollister and Crawford, 1986).

Our data suggest that an inclined shear zone acted as the pathway to drain magmas upward successively from different levels of the stratified lower crust (Fig. 9). These different magmas were emplaced at mid-crustal levels as a single intrusion in a magmatic continuum. The transfer zone between the source region and the emplacement level may be represented by drainage zones similar to the west dipping northwestern satellite and its contiguous shear zone. This is well supported by the location of the two satellites on the outer side of the pluton marginal shear zones. Indeed, such zones, where both anatectic granites and granites of the Rahama Complex are internally sheeted and interlayered with high-grade country rocks, most likely represent the feeding zone of hypothetical upper plutons. The abundance of mafic enclaves in all the satellite granites, interpreted as magmatic precursors, further supports our model. The level reached by the intrusion in the crust is determined by cooling of the magma by its country rocks leading to the solidification of the roof. After this stage, the persistent feeding forces the magma to spread laterally. At mid-crustal levels, such as that of the Rahama Complex, the progressive accumulation of granitic magmas is accommodated in a ductile way by country rocks. Fabrics in the emplacement site still record strike-slip tectonics expressed by the consistent pattern of foliations and lineations over the whole pluton. The non coaxial dextral regime is indicated by the sigmoidal lineation trajectories and the development of incipient plastic to solid-state shear zones in a kinematic continuum.

INFERENCES ON THE REGIONAL PAN-AFRICAN GEODYNAMIC EVOLUTION

The Rahama Complex is dated by an ill-defined Rb-Sr whole-rock age of 668 ± 128 Ma (van Breemen et al., 1977). Similar complexes in Nigeria have yielded more precise younger U-Pb zircon ages, between 638 ± 3 Ma (Dada and Respaut, 1989) and 585 ± 7 Ma (Dada et al., 1989). However, the Nd-Sm model age at 1.92 Ga obtained by Dickin et al. (1991) on a Hbl-Bt monzogranite from Rahama suggests a significant age inheritance from the lower crustal protoliths. Thus, the actual age of the Rahama Complex is more likely around 585 Ma.

The end of the Pan-African evolution of North Central Nigeria belongs to a continuum of thermobarometric conditions in which three episodes are recognised: (1) An early high-pressure episode ($P \approx 800$ - 1000 MPa, $T \approx 650°$-$700°C$) has been recorded by Qtz-Sil-Rt-Grt±Ky assemblages in boudins of a few zones of metapelites (Ferré and Caby, in progress); a similar event is dated around 630±10 Ma in northwest Cameroon (Toteu et al., 1994). (2) A pervasive hydrous migmatitic episode ($P \approx 500$ - 600 MPa, $T \approx 650°$-700 °C) occurred in the upper amphibolite facies as shown by coeval Sil-Crd-Grt±St assemblages; this event, dated by Pb evaporation zircon age on Bt-Grt-Kfs-Qtz leucosomes from Toro at 580 ± 10 Ma (Ferré et al., 1996), is contemporaneous with pluton emplacements. (3) A late greenschist facies episode is characterised by aplite dikes, mylonitic bands and albite + epidote veins; its age, constrained by a Rb-Sr biotite-WR age at 560 Ma in Toro (Ferré et al., 1996), is interpreted as a cooling age at the closure temperature of biotite.

The crustal uplift mechanism responsible for these decreasing P-T conditions can be discussed in the light of structural and thermobarometric data. A frontal thrusting uplift, necessarily associated with crustal thickening, is precluded (i) by the pervasive strike-slip kinematics at the time of plutonism and (ii) by the elevated geothermal gradient (45°C/kilometre) inferred from P-T conditions. Lithospheric extension cannot account either for this uplift because of the absence of extensional tectonics, of high geothermal gradients and of juvenile mantle input.

Therefore, the Rahama pluton, with its mid-crustal level of emplacement and magmatic structures in conformity with the regional structures, is a syntectonic intrusion whose emplacement and solidification occurred in a kinematic continuum during a dextral NNE strike-slip regional deformation by the end of the Pan-African times of Nigeria. Similar syntectonic features are also found in the neighbouring Pan-African plutons of Solli Hills (Ferré et al., 1995) and Toro (Déléris et al., 1996). Numerous Pan-African intrusions of Nigeria probably share the same characters. Recent works further north on the Air and Hoggar Pan-African provinces (Black et al., 1994; Liégeois et al., 1994) show that they result from exotic terrane amalgamation during convergent tectonics from about 730 to 645 Ma. A late Pan-African episode is characterised in these

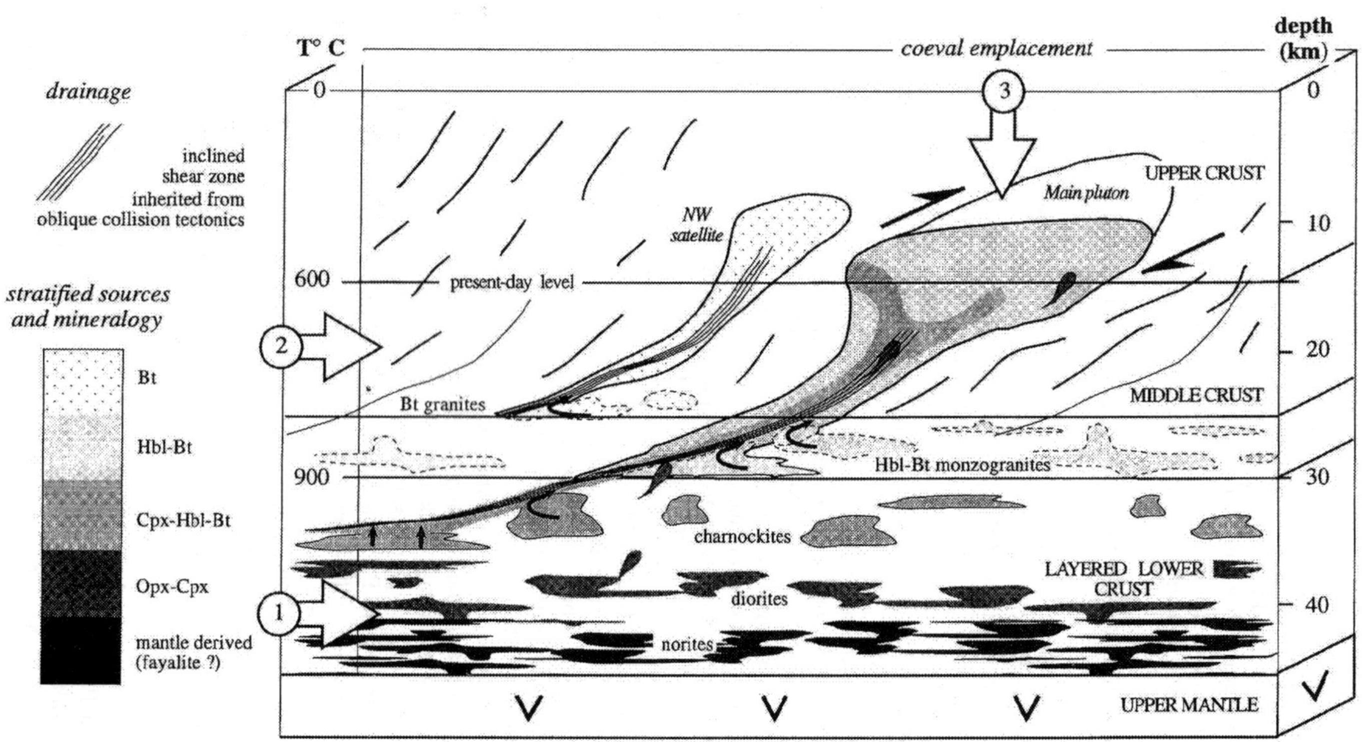

Figure 9. Schematic crustal section showing magma drainage along inclined shear zones: (1) stratified sources of magmas in the layered lower crust; (2) drainage and ascent of magmas along strike-slip inclined active shear zones and (3) coeval emplacement of charnockites and Hbl-Bt monzogranites. Temperatures are those of the regional gradient at the time of emplacement.

provinces by the emplacement in a strike-slip regime of high-K calc-alkaline N-S elongated plutons between 645 and 580 Ma. The evolution of the Rahama area is in agreement with the latter, late stage of the Pan-African history of the Air-Hoggar domain.

CONCLUSION

This petrostructural study of the Rahama Complex gives a model to the zonation of granite-charnockite complexes emplaced at a mid-crustal level and emphasises the role of shear zones in granitic magma transfer. The reverse and continuous petrographic zonation from monzogranites to charnockites is explained by the progressive tapping of increasingly more mafic and deeper magmas from a stratified lower crust. Drainage and upward migration of these magmas are achieved through low-angle shear zones in the lower crust becoming more inclined upward, at the level of emplacement, where they appear as major geological transcurrent boundaries.

The internal structures of the pluton are coeval with a late Pan-African deformation event dominated by a NNE dextral strike-slip tectonics. This event of regional deformation of the country rocks is probably contemporaneous with the emplacement of numerous plutons in northeast Nigeria. These results complement the recent works on other parts of the Pan-African belt in Brazil (Archanjo, 1993) and Hoggar (Djouadi, 1994) where similar dextral strike-slip settings coeval with pluton emplacements are described.

Note 1: AMS raw data are available on request from the first author.

Note 2: Structural data were plotted using Stereonet 4.52 program (Allmendinger, 1992).

Note 3: Mineral abbreviations are from Kretz (1983)

ACKNOWLEDGEMENTS

Jacqueline Déléris, Danladi Ibrahim, Alexander Idornigie and Tunji Solomon provided assistance in the field. Vincent Agboyo, Anne-Marie Roquet, Pierre Lespinasse and Christiane Cavaré-Hester contributed technically. The *Ministère des Affaires Etrangères* who provided logistical support in the framework of the Applied Geological Mapping Project, between the French Embassy in Nigeria and the University of Jos (1990-1994), is gratefully acknowledged. Catherine Suard and Jacques de Monès (French Embassy in Nigeria) personally encouraged this Project. Elf Petrolium Nigeria Ltd. kindly supplied a field vehicle for the duration of this study and is warmly thanked. We are greatly indebted to Louise Corriveau, Renaud Caby and Keith Benn for their constructive review of the paper and subsequent improvements. This contribution is from Centre National de la Recherche Scientifique (UMR n°5563) and University of Toulouse who provided the scientific and technical supports in France.

TRANSCURRENT SHEAR ZONES AND MAGMA EMPLACEMENT IN NEOPROTEROZOIC BELTS OF BRAZIL

Alain VAUCHEZ, Sergio PACHECO NEVES*
and Andréa TOMMASI
Laboratoire de Tectonophysique - ISTEEM - CNRS/
Université de Montpellier II
Pl. Eugène Bataillon - F 34095 Montpellier cedex 5 - France
** also Universidade Federal de Pernambuco,*
Departamento de Geologia, 50000-Recife-PE, Brazil

ABSTRACT

The association of shear zones and magmatism is usually regarded as indicating that partial melting and/or melt transport and emplacement was favored by pre-existing shear zones. However, a detailed examination of a set of examples from Neoproterozoic belts of Brazil allows us to suggest that this interpretation is not unequivocal. In some cases, for instance dyke-swarms, dyke-like plutons and small elongated plutons, a large body of evidence supports shear zone-assisted melt emplacement. In other instances, and especially for large-scale plutons, demonstrative criteria for shear zones predating magma emplacement are lacking, or even support magma emplacement predating shear zone development. Either pluton-enhanced strain localization and shear zone nucleation, or coeval growth of shear zone and granite emplacement, may represent alternative processes resulting in a close association of shear zones and plutons.

INTRODUCTION

In many orogenic belts throughout the world, and independently of the age of the orogeny, an association of magmatic rocks with transcurrent, normal or thrust faults is observed. This led many authors to consider genetic relationships between faults (or shear zones) and magmatism. Several degrees of interaction between shear zones and magma have been suggested and probably all of them may account for some specific examples. Temperature increase due to energy dissipation in lithospheric shear zones (shear-heating) was invoked by Nicolas et al. (1977) to explain the association of magmatism with major shear zones. Brun and Cobbold (1980) have shown that shear heating in crustal shear zones cannot increase the temperature enough to trigger partial

J.L. Bouchez et al. (eds.) Granite: From Segregation of Melt to Emplacement Fabrics, 275–293.

melting. Fleitout and Froidevaux (1980) have shown that for a long lived shear zone rooted in the upper mantle, deforming at geological strain rate, shear heating generated in the uppermost mantle may raise the temperature in the lower crust enough to allow water-present partial melting of metasediments. This process was recently invoked to explain an increase in metamorphic conditions and granitic magmatism, both closely associated with the Red River fault in the Himalayas (Leloup and Kienast, 1993; Leloup et al., 1995). That shear heating may produce water-absent melting, which is more likely in the lower crust (e.g., Clemens and Vielzeuf, 1987), is, however, doubtful. It remains possible that, under favorable circumstances, shear heating may produce limited melting in the crust, and therefore account for small volumes of crustal magmas emplaced in the immediate vicinity of a major shear zone.

Hutton and Reavy (1992), in an attempt to explain "the association of compositionally expanded granitoids with a major mantle component and transcurrent shear zones" observed in the northern part of the British Caledonides, suggested that crustal thickening occurring in transpressional lithospheric shear zones may result in the selective melting of local crustal "roots". In this model partial melting is triggered by the deformation in the shear zones, although it is due to crustal thickening and not to shear heating. This process may certainly account for lower-crust melting. However, melting of crustal roots is a protracted process (England and Thompson, 1986), and magmatism produced through this process should be late- to post-kinematic, and limited to domains that underwent significant thickening through transpression. Shear zone-assisted melt production may therefore only hold for specific cases of crustal magmatism concentrated along lithospheric, usually continental-scale, fault zones.

Whatever the process through which the source rock melts, a large body of evidence indicates that significant volumes of magma ascended in faults and shear zones at different crustal levels, and even that shear zones may contribute in melt extraction and concentration. Faults, and moreover fault arrays, because they frequently display a complex geometry, may provide space for magma emplacement, at least in the brittle crust. Shear zone-assisted magma emplacement is therefore an appealing alternative to diapirism. The result is a clear tendency to regard any spatial association of granite and shear zones as indicative of shear zone-assisted emplacement, although convincing evidence that shear zones predate plutons -and especially large plutons- emplacement is in many cases lacking.

On the other hand, emplacement of igneous bodies introduces a rheological heterogeneity in the crust. Plutons are likely emplaced as almost crystal-free magmas (e.g., Clemens and Mawer, 1992; Brandon et al., 1996) and have therefore a low viscosity compared to solid rocks. If the temperature contrast between the magma and the country rock is moderate (emplacement in the lower to middle crust), plutons may remain molten a long time span before full crystallization, and moreover before thermal equilibrium with the country rock is reached (Davidson et al., 1992; Davidson et al., 1994; Tommasi et al., 1994). In addition, intrusion of magma thermally weakens the surrounding country rocks. The duration of such thermal/rheological heterogeneities may be significant with regard to the rate of deformation processes; they may therefore generate perturbations of the regional strain field and especially favor strain localization and shear zone initiation (e.g., Hollister and Crawford, 1986; Lister and Baldwin, 1993; Neves et al., in press).

Indeed, both shear zone-assisted magma emplacement and magma-enhanced strain localization may exist. To properly interpret pluton emplacement processes, it remains therefore important to discriminate cause and effect, and this appears not to be straightforward. A set of criteria, some of which have been suggested by Paterson (pers. comm. 1995) may help in this analysis. However, only few of them considered alone are conclusive, and a choice between the two solutions should rather result from the fulfilment of several indicative criteria, and the absence of criteria pointing toward the other possibility.

Criteria supporting shear zone-assisted melt transport and emplacement

- Shear zone-assisted extraction of melt in the source region.
- Pluton fed by dyking in the shear zone.
- Magmatic contacts cross-cutting a mylonitic foliation in the country rock.
- Extensive deformation of the country rock showing evidence of pre- and post-emplacement mylonitization.
- Xenoliths of mylonite within the magmatic rock free of solid-state deformation.
- Lesser deformation in the mylonitic igneous rock than in the mylonitic country rock.
- Plutons displaying large aspect-ratios, although they do not show an intense solid-state deformation.
- Magmatism restricted to the vicinity of the shear zone
- Alignment of plutons along the shear zone.

Criteria supporting magma-enhanced strain localization

- Shear zones cross-cutting the country rock metamorphic fabric to which the magmatic fabric of the pluton is parallel and coeval.
- Evidence of emplacement by another process than transcurrent shear zone-assisted emplacement.
- Strike-slip faults end away from the pluton and/or have their greatest displacement near plutons (taking into account that heat advection due to the raising plutons may induce significant changes in deformation mechanisms resulting in enhanced strain localization).
- Evidence that fault orientation is controlled by the spatial distribution of plutons (unexpected bending or splay from a major shear zone).
- Magmatism distributed over a large area independently of the existence of shear zones.

This list of criteria is not exhaustive, other criteria may certainly be found improving this kind of analysis. It is also more appropriate to magmatic rocks emplaced in the medium to lower crust than in the upper crust. Several observations like a progressive transition from a magmatic- to a solid-state deformation are not considered as criteria since they may develop in the same way during the two kinds of evolution considered.

Using this type of approach, we will consider several examples from Neoproterozoic belts of Brazil, for which shear zone-magma interaction can be discussed. Most of them are from the Borborema Province of Northeast Brazil and display evidence supporting both shear zone-assisted melt collecting and magma emplacement, and magma-assisted shear zone nucleation. In the last section, we will consider a syntectonic granite batholith from the Dom Feliciano belt of southeast Brazil that does not satisfactorily fit any of these interpretations.

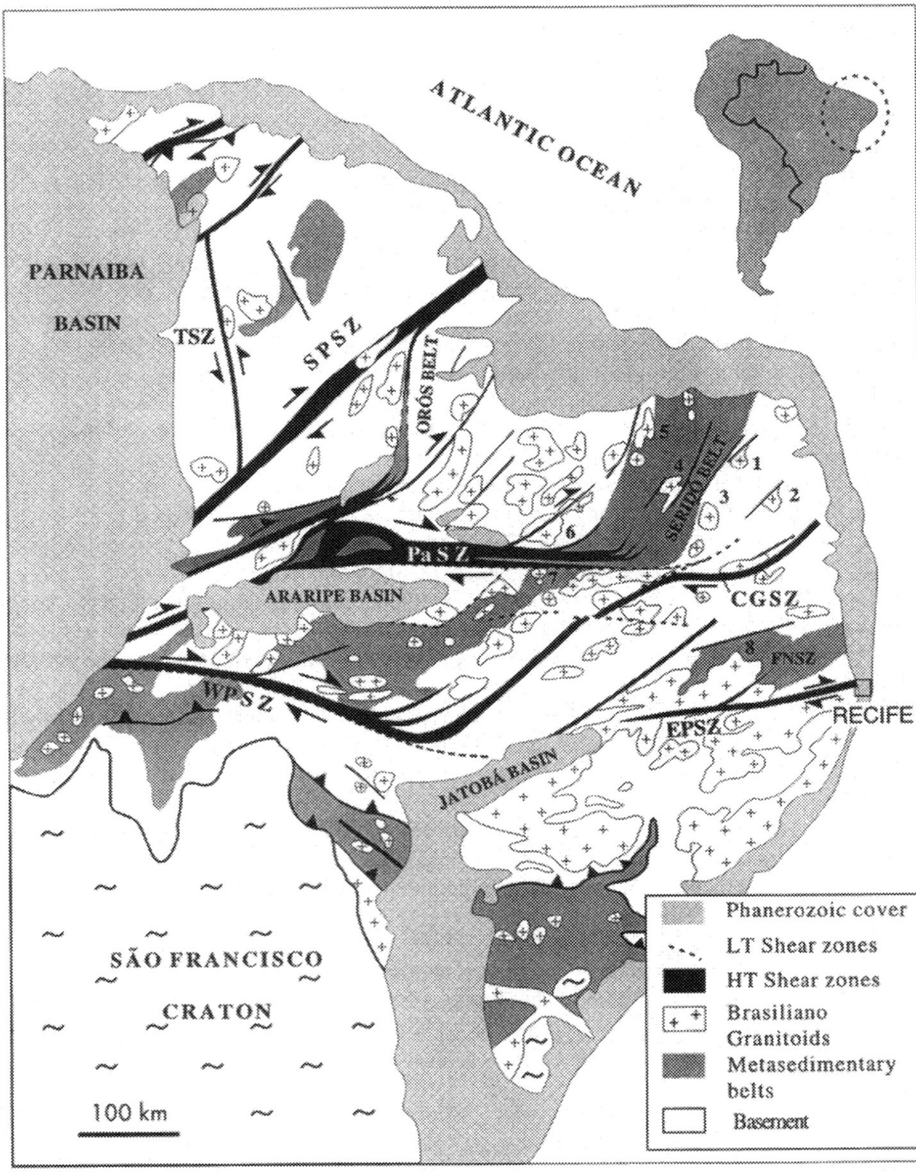

Figure 1. Schematic map showing the Borborema Shear Zone System and brasiliano granites. CGSZ= Campina Grande shear zone; EPSZ= East Pernambuco shear zone; FNSZ= Fazenda Nova shear zone; PaSZ= Patos shear zone; TSZ= Tauá shear zone. Numbers represent plutons refered to in the text: 1= Barcelona; 2= Gameleiras; 3= Picuí; 4= Acarí; 5= São Rafael; 6= Pombal; 7= Emas; 8= Fazenda Nova

GRANITES AND SHEAR ZONES OF THE BORBOREMA PROVINCE, NE BRAZIL

In the Borborema Province of Northeast Brazil, the Neoproterozoic Brasiliano orogeny resulted in the development of an intracontinental network of continental-scale transcurrent shear zones associated with transpressional belts (Vauchez et al., 1995). Northeast-trending dextral strike-slip faults in the northwestern domain and east-trending dextral strike-slip faults associated with north to northeast-trending transpression zones in the southeastern domain form a system kinematically consistent over more than 200,000 km^2 (Fig. 1). The major shear zones are several hundreds kilometres long, and 5 to 25 kilometres wide. They usually display a subvertical foliation bearing a well-developed subhorizontal mineral-stretching lineation and abundant shear criteria. The deformation within the shear zones began under high-temperature (650-700°C) low-pressure (400-500 MPa) metamorphic conditions and was coeval with synkinematic magmatism involving both crustal- and mantle-derived magmas. Continued or cyclic activity of the shear zones under decreasing temperatures down to the greenschist facies is usually restricted to narrow zones developed within or at the southern boundary of HT-mylonitic belts (Vauchez et al., 1995). From numerical modelling, Tommasi et al. (1995) have interpreted the development of the shear zone array as resulting mainly from strain localization enhanced by the presence of lithospheric-scale rheological heterogeneities (a craton and rift basins) within the continental plate during a collision. A similar effect due to the presence of incompletely solidified plutons was suggested by Neves et al. (in press) to explain many smaller-scale (a few tens kilometres in length and 1 kilometre or less in width) shear zones that are associated with plutons. Only few ages have yet been obtained in this area on both magmatic and metamorphic rocks. It is however noteworthy that emplacement of several plutons and peak-metamorphism conditions in mylonites are broadly synchronous (Legrand et al., 1991; Féraud et al., 1993; Leterrier et al., 1994; Monié et al., 1996; Neves, 1996).

SHEAR ZONE-ASSISTED MELT TAPPING IN MIGMATITES

The deformation within the main shear zones of the Borborema Province was coeval with partial melting of gneisses and metasediments of the country rock (Vauchez and Egydio-Silva, 1992; Vauchez et al., 1995). Migmatitic mylonites have been deformed in kilometres-wide transcurrent ductile fault zones, and evidence of melt tapping is observed within numerous small-scale shear zones (Figs. 2 a and b). Melt concentration within the shear zone is indicated by a higher proportion (sometimes 100%) of leucocratic neosome in continuity with the neosome of the migmatite. A transition from "in-situ" melt concentration to small dykes emplaced within the shear zone may be observed and suggests that, after a sufficient volume of melt is concentrated, it may move away from the source region. Melt is frequently thought to be channelled in dilatant shear zones. In opposition with this view, observations from the Seridó belt, the Campina Grande SZ and the West Pernambuco SZ suggest that melt extracted from migmatites tends to concentrate in compressional as well as in extensional small-scale conjugate shear zones.

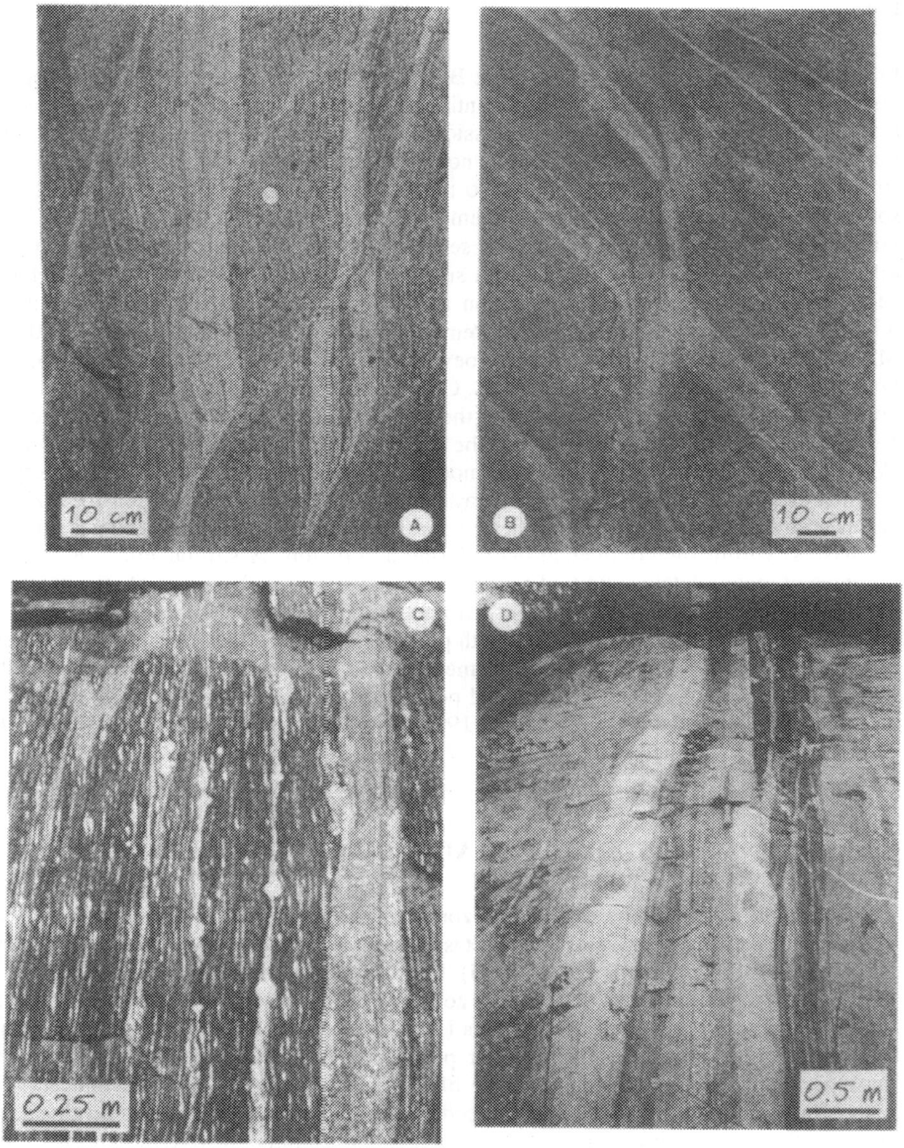

Figure 2. Field examples of shear zone-magma relationships. (A and B) Leucocratic melt collected in small-scale dextral shear zones formed in migmatites. Eastern termination of the West Pernambuco SZ. The migmatitic foliation trends NE-SW and the shear zones EW. Horizontal surfaces; (C) Dyke of leucogranite cross cutting at a small angle the high temperature mylonitic foliation within the Patos shear zone; (D) Mylonitized alternating dykes of porphyritic granite, diorite and aplite in the Fazenda Nova shear zone, illustrating synkinematic magmatic intrusions. Diorite contains K-feldspar xenocrysts suggesting mechanical incorporation that likely began at the magmatic stage.

SHEAR ZONE-ASSISTED MELT EMPLACEMENT

A control of shear zones on magma emplacement is clear for dyke complexes observed in most major shear zones of the Borborema Province, and for dyke-like and elongated plutons of moderate size emplaced within or along the limits of shear zones.

DYKE COMPLEXES IN SHEAR ZONES

Dyke complexes consisting of variably mylonitized to undeformed dykes emplaced parallel to or slightly oblique on the HT-foliation of the mylonites (Fig. 2c) may be observed in most major shear zones (e.g., Arthaud and Caby, 1993; Neves and Vauchez, 1995a; Vauchez et al., 1995). These dyke swarms are particularly well-documented in the left-lateral Fazenda Nova and the right-lateral Eastern Pernambuco shear zones (FNSZ and EPSZ; Fig. 1) where they are composed of up to five different rock types. They consist of successive generations of cross-cutting, light grey to pink leucogranites, porphyritic and equigranular syenites, monzonites, and diorites (Fig. 2d). From field observations it may be inferred that dyke emplacement began with porphyritic syenites and monzonites, which were subsequently intruded by diorites and leucogranites. Leucogranites intruding diorites is a common situation but dioritic dykes may also cross-cut leucogranites, suggesting several episodes of injection of dioritic magma. In some places, dykes may represent almost the whole volume of rock involved in the shear zone.

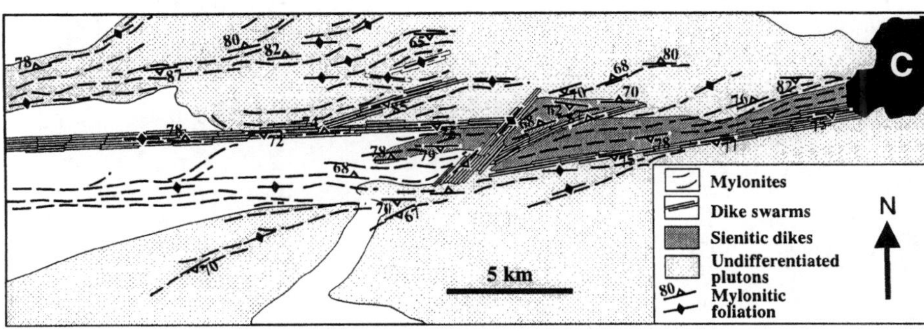

Figure 3. Schematic map showing the orientation of syenitic dykes and dyke swarms in a segment of the EPESZ. In the central part, NE-trending dykes emplaced in a discontinuity between the western and eastern segments. C= Caruaru town.

Voluminous synkinematic magmatism is sometimes associated with local extension due to curvature or discontinuity of the shear zone. Figure 3 illustrates this situation in a segment of the EPSZ west of the city of Caruaru. Westward, this dextral shear zone is composed of three, approximately E-W trending, main mylonitic belts which bound less deformed domains, and E-W trending dyke swarms occur within the mylonite belts. The eastern portion of the shear zone, on the other hand, consists of a single, wider mylonitic zone trending N70°E to N80°E. Large syenitic dykes (up to several hundreds metres wide) and dyke swarms follow the trend of this mylonitic belt. Between the E-W and the

ENE-trending mylonitic belts, dyke swarms are oriented NE-SW. This orientation is consistent with emplacement in an extensional stepover between two disconnected segments, with slightly different orientations, of a right-lateral shear zone.

DYKE-LIKE PLUTONS

Many small plutons displaying a very high shape ratio have been observed in the shear zones of the Borborema province, but most of them have undergone an intense mylonitization and it is impossible to infer whether their shape result from solid-state deformation or from emplacement processes. In a few cases however, dyke-like plutons have been emplaced late during shearing, and their high aspect ratio was essentially developed during emplacement. They are commonly 1 kilometre wide and 10 kilometres long (e. g., the Teresinha and Catinguera granites in the Patos shear zone, Fig. 4) to more than 50 kilometres long (e.g., the Pedra Lisa granite in the Tauá shear zone, Fig. 5).

Figure 4. Landsat image showing the Catinguera (left) and Teresinha (right) dyke-like plutons within the Patos shear zone.

The Pedra Lisa granite, emplaced within the 5-6 kilometre wide left-lateral Tauá shear zone (TSZ; Neves 1989, 1991), provides a good illustration of a dyke-like pluton. It is a homogeneous biotite-amphibole granite several hundred metres to more than one kilometre wide, and continuous over more than 50 kilometres. A pervasive solid-state deformation resulted in the development of S-C mylonites and bands of ultramylonite. A magmatic fabric, defined by amphibole-rich bands, elongate enclaves of microgranite, and by the preferred orientation of amphibole and biotite, is locally preserved. The magmatic lineation is subhorizontal, and tiling of K-feldspar megacrysts is consistent with a left-lateral sense of shear. Xenoliths of the mylonitic country rock are found even in the less deformed granite. Locally magmatic contacts of the granite truncate at low angles the mylonitic banding of the country rock. These observations are consistent with intrusion and subsequent crystallization of the magma in an active shear zone, and imply that emplacement of the Pedra Lisa granite was shear zone-controlled.

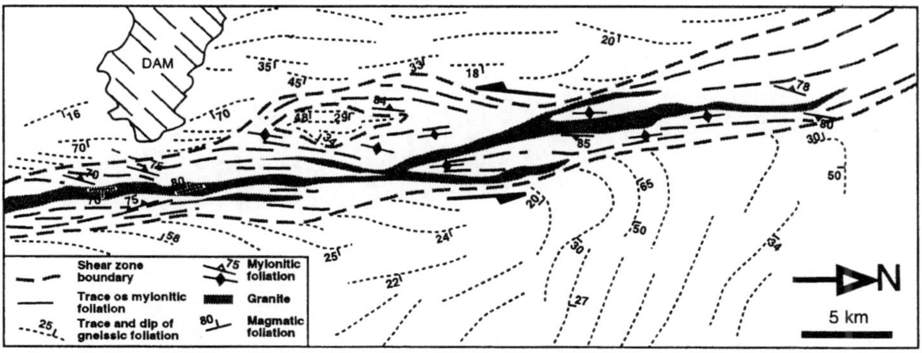

Figure 5. Schematic map of a segment of the Tauá shear zone showing the Pedra Lisa granite. For location see Figure 1.

In summary, except for their large size, plutonic bodies of this type display characteristics quite similar to the synkinematic dykes described previously. They were emplaced in high-temperature shear zones in conformity with the mylonitic foliation. They usually display a lower solid-state deformation than the surrounding mylonites, suggesting that they underwent only part of the deformation. When the magmatic fabric was retained, it is consistently parallel to the mylonitic fabric. Some evidence supporting emplacement in an already deforming shear zone, such as magmatic contacts locally cross-cutting mylonitic foliation or xenoliths of mylonite in undeformed granite, is usually observed.

LATE-KINEMATIC PLUTONS

At the southern boundary of the 10 kilometres-wide West Pernambuco shear zone (Fig. 1) several kilometric-scale peralkalic plutons are present (Fig. 6; Vauchez and Egydio-Silva, 1992). They are slightly elongated, and magmatic contacts truncate the high-temperature mylonitic foliation, showing that these plutons were emplaced within a pre-existing high-temperature mylonitic belt. Their cores are free of solid-state deformation, but display evidence of pre-full crystallization (or transitional) deformation of the suspension at high crystal content (Vauchez and Egydio-Silva, 1992) such as undeformed interstitial quartz associated with local plastic deformation of plagioclase, attributed to neighbour interaction in the deforming magma, and lattice preferred orientation of plagioclase (An_{10}) characteristic of rotation of rigid particles. Close to their boundaries, most of these plutons display evidence of a faint solid-state deformation under greenschist facies metamorphic conditions, due to a late low-temperature reworking of the southern boundary of the West Pernambuco shear zone.

Field relationships therefore support that these small peralkalic bodies have been emplaced within an already existing and probably still deforming mylonite zone, however their full crystallization occurred during a period of relative inactivity between the high- and the low-temperature deformations. These observations, together with the systematic location of the plutons at the shear zone-country rock boundary strongly suggest that the shear zone has acted as a conduit for the magma.

Figure 6. Landsat TM image of the West Pernambuco Shear Zone showing the WNW-ESE mylonite belt that marks the fault and peralkalic plutons (arrowed) emplaced within the HT mylonites along the southern boundary of the shear zone.

PREFERENTIAL NUCLEATION OF SHEAR ZONES IN CRYSTALLIZING PLUTONS

There is a large body of evidence in the Borborema Province that shear zones have favored migration and emplacement of melt from different lithospheric levels as dykes or small plutons. For large plutons, a control of transcurrent faulting on magma emplacement is far from obvious. Evidence that the shear zone predates granite emplacement is frequently lacking (Archanjo et al., 1992; Archanjo et al., 1994; Olivier and Archanjo, 1994). Indeed, for several large plutons of the Borborema Province an emplacement predating strike-slip motion is more likely. A detailed study of the Caruaru magmatic complex has led us to conclude that the formation of shear zones was favored by the presence of incompletely crystallized plutons in the crust. This model may hold for several other plutons in the Borborema Province.

PLUTONS AND SHEAR ZONES IN THE CARUARU AREA

In the Caruaru area, the left-lateral Fazenda Nova shear zone (FNSZ) and the right-lateral East Pernambuco shear zone (EPSZ) are spatially associated with the Toritama pluton and the Fazenda Nova and Serra da Japeganga batholiths (Fig. 7). The Toritama pluton is a shoshonitic syenitic pluton (Guimarães, 1989). The Fazenda Nova/Serra da Japeganga complex is a high-K calc-alkaline association comprising a coarse-grained granites (Serra da Japeganga batholith) at the top, overlying porphyritic granites and diorites (Fazenda Nova batholith) that form the base of the complex. The country rock include migmatites, micaschists, para- and orthogneisses usually displaying a gently-dipping metamorphic foliation.

Away from the shear zones, the Toritama pluton, Fazenda Nova and Serra da Japeganga batholiths display gently- to moderately-dipping magmatic foliations broadly parallel to the low-angle metamorphic foliation in the country rock. The magmatic lineation, although variable, trends dominantly northwestward. Approaching the shear zones, the magmatic foliation and the magmatic lineation are progressively rotated into parallelism with the subvertical foliation and subhorizontal lineation of the mylonite zones.

Considering the gradational contact between the coarse-grained and porphyritic granites, the presence of K-feldspar xenocrysts in the diorites (indicating intrusion during crystallization of the porphyritic granite), and the alignment of chemical analyses from the coarse-grained granite, the porphyritic granite and the diorite in Harker diagrams, Neves and Vauchez (1995a) suggested that: (1) the porphyritic granite was produced by homogeneous mixing of a felsic magma (now represented by the coarse-grained granite) with a dioritic magma; (2) the stratification of the complex reflects the structure of a magma chamber with the felsic member at the top, the mafic member at the base and the hybrid magma in-between; and (3) during crystallization of the porphyritic granite, new mafic batches were injected into the batholith and mingled with the hybrid magma, resulting in texturally and compositionally heterogeneous dioritic rocks. Homogeneous mixing requires almost crystal free magmas; it is therefore likely that a long time span elapsed between completion of mixing and attainment of the large crystal fraction at which the magmatic foliation formed. Over most of the magma chamber, stratified magmas have been "frozen-in" by crystallization without major perturbation. In a few domains the magmatic fabric differs from that of the magma chamber and is related with shear zones. Together with a transition from magmatic to solid-state fabrics, this indicates that magma full crystallization and shear zone development are coeval.

PRE-SHEAR ZONES EMPLACEMENT OF THE CARUARU MAGMATIC COMPLEX

None of the criteria that may support shear-zone assisted magma emplacement is fulfilled. Dykes observed in the batholith were emplaced late. Magmatic contacts crosscutting mylonites have not been observed for the different plutons. The numerous xenoliths of country rock observed near the EPSZ never display a mylonitic fabric. On the

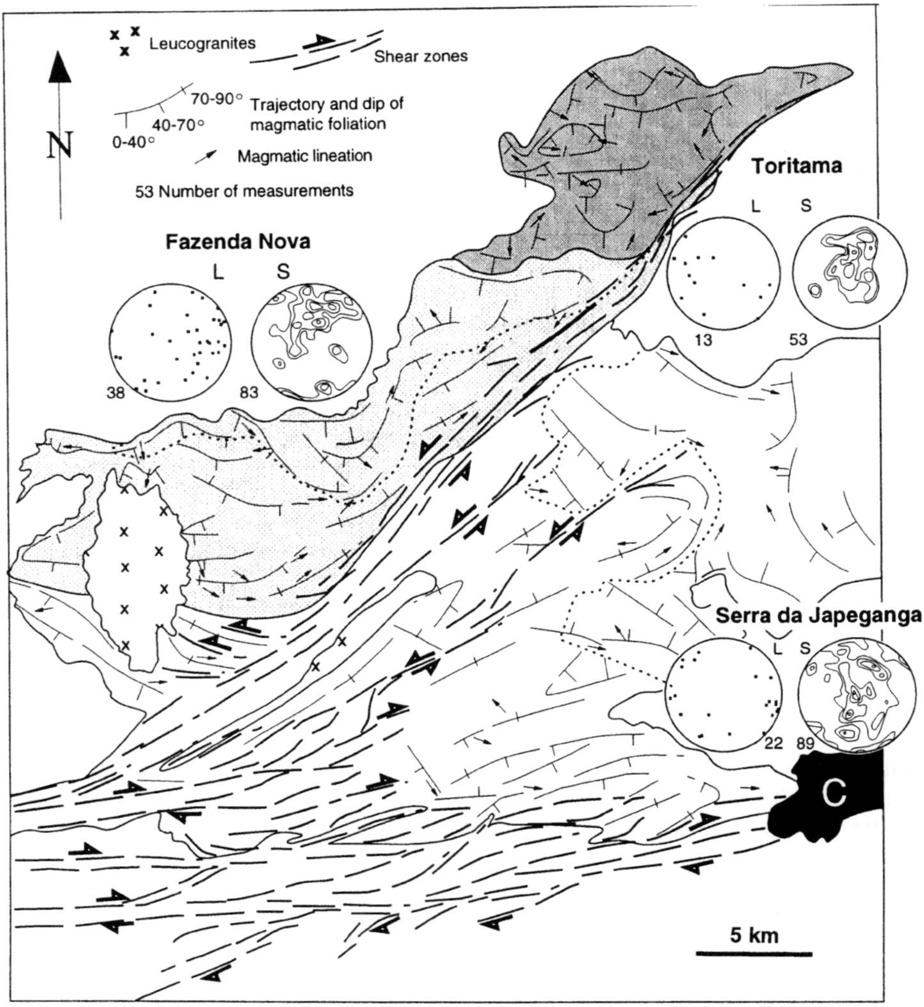

Figure 7. Schematic structural map and stereographic plots of magmatic lineations (L) and foliations (S) of plutons in the Caruaru area. The dotted line in the Fazenda Nova batholith separates diorites (at the northern side) from the porphyritic granite. The area to the east of the dotted line in the Serra da Japeganga batholith is dominated by gently- to moderatly dipping magmatic foliations striking at high angles to the orientation of the shear zones. C - Caruaru town.

other hand, most observations document the deformation by transcurrent shear zones of an already stratified magma chamber containing crystal-rich magmas. The FNSZ offsets the magmatic contacts between the porphyritic and the coarse-grained granite or the diorite (Fig. 7). The mylonitic deformation is mostly localized inside the plutons or at the pluton-country rocks boundary. The magmatic foliation generally parallels the regional flat-lying metamorphic foliation of the country rock, which is cross-cut by the shear zone. Mylonites derived from the country rock never display evidence of polyphase deformation what would be expected if the shear zone pre-dated granite emplacement and was reactivated after the magma was largely crystallized. Probably the most demonstrative feature is that the FNSZ is almost restricted to the batholith; it splays off the East Pernambuco Shear Zone within the Caruaru batholith and ends in the country rock a few kilometres beyond the boundary of the Toritama syenite. Similarly, the EPSZ narrows and terminates westward less than 50 kilometres away from the magmatic complex.

STRAIN LOCALIZATION AND CRYSTALLIZATION TIME

Although shearing may occur in magmas with low crystal fraction, strain localization and development of shear zones is an unlikely process, because their behavior is close to Newtonian. In crystallizing magmas, mechanical interaction between crystals probably begins at crystal fractions as low as 30 vol.% and becomes significant for crystal fractions higher than 40-50 vol.% (e.g., Fernandez and Gasquet, 1994; Lejeune and Richet, 1995; Arbaret et al., in press). Above this threshold the crystal mush is in the submagmatic state (Paterson et al., 1989; Nicolas, 1992; Paterson and Tobisch, 1992) which extends to melt fractions of 20-30 vol.% the transition to solid-state behavior (Nicolas et al., 1993). Nicolas (1992) suggested that strain localization starts in the submagmatic stage and increases with increasing crystal fraction. Consequently, at constant imposed strain rate, during crystallization magmatic shear zones progressively accommodate more strain and the magma outside the shear zones less. The crystallization of the remaining melt occurs in a narrow temperature interval because it approaches an eutectic composition. A solid-state behavior is therefore expected to occur only at temperatures close to the solidus.

The presence of amphibole and biotite as the main mafic minerals in all rock types indicates that they crystallized from water-rich magmas. For hydrous dioritic to granitic melts solidus temperatures differ only by a few tens of degrees and, at medium pressures (around 500 MPa), range from approximately 640°C to 670°C for andesitic to leucogranitic compositions (Wyllie, 1977; Green, 1982; Whitney, 1988). A temperature of about 700°C may be therefore considered as reasonable for a transition to the sub-solidus state. Considering that at the end of the mixing process the magmas were almost crystal free, we may estimate the time needed to reach a crystal fraction favoring strain concentration.

The main factors controlling the rate of crystallization in magma chambers are the temperature of the country rock, the initial temperature of the magma, the shape and dimensions of the magma chamber, and the mechanism of heat transfer (conduction or convection). To estimate crystallization time spans, we used Peacock's (1990) numerical simulation which assumes a tabular geometry for the intrusion, identical temperatures in the country rock at both sides of the intrusion, and one-dimension heat transfer by conduction. We modelled the cooling of the studied granites assuming a sill-like shape (supported by the large extension of the batholith and sub-horizontal magmatic contacts and foliations), a thickness of 5 kilometres, initial temperatures of 800°C and 850°C for the magmas (Neves and Vauchez, 1995b) and 600°C and 650°C for the country rock (Neves and Vauchez, 1995a).

Time-temperature curves were calculated at the centre of the intrusion and at 500 metres from the contact with the wall rock. With increasing time, these two curves get closer suggesting temperature homogenization across the intrusion. For an initial magmatic temperature of 800°C, the core of the pluton requires 0.7 to 1.6 Ma to attain a solid-state behavior, depending on the country rock temperature, with only slightly faster crystallization close to the margins (Fig. 8: dashed line). For an initial magmatic temperature of 850°C a much larger time is necessary; if the temperature of the country rock is set at 650°C the core of the intrusion begins to behave as a solid only after 4 Ma.

The results of this simple conductive model represent minimum estimates, especially because dioritic dykes continued to add heat to the complex during all its crystallization history (Neves and Vauchez, 1995b). Indeed, they support a large time lag (a reasonable mean estimate is 2 My) between the end of magma mixing and the deformation of the magmas at high crystal content, then in the solid-state. It is worth emphasizing that in open magmatic systems the time of crystallization may be significantly larger than implied by the conductive model. Detailed isotopic studies, for instance, have shown that subvolcanic silicic magma systems may be maintained for several hundreds of thousand years (Christensen and DePaolo, 1993) and even up to 1.1 My (Bogaard and Schimick, 1995) and still display low crystal contents, provided basaltic influxes continue to heat the base of the system.

CONSEQUENCES FOR THE BORBOREMA SHEAR ZONES

These data point to an evolution in which a magma chamber was formed, probably during an early tectonic (extensional?) event responsible for the development of the low-angle regional foliation. Mixing occurred while the different magmas injected in the chamber were almost crystal-free. During cooling of the magma chamber, a sub-horizontal magmatic foliation formed parallel to the magmatic contacts. After emplacement of the plutons, but before full crystallization of the magmas in the chamber, the regional deformation responsible for the formation of the shear zone system in the Borborema Province began to be active. The small volume of mylonite, the narrowing and the termination of the FNSZ and the EPSZ away from the magmatic complex point to a strain concentration in the plutons. The material in the magma chamber, still containing a large amount of melt, was able to accommodate a larger strain than the country rock. This may have favored flow instability and strain localization in the plutons. Shear zones may have nucleated preferentially where the largest rheological gradients existed, i.e., at boundaries between plutons and country rock, and between magmas of contrasted composition. In these domains, strain may have preferentially localized into most anisotropic rocks due to magma flow along the main discontinuities, and by a protracted transitional stage (Nicolas, 1992; Nicolas et al., 1993). Enhancement of strain localization by crystallizing magmas may also be related to plutons contracting more than the country rock during cooling. In polyphase ceramics for instance, when particles embedded in a matrix display a larger expansion coefficient than the matrix, i.e., contract more than the matrix during cooling, crack propagation is significantly affected by the particles and tends to curve and follow the particle-matrix boundary (e.g., Davidge, 1979).

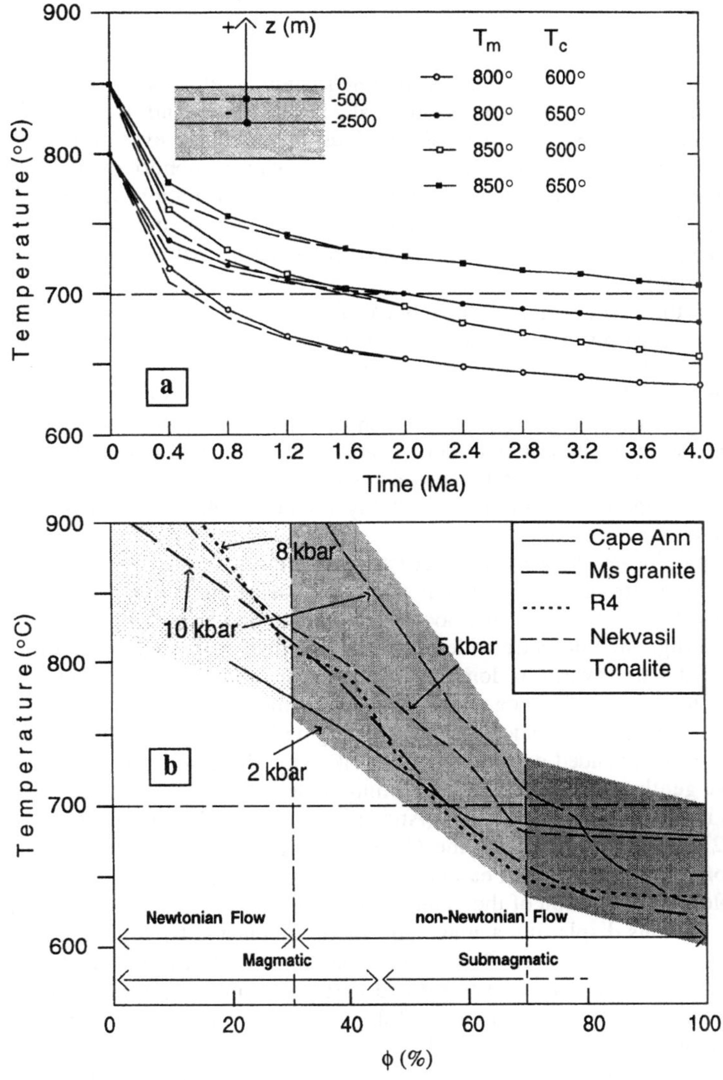

Figure 8. (a) Thermal modeling of a cooling, 5 km-thick tabular pluton emplaced into country rocks at temperatures (T_c) of 600°C and 650°C (after Peacock, 1990). T_m - Initial magmatic temperatures. (b) Temperature versus crystal fraction ϕ in experimentally crystallizing granitoid melts. Cape Ann: Cape Ann granite (Whitney, 1988); Ms granite: muscovite granite (Wyllie, 1977); R4: synthetic granite (Whitney, 1988); Nekvasil: rapakivi granite (Nekvasil, 1991); Tonalite: tonalite (Wyllie, 1977). Curves have been constructed using data from the original references. Light- to dark-grey shades mark Newtonian, Bimgham, and solid state-like rheological behaviors respectively. Appropriate fields for Newtonian and non-Newtonian flow, versus magmatic and submagmatic stages are shown.

COMPARISON WITH PLUTONS NORTH OF THE PATOS-CAMPINA GRANDE SHEAR SYSTEM

North of the Patos-Campina Grande shear zone system, a set of plutons has been studied in detail (Archanjo et al., 1992; Archanjo et al., 1994; Olivier and Archanjo, 1994) and their contrasted situations may represent various stages of the evolution suggested for the Caruaru magmatic complex. Most plutons display a subhorizontal to moderately dipping magmatic foliation consistent with the solid-state foliation in the country rock. Some bodies do not display evidence of strain localization in transcurrent shear zones, whereas others show a moderate to dominant influence of them on the development of the magmatic fabric.

Plutons of the first group may have been fully crystallized, or at least at an advanced stage of crystallization, when shear zones began to form, or may have merely escaped strain localization in strike-slip faults. This is exemplified by the Picuí pluton (Archanjo et al., 1992), which displays a pervasive magmatic foliation gently to moderately dipping eastward and a NS to NE-trending magmatic lineation, concordant with the regional fabric in the country rock. Emplacement of this pluton is interpreted as being coeval with an episode of tangential displacement parallel to the current strike of the regional metamorphic foliation, probably predating the development of the shear zone network.

The others plutons show a clear, but variable, imprint of shear zone-related deformation. First, there are plutons in which magmatic shear zones are present but did not evolve to mylonitic belts. The best documented is the São Rafael pluton which involves a NNW-trending, one kilometre wide belt of subvertical magmatic foliations separating northeastern and southwestern domains of flat-lying foliations (Archanjo et al., 1992). In that case there is no influence of shear zone deformation on magma emplacement, and the shear zone undoubtedly nucleated in the magma before its complete crystallization. Other plutons are bounded or cross-cut by strike-slip shear zones involving mylonites, but preserve gently to moderately dipping magmatic foliations away from the mylonitic belts. This is the case of the Pombal (Archanjo et al., 1994) and Gameleiras (Archanjo et al., 1992) plutons. Similarly to the Caruaru complex, these plutons display magmatic shear zones that reworked an earlier low-angle magmatic foliation, continued to grow after complete solidification of the magma, and propagated in the country rock.

Finally, plutons displaying a magmatic fabric dominated by high-angle to subvertical foliations may be observed. The best example is the Acarí pluton (Archanjo et al., 1992) which is bounded southeastward and northwestward by NE-trending shear zones, and displays a NNE-trending magmatic foliation bearing a gently to moderately SW-plunging magmatic lineation parallel to the stretching lineation in the shear zones. Radiometric dating points to a late emplacement of the Acarí granite, coeval with the deformation in the Borborema shear zone system (Féraud et al., 1993; Leterrier et al., 1994). This led Archanjo et al. (1992) to consider the Acarí granite as an example of shear zone-assisted emplacement. However, the overall steeply dipping magmatic foliation trends more N-S than the orientation of the solid-state fabric in the bounding shear zones. This may agree with a pluton either emplaced in a compressive domain or predating shear zone development, and in this latter case the pluton was still incompletely crystallized when affected by the transcurrent deformation.

SIMULTANEOUS PROPAGATION OF SHEAR ZONES AND MELT?

There is a set of examples for which neither shear zone-assisted pluton emplacement nor magma-assisted shear zone nucleation can satisfactorily account for the evidence. In several orogenic belts, very elongate plutons (shape ratio frequently larger than 10:1) have been described. Their size may reach several hundreds kilometres in length and a few tens of kilometres in width. They usually display a magmatic fabric consistent with non-coaxial flow and are partially reworked in a regional-scale solid-state shear zone in which the mylonitic fabric parallels the magmatic fabric of the pluton. However, although their shape, the well-organized magmatic fabric and the existence of a mylonite belt point toward magma emplacement within a pre-existing shear zone, evidence that the shear zone predates magma emplacement is totally lacking.

A good example where such relationships are observed is the Dorsal de Canguçu shear zone, a lithospheric-scale left-lateral transcurrent shear zone that accommodates a large part of the orogen-parallel deformation in the Neoproterozoic Dom Feliciano belt of SE Brazil (Tommasi et al., 1994). This shear zone runs from Southeast Uruguay to South Brazil over hundreds of kilometres (Fig. 9) and roughly marks the limit between magmatic arc and marginal basin rocks. A closer examination of the shear zone and its surroundings reveals that the shear zone is almost entirely developed within granitoids belonging to two successive generations of magmas of different composition: early coarse-grained granodiorites, which are intruded by peraluminous leucogranites. These granitoids form elongate plutons, defining a narrow (10 to 20 km) magmatic line that coincides with the shear zone (Fig. 9). The deformation history, as inferred by Tommasi et al. (1994) from the macro- and microstructures of the granitoids, reflects the influence of the emplacement of large volumes of magma on the mechanical behavior of the shear zone. Shearing is initially localized within the partially crystallized granodiorite, where it is accommodated by viscous flow. After total crystallization, strain localization within the hot granodiorite remain easier than in the colder country rocks. Emplacement of the leucogranite was accompanied by further strain localization and development of a magmatic fabric within the pluton. After crystallization of the leucogranite, shearing continued under decreasing temperatures, promoting the development of progressively thinner mylonitic zones. The small extent of the low-temperature deformation (lower greenschist facies conditions) suggests that shearing ended roughly when the leucogranite reached a thermal equilibrium with the host rock.

The geochemical evolution of the magmatism in the Canguçu shear zone of the Dom Feliciano belt may be explained by a vertical progression of partial melting within the crust. The granodiorite probably results from melting of the mantle and/or lower crust, whereas the leucogranites are produced by partial melting in the middle crust. Heat advection due to percolation of magmas within a shear zone should favor a vertical progression of partial melting towards the middle crust. Moreover, the large extent of synkinematic magmatism advocates that this shear zone has a lithospheric extension, and therefore may have represented a major channel for magma production and emplacement.

The pre-existence of the shear zone relative to the first magmatic batch cannot be unambiguously proved. The elongate shapes of the granodioritic plutons and their alignment within the shear zone, together with the pervasive and well-organized magmatic fabric, suggest that the shear zone predates the emplacement of the granodiorite. On the other hand, the shear zone is almost entirely developed within granitoids. Mylonitic country rocks are observed only in areas with smaller volumes of granitoids, and even in these areas, the country rock deformation is restricted to decametre-scale late shear zones formed under greenschist facies conditions. Moreover, when not obscured by late granitic intrusions or Cretaceous brittle faults, contacts between the granitoids and country rocks are igneous, and do not intersect any mylonitic fabric of the country rocks. Finally, xenoliths of country rock display a mylonitic fabric only when they have been mylonitized in solid-state shear zones affecting also the granitoids.

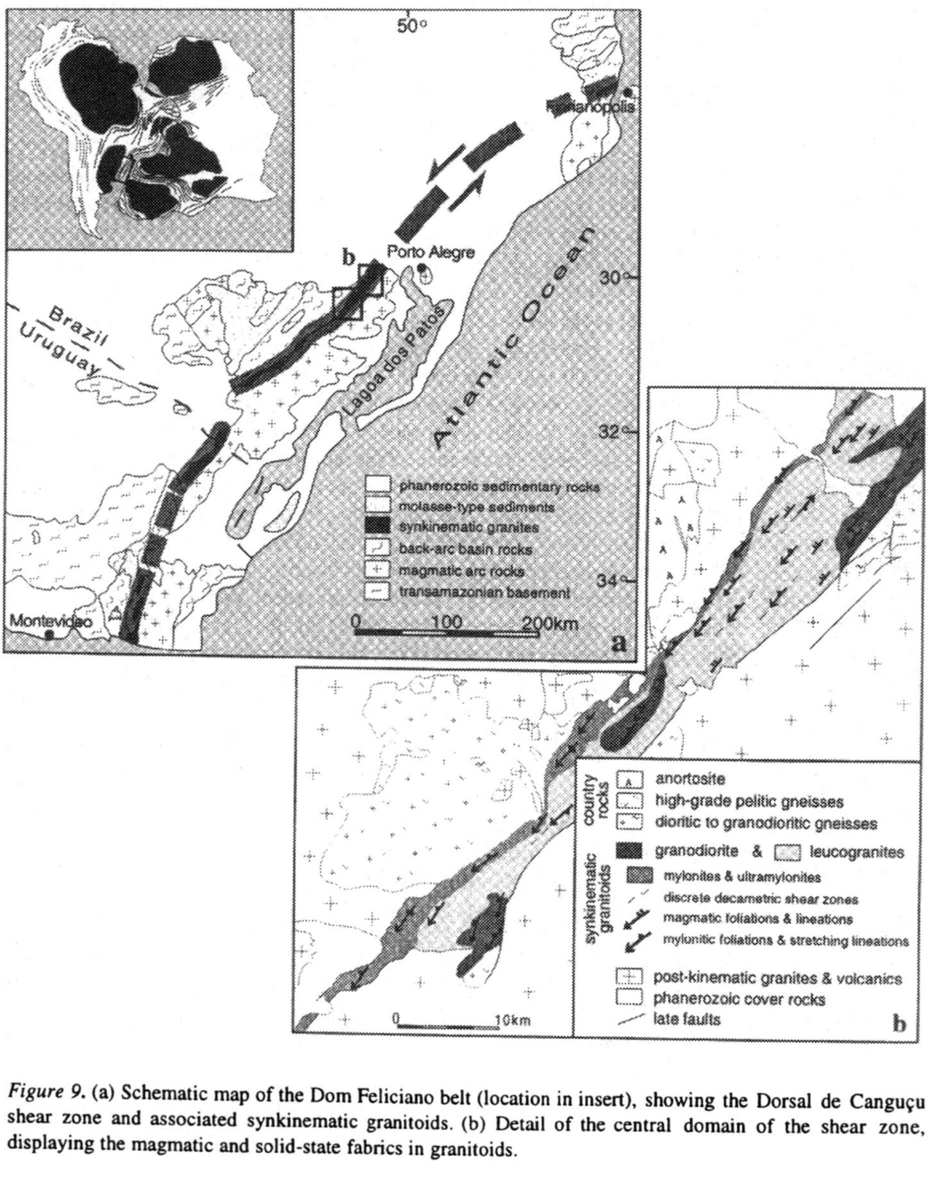

Figure 9. (a) Schematic map of the Dom Feliciano belt (location in insert), showing the Dorsal de Canguçu shear zone and associated synkinematic granitoids. (b) Detail of the central domain of the shear zone, displaying the magmatic and solid-state fabrics in granitoids.

An easy way to reconcile these apparently contradictory observations is to consider that shear zone growth and magma emplacement occurred simultaneously (Hutton, 1988). There, concentration of melt, and shear zone propagation enhance each other. This requires that the growth rate of the shear zone and the emplacement rate of magma are of the same order of magnitude, a condition that may be fulfilled if magma is continuously added in the form of successive batches into the system, maintaining a high fluid pressure and a high temperature by heat advection. Independently of which is the starting process, magmatism or deformation, magma emplacement promotes a significant softening within a restricted domain of the crust, leading to strain localization. Since for large plutons emplaced in the middle to lower crust, cooling is generally slow, this softening may persist during large time spans, allowing the accommodation of significant displacements within a shear zone. Within this shear zone, segregation of melt, leading to emplacement of successive generations of magmas, may extend this time span.

CONCLUSION

From the different examples considered in this study, spatial and temporal association of magmas and shear zones appear to result from various processes. A distinction between the various processes that may be responsible for the association of shear-zones and magmatic bodies is not straightforward. Shear zones may favor magma upwelling and emplacement, at least for bodies of moderate volume as dykes, dyke-like plutons and small plutons. On the other hand, shear zones may nucleate in largely, but incompletely crystallized plutons. The São Rafael granite, with its one kilometre wide magmatic shear zone, or the Mont-Louis-Andorra granite in which a set of magmatic shear zones formed when the magma was in the sub-magmatic state (Bouchez and Gleizes, 1995) are demonstrative of this process. From the example of the Caruaru magmatic complex we suggest that magmatic shear zones may continue to deform during the transition from the submagmatic to the solid states, producing mylonite belts, and may propagate into the country rock to form regional-scale ductile faults. Finally, considering that shear zones and magmas exert a mutual influence, a coeval development of shear zone and emplacement of magmas may be an adequate model in collisional continental zones, especially in the mature stage of the collision, when partial melting of the crust and orogen-parallel transcurrent faulting dominate the orogenic evolution.

ACKNOWLEDGEMENTS

This study was made possible thanks to funding by the Conselho Nacional de Desenvolvimento Científico e Tecnológico (CNPq, Brazil), the European Community and the Centre National de la Recherche Scientifique (CNRS, France). We are indebted to C. Archanjo, J.L. Bouchez, R. Caby, A. Nicolas, and S. Paterson for helpful discussions and suggestions. Special thanks are due to L.A.D. Fernandes and A.N. Sial who largely originated our studies of granites in southeast and northeast Brazil respectively. C. Archanjo and J. Grocott have kindly and efficiently reviewed this paper and hence significantly contributed to its improvement.

OBLIQUE DIAPIRISM OF THE YAKUSHIMA GRANITE IN THE RYUKYU ARC, JAPAN.

Ryo ANMA
Hans Ramberg Tectonic Laboratory, Institute of Earth Sciences, Uppsala University, Norbyvägen 18B, S-75236 Uppsala, Sweden

ABSTRACT

Field observations and structural analyses of the orientations of ductile flow fabrics and brittle fractures are used to infer diapiric rise of the Yakushima pluton into epizonal Paleogene sediments of an accretionary prism in Miocene times. Inside the pluton, preferred orientations of orthoclase megacrysts define a weak primary flow fabric. Shape and intensity of fabric ellipsoids derived from patterns in the alignment of the megacrysts reveal a circulation cell within the domal structure of the Yakushima granite. The asymmetry of this circulation pattern suggests that the pluton rose obliquely upward to the southeast, toward the Ryukyu trench. Fold axes in the ductile strain aureole of the pluton are distorted into conformity with the pluton's shape in the direction of magma ascent. Late aplite sheets, whose orientations record the local paleostress pattern, were emplaced into a concentric fracture system in and around the pluton. The aplites indicate the lifting, and tangential spreading with concentrically disposed intermediate stress axes, of a brittle roof above a buoyant ellipsoidal body of residual magma. Comparable with strain patterns in theoretical models of a viscous sphere rising in ductile surroundings, the deformation pattern in and around the pluton is attributed to the final upsurge of oblique diapiric emplacement.

INTRODUCTION

The processes of intrusion (ascent and emplacement) of magma into crustal environments are complicated by conditions and material properties changing rapidly over a short period (Paterson and Fowler, 1993). There are two primary concerns when interpreting structural observations of a granite intrusion. First are endogenic aspects of the intrusion process and how any deformation related to the emplacement phase (laccolithic growth, ballooning, late diapiric upsurge and/or lateral diapiric widening) overprints features related to the ascent phase of the magma (diapiric rise, stoping, ascent along fractures or a conduit). The second concern is how space is created for granite plutons and how endogenic plutonic buoyancy forces interact with exogenic tectonic forces.

J.L. Bouchez et al. (eds.) Granite: From Segregation of Melt to Emplacement Fabrics, 295–318.

Overprinting of emplacement-related structures on ascent-related fabrics has been recognized by several researchers. The well-known difficulty in distinguishing whether fabrics in plutons with circular planform have formed during diapiric rise of magma, during late diapiric lateral expansion, or through ballooning, has led to debate about the intrusion mechanism of many plutons (Cannibal Creek granite: Bateman, 1985; Paterson, 1988; Davis, 1993; Godin, 1994; Ardara granite: Holder, 1979; Sanderson and Meneilly, 1981; Vernon and Paterson, 1993). Indeed, flows of residual fluid in the pluton´s core may distort the primary flow structure (Cruden, 1990). Hutton (1992) argued that features resulting from pluton ascent may be entirely overprinted by features related to the emplacement phase. These arguments, that structures at the emplacement level may have nothing to do with the ascent mechanism, have discouraged subsequent studies of the ascent mechanism (Hutton, 1992).

Many recent studies of meso- and epizonal intrusions have concentrated on how emplacement mechanisms are related to regional tectonic forces (Brun and Pons, 1981; Hutton, 1982, 1988a,b and 1992; Castro, 1986; Guineberteau et al., 1987; Lagarde et al., 1990; Hutton et al, 1990; Aranguren and Tubia, 1992; Morand, 1992; Tikoff and Teyssier, 1992; Karlstrom et al., 1993). Most of these studies explain the space problem by invoking tectonic forces in the brittle crust. However, there are a number of plutons with circular planforms for which the space problem cannot be explained in terms of passive intrusion into tectonic cavities.

As often described (Paterson et al., 1991 and references therein), many circular plutons have (1) neither visible foliation nor clear compositional zoning, (2) few if any xeno-blocks of country rocks, (3) sharply discordant margins that truncate wall rock structures and (4) narrow contact aureoles with static fabrics. A number of studies have suggested forceful intrusion (i.e. active diapirism and ballooning) to explain these circular plutons, but Paterson and Fowler (1993) argue that the deformation of the country rocks in these cases is inadequate for such mechanisms to explain these. Stoping was postulated in some cases (Paterson and Fowler, 1993) but there are insufficient numbers of xenoliths for stoping to be a credible explanation in most cases. The possibility that granite magma invaded a fracture system and then formed an elliptical pluton at the emplacement level has been discussed (Brun et al., 1990; Clemens and Mawer, 1992; Petford et al., 1993) but few have shown convincing field evidence of deep fracture fillings (Guineberteau et al., 1987; Hutton, 1992; Fowler, 1994). There is no convincing single explanation for the intrusion mechanism of "mystery plutons" (Paterson et al., 1991). On the contrary, circular plutons are expected to preserve primary magmatic structure in any suitable strain marker because their endogenic interiors are isolated from exogenic tectonic overprints.

This work is based on field data from the epizonal Yakushima granite in Southwest Japan. Unlike examples cited by Paterson et al. (1991), the Yakushima granite has a weak magmatic fabric defined by shape preferred orientation of euhedral orthoclase megacrysts. Orientations, symmetries and intensities of the magmatic fabric, together with orientations of brittle fractures were mapped in detail inside the Yakushima granite as well as deformation patterns in the surrounding wall rocks. The field results are compared with strain patterns obtained by analogue modeling in order to test the hypothesized oblique diapirism as the intrusion mechanism for this elliptical pluton.

TECTONIC SETTING

Plutonic rocks in the Outer Zone of Southwest Japan (see Fig. 7 of Anma and Sokoutis, this volume) are all dated at 14±1 Ma (Middle Miocene; Shibata, 1978) using the K-Ar method and are conveniently divisible into two groups: S-type granites along the trench side, and I-type granites along the back-arc side (Nakada and Takahashi, 1979). Among all the contemporaneous granites, the 14 Ma S-type Yakushima granite (Fig. 1) is a large (23 km x 23 km) pluton located closest to the Ryukyu trench-Nankai trough axis. It forms an island 70 km south of the southern tip of Kyushu island in Southwest Japan, near the cusp where the Ryukyu island arc meets the Japan arc. The southwest trending axis of the Ryukyu trench swings toward the east-northeast trend of the Nankai trough at a point 170 km east of the Yakushima granite. Deep seismicity (Yamaoka et al., 1986) suggests that the subducting Philippine Sea plate dips 30-40° beneath Yakushima, increasing to 50° further west.

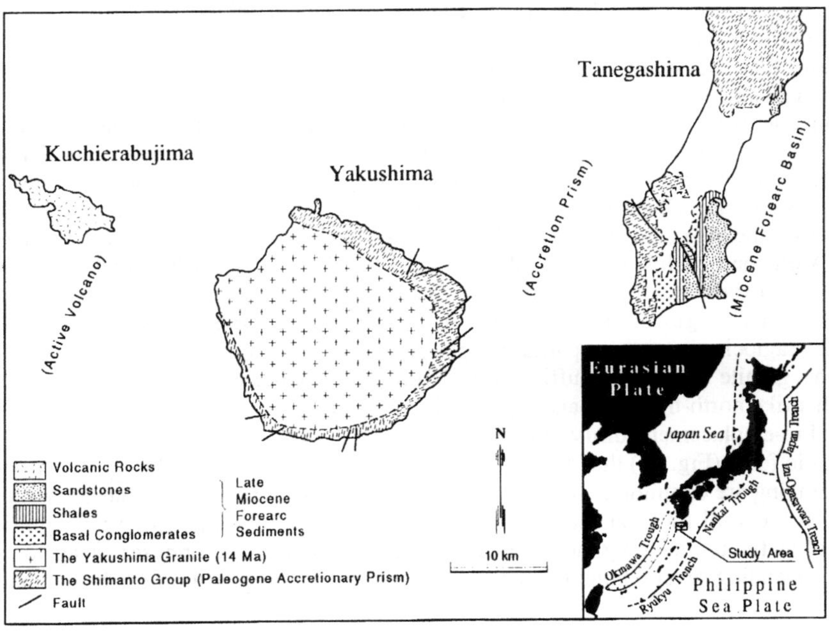

Figure 1. Tectonic setting and geology of the Yakushima granite. Patterns used for the Shimanto group and Miocene shales are parallel to strikes of bedding planes. On the large scale map view, the structures in the Shimanto group are truncated by the granite contact with its smoothly circular planform. A chain of active volcanoes, structures in the Shimanto group, and the Miocene forearc basin that extends north-northeast parallel the Ryukyu trench (insert).

In the Early Miocene, the western margin of the Pacific plate began to retreat eastward from its contact with the Eurasian plate. At the same time, the Philippine Sea plate started to migrate north-eastward along the accreting tip of the Cretaceous-

Paleogene sediments to take the place of the Pacific plate between 20 and 12 Ma. The current configuration of the Ryukyu and Southwest Japan arc systems dates from the Late Miocene (Jolivet et al., 1989; Taira and Ogawa, 1991). Therefore, the subduction of the Philippine Sea plate must have had some affect on the Middle Miocene Yakushima granite in the Ryukyu arc (Stein et al., 1994). However, it is not clear whether the Miocene granites in the Ryukyu and Southwest Japan arcs are relics of magmatism related to subduction of the Pacific or the Philippine Sea plate.

A contemporaneous Middle to Late Miocene forearc basin extends north-northeast from a distance of 40 km east of the Yakushima granite. This Miocene basin contains basal conglomerates overlain by shales with fossils indicating a mangrove-like environment beneath sandstone dipping 20° to the east. This implies that the Miocene forearc basin deepened during the emplacement of the Yakushima granite.

Arc-parallel Miocene dikes (16-15 Ma; Daishi et al., 1987) in the northern and the central domains of the Ryukyu arc (Kizaki, 1986) imply northwest-southeast dilation (perpendicular to the arc) in Miocene times. After the Middle Miocene granite intrusions, the eastern marginal zone of the Okinawa trough filled with Late Miocene sediments (Kimura, 1990). A chain of active volcanoes now parallels the trench axis behind the Miocene magmatic arc and is in line with the Okinawa trough. The overall picture suggests that the study area has been subjected to extension normal to the Ryukyu trench axis since Miocene time (Anma and Sokoutis, this volume).

Outside its smoothly circular planform, the Yakushima granite has a narrow, about 3 km-wide, contact aureole in the clastic sediments of the Shimanto group in the Paleogene accretionary prism. Cordierite and biotite isograds encircle the pluton (Ishikawa, 1964; Sato and Nagahama, 1979) but rocks of the highest metamorphic grade (andalusite-bearing cordierite-biotite hornfels) are found only north and west of the pluton (Anma and Nishikawa, unpublished data). Rocks unaffected by contact metamorphism are exposed along the north and east coasts. There are no diagnostic minerals which indicate regional metamorphism in them; clay minerals are well preserved and clastic plagioclase grains are partially replaced by carbonate aggregates.

The granite contact is sufficiently discordant with bedding in its country rocks, which strike north-northeast and dip steeply in both directions, to be clearly post- accretion. Pre-emplacement thrust anticlines (Fig. 2g) with south-eastward vergence and rhythmic folds (Fig. 2h) that developed in the flysch-like sediments during accretion and off-scraping are common in the northern part of the island. The granite cuts across structures in the wall rocks at the map scale (Fig. 1) and there seems to be no significant control on the intrusion by these pre-emplacement structures. Apophyses of up to 2 m-wide, extending several hundred meters from the pluton, suggest the possibility that the

Figure 2. Structures and fabric elements in the Yakushima granite. (a) A view of a 4 m^2 outcrop. Patterns in alignment of orthoclase megacrysts (white rectangular objects) define the orientation, shape and intensity of the magmatic flow. Note that enclaves (dark ovoids) subparallel a weak shape preferred orientation of megacrysts. (b) Typical tabular shape of orthoclase megacryst with penetration type Carlsbad twin and an aspect ratio of about 3 :3 :1. Textures in the Yakushima granite (c-f, at the same scale) indicate that the fabric is mainly due to magma flow. (c) Straight grain boundary between plagioclase grains and (d) groundmass of anhedral orthoclase (centre), myrmekite in the plagioclase (top centre), and quartz grains with moderate unduratory extinction demonstrate typical magmatic texture without any trace of solid-state deformation. (e) Intensive undulatory extinction of quartz grains, and (f) deformed biotite flakes (top right) accompanied by localized grain size reduction along grain boundaries, are the exceptionally rare indications of solid-state flow seen in the eastern part of the pluton, where oblate-type strain was measured.

Figure 2 (continued).
Pre-emplacement structures in the Shimanto group (g, h). (g) An outcrop view (5m x 7m) showing thrust (arrow) accompanied by thrust anticlines. (h) Flysch-like appearance and rhythmic folds (there is a distance of about 5 m from end to end of the syncline) in rocks of the Shimanto group. Orientations of bedding planes and fold axes were measured for analysis shown in figure 12.

pluton has ascended by stoping. However, only two large blocks of country rock (200 m x 50 m) are exposed within 500 m of the pluton contact. All field observations suggest that the hot granite with low viscosity was able to rise considerable distance to intrude cool and brittle sedimentary rocks at shallow depth.

FIELD EVIDENCE FOR OBLIQUE DIAPIRISM

PRIMARY FLOW FABRIC INSIDE THE PLUTON

The Yakushima granite contains 7-8 modal percent euhedral orthoclase megacrysts (Fig. 2a). These are up to 14 cm across and are interpreted to date from the earliest phase of crystallization (Kawachi and Sato, 1978). The high topographic relief of the island (1935 m above sea level) provides three dimensional outcrops where the megacrysts in turn provide an excellent marker for measuring the orientation, shape and intensity of the primary flow fabric.

Megacrysts with penetration Carlsbad twins are very common and tend to have a tabular shape with a mean aspect ratio of about 3 :3 :1 (Fig. 2b). In the absence of either plastic or brittle deformation of the megacrysts, these rigid discrete particles are assumed to have been aligned by rigid body rotation during the flow of the surrounding magma. No shear sense indicator, such as particle interference or strain shadows (Pons et al., 1995; Passchier and Trouw, 1996, p. 124-127), has been observed.

Each megacryst is in contact with hundreds of smaller crystals (≈5 mm across) in the groundmass. Around 90% of our 100 thin sections show purely magmatic textures (Figs. 2c and 2d) and solid-state deformation in the ground-mass affected re-orientation of the megacrysts in only a few samples (Figs. 2^{e} and 2f) collected along the eastern margin of the pluton where strong oblate strain was measured. Therefore, after full-crystallization, the megacrysts might have not rotated substantially as rigid bodies and the patterns of shape preferred orientation are mainly attributed to magmatic flow. Because rotating rigid particles are expected to have short strain memories, these patterns are likely to record only the last increments of ductile flow (Talbot and Jackson, 1987; Cruden, 1990).

METHODOLOGY OF FABRIC ANALYSIS

A number of directions of {010} plane of the orthoclase megacryst flakes (N=4 to 15) were measured from each outcrop to determine the pattern of local alignment of the megacrysts. This preferred orientations of megacrysts define a weak primary flow fabric in each outcrop inside the pluton. The flow fabric was used to generate an S-pole configuration on a Schmidt net for each square kilometre of the pluton. S-pole concentrations were used to map the orientation, and to define the shape (K-value) and intensity of fabric alignment resulting from flow in each domain. Eigen values $S_1 \geq S_2 \geq S_3$ and corresponding vectors (e_1, e_2 and e_3) were calculated for each domain using the plotting program «Stereonet» developed by Allmendinger (1988). The K-values were calculated from eigen values using the following formula modified from Woodcock (1977):

$$K = \ln (S_2/S_3)/\ln (S_1/S_2)$$

Great circle concentrations of S-poles (Fig. 3a) imply prolate-type fabric ellipsoids (X>Y≈Z) with K>>1 due to constrictive flow or rigid rotation about an axis during shear. S-pole patterns forming point maxima (Fig. 3b) imply K≈0 (oblate-type fabric ellipsoid; X≈Y>Z) and flattening flow. The intensity of fabric alignment was defined by:

$$I = \ln [(S_1+S_2)/2S_3]$$

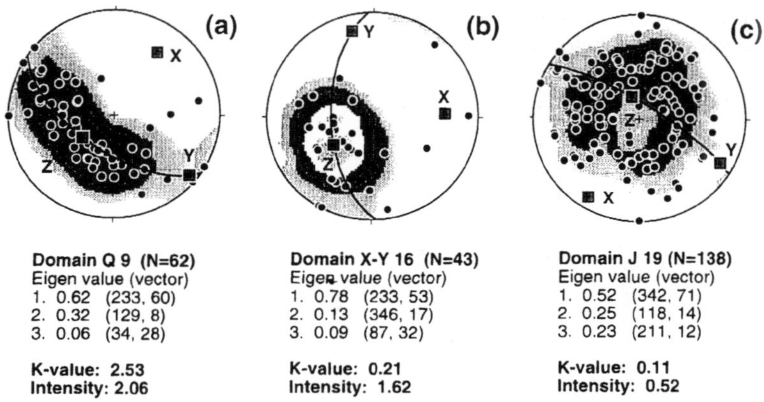

Figure 3. Yakushima granite: three selected lower hemisphere stereonets of field data. Localities are indicated in figure 5c. Solid circles: S-poles to {010} of orthoclase megacrysts. Boxes: eigen vectors (e_1=Z, e_2=Y and e_3=X). The eigen vector directions give approximate directions of principal axes of domain fabric ellipsoids (X>Y>Z): the eigen vector (e_1) connected with the largest eigen value (S_1) corresponds to Z axis (shortening axis) of the domain ellipsoid. In contrast, (e_3, S_3) corresponds to the X axis of the domain ellipsoid. X represents either the axis of rotation of rigid flakes due to shear normal to X, or the axis of constrictive co-zonal alignment of flakes (Fig. 9). Great circle concentration (a) implies a prolate type fabric ellipsoid whereas point maxima concentration (b) implies an oblate type ellipsoid. A low concentration of poles on the stereonet (c) implies a low intensity of fabric alignment.

Examples of field data are shown in figure 3. Comparison of three lower hemisphere Schmidt net projections with different configurations contoured according to the Kamb method shows that: (1) the directions of Z and Y axes become less reliable with increasing K-value; (2) the directions of Y and X axes become less reliable with decreasing K-value; and (3) the shape of the domain ellipsoids (K-value) and the directions of their axes become less reliable with decreasing strain intensity. Figure 3c also illustrates the problem of measuring strikes of planes using a magnetic compass: horizontal planes are difficult to measure and lead to a deficit of poles around the vertical axis. However, the contouring tends to fill such cavities in the diagram and provides patterns sufficiently good to allow subsequent structural analysis.

The reliability of individual domain fabric ellipsoids relate to the number of measurements (N) used for the calculation. The intensity of fabric alignment becomes infinite when N is one or two. A plot against increasing N (Fig. 4) shows that the intensity of fabric alignment first decreases then reaches a plateau when N≈35. In this study, we used domain ellipsoids calculated from more than 20 measurements to interpret patterns of shapes and intensities (Fig. 5c). There was no significant change in scattering fields on Flinn plots in the range from N = 20 to N=138 (Fig. 7).

An obvious question remains: whether configurations of poles to {010} of orthoclase crystals measured within 1 km^2 domains represent the general fabric ellipsoid due to homogeneous magmatic flow at the considered scale, or instead, reflects the fabric heterogeneity over a given area. The Yakushima granite is extremely homogeneous in every outcrop and no deformation feature denoting strain heterogeneity was evidenced on outcrop scale. One square kilometre domains within the 400 km^2 exposure area of the Yakushima granite are considered small enough to represent the general fabric ellipsoids for any arbitrary area. Nevertheless, it should be remembered that the results of the flow fabric analyses shown below are only first-order approximations.

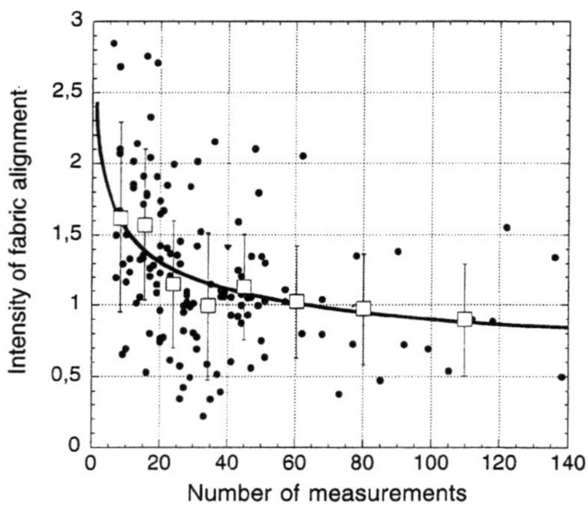

Figure 4. Yakushima granite: plot of fabric alignment intensities (solid circles) against the number of measurements indicating the minimum number of measurements necessary to obtain reliable fabric ellipsoids. Average intensity of fabric alignment (boxes and a fitted curve) decreases rapidly until the number of measurement becomes around 35. Vertical bars: standard deviations (1σ).

FLOW FABRIC ANALYSIS

The results of 129 analyses (≈ 4500 measurements) of the primary flow fabric are shown in a map of the directions of XY planes (Fig. 5a), directions of X axes (Fig. 5b) and shapes and intensities of domain ellipsoids (Fig. 5c; N≥20). Variations in orientation and shape of domain ellipsoids between neighbouring domains appear sufficiently smooth to push forward the following interpretation (Figs. 5 and 6):

(1) The overall orientations and shapes of the domain fabric ellipsoids suggest that the Yakushima granite exposed the crest of a domal structure (Figs. 5a and 6);

(2) Strong oblate-type domain ellipsoids were obtained only in the eastern marginal zone of the pluton (Fig. 5c), where intense unduratory extinctions of quartz grains and weak grain boundary recrystallization were observed (Fig. 2[e], f);

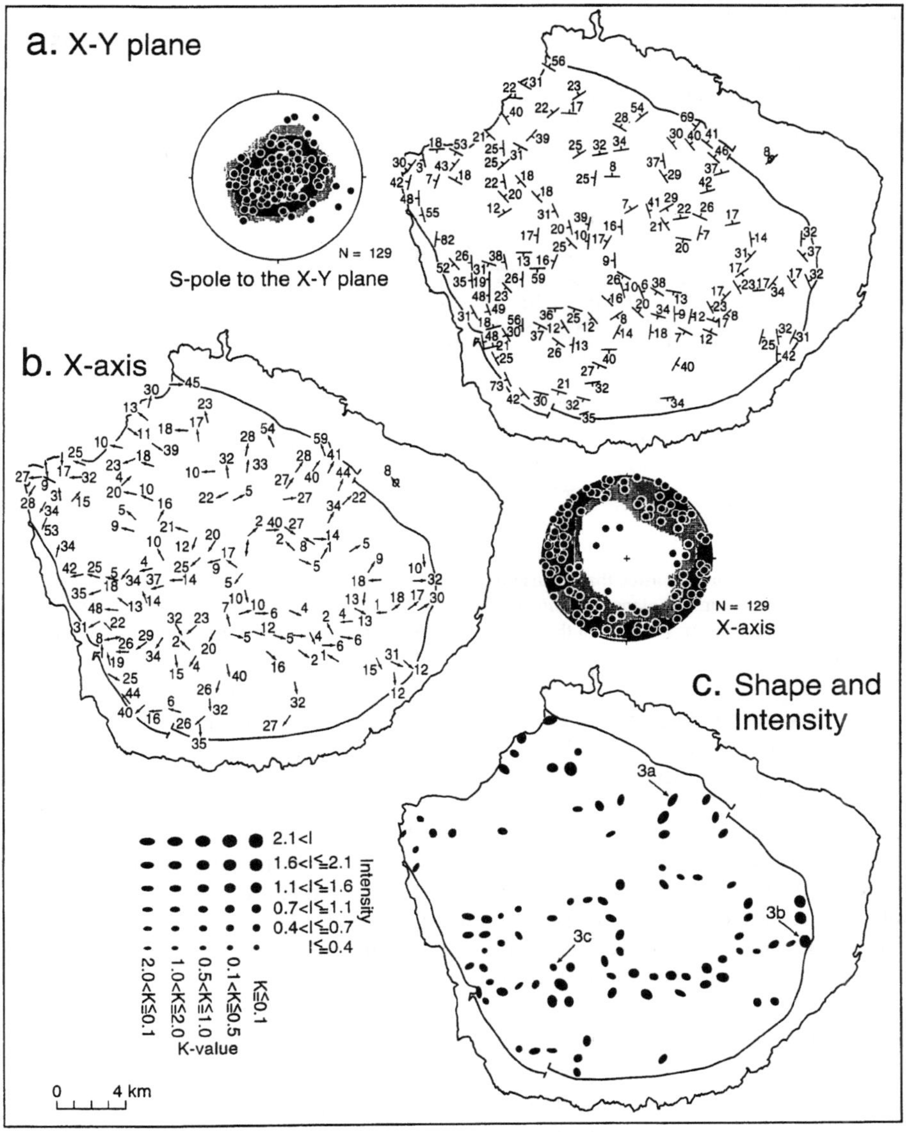

Figure 5. Domain fabric ellipsoids in the Yakushima granite: (a) dips (in degrees) and strikes of XY planes of fabric ellipsoids in 129 domains. (b) Plunges of X axes of domain ellipsoids. (c) Shape and intensity of domain fabric ellipsoids (N ≥ 20) on 90 selected stereonets. Stereonets: (a) dominant horizontal XY planes (dots: poles to planes) implying that the pluton exposed the upper part of its domal structure. Dominant horizontal NW-SE trend of X (b) implies a symmetry axis having this direction.

(3) The XY planes (Fig. 5a) dip outwards and the X axes (Fig. 5b) plunge outwards, in particular, at the northeastern and southwestern margins where X axes display a constant northeast-southwest trend;

(4) Magmatic flow fabric is generally symmetric about a centreline trending northwest-southeast (Fig. 6), and

(5) Inward dipping XY planes of domain ellipsoids are concentrated along the centreline of the pluton (Figs. 5a and 6).

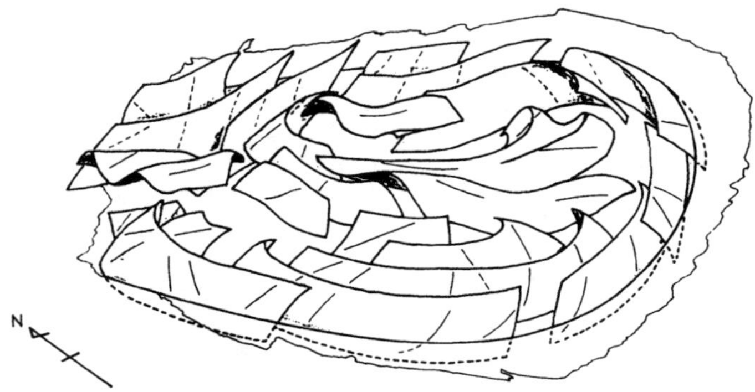

Figure 6. Circulation pattern inside the Yakushima pluton: oblique view from the southwest. Curved surfaces are extrapolated from XY planes of domain ellipsoids. Thin lines on the surface represent the direction of the X axes. The overturn structure in the centre is shown as anticlinal folds surrounding the central zone of upwelling.

The outward-trending prolate domain ellipsoid (Fig. 5c), typical of the northeastern and southwestern peripheries of the pluton, is not the pattern expected in either simple ballooning (point dilation) of magma plutons (Ramsay, 1989) or laccolith inflation (see Fink, 1983). The Flinn plot of figure 7 shows diverse shapes of the domain ellipsoids scattering over both the prolate and oblate fields (compare with shape fabrics in a salt diapir *in* Talbot and Jackson, 1987, fig. 11), whereas in ballooning plutons, they are expected to stay in the oblate field. Therefore, of the possible processes of emplacement, only diapirism appears to be likely (also, see discussion). The flow structure inside the Yakushima pluton is now examined through the pattern of internal deformation theoretically expected in an active buoyant diapir.

INTERNAL DEFORMATION DUE TO BUOYANT DIAPIRISM: THEORY

Whenever a fluid mass moves within an ambient material (brittle or ductile), friction at the contact results in internal circulation whose pattern reflects the mechanism of intrusion. Active buoyant diapirism is where viscous drag of the ductile surroundings circulates the fluid inside a rising body of molten granite. Understanding the principle behaviour of the stress-strain field requires keeping the dynamics as simple as possible (Schmeling et al., 1988). Viscous spheres rising through viscous surroundings were chosen as a simple analogy here, since (1) the Yakushima granite is nearly circular in plan view and (2) high ductility is expected beneath the hot island arc: the lithospheric mantle is expected to be weakened by the heat from underlying asthenosphere and the brittle crust must have been thin (Shimamoto, 1993).

The dynamics of the simplified system is well known (Ramberg, 1981; Talbot and Jackson, 1987; Schmeling et al., 1988; Cruden, 1990). The usual reference frame centred on the rising diapir will be used here. The internal deformation of the rising diapir is characterized by uniaxial constriction up the vertical centre-line; this changes into axial vertical flattening at the crestal stagnant point, followed by non-coaxial flattening along the outer contact surface of the diapir (Dixon, 1975; Talbot and Jackson, 1987; Schmeling et al., 1988; Cruden, 1988; Cruden, 1990). Cruden (1990) calculated the magnitude and K-values for finite deformation of fluid volumes in a rising viscous sphere as a function of travel distance, and found that elongation normal to the ascent direction takes place inside the equator of the diapir after rising several radii (fig. 5 of Cruden, 1990). This suggests that the principal axes of marginal strain ellipsoids swap during diapiric rise of a sphere.

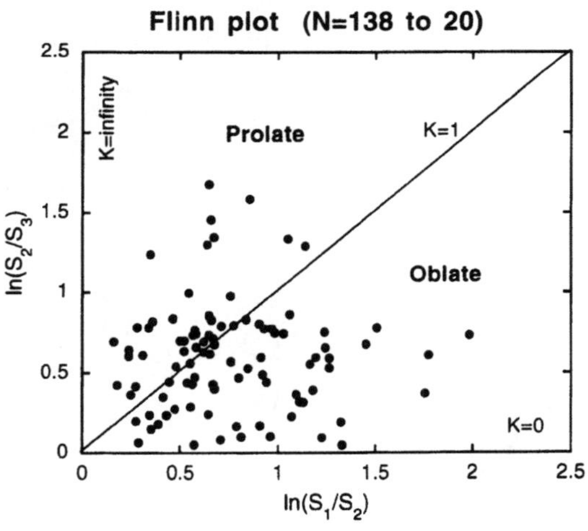

Figure 7. Yakushima granite: Flinn plot. Ninety selected domain ellipsoids (number of measurements = 20 to 138) show a wide scatter of shapes (K), ranging from 0.06 to 6.9, and intensities, from 0.21 to 2.84. Prolate-type fabrics, common here, are unexpected in ballooning plutons.

The experiments of figure 8 illustrate deformation around a spherical diapir, and some of the points raised by previous workers : (1) strong uniaxial flattening above the crestal stagnant point, and (2) shear accompanied by lateral extension around the equator (Fig. 8d). The same type of deformation is expected just inside the upper hemisphere of the diapir contact.

Such theoretical and empirical studies were concentrated on determining finite strains in flowing fluid masses. The effects of the incremental deformation within flowing fluids on the rotation of suspended rigid bodies present a different problem (Owens, 1973; Marre, 1986, p. 13-25; Melka et al., 1992). It is generally accepted that tablet-shaped flakes will rotate toward parallelism with the bulk flattening plane during axial flattening, and around the axis of bulk extension during constriction (Fig. 9). Fernandez and Laporte (1991) reported that shape preferred orientation of flakes with monoclinic symmetry about a plane normal to the rotation axis, could be produced during non-coaxial flattening. Flakes rotate into parallelism with the shear plane at a critical shear strain and continue to rotate if shear continues (see also Ildefonse et al., 1992a, b). This monoclinic symmetry is due to weak alignment of populations of the rotating grains during progressive shear. This rotation model is relevant to the patterns in shape preferred orientation of orthoclase flakes in the Yakushima granite, because (1) the megacrysts are discrete rigid tablet-shaped particles with little interaction between them, and (2) the shape and aspect ratio of the megacrysts is nearly constant at about 3 :3 :1.

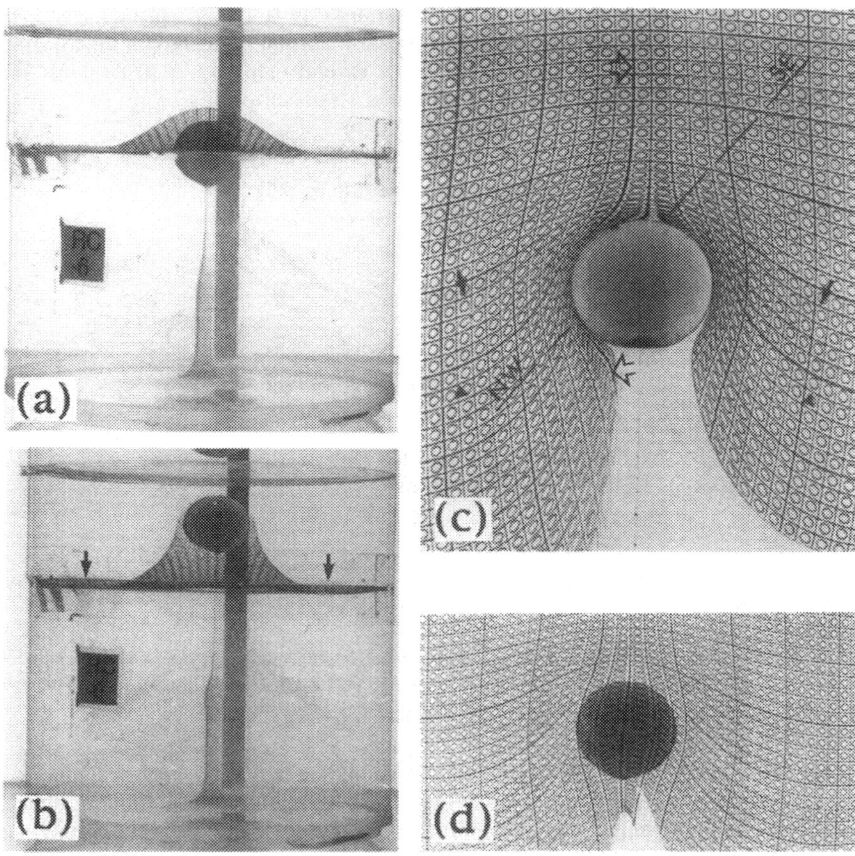

Figure 8. Experiments demonstrating progressive 3D deformation around a viscous sphere during its diapiric rise. The sphere, consisting of Hyvis 2000, a low density (ρ=945 kg.m^{-3}) low viscosity (μ=9.2 x 10^3 Pa.s; Koyi, 1991) polybutene, rose to the free surface through poly-dimethyl siloxane: PDMS (SGM36); with ρ=970 kg.m^{-3}, μ=5 x 10^4 Pa.s (Weijermars, 1986). Horizontal (a, b) and vertical grids (c, d) are printed in the PDMS to visualize deformation around the sphere. Diameter of the rising sphere is 18 mm. Arrows in (b) indicate the far-field material transport. Line in (c) shows interpreted erosion level of the Yakushima granite. Horizontal markers indicated by arrows and triangles in (c) correspond to the deformation stages shown in (a) and (b) respectively. A vertical marker indicated by open arrows corresponds to the position of a grid slightly off-profile (d).

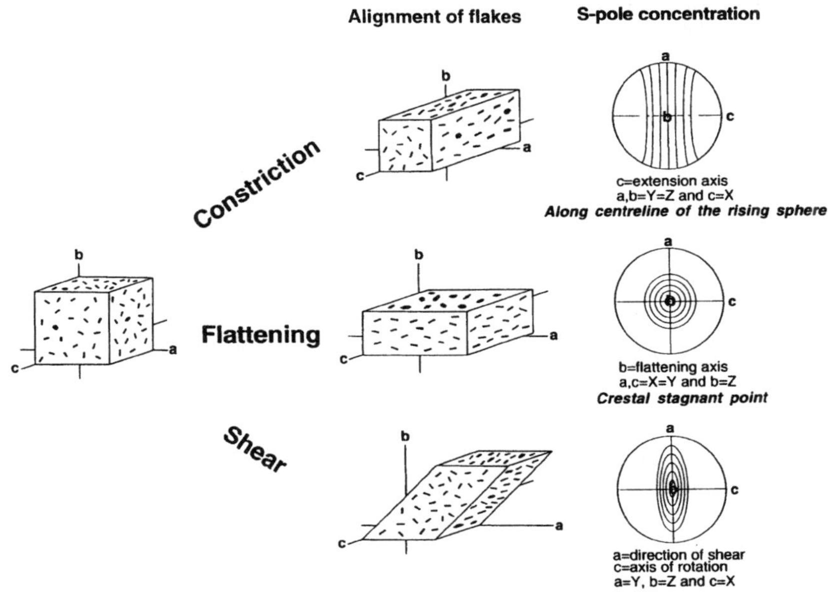

Figure 9. Schematic diagram showing alignment of suspended flakes during the deformation of their ambient fluid. Uniaxial constriction (upper) expected along the centre-line of the diapir, results in a co-zonal alignment of the flakes about the axis of constriction (c-axis). Resulting fabric ellipsoid has its longest axis (X) parallel to the c-axis. The other two principal axes are of equal length. During flattening (middle) at the crestal stagnant point, flakes will become parallel to each other resulting in a fabric ellipsoid with the short Z axis parallel to the flattening axis b. Shear strain (bottom) will lead to a monoclinic symmetry in flake alignment. X corresponds to the axis of rotation of flakes while the intermediate axis Y parallels the direction of shear (a-axis). The short axis (Z) becomes the pole to the shear plane.

In the domains of vertical constriction exposed along the centre-line of the rising model sphere, the long axes of domain ellipsoids (X axes = axes of co-zonal alignment of the flakes) are parallel to the flow direction (Fig. 9). By contrast, in the domains of non-coaxial flattening around the equator of the sphere, the X axes become axes of rotation of the flakes during shear (Fig. 10). This means that the X axes of ellipsoids are parallel to flow up the centre-line, but perpendicular to it in the marginal equatorial zone (Fig. 11a). Volumes of fluid circulating in a spherical diapir will undergo horizontal extension inside the equator (Fig. 8d) as they are moved by upwelling along the core to marginal flow, which is "downward" relative to the equator (see Fig. 11a). Rigid flakes suspended in these volumes of fluid are expected to become aligned along an axis parallel to the equator (Fig. 10). This effect is most pronounced along the particle path of a megacryst travelling from the centre of the rising sphere to its boundary zone. It diminishes as the particle path approaches the stagnant circular axis of the torus of circulation (Fig. 11a; see also Talbot and Jackson 1987). There, the strain is essentially zero and only rigid body rotation occurs. Horizontal lineations parallel to the granite contact consistent with this picture have been reported from the Criffel pluton in southwest Scotland by Courrioux (1987), and from the Bação Complex, southeast Brazil, by Hippertt (1994). Kizaki (1972) reported rotated paleosomes due to shear along the contact in the Oshirabetsu migmatite dome.

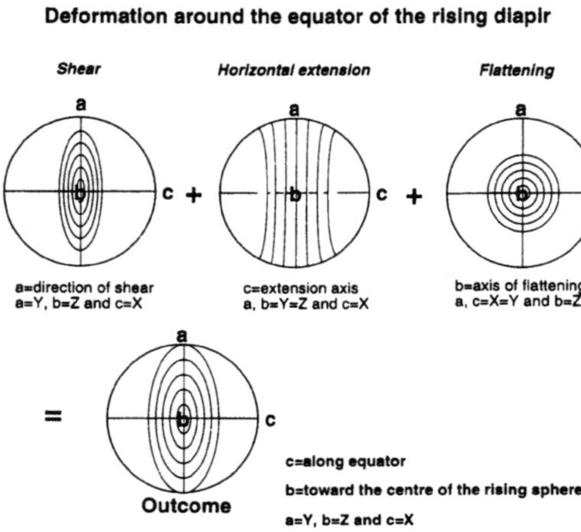

Figure 10. Diagram showing alignment of suspended flakes in the fluid flowing "downward" behind the equator of a rising diapir. Deformation of fluid around the equator inside the diapir is characterized by the combination of shear parallel to the ascent direction (Fig. 8c), horizontal extension parallel to the equator (Fig. 8d) and flattening parallel to the diapir contact (Fig. 8a,b). The resulting fabric alignment around the equator of the diapir is characterized by a weak monoclinic symmetry.

COMPARING THEORY WITH THE FIELD DATA: AN INTERPRETATION

A mature diapir that has risen an appreciable distance can be assumed to have only a single, simple circulation cell. For the Yakushima granite, this assumption is reasonable because (1) there is no significant change in lithology that could have decoupled internal circulation (Weinberg, 1992), and (2) the pluton is isolated and homogeneous, so there is no poly-diapiric interference (Stephansson, 1975; Weinberg and Schmeling, 1992).

By comparing the field data with the simplified dynamic model, the magmatic flow pattern of the Yakushima granite can be attributed to buoyant diapiric rise of a molten granite sphere along a track oblique and upward to the southeast. This is because (1) the strongest oblate-type domain fabric ellipsoids were observed only along the eastern periphery of the pluton (Fig. 5c), and not in its centre; this suggests the presence of a "crestal" stagnant point nearby; (2) X axes of domain fabric ellipsoids plunge outward at the periphery and vary smoothly in trend tracing slight curves (Figs. 5b and 6); this implies a non-vertical axis of internal circulation; and (3) assuming that the equatorial prolate fabric ellipsoids were aligned by shear along the granite contact, the prolate fabrics around the periphery could be attributed to shear normal to the X axes of the domain ellipsoids (compare Figs. 5c and 11b).

The presence of the hypothesized "crestal" stagnant point at the eastern margin of the pluton, together with horizontal shear inferred from the directions of the prolate ellipsoids along the northern contact, implies that the direction of intrusion was almost horizontal toward the southeast in this area. Along the southwestern contact, outward plunging prolate ellipsoids are locally transformed into contact-parallel prolate ellipsoids. This agrees with the theoretically predicted axial swapping: inner prolate ellip soids are parallel to the flow direction, whereas outer prolate ellipsoids are aligned normal to it. Therefore, the direction of internal circulation in the southwestern part of the pluton must have been steeper (≈30° from vertical) than that in the northern domains. Prolate ellipsoids in the northwestern domains could be explained similarly, by axial swapping during northwestward flow. This localized axial swapping is also observed in the northernmost part of the pluton, where oblate domain ellipsoids to the south are transformed into prolate ellipsoids, with their X axes plunging obliquely toward the granite contact. In this area, the inferred direction of internal circulation is toward the north.

Further comparisons between theory and field data show that (4) the inward dipping XY planes of domain ellipsoids at the southeast end of the centre-line (Fig. 11b) could be interpreted as an overturn structure in an internal circulation cell; (5) relatively strong prolate fabrics with X axes sub-parallel to the mid-line of the pluton could be attributed

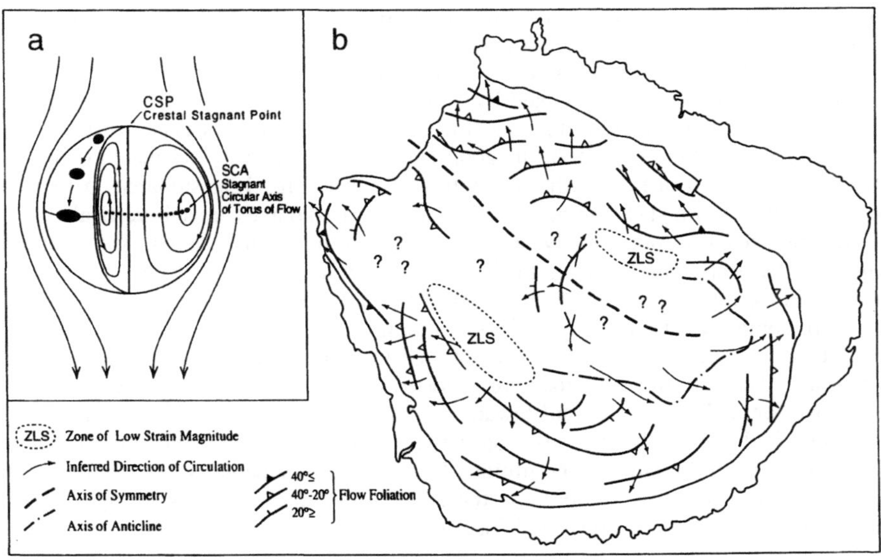

Figure 11. (a) Flow pattern inside a fluid sphere rising through a denser ambient viscous fluid (Schmeling et al., 1988; Cruden, 1990). Strong flattening along a vertical axis will take place at the crestal stagnant point (CSP). On the other hand, only rotational strain is expected around the stagnant circular axis (SCA). Circular passive markers carried from the crest of the sphere to its equator deform to ellipses parallel to the equator. (b) A possible interpretation of the flow structure inside the Yakushima pluton. Zones of low strain (ZLS) coincide with the SCA. Inward dips in figure 5a imply circulation due to diapiric rise of the pluton. A general mirror symmetry about the southeast trending midline of the elliptical granite is attributed to southeastward oblique rise of the pluton. Thin arrows indicate the local direction of internal circulation in accordance with the interpretation discussed in text; they do not correspond to the X-axes of the domain ellipsoids (Fig. 5b).

to the central upwelling; and (6) the stagnant circular axis of the torus of flow is recognizable as zones of low strain (Fig. 11b).

The zone of central upwelling is located southeast from the geometrical centre of the pluton implying an asymmetrical toroidal convection cell leaning to the southeast (Figs. 6 and 11b). Together with the inferred direction of circulation being oblique to the margin (Fig. 11b), this asymmetry suggests that the pluton rose toward the trench axis to the southeast. The magma followed a horizontal path in the northeastern periphery, and ≈30° from vertical in the southwestern periphery. This is probably because magma introduced in the northwest followed a south-easterly directed path along the centre-line before turning sideways to the northeast just short of the eastern periphery (Fig. 11b).

SYN-EMPLACEMENT WALL ROCK DEFORMATION

Syn-emplacement wall rock deformation is marked by regional re-folding (or uplift) of pre-emplacement structures (see Tectonic Setting) and by spaced foliations sub-parallel to the granite contact.

Far-field deformation marked by the distortion of bedding planes and fold axes in the surrounding rocks of the Shimanto group (Fig. 2g,h) also provides clues to the intrusion mechanism of the Yakushima granite. Bedding planes and fold axes were measured in 25 sub-areas (Fig. 12a) divided according to lithology in every stratigraphical unit recognized by Hashimoto (1956) at different distances from the granite. Directions of fold axes from each area are obtained by fitting a great circle to the configuration of poles to the bedding planes. Fold axes (Fig. 12a: open boxes) measured in eight sub-areas along the northeast coast are in good agreement with those calculated (Fig. 12a: solid boxes). Data from the south and southwest coast (Fig. 12a: sub-areas p to x) are taken from Nishikawa (1988, unpublished data), where fold axes are sub-vertical.

The fold axes were presumably horizontal and parallel to the trench axis when they formed in the accretionary prism. In the deformation aureole, they were then distorted towards conformity with the pluton contact and its inferred southeast direction of ascent (Fig. 12b; compare with distortion of horizontal marker in Fig. 8d). In the northeast of the pluton, where horizontal shear was inferred just inside the pluton, original north-northeast fold axes in the wall rock are distorted toward the inferred southeast emplacement direction and warped upward toward conformity with the granite contact. In the southwest, fold axes in the wall rock are upwarped more steeply, reflecting the inferred local emplacement direction of the pluton. However, this scheme is not obvious along the east coast. This might be because the residual magma intruded sideways from the eastern tip of the pluton and distorted further the wall rock structures (see previous section).

The near-field ductile deformation (Paterson and Fowler, 1993) due to emplacement of the Yakushima granite is concentrated in a 200 m wide contact zone where early quartz veins are sheared and folded into parallelism with the spaced foliation (Fig. 13a). This implies a very narrow aureole of high ductile strain (Brun et al., 1991) resulting from thermal (Marsh and Kantha, 1978) and/or strain-rate softening (Weinberg and Podladchikov, 1994) within about 0.02 radii of the granite contact.

The total width of the deformation aureole cannot be determined because it extends beyond the coastline. However it is wider than 1/2 radius of the pluton. England (1992) reported similar patterns of re-folding and uplift of the wall rocks during diapiric emplacement of the Arran granite. Upwarped fold axes could be attributed to either laccolithic growth or ballooning. However, neither of these mechanisms can explain the southeastward distortion of the fold axes in the wall rocks of the Yakushima pluton. Instead, the previously hypothesized oblique rise of a buoyant diapir would better explain the observed distortion.

OTHER EVIDENCE OF OBLIQUE DIAPIRISM

Rocks with the highest metamorphic grade are asymmetrically distributed around the pluton and are found only in the north and west of the island (see Tectonic Setting). This observation alone suggests either oblique diapirism or post-intrusive southeastward tilting of the Yakushima massif. However, rocks away from the thermal contact aureole do not show any difference in regional metamorphic grade, an observation that rules out the tilting scenario (see Discussion). Volcanic breccias with possible fragments of the Yakushima granite occur in the Pleistocene flanks of Kuchierabujima, an active volcano 20 km northwest of Yakushima (Fig. 1). Daishi et al. (1986) used the zircon fission track method to date these fragments at 15.0 ± 0.5 Ma. This suggests that the stem of the 14 Ma Yakushima granite pluton underlies this volcano, a geometry which again is consistent with southeastward oblique rise of the granite.

EFFECT OF FINAL EMPLACEMENT

Deformation during the final emplacement of the Yakushima granite is recorded by late aplites and extensional quartz veins (Fig. 13c-e) forming a concentric system of brittle sheet intrusions in and around the pluton (Fig. 14a). The fibrous extensional quartz veins are associated with tin-tungsten mineralization and are distinct in texture (Fig. 13c,d) from older folded quartz veins in the wall rocks. These sheets and veins are most abundant around the granite contact zone and become less dense both inward and outward from the contact.

Each sheet reflects the orientation of the local paleostress field and the concentric sheet pattern indicates tangential dilation around an ellipsoidal body of residual buoyant fluid with circum-tangential intermediate axes that would extend an arbitrary horizontal marker $\leq 0.5\%$ (sheets and veins sum to widths ≈ 50 cm in every 100 m). Figure 14b reconstructs the stress field geometry in and around the granite, coeval with the emplacement of the aplite and quartz veins. Radial profiles display a paleostress field concordant with the emplacement mechanism mentioned in the previous sections. In the southeast and south of the pluton, the dilation axes (σ_3) parallel the granite contact (profiles C-D and C′-E in Fig. 14b). In contrast, in profiles across the inferred southeast direction of emplacement (profiles C-A, C-B and F-F′ in Fig. 14b), the concentric sheets display an inward-steepening fan-like pattern changing into an outward-steepening fan-like pattern outside the granite contact. The trace of the inflection point of reconstructed paleo-extension axes lies just inside the granite contact. Magma pressure is inferred to have been higher inside the limit defined by the inflection point than outside. This pattern indicates that the stiffened granite crust, together with the thermally weakened wall rocks, were plastically upwarped during the final emplacement of the residual magma of the Yakushima granite.

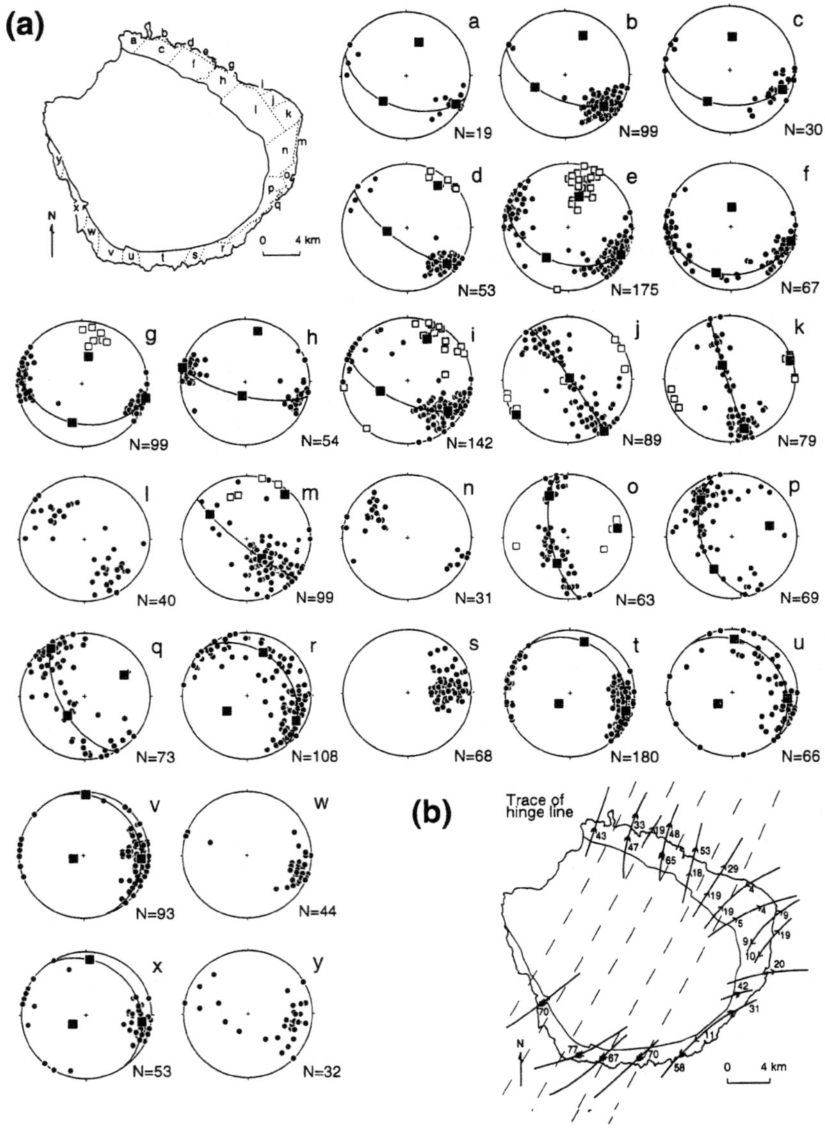

Figure 12. Emplacement of the Yakushima granite: deformation pattern in the wall rocks. (a) Great circle concentration of poles to bedding (solid circles) in the Shimanto group. Calculated fold axes (solid boxes) sub-parallel measured fold axes (open boxes). Stereonets p to x: data from route map by Nishikawa, (1988, unpublished data). (b) Hypothesized original horizontal arc-parallel directions of fold axes in the Shimanto group (dashed lines) are distorted into conformity towards the contact shape and inferred ascent direction of the pluton (solid lines).

Since compression directions, parallel to the directions of the dikes in profiles, are perpendicular to the granite contact in profiles C-D and C′-E (Fig. 14b), the overall orientation of the paleostress field implies that magma pressure during final emplacement was higher in the south and southeast (compare strain field in Fig. 8c with calculations by Phillips, 1974). Profile G-G′ (Fig. 14b) also shows a simple concentric pattern for σ_3, probably because the last dregs of magma entered the pluton through this profile.

Homogenization temperatures of fluid inclusions in quartz veins show that they developed after the pluton had stiffened as it cooled down to ≈300°C (MITI, 1992). At this stage, the rheological contrast between the granite and its country rocks was probably negligible, since fractures cut across the contact without change in either style or orientation.

A uniformly cooling granite mass shrinks with an average volumetric residual strain of about -7×10^{-3} when it cools from 300°C to 0°C (Savage, 1978). With this in mind, assuming 0.5% extension around the sphere of residual buoyant fluid, then 5×10^{-3} dilation due to the intrusion of aplite dikes and quartz veins compensates the expected

Figure 13. Quartz veins in the wall rocks of the Yakushima pluton. (a) Early veins (arrow) in the wall rocks flattened and rotated into parallelism with the spaced cleavage in a zone extending a few hundred metres from the granite contact. (b) Boudins and pinch-and-swell structures are common in the country rocks but probably predate emplacement (thickness of the sandstone units is about 20 cm). (c) Extensional quartz veins measured for analyses of paleostress field during the final emplacement of the pluton. (d) Fibrous texture indicates dilation of quartz veins in the sandstone; it is distinctively different from older finer-grained quartz veins. (e) Thicker quartz veins near the pluton show a primary en-échelon relationship.

shrinkage of the Yakushima granite due to cooling. The total dilation during and after the emplacement of the concentric sheets is therefore in the order of 10 x 10^{-3}, and thus is far too small to disturb the primary flow pattern. Also, note that the magnitude of the residual magma flow mentioned in this section should be significantly smaller than those of previous sections. Nevertheless, the geometry of the concentric sheet system is strongly linked to the pluton shape. The overall picture can be accounted for in terms of late diapiric oblique upsurge, due to differential loading between the inclined diapiric tail and the nearly horizontal diapiric head (Fig. 15, insert).

DISCUSSION

COMPARISON WITH OTHER MODELS

Lateral emplacement of a thin sheet of magma, for instance, the Lebel stock reported by Cruden and Launeau (1994), due to continual magma charge from a conduit (e.g. the Smartville Complex; Beard and Day, 1988) or from fractures (the Kensington pluton; Corriveau and Leblanc, 1995) could be a feasible explanation for the prolate domain ellipsoids found along the periphery of the Yakushima granite. Both Cruden and Launeau (1994), and Corriveau and Leblanc (1995) used the anisotropy of magnetic susceptibility (AMS) method to reveal magmatic flow patterns. Both these studies found marginal magnetic lineations parallel to the granite contact. The marginal AMS ellipsoids are rather flattened in such laterally emplaced plutons (Corriveau and Leblanc, 1995) or in asymmetrically expanded plutons (the Flamanville pluton, Brun et al., 1990), unlike in the Yakushima pluton where marginal domain ellipsoids are prolate with X axes directed toward the granite contact in map view.

Brun et al. (1990) and Bateman (1985) interpreted patterns of horizontal radial shortening of the wall rocks around plutons as due, respectively, to the lateral expansion of the Flamanville granite and in-situ ballooning of the Cannibal Creek pluton. There is no evidence of such horizontal shortening that push ambient rocks horizontally away from the contact of the Yakushima pluton. Instead, the pluton lifted the wall rocks and extended any arbitrary horizontal markers which predate the granite intrusion (previous sections). Furthermore, competent sandstones in the country rocks display boudins and/or pinch-and-swell structures (Fig. 13b) with steep axes implying horizontal extension. However, the arc-parallel pull-apart features in the Shimanto group are most likely due to regional deformation preceding the granite emplacement. Similarly, thrusts having the same vergence as the ascent direction inferred here for the Yakushima granite, are attributed to the accretion that preceded emplacement of the Yakushima granite. There is no definite evidence supporting granite emplacement along thrust faults at the present erosion level.

Bouillin et al. (1993) attributed a stable magnetic lineation in the 6 Ma Monte Capanne pluton of Elba, Italy, to laccolithic emplacement during regional extension parallel to the lineation. Since the direction of extension in Yakushima island should have been normal to the trench axis in Miocene time (i.e. northwest-southeastward, see Tectonic Setting), this mechanism does not account for the prolate domain ellipsoids observed in the northeast and southwest borders of the pluton. Moreover, intense oblate

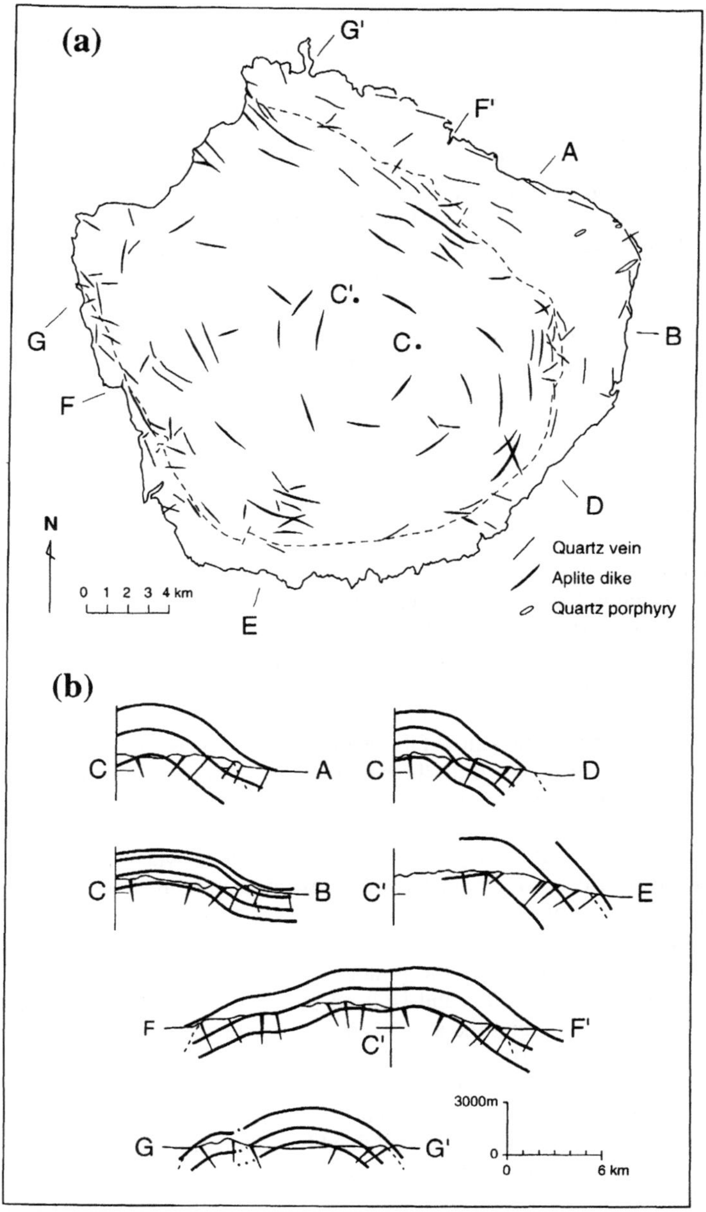

Figure 14. (a) Aplite sheets and late fibrous quartz veins form a sheet system concentric in planview and radial in sections. (b) Radial profiles indicate tangential extension around an ellipsoidal body accompanied by lift (see text).

domain ellipsoids are measured along the eastern margin of the pluton, where the strongest shear would be expected if the emplacement had been controlled by regional extension. Horizontal extension does occur in the Ryukyu subduction zone (see Tectonic Setting), but the strain-rate due to active diapiric rise was likely much faster than the tectonic strain rate.

Stoping followed by in-situ thermal convection (Cruden, personal communication) due to cooling of magma could be a mechanism for the emplacement of the Yakushima pluton. High heat flow beneath the island arc implies a steep geothermal gradient, and therefore, enough heat from beneath to initiate thermal convection inside the emplaced pluton. However, such thermal convection would presumably have been about a vertical axis, unlike the circulation pattern inferred within the Yakushima pluton. Thermal convection is therefore likely only if the whole area was tilted by ≥20° southeastward after any thermal convection ceased. This would imply that the northwestern part of the currently exposed pluton was about 8 km to 10 km deeper than the southeastern part. Then, assuming a geothermal gradient of 30°/km, the regional metamorphic grade in the northwest of the pluton would have been around 300°C and 200 MPa higher than that in the southeast, a difference not observed in regional metamorphic grades (see Tectonic Setting). By the same reasoning, models invoking a tilted laccolith or sill are unlikely.

ASCENT AND EMPLACEMENT OF THE YAKUSHIMA GRANITE

Emplacement by oblique diapirism

Since the emplacement mechanisms mentioned above do not satisfactorily explain field data from the Yakushima pluton, the author favours oblique diapirism as the emplacement mechanism for the Yakushima granite. Throughout the ascent of the magma and its final upsurge, the Yakushima granite seems to have risen obliquely to the southeast (Fig. 15). Although the overprint of the final emplacement phase (below 300°C) is far too small to disturb the primary flow fabrics of the ascent phase, it is clearly recorded by the concentric sheet intrusions. Both the pattern of the early shape preferred orientations of the megacrysts and the late concentric sheets can be explained by the diapiric rise of the pluton. Any solid-state flow between the temperature of full crystallization and 300°C might have a similar flow pattern. Unlike in plutons where ballooning has been proposed as a mechanism of emplacement, buoyant diapirism appear to be the most important emplacement mechanism of the epizonal Yakushima granite.

Although the fold axes in the wall rocks are clearly distorted toward conformity with the granite contact, discordance of the granite contact with the bedding in the country rocks, apophyses from the granite, and isolated blocks of country rocks within it, raise the possibility of stoping being the ascent mechanism of the pluton. Unlike laboratory experiments, surface tension is sufficiently small in large natural plutons (Weinberg, 1993) that apophyses can develop by intruding along radial fractures. Subsequent ascent may isolate a few xeno-blocks in the pluton. Discordance between the contacts and structures in the country rocks may be explained by adding non-Newtonian rheologies to the simplified diapiric system. Because the viscosity of power-law fluids decreases with increasing strain-rate, a power-law diapir in a power-law crust may develop a very narrow marginal zone of pronounced weakness (Weinberg, 1994; Weinberg and Podladchikov, 1994). Sliding along such a zone of weakness may aid piercement so that a diapiric pluton becomes discordant to structures in its wall rocks.

Endogenic buoyancy versus regional tectonic force

If the granite had risen by its buoyancy force alone, presumably it would have risen vertically. The obliquity of the inferred track along which the Yakushima granite rose implies that the intrusion was affected by other tectonic forces. Models accounting for the oblique rise of plutons from a subduction zone are discussed in the following chapter by Anma and Sokoutis (this volume).

In the brittle upper crust, where the influence of tectonic forces on buoyant diapirs is expected to be greater, regional paleostresses are recorded by fracture systems in and around the pluton. The concentric planform of the sheet system implies that even the last brittle fractures were controlled by the emplacement of the Yakushima granite and that regional tectonic overprinting was negligible. Shigeno (1980) attributed the formation of joints in Miocene intrusions within the Shimanto group to horizontal tension at shallow levels during regional uplift and erosion. Whereas earlier studies have emphasized intrusion into tectonic cavities, the Yakushima granite appears to have made its room by its own endogenic forces.

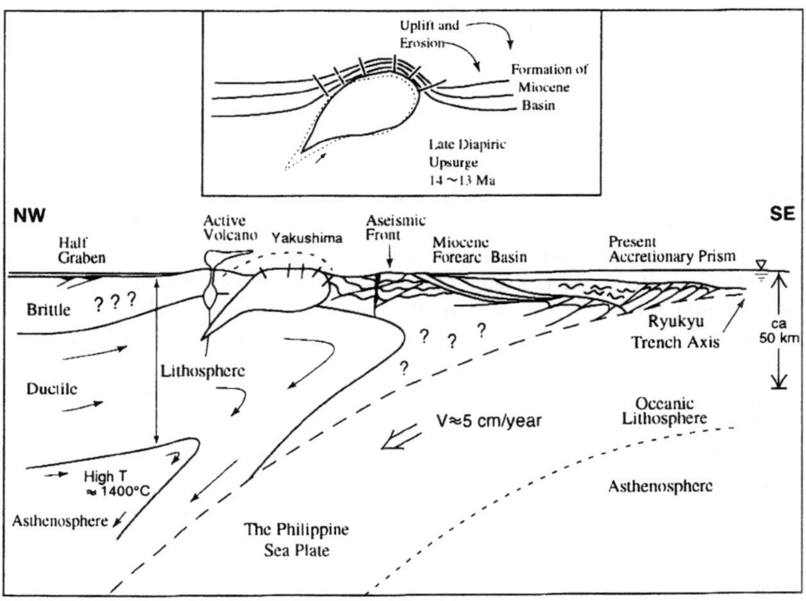

Figure 15. NW-SE profile through the Yakushima area. The pluton rose obliquely towards the southeast throughout the intrusion process. The stem of the pluton underlies the neighbouring active volcano. The contemporaneous Miocene forearc basin is tilted eastward. Sources: Nash (1979), Honda (1985), Letouzey and Kimura (1986), Yamaoka et al. (1986), Iwasaki et al. (1990), and Shimamoto (1993).

Depth of emplacement

None of the model diapirs with Newtonian viscosity distorted their viscous overburden sideways beyond one diameter (Fig. 8; Cruden, 1988; Schmeling et al., 1988). This suggests that the Yakushima pluton could have been emplaced at any depth less than its 20 km diameter. However, regional metamorphism indicates even shallower depths of exposure; these can be rationalized by assuming that the Yakushima pluton lifted the overlying surface. A depth of emplacement of ≤ 14 km would be more consistent with the contact metamorphic grade and the pressure conditions inferred from the normative composition of the Yakushima granite (≈5 kb; Sato and Nagahama, 1979) and the pattern of upwarping attributed to its final emplacement (Fig. 15, insert).

Diapirism in a brittle upper crust

The question remains as to whether fluid diapirs can penetrate the brittle upper crust due to their buoyancy force alone. Flow fabrics inside the Miocene Yakushima pluton, distortion of structures in its country rocks, and concentric sheet intrusions in both, indicate that buoyant diapiric rise was the most important mechanism of emplacement of the granite. Horizontal extension induced by slab subduction (Anma and Sokoutis, this volume) and magma rising along the subduction zone would weaken the upper crust. The high heat flow in the island arc may additionally soften the already weakened crust. Then, the 400 km^2 pluton could lift the surface by distorting it around a ductile aureole. The presence of arc-parallel Miocene dikes and Miocene propagation of the Okinawa trough support the finding of horizontal extension normal to the trench axis.

CONCLUSION

The distinctive feature of the Yakushima granite is a toroidal circulation pattern inside a circular planform with central upwelling along the oblique rise axis. Rotation of megacrysts about an axis parallel to the inferred oblique equator of the pluton is attributed to shear along the granite contact. Zones of low strain are recognized around the stagnant axis in the circular torus of flow. Externally, all the structures in the strain aureole (> 0.5 radius wide) indicate tangential elongation with tangential intermediate axes around a buoyant ellipsoid.

Evidence suggesting the oblique rise of the Yakushima pluton can be summarized as follows: (1) the asymmetry of the toroidal convection cell implying a central zone of upwelling leaning to the southeast; (2) symmetrically re-folded wall rock structures distorted toward a direction parallel to the central upwelling; (3) fragments of the Yakushima granite occuring in the next volcano to the northwest; and (4) the thermal metamorphic grade being highest in the north and west of the pluton.

ACKNOWLEDGEMENTS

I would like to express my gratitude to Prof. Christopher Talbot for discussion and critical reading of this work. I also appreciate thorough reviews given by Dr. A. R. Cruden and Dr. J. M. Tubia; their suggestions and criticism were helpful towards improving the final version of this paper. I would like to thank Prof. Kizaki of Ryukyu University who initiated and supervised this project during my BSc and MSc studies. Mr. Nishikawa mapped the structures in the country rocks in the south of Yakushima island. His data were processed by the author at Uppsala University. I thank Dr. Kawano of Saga University for field assistance and discussion. My best wishes to Mr. Y. Maki of Kusugawa for lending me his cabin during my 7-months field work. I also wish to thank Drs. Sokoutis, Koyi and Weinberg for their assistance, ideas and discussions. My thanks go to Einar Meland for preparation of the experimental equipment, Christina Wernström for preparation of figures and Bertel Glös and Christer Bäck for their technical assistance while taking photographs. This work was supported by a study grant from Uppsala University.

EXPERIMENTAL PLUTON SHAPES AND TRACKS ABOVE SUBDUCTION ZONES

Ryo ANMA and Dimitrios SOKOUTIS
Hans Ramberg Tectonic Laboratory, Institute of Earth Sciences, Uppsala University, Norbyvägen 18B, S-75236 Uppsala, Sweden

ABSTRACT

Field data suggest that the epizonal Yakushima granite rose obliquely into the Ryukyu arc of Japan (Anma, this volume). Analogue models test the hypothesis that the oblique rise of this buoyant diapir was due to drag by the stiff oceanic plate sinking obliquely beneath it. A simplified model of a solid plate sinking obliquely along its length into a ductile medium induces wedge flow in the overlying asthenosphere. Buoyant inclusions in the model lithospheric/asthenospheric wedge trace possible tracks of plutons above subduction zones. The inclusions rise obliquely toward the trench axis in the nearby wedge where rotational shear is greatest, leaving an inclined diapir tail. By contrast, near the surface away from the trench axis, where arc-normal horizontal extension prevails, buoyant inclusions become tabular in shape and are carried toward the trench. Overall, the observations account for the geometries of the Yakushima and other Miocene granites in Southwest Japan.

INTRODUCTION

Anma (this volume) attributes the fabrics and structures in the Yakushima granite pluton to the forceful oblique rise of a buoyant diapir toward the trench axis (Fig. 1). Had the granite risen by buoyancy alone, it would have probably risen vertically. Therefore, the oblique ascent trajectory of this island arc granite is presumed to be due to a tectonic effect of some kind. The likely cause lies beneath it: the subducting oceanic slab, the most distinctive tectonic feature associated with an island arc.

Magmas associated with subduction zones are expected to be generated near the boundary between the subducting oceanic slab and the overriding wedge of sub-continental lithosphere (referred to as the Wadati-Benioff, or WB zone). They are generated by a flux of hydrous fluids (Arculus, 1994) liberated through dehydration of hornblende, mica and/or clay minerals in the subducting lithosphere (Ringwood, 1974; Tatsumi and Eggins, 1995).

J.L. Bouchez et al. (eds.) Granite: From Segregation of Melt to Emplacement Fabrics, 319–334.

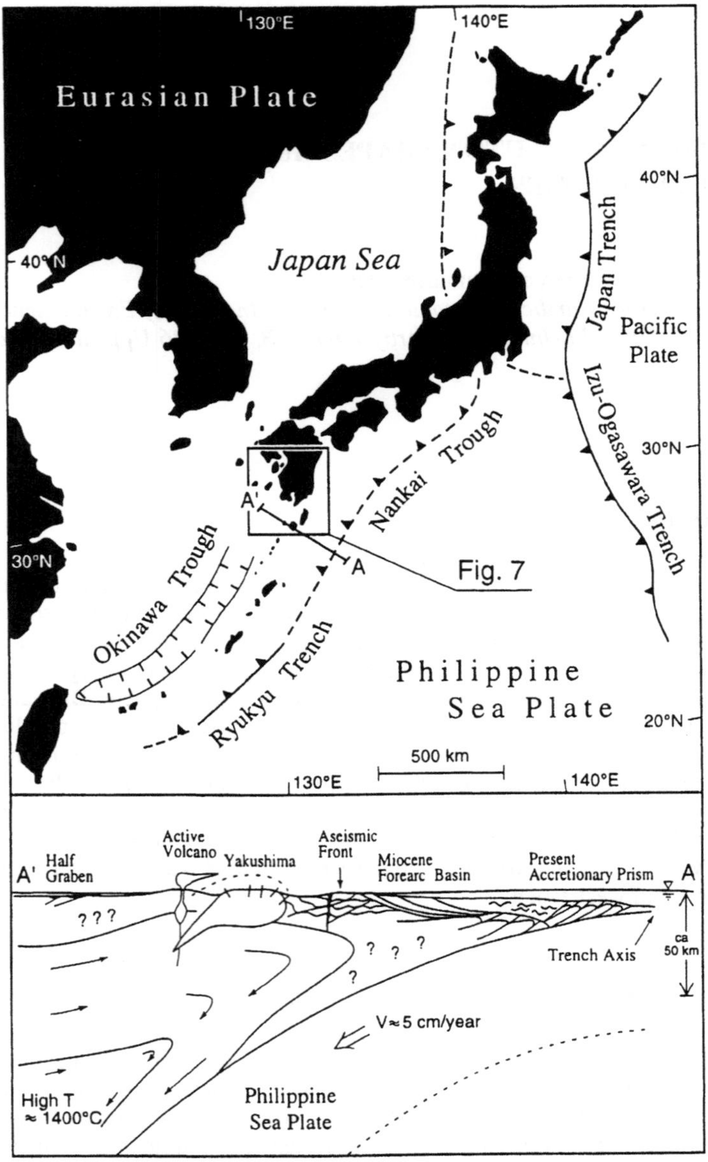

Figure 1. Tectonic setting of the Ryukyu and Southwest Japan arcs. The Yakushima granite (on profile A-A´), located at the northern tip of the Ryukyu island arc, appears to have risen obliquely toward the trench axis (see Anma, this volume). The chain of active volcanoes in the north (profile) and the central axis of the Okinawa trough in the south (map) mark horizontal extension perpendicular to the trench axis. The aseismic front marks the boundary of the ductile lithospheric wedge beneath the island arc. The trace of the aseismic front is not clear in the Ryukyu-Southwest Japan arc systems ; it is here extrapolated from Northeast Japan (Yoshii, 1979).

Current general thinking is that S-type granites like the Yakushima granite (Shibata and Ishihara, 1979) form due to partial melting of the continental crust (Chappell and White, 1974) rather than by differentiation of lithospheric mantle melts. In previous models (Ringwood, 1974; Sakuyama, 1983; Weinberg and Podladchikov, 1994) arc granitoids were attributed to a process where mafic diapirs segregated above the subducting slab, rose through the lithospheric mantle, and collected beneath the continental crust which was partially melted by them. Here, a different possibility is considered: that some S-type granites might originate where crustal rocks are dragged downward along a subducting plate, melt, and then rise as buoyant diapirs. Sediments trapped by the subducting oceanic slab (Hilde, 1983) might melt to generate granitic magma at depths shallower than 110 km beneath the volcanic front (Tatsumi and Eggins, 1995; Anma and Kawano, in preparation). Ductile lower crustal rocks may be dragged down to depths at which their melting is triggered by dehydration of the oceanic slab.

Buoyant diapirs can rise obliquely from passive but inclined tabular or wedge-shaped source regions in systems with inclined top or bottom boundaries (Talbot, 1977; Marsh, 1979). Wedge flow induced in the fluid above a subducting slab (McKenzie, 1969; Toksöz and Hsui, 1978) may also influence the ascent path of granite plutons (Olson and Singer, 1985; Tatsumi et al., 1986; Spiegelman and McKenzie, 1987).

This study examines the mechanism responsible for the oblique ascent of the island arc Yakushima granite using some of the dynamics suggested by previous workers. Analogue models were used to analyse: (1) the patterns of wedge flow induced by a rigid plate sinking obliquely along its length; (2) the influence of this wedge flow on buoyant inclusions rising from the sinking slab; and (3) the influence of wedge flow on the tracks of buoyant inclusions rising from starting points above the subduction zone. The patterns of wedge flow observed in our simplified subduction models can explain the mechanism of back-arc spreading and the observed geometries of Miocene granites in the Southwest Japan.

PRINCIPLES OF MODELING

MODELING STRATEGY, MATERIALS AND DESIGN

Intra-continental earthquakes occurring at depths of less than 15 km together with the steep geothermal gradient beneath the magmatic arc of Northeast Japan imply that the arc lithosphere is hot and weak and that the strength profile above subduction zones can be simplified into two layers: thin brittle crust overriding ductile lithosphere (Shimamoto, 1993). Iwasaki et al. (1990) used wide angle seismic profiling to show that the total thickness of the crust above the Ryukyu subduction zone was about 25 km (Fig. 1: profile). Therefore, to understand the principal behaviour of the stress-strain field in a subduction system, the thin brittle crust above the subduction zone was excluded from our dynamic model which consisted of a simplified two-layer system: a viscous asthenosphere/lithosphere with a rigid oceanic plate sinking through it.

A rigid 6 mm-thick plexiglass plate was used to represent the stiff subducting oceanic plate in the model. This is because field data indicate that the Pacific plate is significantly stiffer than the overriding wedge-shaped ductile lithosphere/asthenosphere, where a zone of anomalous seismic intensity along the eastern coast of Northeast Japan implies that the subducting Pacific plate acts as a wave guide with high seismic velocities (Utsu, 1967; Utsu and Okada, 1968; Shimamura and Asada, 1975). However, the Philippine Sea plate, above which the Yakushima granite rose in Southwest Japan, is younger and hotter and is therefore likely to be softer than the Pacific plate. Nevertheless, deep seismicity along the Philippine Sea plate (Shiono et al., 1980; Kimura et al., 1983) implies that this oceanic slab is also considerably stiffer than the overriding ductile lithosphere/asthenosphere.

In all our models, the ductile lithosphere/asthenosphere is represented by a transparent viscous fluid, SGM36, a poly-dimethyl siloxane manufactured by Dow Corning Inc., with density $\rho = 970$ kg.m^{-3} and Newtonian viscosity $\mu = 5 \times 10^4$ Pa.s below shear strain rates of 3×10^{-3} s^{-1} (Weijermars, 1986).

Tracks of plutons above the subduction zones were studied by inserting less viscous buoyant inclusions in this simplified two-layer system. Two types of model materials were used for the plutons: (1) Hyvis 2000, a low density ρ=945 kg.m^{-3}, low viscosity μ=9.2 x 10^3 Pa.s (Koyi, 1991) polybutene manufactured by BP Chemicals Ltd. and, (2) air bubbles (ρ = 1.3 kg.m^{-3}, μ = 1.8 x 10^{-5} Pa.s).

Patterns of deformation induced by the subducting slab and rising inclusions were visualized by printing, on and inside the SGM36, a grid of orthogonal lines and circles of passive carbon produced by the « unbaked photocopy method » (Dixon and Summers, 1985) modified by Treagus and Sokoutis (1992). The buoyant inclusions and the grid of strain markers were inserted along the mid-plane of the model during construction, well away from the boundary effects caused by the front and back plexiglass walls.

Throughout the experiments (Figs. 2-5), the plexiglass plate sank obliquely through SGM36 by its own weight in a rigid 22cm x 14cm x 14 cm vessel with an open top. The weight of the heavy plate was neutralized as the plate sank and the shear strain-rate in the surrounding fluid decreased and ceased altogether when the plate reached the rigid bottom boundary. The plate sank 150 mm in approximately 24 hours. No slip occurred along any of the boundaries between viscous SGM36 and the plexiglass. The dip of the sinking plate was restricted to 35°, the angle at which the Philippine Sea plate subducts beneath the Yakushima granite (Yamaoka et al., 1986; Jarrard, 1986).

LIMITATIONS OF THE MODELS

The models simplify the natural dynamics. There is no roll back of the hinge line of subduction (Kincaid and Olson, 1987). The trailing edge of the subducting slab does not pull a horizontal plate behind it (Jacoby, 1976; Jacoby and Schmeling, 1981; Shemenda, 1994). The subducting slab in the model cannot sink through the rigid bottom boundary which scales approximately to the 670 km mantle discontinuity (Zhong and Gurnis, 1995 and reference therein). The side boundaries artificially limit the wedge flows above and below the sinking slab. The models also start at an arbitrary stage in the sinking of a slab. Despite all these limitations, the models still answer some of the questions concerning buoyant magmatic diapirs rising above subduction zones.

EXPERIMENTS

WEDGE FLOW INDUCED BY SLAB SUBDUCTION

The pattern of wedge flow due solely to the oblique subduction of a rigid plate sinking along its length is demonstrated in figure 2. The passive carbon marker-grid was printed to visualize flow along the vertical midplane of the working fluid (SGM36). This model has no characteristic length since it involves only a single fluid bound by rigid walls with an open top. The depth of the « trench » that appeared in the surface of the model decreased from an early maximum of 12 mm to 5 mm at the end of the run; this increased the dihedral angle between the slab surface and model fluid as the strain-rate decreased. This model «trench» does not correspond to position and depth of natural trenches, filled by thick sediments. The dihedral angle of the model trench (≈90°) is much smaller than that in nature (≈170°). The shape of the model-trench is rather similar to the wedge-shaped ductile region beneath island arcs illustrated by, for instance, Shimamoto (1993). The position of the inner wall of the model trench approximately corresponds to the position of the aseismic front (profile in Fig. 1).

Intensity, orientation and translation of the flow induced in the viscous fluid by the sinking slab are mapped using the distortions of the rectilinear grid and circular markers (Anma and Sokoutis, in preparation). The SGM36 bound to the no-slip boundary was carried downward with the sinking slab. Vertical marker lines show horizontal flow toward the trench beneath the top free surface implying that the model «back-arc basin» is subjected to horizontal extension normal to the trench axis. The most intense horizontal extension of the top surface overlies the upwelling above the leading edge of the subducting plate and migrates laterally with it, showing « non steady state flow ». Maximum rotational strain was at the trailing edge of the wedge which is dragged down by the sinking slab.

INFLUENCE OF THE SLAB SUBDUCTION ON BUOYANT DIAPIRS SEGREGATED ALONG THE WB ZONE

Ascent velocity of granite diapirs versus tectonic convergence rate

The geometry, trajectory of ascent, and patterns of circulation in and around any diapir rising from a subduction zone will be influenced by wedge flows like those seen in figure 2. The effects of subduction on the shape and track of buoyant inclusions rising from the Wadati-Benioff (WB) zone are considered here.

There is no direct measurement of the rate of melt segregation, and rise of diapiric plutons from WB zones in nature. Knowing the velocity of the subducting slab but not that of the diapirs raises a scaling problem in the model. Ascent velocities used for granites in numerical and analogue experiments were estimated using various approaches. Paterson and Tobisch (1992) used an average rate of ascent for andesite and basalt diapirs of 2 m/year. Calculations by Mahon et al. (1988) assuming reasonable material properties suggested that a diapir 5 km in diameter could rise in a temperature-varying medium with velocities that vary by four orders of magnitude but with an avera-

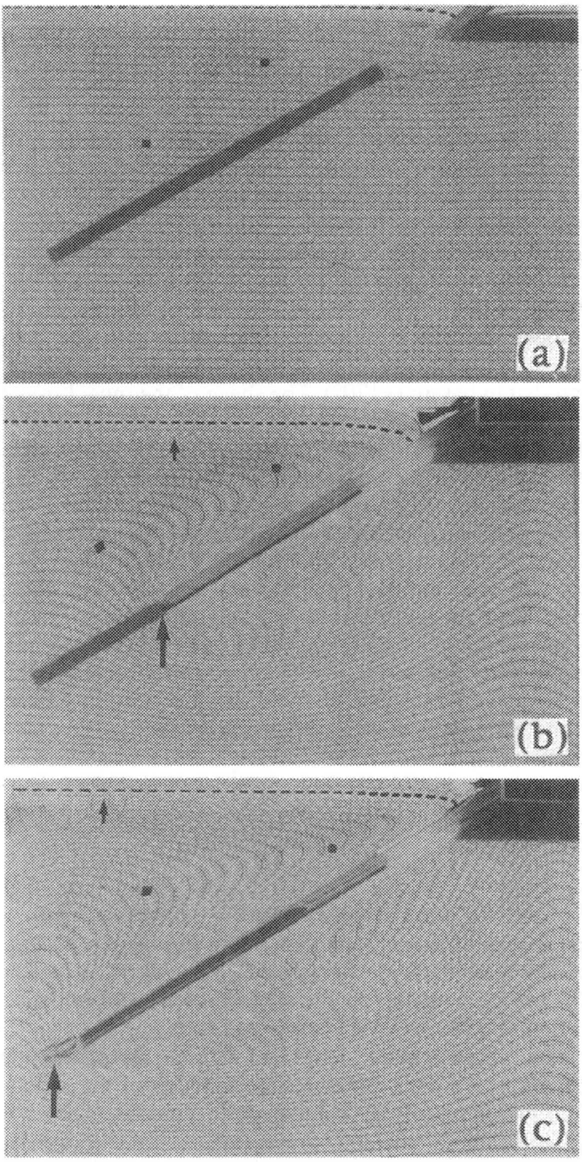

Figure 2. Patterns of wedge flow induced by a rigid subducting slab seen in a grid printed along the vertical mid-plane of the model. The dashed line represents the position of the top free surface of SGM36 (ρ = 970 kg.m^{-3}, μ = 5 x 10^{4} Pa.s). Markers above the dashed line are reflections from the top surface. Maximum horizontal elongation of top free surface (indicated by thin arrow) lies above the leading edge of the subducting plate (thick arrow) and migrates with it. The solid circle and square are inserted to ease visualization of movement paths of particular elements.

ge velocity of 6-12 m/year. The velocities decreased when the diapir reached cold crust, and fell to zero at the final emplacement level.

The tectonic convergence rate between the plates on either sides of a subduction zone can be related to the ascent rate of granite plutons rising from the WB zone. The recent convergent rate between the Eurasian plate and the Philippine Sea plate has been estimated at 4.9 x 10^{-2} m/year in the northern domain and 7 x 10^{-2} m/year in the southern domain of the Ryukyu arc (Letouzey and Kimura, 1986; Uyeda, 1982). Therefore, the ascent velocity of plutons ($V_{inclusion}$) appears to be one to three orders of magnitude faster than the rate of local subduction (V_{slab}) in the deep crust and mantle where the viscosities of rocks are lower and the density contrast between the rising diapirs and their ambient rocks are greater. However, the effect of wedge flow increases in significance as $V_{inclusion}$ decreases toward emplacement levels.

The scaling problem is therefore, further simplified by modeling two extreme cases: 1) where the ascent velocity of a diapir ($V_{inclusion}$) >> the sinking velocity of the oceanic slab (V_{slab}), and 2) $V_{inclusion} \approx V_{slab}$. The natural phenomenon is likely to occur somewhere between these two extremes.

Geometry of the buoyant source layer

Hyvis 2000, stained pale blue, was used for the buoyant inclusion in both models described below because it sticks to the rigid plexiglass plate and is thus dragged down with the sinking plate. The Hyvis 2000 was poured onto the top surface of the SGM36 and was dragged with the sinking slab when V_{slab} >> $V_{inclusion}$ despite its positive buoyancy. After the buoyant inclusion was covered by the surrounding SGM36, V_{slab} was controlled to achieve the above mentioned velocity contrast. In both cases, the source layer was an oblique ellipse stuck to the toe of the sinking slab. Since the generation of magma in subduction zones is controlled by the dehydration reaction of hydrous minerals, and since dehydration depends on pressure (Tatsumi and Eggins, 1995), the geometry of natural magma sources is likely to be linear (Marsh and Carmichael, 1974). However, even linear source layers should develop bulges with a wavelength due to the Rayleigh-Tayler instability (Rönnlund, 1991). In this case, the elliptical initial source layer in our experiments models a single bulge in a linear source region.

Case 1: $V_{inclusion} >> V_{slab} = 0$

To simulate this extreme case (Fig. 3), the buoyant inclusion was first dragged downward with the sinking slab to the base of the model. As the rigid slab sank, the top edge of the « magma » bulged upward but did not separate from the slab until the slab stopped (Fig. 3b) at the bottom of the box at which point the inclusion started to rise due to its positive buoyancy. Note that the obliquely curved stem left beneath the diapir at the end of the run (Fig. 3c) does not trace the trajectory of diapiric rise. The elliptical pool of « magma » dragged beneath the surface developed a peripheral levee which fed a bulge at its highest point (Fig. 3a, b). The levee on either side of the centreline of the elliptical source layer coalesced into the stem feeding the rising bulge. The bulge of « magma » developed into an asymmetric diapir by growing faster on the side farthest from the no-slip top of the oblique slab (Fig. 3c). After the stage shown in figure 3c, the stem pinched and parted just below the spherical diapir which was rising faster than the stem. The asymmetry of the diapir was derived from the oblique source layer and remained stable even after rising a few body radii.

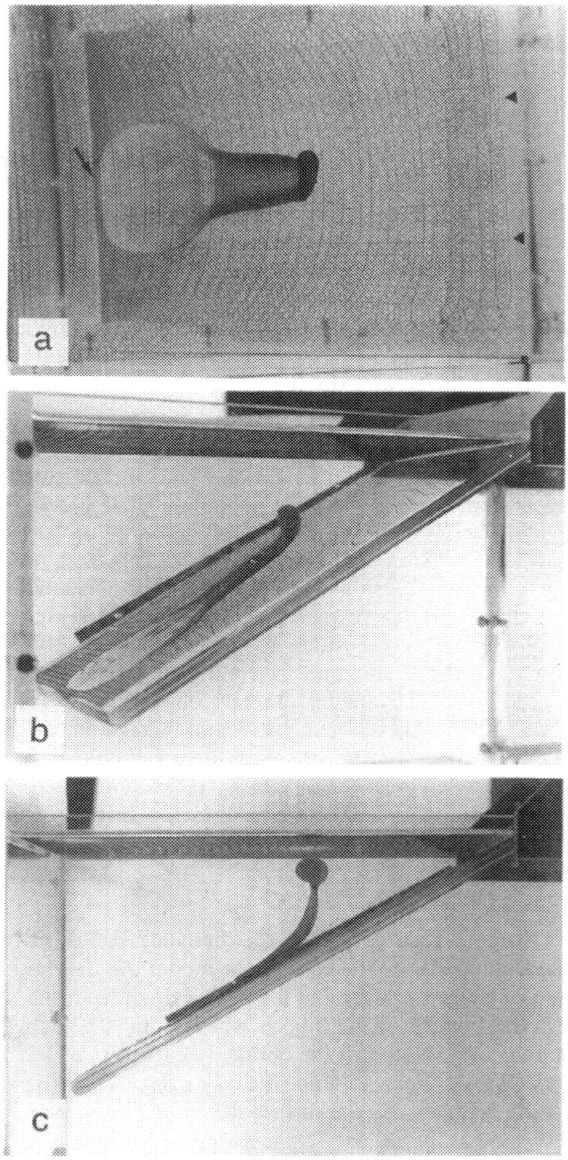

Figure 3. Influence of wedge flow on rising diapir segregated from the WB zone ($V_{inclusion} >> V_{slab}$). The buoyant blue Hyvis ($\rho = 945$ kg.m^{-3}, $\mu = 9.2 \times 10^3$ Pa.s) was first poured onto the top free surface and was then dragged downward with the sinking slab. After the slab reached the bottom, the buoyant pool developed an ellipsoidal bulge at its top edge (a, b) and rose vertically (c). Note that strongest trench-normal horizontal extension (arrow) lies immediately above the leading edge of the sinking plate (a). Triangles indicate the trench axis.

Case 2: $V_{inclusion} \approx V_{slab}$

In another experiment, V_{slab} was slowed just after the buoyant Hyvis 2000 was covered by the SGM36 such that $V_{inclusion} \approx V_{slab}$. This time, the buoyant inclusion reached the top free surface before the slab reached the base of the model, without ever having escaped from the drag of the sinking slab. The model magma pooled in the upper part of the buoyant layer with an ellipsoidal axis parallel to the trench axis (Fig. 4b). Therefore, we may conclude that diapirs affected by wedge flow are expected to rise obliquely and become elongated parallel to the trench.

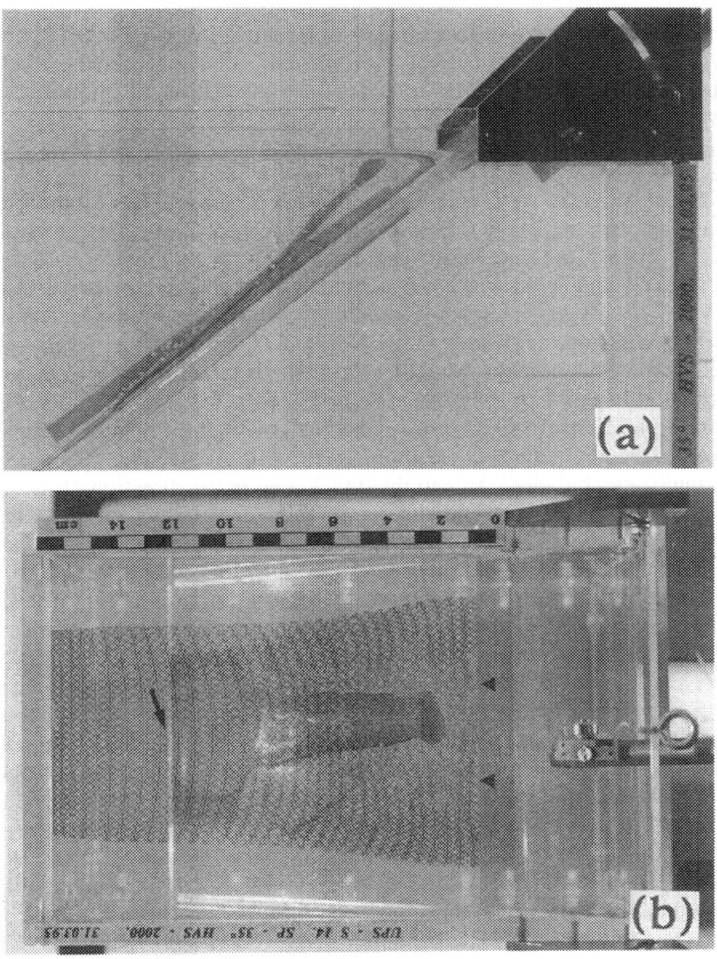

Figure 4. Influence of wedge flow on diapirs rising from the WB zone ($V_{inclusion} \approx V_{slab}$). When $V_{inclusion} \approx V_{slab}$, the buoyant inclusion was dragged into a thin oblique sheet (a; side view, b; top view). Symbols: same as in figure 3.

This experiment illustrates an essential aspect regarding subduction zone magmatism: there must be enough time for a bulge of sufficient volume to segregate for the buoyancy force to overcome the downward drag. This temporal aspect must be taken into account in future considerations concerning subduction zone magmatism.

RISE OF BUOYANT INCLUSIONS STARTING ABOVE A SUBDUCTION ZONE

Buoyant inclusions rising in the ductile lithosphere above a subduction zone were modelled by injecting air bubbles into the SGM36. Different sized air bubbles (Fig. 5a) were used to mimic the effect of slab subduction on granites starting above the WB zone (Figs. 5b-f). The scaling in this model is less realistic than the other models because small air bubbles have high surface tension and contrast significantly in viscosity and density with the enclosing SGM36. Nevertheless, this experiment illustrates the potential influence of wedge flow on the tracks and geometries of plutons rising in the lower crust.

$V_{inclusion}/V_{slab}$ varies with the size of each bubble. The ratio varies from $V_{inclusion} \geq V_{slab}$ for large bubbles away from the slab, to $V_{inclusion} \leq V_{slab}$ for small bubbles close to the slab. We propose that, wherever granites form in relation to a subduction zone, they are likely to follow tracks similar to those illustrated in figure 6. The tracks of bubble paths reconstructed from serial photographs are loops having different shapes depending on where the bubbles started relative to the subducting plate (Fig. 6). $V_{inclusion}/V_{slab}$ changes continuously along the loop and depends on the interaction between diapir buoyancy and slab drag.

Air bubbles located near the leading edge of the subducting slab were pushed away when subduction began (Fig. 5b). After being expelled by the sinking slab, large air bubbles having sufficient buoyancy to overcome downward drag rose obliquely toward the trench axis, leaving a diapiric tail (bubble A in Fig. 5 and Fig. 6).

In the final stage, bubble A bulged the trench wall in the direction of its ascent (Fig. 5g; Fig. 6: step A9). Profiles reconstructing the three dimensional shape of bubble A at step A9 are shown (Fig. 6, insert). Although bubble A surfaced with a circular planform (Fig. 5g), in profile, it had a flat bottom and a tail parallel to the slab (Fig. 6, insert).

Small bubbles with insufficient buoyancy to escape downward drag from the sinking slab did not reach the free surface (bubble C in Figs. 5 and 6). Serial profiles show that bubble C rotated clockwise from step 3 to step 4 (Fig. 6).

Further, shallow air bubbles rising far from the trench axis underwent horizontal extension perpendicular to the trench axis, becoming tabular in section (bubble B in Figs. 5 and 6). Such inclusions became elliptical in plan view (bubbles B and D in Fig. 5g) with elongation axes perpendicular to the trench axis during translation, first away from the trench then toward it. This is because air bubbles are highly deformable compared with the ambient SGM36. In detail, bubble B formed an asymmetrical bulb above a small asymmetric tail distal to the trench (Fig. 6 step 6). The tail shortened between steps 7 and 10 (Fig. 6). This demonstrates the possibility that shallow intrusions rising well away from the trench can have an asymmetric zoning with wider zones on the trench side.

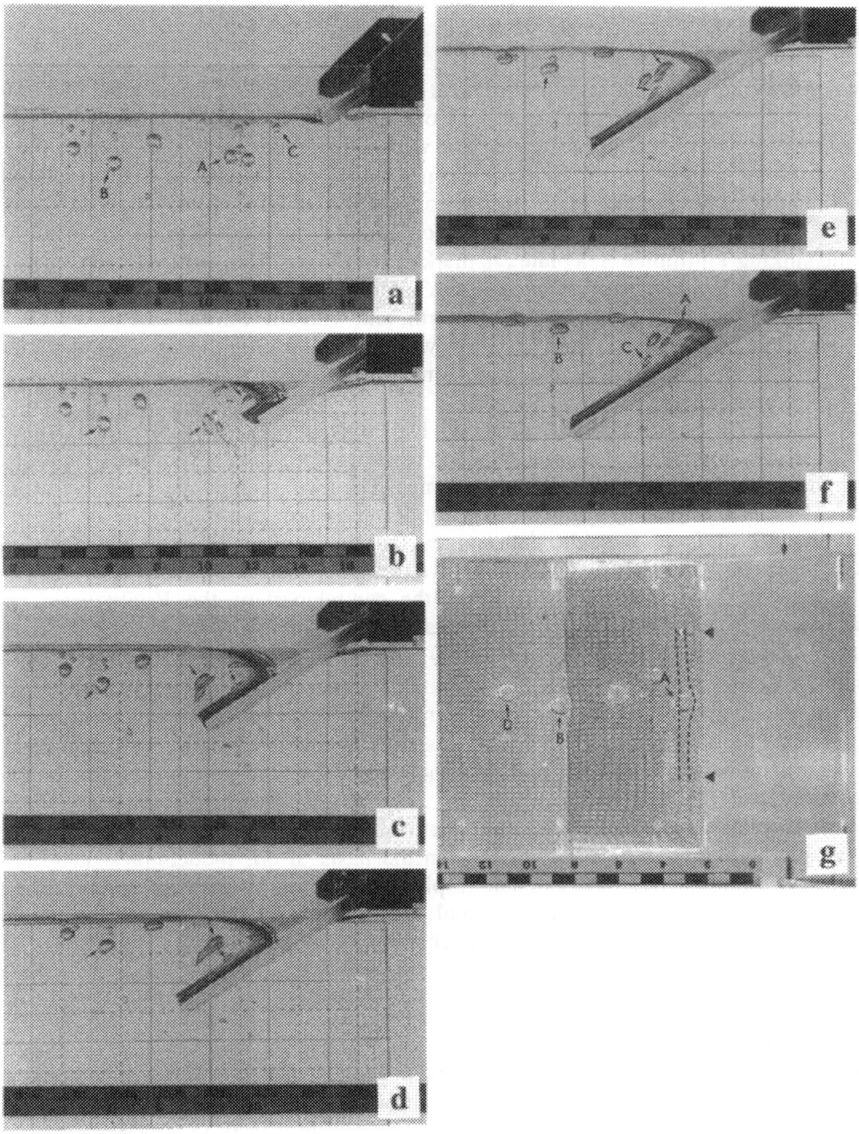

Figure 5. Influence of wedge flow on the geometry and track of buoyantly rising air bubbles. Photographs (a) to (f) were taken at 0 min., 16 min., 52 min., 89 min., 140 min. and 201 min. after the start of the run. Note that the positions of inclusions A and C at the beginning (a) of the run had switched by the end (f). Photograph from above (g) was taken 222 min. after start. Bubble A has an inclined tail (f), although it looks circular in plan view (g) where it distorts horizontal markers (g ; dashed) towards its ascent direction. Inclusion B and D became elliptical with elongation axes perpendicular to the trench axis in plan view (g).

APPLICATION

REGIONAL SCALE

The subduction zones along the northwest margin of the Philippine Sea plate provide opportunities to examine the results of our experiments on wedge flow and pluton emplacement above subduction zones. The wedge flow induced by slab subduction in the model shown in figure 2 suggests that a zone of horizontal extension migrates above the leading edge of the sinking plate. This is consistent with field observations and previous reports on the Ryukyu subduction zone, including: (1) juxtaposition of the leading edge of the sinking Philippine Sea plate (as marked by the furthest limit of deep seismicity: Shiono et al., 1980; Kimura et al., 1983; Yamaoka et al., 1986; Uyeda, 1991) with the chain of current volcanic activity in the north, and with the central axis of the Okinawa trough in the south (Fig. 1), and (2) north-westward migration of this zone of horizontal extension above the Ryukyu subduction zone after Miocene times (see Tectonic Setting *in* Anma, this volume).

Arc-parallel, 15-16 Ma dikes (Fig. 1; Daishi et al., 1987; Taneda and Kinoshita, 1972) in the northern and central Ryukyu arc imply northwest-southeast dilation, i.e. perpendicular to the arc axis in Miocene times. The formation of the Okinawa trough is marked by late Miocene sediment infill along an eastern marginal zone of the trough (Kimura, 1990). Paleomagnetic data suggest that the southern Ryukyu block rotated 25° clockwise relative to Eurasia as the Okinawa trough opened between 10 and 6 Ma (Miki et al., 1990; Miki, 1995). By contrast, the central part of the Ryukyu arc opened without any rotation (Miki, 1995). The rapid rifting of the Okinawa trough started in early Pleistocene (1.5-2 Ma; Kimura, 1990) and is still active (Kodama et al., 1995). A chain of active volcanoes parallels the axis of the northern Ryukyu arc. Volcanic activity can then be inferred along the outer margin of the Okinawa trough further to the southwest (Kato et al. 1992; Fig. 1). Thus, the zone of horizontal northwest-southeast extension appears to have migrated from the southeast to the northwest after the Middle Miocene.

Arc-normal extension was previously attributed to suction induced by the roll back of the subducting slab (Elsasser, 1971). On the basis of our models presented here, we propose that both back-arc extension, plus migration of the extension zone, and upwelling of mantle material (or thermal diapir; Karig, 1971) can be attributed to a single action; oblique subduction of the oceanic lithosphere alone.

EXAMINING INDIVIDUAL PLUTONS

Experimental observations in figures 5 and 6 suggest that the shape of plutons above the subduction zones depends on their distance from the « model » trench axis. Note that the position of the model trench corresponds to the position of the aseismic front shown in the profile of figure 1, rather than to the natural sediment-filled trench axis. Buoyant bubbles starting near the model trench rose obliquely toward the trench axis leaving an inclined diapir tail (Fig. 6: bubble A). Shallow bubbles starting far from the trench axis became tabular with their longest axis oriented horizontally (Fig. 6: bubble B). This finding agrees with field observations of the Yakushima granite (Anma, this volume) and with previous reports of the shapes of other contemporaneous Miocene plutons in the Outer Zone of Southwest Japan (Shibata, 1978; Fig. 7).

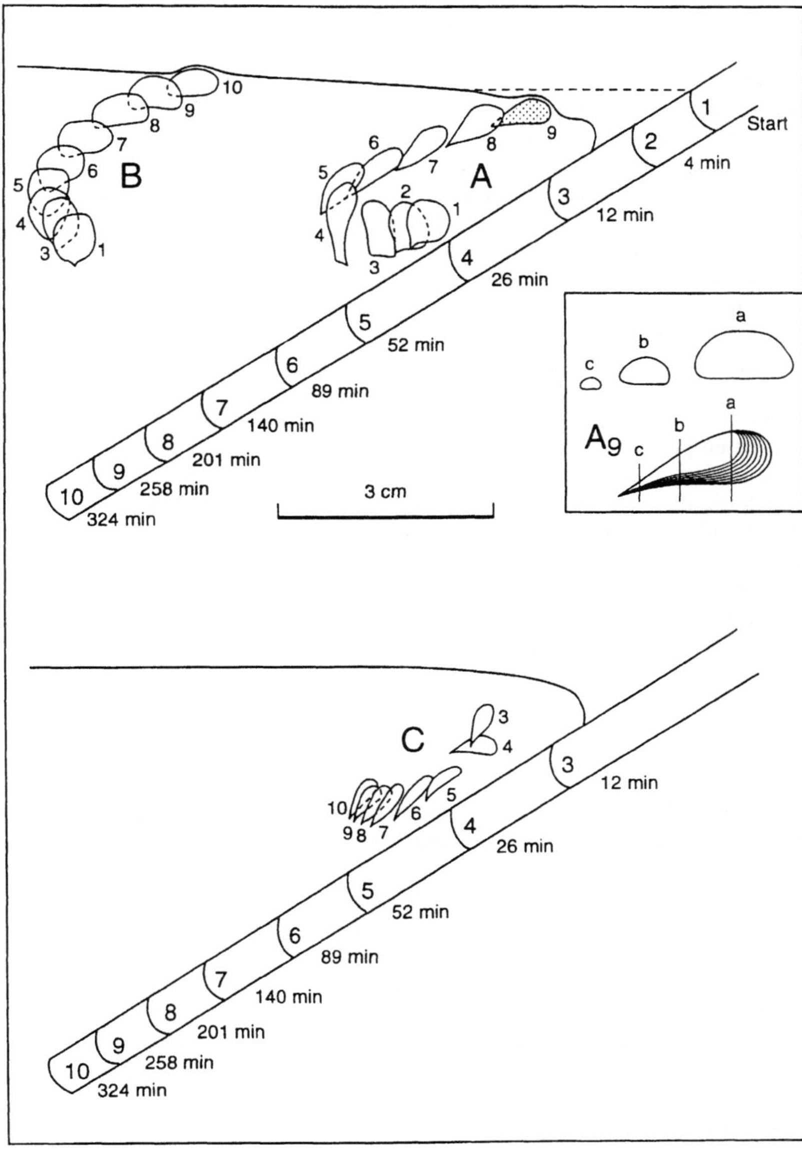

Figure 6. Looped tracks of air bubbles A, B and C in figure 5. Steps 1 to 10 are the positions of each inclusion and the leading edge of the subducting plate at 0 min., 4 min., 12 min., 26 min., 52 min., 89 min., 140 min., 201 min., 258 min. and 324 min. after the start of the run. Insert : profiles reconstructing the three dimensional shape of bubble A at step 9.

The Yakushima granite is inferred to have risen obliquely in the Ryukyu arc along a track inclined upward toward the trench axis. It is located close to the trailing edge of the Miocene forearc basin (≈30 km from the aseismic front) and is comparable in relative location to bubble A of the experiment (Figs. 5 and 6). This bubble ascended along a loop and bulged the inner trench wall (Fig. 5g) during the emplacement stage. Bubble A surfaced with a circular planform (Fig. 5g) and with an inclined tail oblique to the trench (Fig. 6, insert) as inferred for the Yakushima granite (Anma, this volume). Whether such S-type granites are generated in the lower crust or rise directly from the WB zone, the influence of wedge flow increases significantly as they slow on entering the crust. Then, the wedge flow will drag the diapiric tail parallel to the subducting plate. Similarities between shape of bubble A and field observations raise an additional possibility: that the Yakushima granite has a flat bottom shallower than expected.

Unlike the Yakushima granite, the S-type Minami-Osumi pluton (Fig. 7) has an elliptical planform elongate parallel to the trench (Fig. 1). This may be because it is bigger, hence had a higher $V_{inclusion}$ and rose along a steeper ascent path. Alternatively, it may have been affected more by the slab drag and became elliptical as seen in figure 4. Arc-normal horizontal extension could also account for this geometry.

Using the surface geometry of the I-type Okueyama granite, the width of its thermal aureole and borehole data, Takahashi (1986) estimated the shape of this pluton (Fig. 7, insert) to be elliptical and elongate normal to the trench axis in plan view and tabular-shaped in section. A map by Nakada et al. (1988) shows the Ichifusayama pluton to be elliptical with its longest axis parallel to that of the Okueyama pluton. Concentric zoning within the Ichifusayama pluton appears to be asymmetric (Fig. 7, insert) despite its southeastern block being elevated by a Quaternary fault (Nakada et al., 1988). The geometry and asymmetry of the Okueyama and Ichifusayama plutons and their locations at about 250 km from the trench axis (≈ 110 km from the aseismic front), are comparable to bubbles B or D of figure 5.

CONCLUSIONS

Our experiments demonstrate that subduction of a cold and stiff oceanic plate causes (1) asymmetry in the shapes of magma plutons rising above the subduction zone, and (2) obliquity in their ascent paths. The experiment shown in figure 3 suggests that plutons acquire an asymmetric shape due to the obliquity of the source layer and drag along the top of the sinking slab. The effect of these factors will increase when the angle of subduction is steeper, and rising diapirs will then become more asymmetric.

Obliquity of the ascent trajectory in asymmetric plutons is attributed to wedge flow induced by viscous drag from the sinking lithospheric plate (Fig. 2). Another effect which might lead to oblique diapiric rise is the wedge-shaped temperature and rheology distribution (Fig. 1) predicted by thermal modeling of subduction zones (Toksöz and Hsui, 1978). The shape of the curved ascent path depends on $V_{inclusion}/V_{slab}$. In the deep crust and/or mantle, $V_{inclusion}$ may be one to three orders larger than V_{slab}. Therefore, plutons rising from the WB zone are likely to follow almost vertical tracks and their stems will be inclined as in figure 3. However, for a diapir to rise rather than be dragged downward into a thin oblique sheet (Fig. 4), enough time needs to pass for a bulge of sufficient volume to segregate for its buoyancy to allow escape from downward drag.

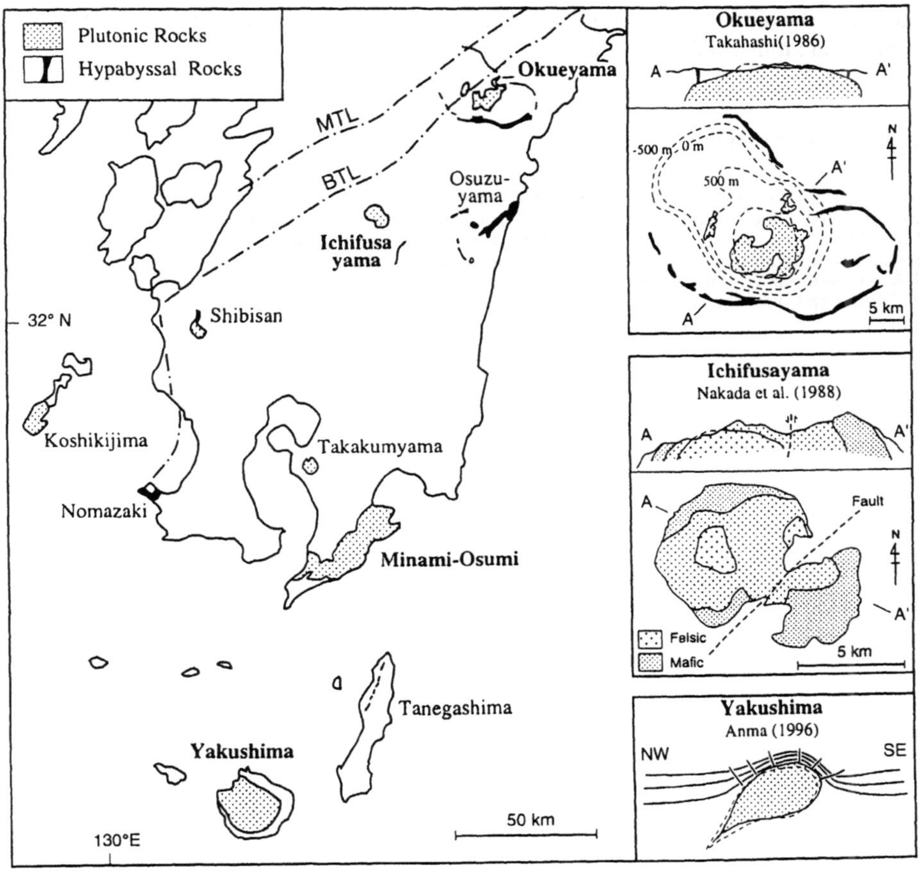

Figure 7. Distribution of Miocene plutons in the Kyushu area, the Outer zone (south of MTL) of Southwest Japan. Position of this map is indicated in figure 1. The axes of the Ryukyu trench - Nankai trough are located southwest from this map area and almost parallel the BTL. MTL : Median Tectonic Line, BTL: Butsuzou Tectonic Line. S-type plutons: the Yakushima, Minami-Osumi and Takakumayama plutons. I-type plutons: the Koshikijima, Shibisan, Ichifusayama and Okueyama plutons (Nakada and Takahashi, 1979). S-type plutons located near the trench axis exhibit either an asymmetrical circulation pattern (the Yakushima granite) or an elliptical elongation parallel to the trench axis (Minami-Osumi pluton). By contrast, the I-type Okueyama (Takahashi, 1986) and Ichifusayama plutons (Nakada et al., 1988) located further from the trench axis are elongate perpendicular to the trench axis (see text). These shapes, in map view, with respect to distance from the trench favorably compare with those of air bubbles in our simplified model (Figs. 5 and 6).

$V_{inclusion}$ slows down significantly in the middle or upper crust. Wedge flow will then influence buoyant rise, resulting in the oblique rise of epizonal diapirs in magmatic arcs (Figs. 5 and 6). Diapirs starting in the region of high rotational strain near the trench rose obliquely toward the inner wall of the model trench leaving inclined tails trailing along their tracks. By contrast, diapirs starting in the region of horizontal extension above the leading edge of the sinking slab rotated to horizontal-tabular shapes as they were translated toward the trench.

ACKNOWLEDGEMENTS

We would like to express our gratitude to Prof. Christopher Talbot, Drs. H. Koyi and M. Watkeys for discussion and critical reading of this work. We appreciate careful reviews given by Drs. A. R. Cruden and J. M. Tubia. Our thanks go to Einar Meland for experimental equipment, Christina Wernström for preparation of figures and Bertel Glös and Christer Bäck for technical assistance with taking photographs. R.A. received support through a study grant from Uppsala University and research aid granted by the 29th IGC Wind-up Committee. D.S. received support from the Swedish Natural Science Research Council.

BIBLIOGRAPHY

Affaton, P., Rahaman, M.A., Trompette, R., and Sougy, J. (1991) The Dahomeyide orogen: tectonothermal evolution and relationships with the Volta Basin, in R.D Lécorché and J.P. Dallmeyer (eds.), *The West African orogens and circum-Atlantic correlatives*, I.G.C.P., Special Publication, Berlin: Springer Verlag, pp. 107-122.

Ague, J.J. and Brimhall, G.H. (1988) Magmatic arc asymmetry and distribution of anomalous plutonic belts in the batholiths of California: effects of assimilation, crustal thickness, and depth of crystallization, *Geol. Soc. Am. Bull.* **100**, 912-927.

Ajibade, A.C. and Wright, J.B. (1989) The Togo-Benin-Nigeria Shield: evidence of crustal aggregation in the Panafrican Belt, *Tectonophysics* **165**, 125-129.

Allard, B. and Benn, K. (1989) Shape preferred orientation analysis using digitized images on a microcomputer, *Computer and Geosciences* **15**, 441-448.

Allmendinger, R.W. (1988) Stereonet plotting program. v.4.5., Allmendinger, Cornell Univ.

Allmendinger, R.W., Hauge, T.A., Hauser, E.C., Potter, C.J., and Oliver, J. (1987) Tectonic heredity and the layered lower crust in the Basin and Range Province, western United States, in M.P. Coward, J.F. Dewey, and P.L. Hancock, *Continental extensional tectonics, Geol. Soc. Special Publication*, London, pp. 223-246.

Améglio, L., Vigneresse, J.L., Darrozes, J., and Bouchez, J.L. (1994) Forme du massif granitique du Sidobre (Montagne Noire, France): sensibilité de l'inversion des données gravimétriques au contraste de densité, *C. R. Acad. Sci. Paris* **319**, 2, 1183-1190.

Améglio, L., Vigneresse, J.L., and Bouchez, J.L. (1997) Granite pluton geometry and emplacement mode inferred from combined fabric and gravity data, in J.L. Bouchez, D.H.W. Hutton and W.E. Stephens (eds.), *Granite: from segregation of melt to emplacement fabrics*, Kluwer Academic Publishers, Dordrecht, pp. 199-214.

Amice, M. (1990) Le complexe granitique de Cabeza de Araya (Extremadure, Espagne): zonation, structures magmatiques et magnétiques, géométrie. Discussion du mode de mise en place, *Unpubl. Thesis, Univ. Toulouse*, France, 250 p.

Amice, M., Bouchez, J.L., Aranguren, A., Alvarez, F., and Vigneresse, J.L. (1991) El batolito granitico de Cabeza de Araya (Extremadura): comparacion de sus estructuras magmaticas y magneticas, *Boletin Geologico y Minero* **102/3**, 455-471.

Ananaba, S.E. and Ajakaye, D.E. (1987) Evidence of tectonic control of mineralization in Nigeria from lineament density analysis: a Landsat study, *Int. J. Remote Sensing* **810**, 1445-1452.

Anderson, E.M. (1951) The dynamics of faulting, Edinburgh, Oliver & Boyd, 206 pp.

Anma, R. (1997) Oblique diapirism of the Yakushima granite in the Ryukyu arc, Japan, in J.L. Bouchez, D.H.W. Hutton and W.E. Stephens (eds.), *Granite: from segregation of melt to emplacement fabrics*, Kluwer Academic Publishers, Dordrecht, pp. 287-311.

Anma, R. and Sokoutis, D. (1997) Experiments on pluton shapes and tracks above subduction zones, in J.L. Bouchez, D.W.H. Hutton and W.E. Stephens (eds.), *Granite: from segregation of melt to emplacement fabrics*, Kluwer Academic Publishers, Dordrecht, pp. 313-328.

Annor, A.E. and Freeth, S.J. (1985) Thermo-tectonic evolution of the Basement Complex around Okene, Nigeria, with special reference to deformation mechanism, *Precamb. Res.* **28**, 269-281.

Apalategui, O., Eguiluz, L., and Quesada, C. (1990) Ossa Morena Zone: structure, in R.D. Dallmeyer and E. Martínez García (eds.), *Pre-Mesozoic Geology of Iberia*, Springer-Verlag, Berlin, pp. 280-291.

Aranguren, A. and Tubia, J.M. (1992) Structural evidence for the relationship between thrusts, extensional faults and granitic intrusions in the Variscan Belt of Galicia (Spain), *J. Struct. Geol.* **14**, 1229-1237.

Aranguren, A., Cuevas, J., and Tubia, J.M. (1995) Composite magnetic fabric in S-C mylonites from a granodioritic pluton, *Terra Abstract* **7**, 142.

Aranguren, A., Larrea, F., Carracedo, M., Cuevas, J., and Tubia, J.M. (1997) The Los Pedroches batholith (Southern Spain): polyphase interplay between shear zones in transtension and setting of granites, in J.L. Bouchez, D.H.W. Hutton and W.E. Stephens (eds.), *Granite: from segregation of melt to emplacement fabrics*, Kluwer Academic Publishers, Dordrecht, pp. 215-229.

Arbaret, L. (1995) Orientation préférentielle de forme dans les magmas: modélisation analogique 3D en cisaillement simple, *Unpubl. Thesis Univ. Blaise-Pascal*, Clermont-Ferrand, 264 pp.

Arbaret, L., Diot, H., Bouchez, J.L., and Launeau, P. (1995) Three dimensional shape preferred orientation of particles in experimental simple shear flow: AMS fabric and image analysis, *Journ. Czech. Geol. Soc.* **40/3**, B57.

Arbaret, L., Diot, H., and Bouchez, J.L. (1996) Shape fabrics of particles in low concentration suspensions: 2D analog experiments and application to tiling in magma, *J. Struct. Geol.* **18**, 941-950.

Arbaret, L., Diot, H., Bouchez, J.L., Lespinasse, P., and de Saint-Blanquat, M. (1997) Analogue 3D simple shear experiments of magmatic biotite subfabrics, in J.L. Bouchez, D.H.W. Hutton and W.E. Stephens (eds.), *Granite: from segregation of melt to emplacement fabrics*, Kluwer Academic Publishers, Dordrecht, pp. 129-143.

Archanjo, C.J. (1993) Fabriques de plutons granitiques et déformation crustale du Nord-Est du Brésil: une étude par l'anisotropie de la susceptibilité magnétique de granites ferromagnétiques, *Unpubl. Thesis Univ. Paul-Sabatier,* Toulouse, 167 pp.
Archanjo, C.J., Olivier, P., and Bouchez, J.L. (1992) Plutons granitiques du Seridó (NE du Brésil); écoulement magmatique parallèle à la chaîne révélé par leur anisotropie magnétique, *Bull. Soc. géol. France* **163**, 509-520.
Archanjo, C.J., Bouchez, J.L., Corsini, M., and Vauchez, A. (1994) The Pombal granite pluton: magnetic fabric, emplacement and relationships with the Brasiliano strike-slip setting of NE Brazil (Paraíba State), *J. Struct. Geol.* **163**, 323-335.
Archanjo, C.J., Launeau, P., and Bouchez, J.L. (1995) Magnetic fabric versus magnetite and biotite shape fabrics of the magnetite-bearing granite pluton of Gameleiras (Northeast Brasil), *Phys. Earth Plan. Inter.* **89**, 63-75.
Arculus, R.J. (1994) Aspects of magma genesis in arcs, *Lithos* **33**, 189-208.
Arzi, A.A. (1978) Critical phenomena in the rheology of partially melted rocks, *Tectonophysics* **44**, 173-184.
Atlan, H. (1992) *L'organisation biologique et la théorie de l'information*, Hermann, Paris, 299 pp.
Audrain, J., Amice, M., Vigneresse, J.L., and Bouchez, J.L. (1989a) Gravimétrie et géométrie tri-dimensionelle du pluton granitique de Cabeza de Araya (Extrémadure, Espagne), *C. R. Acad. Sci. Paris* **309**, 2, 1757-1764.
Audrain, J., Vigneresse, J.L., Cuney, M., and Friedrich, M. (1989b) Modèle gravimétrique et mise en place du complexe granitique hyperalumineux de Saint Sylvestre (Massif Central français), *C. R. Acad. Sci. Paris* **309**, 2, 1907-1914.
Balk, R. (1937) Structural behaviour of igneous rocks, *Geol. Soc. Am. Memoir* **5**, 117 pp.
Baker, D.R. and Vaillancourt, J. (1995) The low viscosities of F+H_2O-bearing granitic melts and implications for melt extraction and transport, *Earth Planet. Sci. Lett.* **132**, 199-211.
Batchelor, R.A. and Bowden, P. (1986) Major and trace element analyses of volcanic and sub-volcanic igneous rocks from the Nigeria-Niger anorogenic province, *Overseas Development Administration*, London, Research Scheme R2679, Supplement B.
Bateman, R. (1985) Aureole deformation by flattening around a diapir during in situ ballooning: the Cannibal Creek Granite, *J. Geol.* **93**, 293-310.
Bateman, P.C. (1992) Plutonism in the central part of the Sierra Nevada Batholith, California, *U.S. Geol. Surv. Prof. Pap.*, 1483.
Bateman, P.C. and Chappell, B.W. (1979) Crystallization, fractionation, and solidification of the Tuolumne Intrusive series, Yosemite National Park, California, *Geol. Soc. Am. Bull.* **90**, 465-482.
Beard, J.S. and Day, H.W. (1988) Petrology and emplacement of reversely zoned gabbro-diorite plutons in the Smartville complex, *J. Petrol.* **29**, 965-995.
Beere, W. (1975) A unifying theory of the stability of penetrating liquid phases and sintering pores, *Acta Metall.* **23**, 131-138.
Bell, F.G. (1983) *Engineering properties of rocks and soils,* 2nd edition, Butterworths, London, 149 pp.
Benn, K. (1994) Overprinting of magnetic fabrics in granites by small strains: numerical modelling, *Tectonophysics* **233**, 153-162.
Benn, K. and Allard, B. (1989) Preferred mineral orientations related to magmatic flow in ophiolite layered gabbros, *J. Petrol.* **30**, 925-946.
Benn, K., Rochette, P., Bouchez, J.L., and Hattori, K (1993) Magnetic susceptibility, magnetic mineralogy and magnetic fabric in a late Archean granitoid-gneiss belt, *Precamb. Res.* **63**, 59-81.
Bergantz, G.W. (1995) Changing techniques and paradigms for the evaluation of magmatic processes, *J. Geophys. Res.* **100**, B9, 17603-17613.
Bergantz, G.W. and Dawes, R. (1994) Aspects of magma generation and ascent in continental lithosphere, in M.P. Ryan (ed.), *Magmatic systems*, Academic Press, San Diego, Calif, pp. 291-317.
Bernier, S., Bouchez, J.L., and Rochette, P. (1987) Anisotropie de la susceptibilité magnétique du granite de Beauvoir (sondage d'Echassières, Massif Central français), *C. R. Acad. Sci. Paris* **305**, 1167-1173.
Berthé, D., Choukroune, P., and Jégouzo, P. (1979) Orthogneiss, mylonite and noncoaxial deformation of granites: the example of the South Armorican Shear Zone, *J. Struct. Geol.* **1**, 31-42.
Bhattacharyya, D.S. (1966) Orientation of mineral lineation along the flow direction in rocks, *Tectonophysics* **3**, 29-33.
Bianchi, A., Callegari, E., and Jobstraibizer, P.G. (1970) I tipi petrografici fondamentali del plutone dell' Adamello, Memorie degli Istituti di Geologia e Mineralogia dell'Università di Padova **27**, 1-148.
Birch, F.S. (1979) Magnetic fabric of the Exeter pluton, New Hampshire, *J. Geophys. Res.* **84**, 1129-1137.
Black, R., Latouche, L., Liégeois, J.P., Caby, R., and Bertrand, J.M. (1994) Pan-African displaced terranes in the Tuareg shield (central Sahara), *Geology* **22**, 641-644.
Blake, S. and Campbell, I.H. (1986) The dynamics of magma-mixing during flow in volcanic conduits, *Contr. Mineral. Petrol.* **94**, 72-81.
Blake, S. and Koyaguchi, T. (1991) Insights on the magma mixing model from volcanic rocks, in J. Didier and B. Barbarin (eds.), *Enclaves and granite petrology*, Elsevier, Amsterdam, pp. 403-413.
Blank, J., Stolper, E.M, and Carrol, M.R. (1993) Solubilities of carbon dioxide and water in rhyolitic melt at 850°C and 750 bars, *Earth Planet. Sci. Lett.* **119**, 27-36.
Blumenfeld, P., Mainprice, D., and Bouchez, J.L. (1986) C-slip in quartz from subsolidus deformed granite, *Tectonophysics* **127**, 97-115.

Blumenfeld, P. and Bouchez, J.L. (1988) Shear criteria in granite and migmatite deformed in the magmatic and solid states, *J. Struct. Geol.* **10**, 361-372.

Blundy, J.D. (1989) The geology of the southern Adamello Massif, Italy, *University of Cambridge Ph.D. thesis*, 221 pp.

Blundy, J.D. and Sparks, R.S.J. (1992) Petrogenesis of mafic inclusions in granitoids of the Adamello Massif, Italy, *J. Petrol.* **33**, 1039-1104.

Bogaard, P. and Schimick, C. (1995) $^{40}Ar/^{39}Ar$ laser probe ages of Bishop Tuff quartz phenocrysts substantiate long-lived silicic magma chamber at Long Valley, United States, *Geology* **23**, 759-762.

Borradaile, G.J. (1988) Magnetic susceptibility, petrofabrics and strain, *Tectonophysics* **156**, 1-20.

Borradaile, G.J. (1991) Correlation of strain with Anisotropy of Magnetic Susceptibility (AMS), *Pageoph.* **135**, 15-29.

Borradaile G.J., Keeler, W., Alford, C., and Sarvas, P. (1987) Anisotropy of magnetic susceptibility of some metamorphic minerals, *Phys. Earth Plan. Inter.* **48**, 161-166.

Borradaile, G.J. and Puumala M.A. (1989) Synthetic magnetic fabrics in a plasticene medium, *Tectonophysics* **164**, 73-78.

Borradaile, G.J. and Werner, T. (1994) Magnetic anisotropy of some phyllosilicates, *Tectonophysics* **235**, 223-248.

Bott, M.H.P. (1962) A simple criterion for interpreting negative gravity anomalies, *Geophysics* **27**, 376-381.

Bottinga, Y. and Weill, D.F. (1972) The viscosity of magmatic silicate liquids: a model for calculation, *Am. J. Sci.* **272**, 438-475.

Bouchez, J.L. (1997) Granite is never isotropic: an introduction to AMS studies of granitic rocks, in J.L Bouchez, D.H.W. Hutton and W.E. Stephens (eds), *Granite: from segregation of melt to emplacement fabrics*, Kluwer Academic Publishers, Dordrecht, pp. 95-112.

Bouchez, J.L, Guillet, P., and Chevalier, F. (1981) Structures d'écoulement liées à la mise en place du granite de Guérande (Loire-Atlantique, France), *Bull. Soc. géol. France* **7-XXIII (4)**, 387-399.

Bouchez, J.L., Bernier, S., Rochette, P., and Guineberteau, B. (1987) Log des susceptibilités magnétiques et anisotropies de susceptibilité dans le granite de Beauvoir: conséquences pour sa mise en place, *Géologie de la France*, Mém. GPF, BRGM Publ., Orléans **2-3**, 223-232.

Bouchez, J.L. and Diot, H. (1990) Nested granites in question: contrasted emplacement kinematics of independent magmas in the Zaër pluton, Morocco, *Geology* **18**, 966-969.

Bouchez, J.L., Gleizes, G., Djouadi, M.T., and Rochette, P. (1990) Microstructure and magnetic susceptibility applied to emplacement kinematics of granites: the example of the Foix pluton (French Pyrenees), *Tectonophysics* **184**, 157-171.

Bouchez, J.L., Delas, C., Gleizes, G., Nédélec A., and Cuney, M. (1992) Submagmatic microfractures in granites, *Geology* **20**, 35-38

Bouchez, J.L. and Gleizes, G. (1995) Two-stage deformation of the Mont-Louis-Andorra granite pluton (Variscan Pyrenees) inferred from magnetic susceptibility anisotropy, *J. Geol. Soc. London* **152**, 669-679.

Bouillin, J.P., Bouchez, J.L., Lespinasse, P., and Pêcher, A. (1993) Granite emplacement in an extensional setting: an AMS study of the magmatic structures of Monte Capanne (Elba, Italy), *Earth Planet. Sci. Lett.* **118**, 263-279.

Boutin R., Montigny R., and Thuizat R. (1995) Chronologie K-Ar et ^{39}Ar-^{40}Ar du métamorphisme et du magmatisme des Vosges: comparaison avec les massifs varisques avoisinants, *Géologie de la France* **1**, 3-25.

Bowen, N.L. (1948) The granite problem and the method of multiples prejudices, in J. Giluly and coll. (eds.) *Origin of granite*, *Geol. Soc. Am. Mem.* **28**, pp. 79-90.

Bowden, P., Black, R., Martin, R.F., Ike, E.C., Kinnaird, J.A., and Batchelor, R.A. (1987) Niger-Nigerian alkaline ring complexes: a classic example of African Phanerozoic anorogenic mid-plate magmatism, in J.G. Fitton and B.G.J. Upton, *Alkaline Igneous Rocks*, Oxford, *Geol. Soc. Sp. Pub.*, pp. 357-379.

Brack, P. (1983) Multiple intrusions: examples from the Adamello batholith (Italy) and their significance on the mechanism of intrusion, *Memorie della Società Geologica Italiana* **26**, 145-157.

Brack, P. (1984) Geologie der Intrusiva und Rahmegesteine des Südwest-Adamello (Nord-Italien), Eidgenössischen Technischen Hochschule, *Zürich, Switzerland, PhD Dissertation #7612*, 253 pp.

Brandeis, G. and Jaupart, C. (1986) On the interaction between convection and crystallization in cooling magma chambers, *Earth Planet. Sci. Lett.* **77**, 345-361.

Brandeis, G. and Jaupart, C. (1987) The kinetics of nucleation and crystal growth and scaling laws for magmatic crystallization, *Contr. Mineral. Petrol.* **96**, 24-34.

Brandon, A.D., Creaser, R.A., and Chako, T. (1996) Constraints on rates of granitic magma transport from epidote dissolution kinetics, *Science* **271**, 1845-1848.

Brenan, J. M. (1993) Diffusion of chlorine in fluid-bearing quartzite: effects of fluid composition and total porosity, *Contr. Mineral. Petrol.* **115**, 215-224.

Brew, D.A. and Ford, A.B. (1981) The Coast Plutonic Complex Sill, southeastern Alaska, in N.R.D. Albert and T. Hudson (eds.), *The U.S. Geological Survey in Alaska: accomplishments during 1979*, U.S. Geological Survey Circular 823-B, pp. 96-99.

Britto, A.M. and Gunn, M.J. (1987) *Critical state soil mechanics via finite elements*, E. Horwood Ltd., 488 pp.

Brown, M. (1994) The generation, segregation, ascent and emplacement of granite magma: the migmatite-to-crustally-derived granite connection in thickened orogens, *Earth. Sci. Rev.* **36**, 83-130.

Brown, M., Averkin, Y.A., and McLellan, E.L. (1995) Melt segregation in migmatites, *J. Geophys. Res.* **100 B8**, 15655-15679.

Brun, J.P. and Cobbold, P.R. (1980) Strain heating and thermal softening in continental shear zones: a review, *J. Struct. Geol.* **2**, 149-158.

Brun, J.P. and Pons, J. (1981) Strain patterns of pluton emplacement in a crust undergoing non-coaxial deformation, Sierra Morena, southern Spain, *J. Struct. Geol.* **3**, 219-229.

Brun, J.P., Gapais, D., Cogné, J.P., Ledru, P., and Vigneresse, J.L. (1990) The Flamanville granite (northwest France): an unequivocal example of a syntectonically expanding pluton, *Geol. J.* **25**, 271-286.

Bryon, D.N., Atherton, M.P., and Hunter, R.H. (1994) The description of the primary textures of "Cordilleran" granitic rocks, *Contr. Mineral. Petrol.* **117**, 66-75.

Bulau, J.R., Waff, H.S., and Tyburczy, J.A. (1979) Mechanical and thermodynamical constraints on fluid distribution in partial melts, *J. Geophys. Res.* **84**, 6102-6108.

Burke, K., Freeth, S.J., and Grant, N.K. (1976) The structure and sequence of geological events in the basement complex of the Ibadan area, Western Nigeria, *Precamb. Res.* **3**, 537-545.

Burnham, C.W. (1979) The importance of volatile constituents, in H.S. Yoder (ed.), *The evolution of the igneous rocks*, Princeton University Press, pp. 1077-1084.

Busby-Spera, C.J. and Saleeby, J. (1990) Intra-arc strike-slip fault exposed at batholithic levels in the southern Sierra Nevada, California, *Geology* **18**, 255-259.

Busch, W., Schneider, G., and Mehnert, K.R. (1974) Initial melting at grain boundaries. Part II: melting in rocks of granodioritic, quartz-dioritic and tonalitic compositions, *N. Jahr. Mineral. Monatsh.* **8**, 345-370.

Butler, R.W.H. and Hutton, D.H.W. (1994) Basin structure and Tertiary magmatism in Skye, NW Scotland, *J. Geol. Soc. London* **151**, 931-944.

Caby, R. (1989) Precambrian terranes of Benin-Nigeria and Northeast Brazil and the Late Proterozoic South Atlantic fit, *Geol. Soc. Am. Special Paper* **230**, 145-158.

Candela, P.A. (1986) The evolution of aqueous vapor from silicate melts: effects on oxygen fugacity, *Geochim. Cosmochim. Acta.* **50**, 1205-11.

Capper, P.L. and Cassie, W.F. (1976) *The mechanics of engineering soils*, 6th ed., Wiley, New York, 376 pp.

Carmichael, I.S.E. (1991) The redox state of basic and silicic magmas: a reflection of their source region? *Contr. Mineral. Petrol.* **106**, 129-141.

Carmichael, I.S.E. and Ghiorso, M.S. (1990) The effect of oxygen fugacity on the redox state of natural liquids and their crystallizing phases, in J. Nicholls and J.K. Russell (eds.), *Modern methods of igneous petrology: understanding magmatic processes, Rev. Mineral.* **24**, pp. 191-212.

Carracedo, M. (1991) Contribución el estudio del batolito de Los Pedroches (Córdoba), *Doctorate Thesis*, Basque Country Univ., 443 pp.

Carracedo, M., Larrea, F.J., Aranguren, A., Cuevas, J., and Tubía, J.M. (1994) La falla de Conquista: una zona de cizalla dúctil transtensiva asociada al emplazamiento del batolito de Los Pedroches, *XVI Reunión de Xeoloxía e Minería*, Laxe, 51-54.

Cassard, D., Feybesse, J.L., and Lescuyer, J.L. (1993) Variscan crustal thickening, extension and late overstacking during the Namurian-Westphalian in the western Montagne Noire (France), *Tectonophysics* **222**, 33-53.

Castaing, C., Triboulet, C., Feybesse, J.L., and Chèvremont, P. (1993) Tectonometamorphic evolution of Ghana, Togo and Benin in the light of the Pan-African/Brasiliano orogeny, *Tectonophysics* **218**, 323-342.

Castro, A. (1985) The Central Extremadura batholith, geotectonic implications (European Hercynian belt): an outline, *Tectonophysics* **120**, 57-68.

Castro, A. (1986) Structural pattern and ascent model in the central Extremadura batholith, Hercynian belt, Spain, *J. Struct. Geol.* **8**, 633-645.

Castro, A., Moreno-Ventas, I., and De la Rosa, J.D. (1990) Microgranular enclaves as indicators of hybridization processes in granitoid rocks, Hercynian belt, Spain, *Geol. J.* **25**, 391-404.

Castro, A., Moreno-Ventas, I., and De la Rosa, J.D. (1991) H-type (hybrid) granitoids: a proposed revision of the granite-type classification and nomenclature, *Earth Sci. Rev.* **31**, 237-253.

Castro, A., De la Rosa, J.D., Fernández, C., and Moreno-Ventas, I. (1995) Unstable flow, magma mixing and magma-rock deformation in a deep-seated conduit: the Gil-Márquez complex, southwest Spain, *Geol. Runds.* **84**, 359-374.

Chappell, B.W. and White, A.J.R. (1974) Two contrasting granite types, *Pacific Geol.* **8**, 173-174.

Chauris, L. and Garreau, J. (1983) Le massif de Plouaret, *C. R. Acad. Sci. Paris* **296**, 1591-1594.

Chen, J.H. and Moore, J.G. (1982) Uranium-lead isotopic ages from the Sierra Nevada batholith, California, *J. Geophys. Res.* **87**, B6, 4761-4784.

Christensen, J.N. and De Paolo, D.J. (1993) Time scales of large volume silicic magma systems: Sr isotopic systematics of phenocrysts and glass from the Bishop Tuff, Long Valley, California, *Contr. Mineral. Petrol.* **113**, 110-124.

Clark, S.P., Jr. (1966) Handbook of physical constants, *Geol. Soc. Am. Memoir* **97**, 587 p.

Clemens, J.D. (1984) Origin and evolution of peraluminous silicic ignimbrite suite: the Violet Town Volcanics, *Contrib. Mineral. Petrol.* **88**, 354-371.

Clemens, J.D. (1984) Water contents of silicic to intermediate magmas, *Lithos* **17**, 273-287.

Clemens, J.D. and Wall, V.J. (1981) Origin and crystallization of some peraluminous (S-type) granitic magmas, *Can. Mineral.* **19**, 111-131.

Clemens, J.D., Holloway, J.R., and White, A.J.R. (1986) Origin of an A-type granite: experimental constraints, *Am. Mineral.* **71**, 317-324.

Clemens, J.D. and Vielzeuf, D. (1987) Constraints on melting and magma production in the crust, *Earth Planet. Sci. Lett.* **86**, 287-306.

Clemens, J.D. and Mawer, C.K. (1992) Granitic magma transport by fracture propagation, *Tectonophysics* **204**, 339-360.

Clemens, J.D., Petford, N., and Mawer, C.K. (1997) Ascent mechanisms of granitic magmas: causes and consequences, in M. Holness (ed.), *Deformation-enhanced fluid flow*, Chapman & Hall, London, in press.

Clemm, P.J. and Fisher, J.C. (1955) The influence of grain boundaries on the nucleation of secondary phases, *Acta Metall.* **3**, 70-73.

Cloos, E. (1946) Lineation: a critical review and annotated bibliography, *Geol. Soc. Am. Memoir* **18**, 122 pp.

Cogné, J.P. and Perroud, H. (1988) Anisotropy of magnetic susceptibility as a strain gauge in the Flamanville granite, NW France, *Phys. Earth Planet. Inter* **51**, 264-270.

Collinson, D.W. (1983) Methods in rock magnetism and paleomagnetism: techniques and instrumentation, Chapman and Hall, London.

Cordell, L. and Henderson, R.G. (1968) Iterative three dimensional solution of gravity anomaly using a digital computer, *Geophysics* **33**, 596-601.

Cooper, R.F. and Kohlstedt, D.L. (1982) Interfacial energies in the olivine-basalt system, in S. Akimoto and M.H. Manghnani (eds.), *High pressure research in geophysics (Advances in Earth and Planetary Sciences 12)* Center for Academic Publication, Tokyo, pp. 217-228.

Cooper, R.F. and Kohlstedt, D.L. (1984) Solution-precipitation enhanced diffusional creep of partially molten olivine-basalt aggregates during hot-pressing, *Tectonophysics* **107**, 207-233.

Cornwell, J. (1995) *Nature's imagination: the frontiers of scientific vision*, Oxford University Press, 212 pp.

Corretgé, L.G. (1971) Estudio petrologico del batolito de Cabeza de Araya (Cáceres), *Unpubl. Thesis Univ. Salamanca, Spain*, 2 vol., 453 p.

Corretgé, L.G. (1983) Las rocas graníticas y granitoides del Macizo Ibérico, in J.M. Comba (ed), *Geología de España, Libro Jubilar J.M. Ríos, Tomo I*, IGME, Madrid, pp. 569-593.

Corriveau, L. and Leblanc, D. (1995) Sequential nesting of magmas in marble, southeastern Grenville Province, Quebec: from fracture propagation to diapirism, *Tectonophysics* **246**, 183-200.

Corry, C.E. (1988) Laccoliths: mechanics of emplacement and growth, *Spec. Pap. Geol. Soc. Am.* **220**, 110 pp.

Courrioux, G. (1987) Oblique diapirism: the Criffel granodiorite/granite zoned pluton (southwest Scotland), *J. Struct. Geol.* **3**, 313-330.

Craig, R.F. (1992) *Soil Mechanics*, 5th edition, Chapman and Hall, 427 pp.

Cruden, A.R. (1988) Deformation around a rising diapir modeled by creeping flow past a sphere, *Tectonics* **7**, 1091-1101.

Cruden, A.R. (1990) Flow and fabric development during the diapiric rise of magma, *J. Geol.* **98**, 681-698.

Cruden, A.R. and Launeau, P. (1994) Structure, magnetic fabric and emplacement of the Archean Lebel Stock, SW Abitibi Greenstone belt, *J. Struct. Geol.* **16**, 677-692.

Cuney, M., Friedrich, M., Blumenfeld, P., Bourguignon, A., Boiron, M.C., Vigneresse, J.L., and Poty, B. (1990) Metallogenesis in the French part of the Variscan orogen. Part I: U-preconcentrations in the pre-Variscan formations: a comparison with Sn, W and Au, *Tectonophysics* **177**, 39-57.

Czmanske, G.K. and Wones, D.R. (1973) Oxidation during magmatic differentiation, Finnmarka complex, Oslo area, Norway, *J. Petrol.* **14**, 349-380.

D'Lemos, R.S., Brown, M., and Strachan, R.A. (1992) Granite magma generation, ascent and emplacement in a transpressional orogen. *J. Geol. Soc. London* **149**, 487-490.

Dada, S.S. and Respaut, J.P. (1989) La monzonite à fayalite de Bauchi (bauchite), nouveau témoin d'un magmatisme syntectonique panafricain au Nord du Nigéria, *C. R. Acad. Sci. Paris* **309**, 887-892.

Dada, S.S., Lancelot, J.R., and Briqueu, L. (1989) Age and origin of the annular charnockitic complex at Toro, Northern Nigeria: U-Pb and Rb-Sr evidence, *J. African Earth Sciences* **9**, 227-234.

Dada, S.S., Briqueu, L., Harms, U., Lancelot, J.R., and Matheis, G. (1995) Charnockitic and monzonitic Pan-African series from north-central Nigeria: trace-element and Nd, Sr, Pb isotopes constraints on their petrogenesis, *Chem. Geol.* **124**, 233-252.

Daines, M.J. and Richter, F.M. (1988) An experimental method for directly determining the interconnectivity of melt in a partially molten system, *Geophys. Res. Lett.* **15**, 1459-1462.

Daishi, M., Hayashi, M., and Kato, Y. (1986) Fission track ages of some Cenozoic acidic intrusive rocks from Ryukyu Islands, *J. Mineral. Petrol. Economic Geol.* **81**, 324-332 (in Japanese).

Daishi, M., Hayashi, M., and Kato, Y. (1987) Radiometric ages of some Cenozoic volcanic rocks from Ryukyu Islands, *J. Mineral. Petrol. Economic Geol.* **82**, 370-381 (in Japanese).

Dall'Agnol, R., Scaillet, B., and Pichavant, M. (1994) Phase relations of amphibole-biotite bearing granites. I. Effects of a_{H2O} under oxidizing conditions, *Terra Nova* **6**, 13.

Daly, L. (1959) Sur l'anisotropie magnétique dans les roches déformées et la nature de leur aimantation rémanente naturelle, *C. R. Acad. Sci., Paris* **248**, 2614-2616.

Daniel, J.M. and Jolivet, L. (1995) Detachment faults and pluton emplacement: Elba Island (Tyrrhenian Sea), *Bull. Soc. géol. France* **166**, 4, 341-354.

Darrozes, J., Moisy, M. Olivier, P., Améglio, L., and Bouchez, J.L. (1994) Structure magmatique du granite du Sidobre (Tarn, France): de l'échelle du massif à celle de l'échantillon, *C. R. Acad. Sci. Paris* **318**, 243-250.

Davidge, R.W. (1979) *Mechanical behaviour of ceramics*, Cambridge University Press, Cambridge, 165 p.

Davidson, C., Hollister, L.S., and Schmid, S.M. (1992) Role of melt in the formation of a deep-crustal compressive shear zone: the MacLaren Glacier metamorphic belt, South Central Alaska, *Tectonics* **11**, 348-359.

Davidson, C., Schmid, S.M., and Hollister, L.S. (1994) Role of melt during deformation in the deep crust, *Terra Nova* **6**, 133-142.

Davies, G.H. (1987) A shear-zone model for the structural evolution of metamorphic core complexes in southeastern Arizona, in: M.P. Coward, J.F. Dewey, and P.L. Hancock, *Continental extensional tectonics, Geol. Soc. Sp. Publ.*, London, 247-266.

Davis, B.K. (1993) Mechanism of emplacement of the Cannibal Creek Granite with special reference to timing and deformation history of the aureole, *Tectonophysics* **224**, 337-362.

De la Rosa, J.D. (1992) Petrología de las rocas básicas y granitoides del batolito de la Sierra Norte de Sevilla, Zona Surportuguesa, Macizo Ibérico, *Unpub. Ph. D. Thesis*, Univ. Sevilla, 312 pp.

Debat, P., Sirieys, P., Déramond, J., and Soula, J.C. (1975) Paléodéformation d'un massif orthogneissique (massif des Cammazes, Montagne Noire Occidentale, France), *Tectonophysics* **28**, 159-183.

Défalque, G., Demaiffe, D., Dumont, P., and Lalieux, P. (1992) Le batholite de "Los Pedroches" (Sierra Morena): études cartographique, pétrographique, géochimique, géochronologique et métallogénique, *Ann. Soc. géol. Belgique* **115**, 77-89.

Del Moro, A., Pardini, G.C., Quericioli, C., Villa, I.M., and Callegari, E. (1983) Rb/Sr and K/Ar chronology of Adamello granitoids, Southern Alps, *Memorie della Società Geologica Italiana* **26**, 285-299.

Déléris, J., Nédélec, A., Ferré, E., Gleizes, G., Ménot, R.P., Obasi, C.K., and Bouchez, J.L. (1996) The Pan-African Toro Complex (Northern Nigeria): magmatic interactions and structures in a bimodal intrusion, *Geol. Mag.* **133**, 18 pp.

Dell'Angelo, L.N., Tullis, J., and Yund, R.A. (1987) Transition from dislocation creep to melt-enhanced diffusion creep in fine-grained granitic aggregates, *Tectonophysics* **139**, 325-332.

Dell'Angelo, L.N. and Tullis, J. (1988) Experimental deformation of partially melted granitic aggregates, *J. Metam. Geol.* **6**, 495-515.

Dickin, A.P., Halliday, A.N., and Bowden, P. (1991) A Pb, Sr and Nd isotope study of the basement and Mesozoic ring complexes of the Jos Plateau, Nigeria, *Chem. Geol. (Isot. Geosci. Section)* **94**, 23-32.

Díez Balda, M.A., Vegas, R., and González-Lodeiro, F. (1990) Central-Iberian Zone: autochtonous sequences, structure, in R.D. Dallmeyer and E. Martínez García (eds.), *Pre-Mesozoic Geology of Iberia*, Springer-Verlag, Berlin, pp. 172-188.

Dingwell, D.B., Scarfe, C.M., and Cronin, D.J. (1985) The effect of fluorine on viscosities in the system Na_2O-Al_2O_3-SiO_2: implications for phonolites, trachytes and rhyolites, *Am. Miner.* **70**, 80-87.

Dingwell, D.B., Knoche, R., and Webb, S.L. (1992) The effect of B_2O_3 on the viscosity of haplogranitic liquids, *Am. Miner.* **77**, 457-461.

Dingwell, D.B., Bagdassarov, N.S., Bussod, N.S., and Webb, S.L. (1993) Magma rheology, in R.W. Luth (ed.), *Experiments at high pressure and applications to the Earth's mantle*, Mineralogical Association of Canada, Nepean, Ontario, pp. 131-196.

Dixon, J.M. (1975) Finite strain and progressive deformation in models of diapiric structures, *Tectonophysics* **28**, 89-124.

Dixon, J.M. and Summers, J.M. (1985) Recent developments in centrifuge modelling of tectonic processes: equipment, model construction, techniques and rheology of model materials, *J. Struct. Geol.* **7**, 83-102.

Djouadi, M.T. (1994) Granites fini-panafricains de type Taourirt (Hoggar, Algérie): une étude structurale de la susceptibilité magnétique et modèles de mise en place du complexe du Tesnou et du massif de Tioueïne, *Unpubl. Thesis*, Univ. Paul-Sabatier, Toulouse, France, 161 pp.

Djouadi, M.T. and Bouchez, J.L. (1992) Structure étrange du granite du Tesnou (Hoggar, Algérie), *C. R. Acad. Sci. Paris* **315**, 1231-1238.

Doblas, M. (1991) Tardi-Hercynian extensional and transcurrent tectonics in Central Iberia, *Tectonophysics* **191**, 325-334.

Donaire, T. and Pascual, E. (1991) Paragénesis metamórficas de contacto en xenolitos de la granodiorita de Los Pedroches (Córdoba, España): evidencia de emplazamiento a alta temperatura, *Geogaceta* **10**, 90-93.

Donath, F.A. (1964) Strength variation and deformational behavior in anisotropic rock, in W.R. Judd (ed.), *State of stress in the Earth's crust*, New York, Elsevier, pp. 281-297.

Druguet, E. and Hutton, D.H.W. (1995) Granitoid emplacement during non-coaxial deformation: an example from a syntectonic migmatite complex, Cap de Creus (Eastern Pyrenees), *Terra Abstracts, Terra Nova* **7**, 141.

Ebadi, A. and Johannes, W. (1991) Beginning of melting and composition of first melts in the system Qz-Ab-Or-H_2O-CO_2, *Contr. Mineral. Petrol.* **106**, 286-295.

Edel J.B. (1982) Le socle varisque de l'Europe moyenne: apports du magnétisme et de la gravimétrie, *Sci. Géol. Bull.* **35**, 207-224.

Egbuniwe, I.G., Fitches, W.R., Bentley, M., and Snelling, N.J. (1985) Late Pan-African syenite-granite plutons in NW Nigeria, *J. African Earth Sciences* **34**, 427-435.

Elasser, W.M. (1958) *The physical foundation of biology,* Pergamon Press, 219 pp.
Ellwood, B.B. and Whitney, J.A. (1980) Magnetic fabric of the Elberton granite, Northeast Georgia, *J. Geophys. Res.* **85**, 1481-1486.
Ellwood, B.B., Hrouda, F., and Wagner, J.J. (1988) Symposia on magnetic fabrics: introductory comments, *Phys. Earth Plan. Inter.* **51**, 249-252.
Elsasser, W.M. (1971) Sea-floor spreading as thermal convection, *J. Geophys. Res.* **76**, 1101-1112.
Emslie R.F. and Stirling J.A.R. (1993) Rapakivi and related granitoids of the Nain plutonic suite: geochemistry, mineral assemblages and fluid equilibria, *Can. Mineral.* **31**, 821-847.
Engebretson, D.C., Cox, A., and Gordon, R.G. (1985) Relative motions between oceanic and continental plates in the Pacific Basin, *Geol. Soc. Amer. Special Pap.*, 206 pp.
England, R.W. (1992) The genesis, ascent, and emplacement of the Northern Arran Granite, Scotland: implications for granitic diapirism, *Geol. Soc. Amer. Bull.* **104**, 606-614.
England, P.C. and Thompson, A. (1986) Some thermal and tectonic models for crustal melting in continental collision zones, in M.P. Coward and A.C. Ries (eds.), *Collision Tectonics, Geol. Soc. Spec. Publ., London,* 83-94.
Erdmer, P and Mortensen, J.K. (1993) A 1200 km long Eocene metamorphic-plutonic belt in the northwestern Cordillera: evidence from southwest Yukon, *Geology* **21**, 1039-1042.
Escuder Viruete, J., Arenas, R., and Martínez Catalán, J.R. (1994) Tectonothermal evolution associated with Variscan crustal extension in the Tormes Gneiss Dome (NW Salamanca, Iberian Massif, Spain), *Tectonophysics* **238**, 117-138.
Etheridge, M.A. (1983) Differential stress magnitudes during regional deformation and metamorphism upper bound imposed by tensile fracturing, *Geology* **11**, 231-234.
Evernden, J.F. and Kistler, R.W. (1970) Chronology of emplacement of Mesozoic batholithic complexes in California and western Nevada, *U.S. Geol. Survey Prof. Paper,* 623.
Fairhead, J.D. and Okereke, C.S. (1987) A regional gravity study of the West African rift system in Nigeria and Cameroon and its tectonic interpretation, *Tectonophysics* **143**, 141-159.
Fara, H.D. and Scheidegger, A.E. (1963) An eigenvalue method for the statistical evaluation of fault plane solution of earthquakes, *Bull. Seism. Soc. Am.* **53**, 811-816.
Faure, M. (1995) Late orogenic Carboniferous extensions in the Variscan French Massif Central, *Tectonics* **14**, 132-153.
Faure, M. and Pons, J. (1991) Crustal thinning recorded by the shape of the Namurian-Westphalian leucogranite in the Variscan belt of the Northwest Massif Central, France, *Geology* **19**, 730-733.
Ferguson, C.C. (1979) Rotation of elongate rigid particles in slow non-Newtonian flows, *Tectonophysics* **60**, 247-262.
Ferguson, J. (1980) Application of information theory in geological data processing, *J. Geol. Soc. Lond.* ***137***, 107-108.
Ferguson, J. (1988) *Mathematics in Geology,* Allen and Unwin, UK, 299 pp.
Fernandez, A. (1981) Une généralisation du modèle de March applicable à l'analyse des orientations préférentielles de forme issues de la déformation coaxiale dans les roches éruptives, *C. R. Acad. Sci. Paris* **293**, sér. II, 1091-1094.
Fernandez, A. (1984) Etude théorique et expérimentale du développement de la fabrique dans les roches magmatiques: application à l'étude structurale des granitoïdes. *Unpubl. Thesis Univ. Clermont-Ferrand.*
Fernandez, A. (1987) Preferred orientation developed by rigid markers in two-dimensional simple shear strain: a theoretical and experimental study, *Tectonophysics* **136**, 151-158.
Fernandez, A. (1988) Strain analysis from shape preferred orientation in magmatic rocks, in C. Talbot (ed.) *Geological kinematics and dynamics, Bull. Geol. Inst. Univ. Uppsala* **14**, 61-67.
Fernandez, A., Febesse, J.L., and Mezure, J.F. (1983) Theoretical and experimental study of fabrics developed by different shaped markers in two-dimensional simple shear, *Bull. Soc. géol. France* **25**, 319-326.
Fernandez, A. and Barbarin, B. (1991) Relative rheology of coeval mafic and felsic magmas: nature of resulting interaction processes and shape and mineral fabrics of mafic microgranular enclaves, in J. Didier and B. Barbarin (eds), *Enclaves and granite petrology,* Elsevier, Amsterdam, pp. 243-275.
Fernandez A. and Laporte, D. (1991) Significance of low symmetry fabrics in magmatic rocks, *J. Struct. Geol.* **13**, 337-347.
Fernandez, A. and Gasquet, D.R. (1994) Relative rheological evolution of chemically contrasted coeval magmas: example of the Tichka plutonic complex (Morocco), *Contr. Mineral. Petrol.* **116**, 316-326.
Fernandez, A., Machado da Silva, G., and Fernández-Catuxo, J. (1995) Modélisation analogique du développement de l'orientation préférentielle 3D dans les roches magmatiques, *C. R. Acad. Sci. Paris* **320**, 1051-1054.
Fernandez, A. and Fernandez-Catuxo, J. (1997) 3D biotite shape fabric experiments under simple shear strain, in J.L. Bouchez, D.H.W. Hutton and W.E. Stephens (eds.), *Granite: from segregation of melt to emplacement fabrics,* Kluwer Academic Publishers, Dordrecht, pp. 145-158.
Fernández, F.J., Cueto, L.A., Larrea, F.J., and Quesada, C. (1990) El plutón de El Guijo: petrología, geoquímica, edad y relación con otras rocas del batolito de Los Pedroches, *Cuad. Lab. Xeol. Laxe* **15**, 89-103.
Fernández-Catuxo, J. (1994) Geología granítica del Macizo del Confurco (Galicia, España), *Unpubl. Thesis Univ. Oviedo.*

Ferré, E., Gleizes, G., Bouchez, J.L., and Nnabo, P.N. (1995) Internal fabric and strike-slip emplacement of the Panafrican granite of Solli Hills, Northern Nigeria, *Tectonics* **14/5**, 1205-1219.
Ferré, E., Déléris, J., Bouchez, J.L., Lar, A.U., and Peucat, J.J. (1996) Pan-African reactivation of contrasted Eburnean and Archean provinces in Nigeria: structural and isotopic data. *J. Geol. Soc. London* **153**, pp. 719-728.
Ferré, E., Gleizes, G., Djouadi, M.T., Bouchez, J.L., and Ugodulunwa, F.X.O. (1997) Drainage and emplacement of magmas along an inclined transcurrent shear zone: petrophysical evidence from a granite-charnockite pluton (Rahama, Nigeria), in J.L. Bouchez, D.H.W. Hutton and W.E. Stephens (eds.), *Granite: from segregation of melt to emplacement fabrics*, Kluwer Academic Publishers, Dordrecht, pp. 253-274.
Fink, J.H. (1983) Structure and emplacement of a rhyolitic obsidian flow: Little Glass Mountain, Medicine Lake Highland, northern California, *Geol. Soc. Amer. Bull* **94**, 362-380.
Fitches, W.R., Ajibade, A.C., Egbuniwe, I.G., Holt, R.W., and Wright, J.B. (1985) Late Proterozoic schist belts and plutonism in Northwest Nigeria, *J. Geol. Soc. London* **142**, 319-337.
Fleitout, L. and Froidevaux, C. (1980) Thermal and mechanical evolution of shear zones, *J. Struct. Geol.* **2**, 159-164.
Flinn, D. (1962) On folding during three-dimensional progressive deformation, *Geol. Soc. London Quaterly Journal* **118**, 385-428.
Fowler, T.J. (1994) Sheeted and bulbous pluton intrusion mechanisms of a small granitoid from southeastern Australia: implications for dyke-to-pluton transformation during emplacement, *Tectonophysics* **234**, 197-215.
Freeman, B. (1985) The motion of rigid ellipsoidal particles in slow flow, *Tectonophysics* **113**, 163-183.
Fritz, H. (1995) Emplacement of granites and exhumation of basement domes during oblique collision: eastern desert (Egypt), *Terra Abstracts, Terra Nova* **7**, 137.
Frost, T.P. and Mahood, G.A. (1987) Field, chemical, and physical constraints on mafic-felsic magma interaction in the Lamarck granodiorite, Sierra Nevada, California, *Geol. Soc. Amer. Bull.* **99**, 272-291.
Gapais, D. (1989) Shear structures within deformed granites: mechanical and thermal indicators, *Geology* **17**, 1144-1147.
Gapais, D. and Barbarin, B. (1986) Quartz fabric transition in a cooling syntectonic granite (Hermitage massif, France), *Tectonophysics* **125**, 357-370.
Gasquet, D. (1992) Mise en évidence d'intrusions emboitées dans le Massif du Tichka (Haut Atlas occidental, Maroc), *C. R. Acad. Sci. Paris* **314(II)**, 931-936.
Gay, N.C. (1966) Orientation of mineral lineation along the flow direction in rocks: a discussion, *Tectonophysics* **3**, 559-564.
Gay, N.C. (1968) Pure shear and simple shear deformation of inhomogeneous viscous fluids. 1. Theory, *Tectonophysics* **5**, 211-234.
Ghosh, S.K. and Ramberg, H. (1976) Reorientation of inclusions by combinaison of pure shear and simple shear, *Tectonophysics* **34**, 1-70.
Glazner, A.F. (1991) Plutonism, oblique subduction and continental growth: an example from the Mesozoic of California, *Geology* **19**, 784-786.
Glazner, A.F. (1994) Foundering of mafic plutons and density stratification of continental crust, *Geology* **22**, 435-438.
Gleizes, G. (1992) Structures des granites hercyniens des Pyrénées de Mont-Louis-Andorre à la Maladeta, *Unpubl. Thesis, Univ. Toulouse*, France, 259 p.
Gleizes, G., Leblanc, D., and Bouchez, J.L. (1991) Le pluton granitique de Bassiès (Pyrénées ariégeoises): zonation, structure et mise en place, *C. R. Acad. Sci. Paris* **312(II)**, 755-762.
Gleizes, G., Nédélec, A., Bouchez, J.L., Autran, A., and Rochette, P. (1993) Magnetic susceptibility of the Mont-Louis Andorra ilmenite-type granite (Pyrenees): a new tool for the petrographic characterization and regional mapping of zoned granite plutons, *J. Geophys. Res.* **98**, B3, 4317-4331.
Godin, P. (1994) Deformation within the Cannibal Creek Pluton and its aureole, Queensland, Australia: a re-evaluation of ballooning as an emplacement mechanism, *J. Struct. Geol.* **16**, 693-707.
Graham, J.W. (1954) Magnetic susceptibility anisotropy: an unexploited petrofabric element, *Geol. Soc. Am. Abstr. Program* **65**, 1257-1258.
Gratier, J.P. (1987) Pressure solution-deposition creep and associated tectonic differentiation in sedimentary rocks, in M.E. Jones and R.M.F. Preston (eds.), *Deformation of sediments and sedimentary rocks*, *Geol. Soc. London Spec. Publ.* **29**, 25-38.
Green, T.H. (1982) Anatexis of mafic crust and high pressure crystallization of andesite, in R.S. Thorpe (ed.), *Andesites: orogenic andesites and related rocks*, John Wiley & Sons, Chischerter, pp. 465-487.
Greene, D.C. and Schweickert, R.A. (1995) The Gem Lake shear zone: Cretaceous dextral transpression in the northern Ritter Range pendant, eastern Sierra Nevada, California, *Tectonics* **14**, 4, 945-961.
Grégoire, V., de Saint-Blanquat, M., Nédélec, A., and Bouchez, J.L. (1995) Shape anisotropy versus magnetic interactions of magnetite grains: experiments and application to AMS in granitic rocks, *Geophys. Res. Letters* **22**, 2765-2768.
Grocott, J., Brown, M., Dallmeyer, R.D., Taylor, G.K., and Treloar, R.J. (1994) Mechanisms of continental growth in extensional arcs: an example from the Andean plate-boundary zone, *Geology* **22**, 391-394.
Grocott, J. and Taylor, G.K. (1995) Structural and magmatic evolution of the Andean convergent plate boundary zone in Triassic to early Cretaceous time, northern Chile, *Terra Abstract, Terra Nova* **7**, 138.

Guillet, P. (1983) Les granites de Guérande et de Plouaret (Massif Armoricain): aspects structuraux et cinématique de leur mise en place, *Unpubl. Thesis, Univ. Nantes*, France, 162 p.
Guillet, P., Bouchez, J.L., and Wagner, J.J. (1983) Anisotropy of magnetic susceptibility and magnetic structures in the Guérande granite massif (France), *Tectonics* **2(5)**, 419-429.
Guillet, P., Bouchez, J.L., and Vigneresse, J.L. (1985) Le complexe granitique de Plouaret: mise en évidence structurale et gravimétrique de diapirs emboités, *Bull. Soc. géol. France* **8**, 503-513.
Guineberteau, B., Bouchez, J.L., and Vigneresse J.L. (1987) The Mortagne granite pluton (France) emplaced by pull-apart along a shear zone: structural and gravimetric arguments and regional implication, *Geol. Soc. Am. Bull.* **99**, 763-770.
Guineberteau, B., Cuney, M., and Carré, J.L. (1989) Structure magmatique et plastique des granites de la Marche Occidentale: un couloir transformant hercynien dans le Nord-Ouest du Massif Central français, *C. R. Acad. Sci. Paris* **309**, 2, 1685-1702.
Hacker, R.H. (1990) Amphibolite-facies-to-granulite-facies reactions in experimentally deformed, unpowdered amphibolite, *Amer. Mineral.* **75**, 1349-1361.
Haken, H. (1983) *Synergetics: an introduction*, Springer, 371 pp.
Hammer, S. (1939) Terrain corrections for gravimeter stations, *Geophysics* **4**, 184 p.
Hand, M. and Dirks, P.H.G.M. (1992) The influence of deformation on the formation of axial-planar leucosomes and the segregation of small melt bodies within the migmatitic Napperby gneiss, central Australia, *J. Struct. Geol.* **14**, 591-604.
Hansmann, W. (1986) U-Pb datierungen am Zirkonen und Thoriten aus einer magmatischen Gesteinssuite am beispiel des südlichen Adamello, Eidgenössischen Technischen Hochschule *Ph.D. thesis*, Zürich, Switzerland, 212 pp.
Hansmann, W. and Oberli, F. (1991) Zircon inheritance in an igneous rock suite from the southern Adamello batholith (Italian Alps), *Contr. Mineral. Petrol.* **107**, 501-518.
Hanson, R.B. and Glazner, A.F. (1995) Thermal requirements for extensional emplacement of granitoids, *Geology* **23**, 3, 213-216.
Hargraves, R.B., Johnson, D., and Chan, C.W. (1991) Distribution anisotropy: the cause of AMS in igneous rocks? *Geophys. Res. Letters* **18**, 2193-2196.
Harris, C. and Bell, J.D. (1982) Natural partial melting of syenite blocks from Ascension Island, *Contr. Mineral. Petrol.* **79**, 107-113.
Harvey, P.K. and Laxton, R.R. (1980) The estimation of finite strain from orientation distribution of passively deformed linear markers: eigenvalue relationship, *Tectonophysics* **70**, 285-307.
Hashimoto, I. (1956) Geological succession and structure of the undated strata in Yakushima, and some facts concerning the Kumage group in the western part of Tane-ga-shima, Kagoshima prefecture, *Rep. Liberal Arts, Kyushu Univ.* **2**, 23-34.
Heller, F. (1973) Magnetic anisotropy of granitic rocks of the Bergell massif (Switzerland), *Earth Planet. Sci. Lett.* **20**, 180-188.
Hernández-Enrile, J.L. (1991) Extensional tectonics of the Toledo ductile-brittle shear zone, Central Iberian Massif, *Tectonophysics* **191**, 311-324.
Hecht, L. Vigneresse, J.L., Spiegel, W., Cuney, M., and Morteani, G. (1994) Shape at depth and magma evolution in the granitic complex of the Fichtelgebirge (Bavaria, Germany), implication on the zoning of granitic plutons, *EOS Trans-Am. Geophys. Union* **75**, 365 p.
Hibbard, M.J. (1987) Deformation of incompletely crystallised magma systems, granitic gneisses and their tectonic implications, *J. Geol.* **95**, 543-561.
Hilde, T.W.C. (1983) Sediment subduction versus accretion around the Pacific, *Tectonophysics* **99**, 381-397.
Hinch, E.J. and Leal, L.G. (1979) Rotation of small non-axisymmetric particles in a simple shear flow, *J. Fluid. Mech.* **92**, 591-608.
Hippertt, J.F. (1994) Structures indicative of helicoidal flow in a migmatitic diapir (Bação Complex, southeastern Brazil), *Tectonophysics* **234**, 169-196.
Hirt, W.H. (1989) The petrological and mineralogical zonation of the Mount Whitney Intrusive Suite, eastern Sierra Nevada, California, *PhD thesis*, U.C.-Santa Barbara.
Hirth, G. and Tullis, J. (1992) Dislocation creep regimes in quartz aggregates, *J. Struct. Geol.* **14**, 145-159.
Hirth, G. and Kohlstedt, D.L. (1995) Experimental constraints on the dynamics of the partially molten mantle: deformation in the diffusion creep regime, *J. Geophys. Res.* **100**, 1981-2001.
Holder, M.T. (1979) An emplacement mechanism for post-kinematic granites and its implications for their geochemical features, in M.P. Atherton and J. Tarney (eds.), *Origin of granite batholiths: geochemical evidence*, Shiva, Kent, pp. 116-128.
Holland, T. and Blundy, J.D. (1994) Non-ideal interactions in calcic amphiboles and their bearing on amphibole-plagioclase thermometry, *Contr. Mineral. Petrol.* **116**, 433-447.
Holliger, K. and Levander, A. (1994) Lower crustal reflectivity modeled by rheological controls on mafic intrusion, *Geology* **22**, 367-370.
Hollister, L.S. and Crawford, M.L. (1986) Melt-enhanced deformation: a major tectonic process, *Geology* **14**, 558-561.
Holloway, J.R. (1976) Fluids in the evolution of granitic magmas: consequences of finite CO_2 solubility, *Geol. Soc. Amer. Bull.* **87**, 1513-1518.
Holloway, J.R. (1987) Igneous fluids, in I.S.E. Carmichael and H.P. Eugster (eds.) *Thermodynamic modeling of geological materials: minerals, fluids and melts, Rev. Mineral.* **17**, pp. 211-233.

Holness, M.B. (1995) The effect of feldspar on quartz-H_2O-CO_2 dihedral angles at 4 kb, with consequences for the behaviour of aqueous fluids in migmatites, *Contr. Mineral. Petrol.* **118**, 356-364.

Holtz, F. and Johannes, W. (1994) Maximum and minimum water contents of granitic melts: implications for chemical and physical properties of ascending magmas, *Lithos* **32**, 149-159.

Honda, S. (1985) Thermal structure beneath Tohoku, Northeast Japan: a case study for understanding the detailed thermal structure of the subduction zone, *Tectonophysics* **112**, 69-102.

Housen, B.A. and van der Pluijm, B.A. (1990) Chlorite control of correlations between strain and anisotropy of magnetic susceptibility, *Phys. Earth Plan. Inter.* **61**, 315-323.

Housen, B.A., Richter, C., and van der Pluijm, B.A. (1993) Composite magnetic anisotropy fabrics: experiments, numerical models, and implications for the quantification of rock fabrics, *Tectonophysics* **220**, 1-12.

Housen, B.A., van der Pluijm, B.A., and Essene, E.J. (1995) Plastic behavior of magnetite and high strains obtained from magnetic fabrics in the Parry Sound shear zone, Ontario Grenville Province, *J. Struct. Geol.* **17**, 265-278.

Howard, K.A., McCaffrey, K.J.W, Wooden, J.L., Foster, D.A., and Shaw, S.E. (1995) Jurassic thrusting of Precambrian basement over Paleozoic cover in the Clipper Mountains, southeastern California, in D.M. Miller and L. Busby, *Jurassic magmatism and tectonics in the North American Cordillera: Boulder, Colorado, Geol. Soc. Am. Special Paper* **229**.

Hrouda, F. (1982) Magnetic anisotropy of rocks and its application in geology and geophysics, *Geophys. Surv.* **5**, 37-82.

Hrouda, F. (1986) The effect of quartz on the magnetic anisotropy of quartzite, *Studia Geophys. Geod.* **33**, 39-45.

Hrouda, F. (1987) Mathematical model relationship between the paramagnetic anisotropy and strain in slates, *Tectonophysics* **142**, 323-327.

Hrouda, F. (1993) Theoretical models of magnetic anisotropy to strain relationship revisited, *Phys. Earth Plan. Inter.* **77**, 237-249.

Hrouda, F. (1994) Mathematical modelling of the behaviour of passive fabric elements and corresponding AMS, in the transpression zone, in H.J. Bunge, S. Siegesmund, W. Skrotzki, and K. Weber (eds.), *Textures of geological materials,* DGM Verlag, Oberursel, pp. 381-392.

Hunter, R.H. (1987) Textural equilibrium in layered igneous rocks, in I. Parsons (ed.), *Origins of igneous layering*, NATO ASI Series, Reidel, pp. 473-503.

Huppert, H.E. and Sparks, R.S.J. (1988) The generation of granitic magmas by intrusion of basalt into continental crust, *J. Petrol.* **29**, 599-624.

Hutton, D.H.W. (1982) A tectonic model for the emplacement of the main Donegal granite, NW Ireland, *J. Geol. Soc. London* **139**, 615-631.

Hutton, D.W.H. (1988a) Igneous emplacement in a shear-zone termination: the biotite granite at Strontian, Scotland, *Geol. Soc. Amer. Bull.* **100**, 1392-1399.

Hutton, D.H.W. (1988b) Granite emplacement mechanism and tectonic controls: inferences from deformation studies, *Trans. Royal Soc. Edinburgh: Earth Sci.* **79**, 245-255.

Hutton, D.H.W. (1992) Granite sheeted complex: evidence for the dyking ascent mechanism, *Trans. Royal Soc. Edinburgh: Earth Sci.* **83**, 377-382.

Hutton, D.H.W., Dempster, T.J., Brown, P.E., and Becker, S.D. (1990) A new mechanism of granite emplacement: intrusion in active extensional shear zones, *Nature* **343**, 452-455.

Hutton, D.H.W. and Ingram, G.M. (1992) The Great Tonalite Sill of Southeastem Alaska and British Columbia: emplacement into an active contractional high angle reverse shear zone, *Trans. Royal Soc. Edinburgh: Earth Sci.* **83**, 383-386.

Hutton, D.H.W. and Reavy, R. J. (1992) Strike-slip tectonics and granite petrogenesis, *Tectonics* **11**, 960-967.

Hutton, D.H.W. and Miller, R.B. (1994) A major dip-slip shear zone associated with emplacement of the Tuolumne Intrusive Suite, central Sierra Nevada batholith, *Geol. Soc. Am. Abstracts with program* **26**, A133.

Hutton, D.H.W and Alsop, G.I. (1996) The Caledonian strike swing and associated lineaments in NW Ireland and adjacent areas: sedimentation, deformation and igneous intrusion patterns, *J. Geol. Soc. London* **153**, 345-360.

Hutton, J. (1795) Theory of the Earth with proofs and illustrations. Facsimile reprint (1993), J. Cramer and H.K. Swann (eds.), Wheldon and Wesly, Codicote (Herts).

Ildefonse, B. and Fernandez, A. (1988) Influence of the concentration of rigid markers in a viscous medium on the production of preferred orientations: an experimental contribution, *Bull. Geol. Univ. Uppsala,* GKD **14**, 55-60.

Ildefonse, B., Launeau, P., Fernandez, A., and Bouchez, J.L. (1992a) Effect of mechanical interactions on development of shape preferred orientations: a two-dimensional experimental approach, *J. Struct. Geol.* **14**, 73-83.

Ildefonse, B., Sokoutis, D., and Mancktelow, N.S. (1992b) Mechanical interactions between rigid particles in a deforming ductile matrix: analogue experiments in simple shear flow, *J. Struct. Geol.* **14**, 1253-1266.

Ildefonse, B. and Mancktelow, N.S. (1993) Deformation around rigid particles: the influence of slip at the particle/matrix interface, *Tectonophysics* **221**, 345-359.

Ildefonse, B., Arbaret, L., and Diot, H. (1997) Rigid particles in simple shear flow: is their preferred orientation periodic or steady state? in J.L. Bouchez, D.H.W. Hutton and W.E. Stephens (eds.), *Granite: from segregation of melt to emplacement fabrics*, Kluwer Academic Publishers, Dordrecht, pp. 177-185.

Ingram, G.M. and Hutton, D.H.W. (1994) The Great Tonalite Sill: emplacement into a contractional shear zone and implications for Late Cretaceous to early Eocene tectonics in southeastern Alaska and British Columbia, *Geol. Soc. Am. Bull.* **106**, 715-728.

Instituto Geográfico Nacional (1976) Mapa gravimétrico de la península Ibérica e Islas Baleares (anomalías de Bouger), 1:500.000, Madrid.

Ishihara, S. (1977) The magnetite-series and ilmenite series granitic rocks, *Mining Geology* **27**, 293-305.

Ishihara, S. (1981) The granitoid series and mineralization, *Econ. Geology*, 75th An. Volume, 458-484.

Ishikawa, H. (1964) The geochemical study of the contact metamorphism associated with the intrusion of the Yakushima Granite, *Bull. College of Education, Kagoshima Univ.* **16**, 10-20.

Iwasaki, T., Hirata, N., Kanazawa, T., Melles, J., Suyehiro, K., Urabe, T., Möller, L., Makris, J., and Shimamura, H. (1990) Crustal and upper mantle structure in the Ryukyu Island Arc deduced from deep seismic sounding, *Geophys. J. Int.* **102**, 631-651.

Jackson, M. (1991) Anisotropy of magnetic remanence: a brief review of mineralogical sources, physical origins and geological applications, *Pure Appl. Geophys*. **136**, 1-28.

Jacoby, V. (1976) Paraffin model experiment of plate tectonics, *Tectonophysics* **35**, 103-113.

Jacoby, V. and Schmeling, H. (1981) Convection experiments and the driving mechanism, *Geol. Rundschau* **70**, 207-230.

Jacques, J.M. and Reavy, R.J. (1994) Caledonian plutonism and major lineaments in the SW Scottish Highlands, *J. Geol. Soc. London* **151**, 955-970.

Jacques, J.M., Hutton, D.H.W., and Reavy, R.J. (1997) Crustal depth control of shear zones and deep lineaments in the Caledonian plutons of the SW Highlands of Scotland, *Geol. Mag.*, in press.

Jarrard, R.D. (1986) Relations among subduction parameters, *Rev. Geophys.* **24**, 217-284.

Jeffery, G.B. (1922) The motion of ellipsoidal particles immersed in a viscous fluid, *Proc. R. Soc. Lond.* **102**, 201-211.

Jelinek, V. (1977) The statistical theory of measuring anisotropy of magnetic susceptibility of rocks and its application, *Geofyzika* n.p. ed., Brno, Czech Republic, 88 pp.

Jelinek, V. (1981) Characterization of the magnetic fabrics of rocks, *Tectonophysics* **79**, 63-67.

Jezek, J. (1994) Software for modelling the motion of rigid triaxial ellipsoidal particles in viscous flow, *Computers & Geosciences* **20**, 409-424.

Jezek, J., Melka, R., Schulmann, K., and Venera, Z. (1994) The behaviour of rigid triaxial ellipsoidal particles in viscous flows: modelling of fabric evolution in a multiparticle system, *Tectonophysics* **229**, 165-180.

Jezek, J., Schulmann, K., and Segeth, K. (1996) Fabric evolution of rigid particles during mixed coaxial and simple shear flows, *Tectonophysics* **257**, 203-220.

John, B.E. and Blundy, J.D. (1993) Thermal and mechanical aspects of granitoid emplacement in the southern Adamello Massif, Italy, *Geol. Soc. Amer. Bull.* **105**, 1517-1541.

John, B.E. and Stünitz, H. (1997) Magmatic fracturing and small-scale melt segregation during pluton emplacement: evidence from the Adamello massif (Italy), in J.L. Bouchez, D.H.W. Hutton and W.E. Stephens (eds.), *Granite: from segregation of melt to emplacement fabrics*, Kluwer Academic Publishers, Dordrecht, pp. 55-74.

Johnson, M.C. and Rutherford, M.R. (1989) Experimentally determined conditions in the Fish Canyon Tuff, Colorado, magma chamber, *J. Petrol.* **30**, 711-737.

Johnson, M.C., Anderson, A.T., and Rutherford, M.R. (1994) Pre-eruptive volatile contents of magmas, in M. Carroll and J.R., Holloway (eds.), *Volatiles in Magmas, Rev. Mineral.* **30**, pp. 281-330.

Jolivet, R., Huchon, P., and Pangin, C. (1989) Tectonic setting of Western Pacific marginal basins, *Tectonophysics* **160**, 23-47.

Jover, O. (1986) Les massifs granitiques de Guéret et du Nord-Millevaches (Massif Central français): analyse structurale et modèle de mise en place, *Unpubl. Thesis, Univ. Nantes*, France, 233 p.

Jover, O. and Bouchez, J.L. (1986) Mise en place syntectonique des granitoïdes de l'Ouest du Massif Central français, *C. R. Acad. Sci. Paris* **303**, 969-974.

Jover, O., Rochette, P., Lorand, J.P., Maeder, M., and Bouchez, J.L. (1989) Magnetic mineralogy of some granites from the French Massif Central: origin of their low-field susceptibility, *Phys. Earth Planet. Inter.* **55**, 79-92.

Julivert, M., Fontboté, J.M., Ribeiro, A., and Conde, A. (1972) Mapa Tectónico de la Península Ibérica y Baleares, 1:1000000, Inst. geol. Min. España, Madrid.

Jurewicz, S.R. and Watson, E.B. (1984) Distribution of partial melt in a felsic system: the importance of surface energy, *Contr. Mineral. Petrol.* **85**, 25-29.

Jurewicz, S.R. and Watson, E.B. (1985) The distribution of partial melt in a granitic system: the application of liquid phase sintering theory, *Geochim. Cosmochim. Acta* **49**, 1109-1121.

Jurewicz, S.R. and Jurewicz, A.J.G. (1986) Distribution of apparent angles on random sections with emphasis on wetting angle measurements, *J. Geophys. Res.* **91**, 9277-9282.

Jurewicz, S.R. and Jones, J.H. (1995) Preliminary results of olivine-metal wetting experiments and the direct measurement of metal phase interconnectivity, *Lunar Planet. Sci.* **XXVI**, 709-710.

Kaczor, S.M., Hanson, G.N., and Peterman, Z.E. (1988) Disequilibrium melting of granite at the contact with a basic plug: a geochemical and petrographic study, *J. Geol.* **96**, 61-78.

Kagami, H., Ulmer, P., Hansmann, W., Dietrich, V., and Steiger, R.H. (1991) Nd-Sr isotopic and geochemical characteristics of the southern Adamello (N. Italy) intrusives: implications for crustal versus mantle origin, *J. Geophys. Res.* **96**, 14331-14346.

Karig, D.E. (1971) Origin and development of marginal basins in the western Pacific, *J. Geophys. Res.* **76**, 2542-2561.

Karlstrom, K.E. (1989) Toward a syntectonic paradigm for granitoid, *EOS* **70**, 762.

Karlstrom, K.E., Miller, C.F., Kingsbury, J.A., and Wooden, J.L. (1993) Pluton emplacement along an active ductile thrust zone, Piute Mountains, southeastern California: interaction between deformational and solidification processes, *Geol. Soc. Am. Bull.* **105**, 213-230.

Kato, Y., Shinjo, R. and Kawano, Y. (1992) Cenozoic igneous activity in the Ryukyu islands, *Exploration of volcanoes and rocks in Japan, China and Antarctica*, Commemorative paper for Prof. Yukio Matsumoto, 39-48 (in Japanese).

Kawachi, Y. and Sato, T. (1978) Orthoclase megacrysts in the Yakushima granite, southern Kyushu, Japan, *N. Jb. Miner. Abh.* **132**, 136-152.

Kerrick, D.M. and Jacobs, G.K. (1981) A modified Redlich-Kwong equation for H_2O, CO_2, and H_2O-CO_2 mixtures at elevated pressures and temperatures, *Am. J. Sci.* **281**, 735-767.

Khan, M.A. (1962) The anisotropy of magnetic susceptibility of some igneous and metamorphic rocks, *J. Geophys. Res.* **67**, B7, 2873-2885.

Kimura, M. (1990) Genesis and formation of the Okinawa trough, Japan, *Geol. Soc. Japan Memoirs* **34**, 77-88 (in Japanese).

Kimura, M., Oue, S., and Kasahara, J. (1983) Seismicity along the Ryukyu arc and its vicinity, *Gekkan Chikyu* **5**, 753-758 (*in Japanese*).

Kincaid, C. and Olson, P. (1987) An experimental study of subduction and slab migration, *J. Geophys. Res.* **92**, 13832-13840.

King, R.F. (1966) The magnetic fabric of some Irish granites, *J. Geol.* **5**, 43-66.

Kirpatrick, J.A. and Ellis, D.J. (1992) C-type magmas: igneous charnockites and their extrusive equivalents. *Trans. Royal Soc. Edimburgh: Earth Sci.* **83**, 155-164.

Kistler, R.W., Chappell, B.W., Peck, D.L., and Bateman, P.C. (1986) Isotopic variation in the Tuolumne Intrusive Suite, central Sierra Nevada, California, *Contr. Mineral. Petrol.* **94**, 205-220.

Kizaki, K. (1986) Geology and tectonics of the Ryukyu Islands, *Tectonophysics* **125**, 193-207.

Kizaki, K. (1972) Configuration of migmatite dome: comparative tectonics of migmatite in the Hidaka Metamorphic belt, *J. Fac. Sci., Hokkaido Univ.* **15**, 157-172.

Knight, M.D. and Walkers, G.P.L. (1988) Magma flow directions in dikes of the Koolau complex, Oahu, determined from magnetic fabric studies, *J. Geophys. Res.* **93**, B5, 4301-4319.

Kodama, K., Tashiro, H., and Takeuchi, T. (1995) Quaternary counterclockwise rotation of south Kyushu, southwest Japan, *Geology* **23**, 823-826.

Kohut, M. and Janak, M. (1995) The Tatra granitoid pluton: an example of intrusion into the Variscan shear zone in the Western Carpathians, *Terra Abstract, Terra Nova* **7**, 141.

Koukoukevlas, L., Pe-Piper, G. and Piper, D.J.W. (1996) Pluton emplacement by wall rock thrusting, hanging-wall translation and extensional collapse: latest Devonian plutons of the Cobequid fault zone, Nova Scotia, Canada, *Geol. Mag.* **133**, 285-298.

Koyaguchi, T. (1987) Magma mixing in a squeezed conduit, *Earth Planet. Sci. Lett.* **84**, 339-344.

Koyi, H. (1991) Mushroom diapirs penetrating overburdens with high effective viscosities, *Geology* **19**, 1229-1232.

Kress, V.C. and Carmichael, I.S.E. (1991) The compressibility of silicate liquids containing Fe_2O_3 and the effect of composition, temperature, oxygen fugacity and pressure on their redox state, *Contr. Mineral. Petrol.* **108**, 82-92.

Kretz, R. (1966) Interpretation of the shape of mineral grains in metamorphic rocks, *J. Petrol.* **7**, 68-94.

Kretz, R. (1983) Symbols for rock-forming minerals, *Am. Mineralogist* **68**, 277-279.

Krohe, A. (1991) Emplacement of synkinematical plutons controlled by transtensional tectonics in the Variscan Odenwald (SW Germany), *Geol. Rundsch.* **80**, 391-409.

Kukowski, H. and Neugebauer, H.J. (1990) On the ascent and emplacement of granitoid magma bodies: dynamic-thermal numerical models, *Geol. Rundsch.* **79**, 227-239.

Kushiro, I. (1976) Changes in viscosity and structure of melt of $NaAlSi_2O_6$ composition at high pressures, *J. Geophys. Res.* **81**, 6347-6350.

Kusznir, N.J. and Park, R.G. (1984) Intraplate lithosphere deformation and the strength of the lithosphere, *Geophys. J. Royal Astron. Soc.* **79**, 513-538.

Lagarde, J.L., Omar, S.A., and Roddaz, B. (1990) Structural characteristics of granitic plutons emplaced during weak regional deformation: examples from late Carboniferous plutons, Morocco, *J. Struct. Geol.* **12**, 805-821.

Lange, R.A. and Carmichael, I.S.E. (1990) Thermodynamic properties of silicate liquids with emphasis on density, thermal expansion and compressibility, in J. Nicholls and J.K. Russel (eds.), *Modern methods of igneous petrology: understanding magmatic processes, Rev. Mineral.* **24**, pp. 25-64.

Lange, R.A., Cashman, K.V., and Navrotsky, A. (1994) Direct measurements of latent heat during crystallization and melting of a ugandite and an olivine basalt, *Contr. Mineral. Petrol.* **118**, 169-181.

Laporte, D. (1994) Wetting behaviour of partial melts during crustal anatexis: the distribution of hydrous silicic melts in polycrystalline aggregates of quartz, *Contr. Mineral. Petrol.* **116**, 486-499.

Laporte, D. and Watson, E.B. (1991) Direct observation of near-equilibrium pore geometry in synthetic quartzites at 600°-800 °C and 2-10.5 kb, *J. Geol.* **99**, 873-878.
Laporte, D. and Provost, A. (1994) The equilibrium crystal shape of silicates: implications for the grain-scale distribution of partial melts, *EOS* **75**, 364.
Laporte, D. and Vielzeuf, D. (1994) Wetting behaviour of partial melts during crustal anatexis: the distribution of hydrous silicic melts in polycrystalline aggregates of quartz, *EOS* **75**, 364.
Laporte, D. and Watson, E.B. (1995) Experimental and theoretical constraints on melt distribution in crustal sources: the effect of crystalline anisotropy on melt interconnectivity, *Chem. Geol.* **124**, 161-184.
Latouche, L., Black, R., and Liégeois, J.P. (1993) Pan-African assembly of terranes in the Tuareg Shield, in Thorweihe and Schandelmeier, *Geoscientific Research in Northeast Africa*, Rotterdam, Balkema, pp. 165-169.
Launeau, P., Bouchez, J.L., and Benn, K. (1990) Shape preferred orientation of object populations: automatic analysis of digitized images, *Tectonophysics* **180**, 201-211.
Launeau, P., and Bouchez, J.L. (1992) Mode et orientation préférentielle de forme des granites par analyse d'images numériques, *Bull. Soc. géol. France* **163**, 721-732.
Launeau, P., Cruden, A., and Bouchez, J.L. (1994) Mineral recognition in digital images of rocks: a new approach using multichannel classification, *The Canadian Mineralogist* **32**, 919-933.
Launeau, P., Cruden, A., and Robin, P.Y. (1995) Magmatic flow analysis using crystal size and shape preferred orientation distributions, EUG8, Strasbourg, *Terra abstracts* **7**, 137.
Le Fort, P. (1981) Manaslu leucogranite: a collision signature of the Himalaya and a model for its genesis and emplacement, *J. Geophys. Res.* **86**, 10545-10568.
Leblanc, D., Gleizes, G., Lespinasse, P., Olivier, P., and Bouchez, J.L. (1994) The Maladeta granite polydiapir, Spanish Pyrenees: a detailed magneto-structural study, *J. Struct. Geol.* **16**, 223-235.
Leblanc, D., Gleizes, G., Roux, L., and Bouchez, J.L. (1996) Variscan dextral transpression in the French Pyrenees: new data from the Pic des Trois-Seigneurs granodiorite and its country rocks, *Tectonophysics* **261**, 331-345.
Lejeune, A. and Richet, P. (1995) Rheology of crystal-bearing silicate melts: an experimental study at high viscosities, *J. Geophys. Res.* **100**, 4215-4229.
Leloup, P.H. and Kienast, J.R. (1993) High temperature metamorphism in a major strike-slip shear zone: the Ailao Shan-Red River, China, *Earth Planet. Sci. Lett.* **118**, 213-234.
Leterrier, J., Jardim de Sá, E., Bertrand, J.M., and Pin, C. (1994) Ages U-Pb sur zircon de granitoïdes "brasilianos" de la ceinture du Seridó (Province Borborema, NE Brésil), *C. R. Acad. Sci. Paris* **318**, 1505-1511.
Letouzey, J. and Kimura, M. (1986) The Okinawa trough: genesis of a back-arc basin developing along a continental margin, *Tectonophysics* **125**, 209-230.
Liégeois, J.P., Black, R., Navez, J., and Latouche, L. (1994) Early and late Pan-African orogenies in the Aïr assembly of terranes (Tuareg shield, Niger), *Precamb. Res.* **67**, 59-88.
Linnen, R., Williams-Jones, A.E., and Martin, R.F. (1992) Evidence of magmatic cassiterite mineralization at the Nong Sua aplite-pegmatite complex, Thailand, *Can. Mineral.* **30**, 739-761.
Lisle, R.J. (1985) The use of the orientation tensor for the description and statistical testing of fabrics, *J. Struct. Geol.* **7**, 115-117.
Lister, G.S. (1981) The effect of basal-prism mechanism switch on fabric development during plastic deformation of quartzite, *J. Struct. Geol.* **3**, 67-75.
Lister, G.S. and Baldwin, S.L. (1993) Plutonism and the origin of metamorphic core complexes, *Geology* **21**, 607-610.
Lockwood, J.P. (1975) Mount Abbot quadrangle, central Sierra Nevada, California: analytic data, *U.S. Geological Survey Prof. paper* **774**-C, 18 p.
Lockwood, J.P. and Lydon, P.A. (1975) Geologic map of the Mount Abbot quadrangle, central Sierra Nevada, California, Geol. Quadrangle Map GQ-1155, *U.S. Geol. Survey.*
Lockwood, J.P. and Moore, J.G. (1979) Regional deformation of the Sierra Nevada, California, on conjugate microfault sets, *J. Geophys. Res.* **84**, B11, 6041-6049.
Longhi, J. and Jurewicz, S.R. (1995) Plagioclase-melt wetting angles and textures: implications for anorthosites, *Lunar Planet. Sci.* **XXVI**, 859-860.
Lotze, F. (1945) Zur Gliederung der Varisziden der Iberischen Meseta, *Geotekt. Forsch.* **6**, 78-92.
Lucas, S.B. and St. Onge, M.R. (1995) Syn-tectonic magmatism and the development of compositional layering, Ungava Orogen (northern Quebec, Canada), *J. Struct. Geol.* **17**, 475-491.
Lupulescu, A. and Watson, E.B. (1994) Granitic melt connectivity at low-melt fraction in a mafic crustal protolith, *EOS* **75**, 585-586.
Lupulescu, A. and Watson, E.B. (1995) Tonalitic melt connectivity at low-melt fraction in a mafic crustal protolith at 10 kb and 800 °C, *EOS* **76**, 299-300.
Maaløe, S. (1985) *Principles of igneous petrology*, Springer-Verlag, Heidelberg.
Maaløe, S. and Wyllie, P.J. (1975) Water content of a granite magma deduced from the sequence of crystallization determined experimentally with water-undersaturated conditions, *Contr. Mineral. Petrol.* **52**, 175-191.
MacLeod, W.N., Turner, D.C., and Wright, E.P. (1971) The geology of the Jos Plateau, *Bull. Geol. Survey of Nigeria*, Geological Survey of Nigeria, vol **1**, 118 pp.

Mahon, K.I., Harrison, T.M., and Drew, D.A. (1988) Ascent of a granitoid diapir in a temperature varying medium, *J. Geophys. Res.* **93**, 1174-1188.
Mahood, G.A. and Cornejo, P.C. (1992) Evidence for ascent of differentiated liquids in a silicic magma chamber found in a granitic pluton, *Trans. Roy. Soc. Edinburgh, Earth Sci.* **83**, 63-69.
Mainprice, D., Bouchez, J.L., Blumenfeld, P., and Tubía, J.M. (1986) Dominant *c*-slip in naturally deformed quartz: implications for dramatic plastic softening at high temperature, *Geology* **14**, 819-822.
Mair, J.A. and Green, A.G. (1981) High resolution seismic reflection profiles reveal fractures zones within a 'homogeneous' granite batholith, *Nature* **294**, 439-442.
March, A. (1932) Mathematische Theorie der Regelung nach der Korngestalt bei affiner Deformation, Z. Kristallogr. **81**, 285-297.
Marre, J. (1986) *The Structural analysis of granitic rocks,* North Oxford Academic, 123 p.
MARS (1977) Radar image - number NC32-7, N and S. Motorola Aerial Remote Sensing, Inc., Phoenix, Federal Department of Forestry, Ibadan, Nigeria.
Marsh, B.D. (1979) Island arc development: some observations, experiments and speculations, *J. Geol.* **87**, 687-713.
Marsh, B.D. (1982) On the mechanics of igneous diapirism, stoping and zone melting, *Am. J. Sci.* **282**, 808-855.
Marsh, B.D. (1984) Mechanics and energetics of magma formation and ascension, in F.R. Boyd. Jr. (ed.), *Studies in Geophysics: explosive volcanism, inception, evolution and hazards*, National Academic Press, Washington D.C., pp. 67-83.
Marsh, B.D. and Carmichael, I.S.E. (1974) Benioff zone magmatism, *J. Geophys. Res.* **79**, 1196-1206.
Marsh, B.D. and Kantha, L.H. (1978) On the heat and mass transfer from an ascending magma, *Earth Planet. Sci. Lett.* **39**, 435-443.
Masuda, T., Michibayashi, K., and Ohta, H. (1995) Shape preferred orientation of rigid particles in a viscous matrix: reevaluation to determine kinematic parameters of ductile deformation, *J. Struct. Geol.* **17**, 115-129.
Masuda, T., Mizuno, N., Kobayashi, M., and Ngoc Nam, T. (1995) Stress and strain estimates for Newtonian and non-Newtonian materials in a rotational shear zone, *J. Struct. Geol.* **17**, 451-454.
Mattern, F. (1996) The Elbe zone at Dresden: a Late Proterozoic pull-apart intruded shear zone, *Z. dt. geol. Ges.*, Stuttgart **147**, 57-80.
Matthews, D.H. (1986) Can we see granites on reflection profiles? *Annales Geophysicae* **5B**, 353-356.
Maury, R.C. and Bizouard, H. (1974) Melting of acid xenoliths into a basanite: an approach to the possible mechanisms of crustal contamination, *Contrib. Mineral. Petrol.* **48**, 275-286.
McCaffrey, K.J.W. (1992) Igneous emplacement in a transpressive shear zone: Ox Mountains igneous complex, *J. Geol. Soc. London* **149**, 221-235.
McCaffrey, K.J.W., Howard, K.A., Bailey, C.M., and Foster, D.A. (1991) Jurassic syntectonic magmatism in the Clipper Mountains, eastern Mojave Desert, California, *Geol. Soc. Am. Abstr. with Programs* **23**, p. A250.
McCaffrey, K.J.W., Petford, N., and Evans, D.J. (1997) Scale invariant geometry of tabular sheet intrusions, *J. Geol. Soc. London.* (in press).
McCarthy, J. and Thompson, G.A. (1988) Seismic imaging of extended crust with emphasis on the western United States, *Geol. Soc. Amer. Bull.* **100**, 1361-1374.
McKenzie, D. (1969) Speculations on the consequences and causes of plate motions, *Geophys. J. R. Astr. Soc.* **18**, 1-32.
McKenzie, D. (1989) Some remarks on the movement of small melt fractions in the mantle, *Earth Planet. Sci. Lett.* **95**, 53-72.
McLellan, E.L. (1988) Migmatite structures in the Central Gneiss Complex, Boca de Quadra, Alaska, *J. Metam. Geol.* **6**, 517-542.
McNulty, B.A. (1995) Shear zone development during magmatic arc construction: the Bench Canyon shear zone, central Sierra Nevada, California, *Geol. Soc. Am. Bull.* **107**, 1094-1107.
Means, W.D. (1990) Kinematic, stress, deformation and material behavior, *J. Struct. Geol.* **12**, 953-971.
Mehnert, K.R., Busch, W., and Schneider, G. (1973) Initial melting at grain boundaries of quartz and feldspar in gneisses and granulites, *N. Jahr. Mineral. Monatsh.* **4**, 165-183.
Melka, R., Schulmann, K., Schulmannová, B., Hrouda, F., and Lobkowicz, M. (1992) The evolution of perpendicular linear fabrics in a synkinematically emplaced tourmaline granite (central Moravia-Bohemian Massif), *J. Struct. Geol.* **14**, 605-620.
Meneilly, A.W. (1982) Regional structure and syntectonic granite intrusion in the Dalradian of the Gweebarra Bay area, Donegal, *J. Geol. Soc. London* **139**, 633-646.
Miki, M. (1995) Two-phase opening model for the Okinawa Trough inferred from paleomagnetic study of the Ryukyu arc, *J. Geophys. Res.* **100**, 8169-8184.
Miki, M., Matsuda, T., and Otofuji, Y. (1990) Opening mode of the Okinawa Trough: paleomagnetic evidence from the South Ryukyu Arc, *Tectonophysics* **175**, 335-347.
Miller, R.B. and Paterson, S.R. (1994) The transition from magmatic to high-temperature solid-state deformation: implications from the Mount Stuart batholith, Washington, *J. Struct. Geol.* **16**, 853-865.
Milligan, G.W.E. and Houlsby, G.T. (1984) *Basic soil mechanics,* Butterworths, London, 132 pp.
Minarik, W.G. and Watson, E.B. (1995) Interconnectivity of carbonate melt at low melt fraction, *Earth Planet. Sci. Lett.* **133**, 423-437.

Mira, M., Ortega, E., and Rodriguez, L. (1987) Hoja y memoria n° 834, San Benito, E: 1/50.000. *Instit. Geol. Min. España*, Madrid.
Mitchell, J.K. (1993) *Fundamentals of soil behaviour*, 2nd edition, John Wiley and Sons, 437 pp.
MITI (Ministry of International Trade and Industry) (1992) Report of potential investigation of rare metal mineral resources: the Yakushima area, MITI, Tokyo, 171 p. (in Japanese).
Moisy, M. (1993) Evolution de la perméabilité et de la circulation des fluides hydrothermaux dans une zone de cisaillement fragile, *Unpubl. Thesis Univ. Grenoble*, 214 p.
Mollier, B. (1984) Le granite de Brâme-St Sylvestre-St Goussaud; ses structures magmatiques; une étude de la distribution de l'uranium à l'échelle du grain, *Unpubl. Thesis Univ. Nantes*, CREGU Mém., Nancy, France **7**, 172 p.
Mollier, B. and Bouchez, J.L. (1982) Structuration magmatique du complexe granitique de Brâme-St Sylvestre-St Goussaud (Limousin, Massif Central français), *C. R. Acad. Sci. Paris* **294 II**, 1329-1334.
Monié, P., Caby, R., and Arthaud, M. (1996) The Neoproterozoic Brasiliano orogeny in Northeast Brazil: $^{40}Ar/^{39}Ar$ and petrostructural data from Ceará, *Precamb. Res.*, in press.
Montel, J.-M. and Vielzeuf, D. (1996) Partial melting of metagreywackes. Part II: composition of minerals and melts, *Contrib. Mineral. Petrol.* (in press).
Morand, V.J. (1992) Pluton emplacement in a strike-slip fault zone: the Doctors Flat Pluton, Victoria, Australia, *J. Struct. Geol.* **14**, 205-213.
Moreno-Ventas, I., Rogers, G., and Castro, A. (1995) The role of hybridization in the genesis of Hercynian granitoids in the Gredos massif. Inferences from Sm/Nd isotopes, *Contr. Mineral. Petrol.* **120**, 137-149.
Musumeci, G. and Pertusati, P.C. (1995) Syntectonic magmatism in a collisional belt: North Victoria Land Antarctica, *Terra Abstract Terra Nova* **7**, 140.
Nadai, A. (1963) Theory of flow and fracture of solids, McGraw-Hill, New York, 705 p.
Nakada, S. and Takahashi, M. (1979) Regional variation in chemistry of the Miocene intermediate to felsic magmas in the Outer Zone and the Setouchi Province of southwest Japan, *J. Geol. Soc. Japan* **85**, 571-582.
Nakada, S., Hirahara, S., and Okamoto, Y. (1988) The Ichifusayama zoned pluton, SW Japan: concentric zoning versus vertical zoning, *Fifth International Symposium on Tin/Tungsten Granites in Southeast Asia and the Western Pacific, IGCP Project* **220**, 104-108.
Naney, M.F. (1983) Phase equilibria of rock-forming ferromagnesian silicates in granitic systems, *Am. J. Sci.* **283**, 993-1033.
Nash, D.F. (1979) The geological development of the north Okinawa trough area from Neogene times to recent, *J. Japanese Assoc. Petroleum Technologists* **44**, 109-119.
Nédélec, A., Paquette, J.L., Bouchez, J.L., Olivier, P., and Ralison, B. (1994) Stratoid granites of Madagascar: structure and position in the Panafrican orogeny, *Geodinamica Acta* **7**, 48-56.
Nekvasil, H. (1988) Calculated effect of anorthite component on the crystallization paths of H_2O-undersaturated haplogranitic melts, *Am. Miner.* **73**, 966-981.
Nekvasil, H. (1991) Ascent of felsic magmas and formation of rapakivi, *Am. Miner.* **76**, 1279-1290.
Neves, S.P. (1996) Etude des relations entre magmatisme et zones de cisaillement lithosphériques: exemple des décrochements de Pernambuco et Fazenda Nova (Etat de Pernambuco, Brésil), *Unpubl. Thesis Univ.* Montpellier II, 243 p.
Neves, S.P. and Vauchez, A. (1995a) Magma emplacement and shear zone nucleation and development in Northeast Brazil (Fazenda Nova and Pernambuco shear zones; state of Pernambuco), *J. S. Am. Earth Sci.* **8**, 289-298.
Neves, S.P. and Vauchez, A. (1995b) Successive mixing and mingling of magmas in a plutonic complex of Northeast Brazil, *Lithos* **24**, 275-299.
Neves, S.P., Vauchez, A., and Archanjo, C.J. (1996) Shear-zone controlled magma emplacement or magma-assisted nucleation of shear zones? Insights from northeast Brazil, *Tectonophysics*, in press.
Nicolas, A. (1986) A melt extraction model based on structural studies in mantle peridotites, *J. Petrol.* **27**, 999-1022.
Nicolas, A. (1989) Principes de tectonique, *Masson Ed.*, 2nd edition, 223 p.
Nicolas, A. (1992) Kinematics in magmatic rocks with special reference to gabbros, *J. Petrol.* **33**, 891-915.
Nicolas, A., Bouchez, J.L., Blaise, J., and Poirier, J.P. (1977) Geological aspects of deformation in continental shear zones, *Tectonophysics* **42**, 55-73.
Nicolas, A., Freydier C., Godard, M., and Vauchez A. (1993) Magma chambers at oceanic ridges: how large? *Geology* **21**, 53-56.
Nicolas, A. and Boudier, F. (1995) Mapping oceanic ridge segments in Oman ophiolites, *J. Geophys. Res.* **100**, 6179-6197.
Oertel, G. (1955) Der Pluton von Loch Doon in Sudscotland, *Geotecktonischer Forshungen* **11**, 1-83.
Okudaira, T., Takeshita, T., Hara, I., and Ando, J. (1995) A new estimate of the conditions for transition from basal <a> to prism [c] slip in naturally deformed quartz, *Tectonophysics* **250**, 31-46.
Olarewaju, V.O. (1978) On the Basement Complex Geology of 1:100,000 Sheet 126, Ririwai, Northern Nigeria. *Bull. Dept. Geol.*, Ahmadu Bello Univ., Zaria, Nigeria **1**, 63-96.
Olgaard, D. L. and Evans, B. (1988) Grain growth in synthetic marbles with added mica and water, *Contrib. Mineral. Petrol.* **100**, 246-260.
Oliver, H.W. (1977) Gravity and magnetic investigations of the Sierra Nevada batholith, California, *Geol. Soc. Am. Bull.* **88**, 445-461.

Olivier, P. and Archanjo, C.J. (1994) Magnetic and magmatic structures of the Emas granodioritic pluton (Cachoeirinha belt, NE Brazil). Relationships with Pan-African strike-slip fault systems, *Tectonophysics* **229**, 239-250.

Olivier, P., de Saint-Blanquat, M., Gleizes, G., and Leblanc, D. (1997) Homogeneity of granite fabrics at the metre and dekametre scales, in J.L. Bouchez, D.H.W. Hutton and W.E. Stephens (eds.), *Granite: from segregation of melt to emplacement fabrics*, Kluwer Academic Publishers, Dordrecht, pp. 113-128.

Olson, P. and Singer, H. (1985) Creeping plumes, *J. Fluid Mech.* **158**, 511-531.

Owens, W.H. (1973) Strain modification of angular density distributions, *Tectonophysics* **16**, 249-261.

Panozzo-Heilbronner, R. (1992) The autocorrelation function: an image processing tool for fabric analysis, *Tectonophysics* **212**, 351-370.

Parsons, T., Sleep, N.H., and Thompson, G.A. (1992) Host rock rheology controls on the emplacement of tabular intrusions: implications for underplating of extended crust, *Tectonics* **11**, 1348-1356.

Parsons, T. and Thompson, G.A. (1993) Does magmatism influence low-angle normal faulting? *Geology* **21**, 247-250.

Passchier, C.W. (1987) Stable positions of rigid objects in non-coaxial flow: a study in vorticity analysis, *J. Struct. Geol.* **9**, 679-690.

Passchier, C.W. and Simpson, C. (1986) Porphyroclast systems as kinematic indicators, *J. Struct. Geol.* **8**, 831-843.

Passchier, C.W. and Sokoutis, D. (1993) Experimental modelling of mantled porphyroclasts, *J. Struct. Geol.* **15**, 895-909.

Passchier, C.W. and Trouw, R.A.J. (1996) *Microtectonics*, Springer, Berlin, 289 p.

Paterson, S.R. (1988) Cannibal Creek Granite: post-tectonic "ballooning" pluton or pre-tectonic piercement diapir? *J. Geol.* **96**, 730-736.

Paterson, S.R. (1989) Are syntectonic granites truly syntectonic? *EOS* **70**, 762.

Paterson, S.R. (1994) Comment on "Dike transport of granitoid magmas" by Petford, N., Kerr, R.C., and Lister, J.R., 1993, *Geology* **22**, 473.

Paterson, S.R., Vernon, R.H., and Tobish, O.T. (1989) A review of criteria for the identification of magmatic and tectonic foliations in granitoids, *J. Struct. Geol.* **11**, 349-363.

Paterson, S.R., Vernon, R.H., and Fowler, T.K. (1991) Aureole tectonics, in D.M. Kerrick (ed.), *Contact metamorphism*, *Reviews in Mineralogy* **26**, pp. 673-722.

Paterson, S.R. and Tobisch, O. (1992) Rates of processes in magmatic arcs: implications for the timing and nature of pluton emplacement and wall rock deformation, *J. Struct. Geol.* **14**, 291-300.

Paterson, S.R. and Fowler, T.K. (1993) Re-examining pluton emplacement processes, *J. Struct. Geol.* **15**, 191-206.

Paterson, S.R. and Vernon, R.H. (1995) Bursting the bubble of ballooning plutons: a return to nested diapirs emplaced by multiple processes, *Geol. Soc. Am. Bull.* **107**, 1356-1380.

Peacock, S.M. (1990) Numerical simulation of regional and contact metamorphism using the Macintosh microcomputer, *J. Geol. Educ.* **38**, 132-137.

Pearce, G.W. and Fueten, F. (1989) An intensive study of magnetic susceptibility anisotropy of amphibolite layers of the Thompson Belt, North Manitoba, *Tectonophysics* **162**, 315-329.

Pelto, C.R. (1954) Mapping of multicomponent systems, *J. Geol.* **62**, 501-511.

Perkins, D. and Vielzeuf, D. (1992) Reinvestigation of fayalite+anorthite=garnet, *Contr. Mineral. Petrol.* **111**, 260-263.

Persikov, E.S. (1991) The viscosity of magmatic liquids: experiment, generalized patterns. A model for calculation and prediction. Applications, In L.L. Perchuck and I. Kushiro (eds.) *Advances in Physical Chemistry*, Springer Verlag, New York, pp. 1-40.

Pesquera, A. and Pons, J. (1990) Le pluton hercynien de Aya (Pyrénées basques espagnoles): structure, mise en place et incidences tectoniques régionales. *Bull. Soc. géol. France* **8-VI(1)**, 13-21.

Petford, N. (1993) Porous media flow in granitoid magmas: an assessment, in D.B. Stone and S.K. Runcorn (eds.) *Flow and creep in the solar system: observations, modeling and theory*, Kluwer Academic Publisher, Dordrecht, pp. 261-286

Petford, N. (1995) Segregation of tonalitic and trondhjemitic melts from the continental crust: the mantle connection, *J. Geophys. Res.* **B100**, 15735-15743.

Petford, N. (1996) Dykes or diapirs? *Trans. Roy. Soc. Edinburgh: Earth Sci.* **87**, 105-114.

Petford N., Kerr, R.C., and Lister, J.R. (1993) Dike transport of granitoid magmas, *Geology* **21**, 845-848.

Pfiffner, O.A. and Ramsay J.G. (1982) Constraints on geological strain rates: arguments from finite strain states of naturally deformed rocks, *J. Geophys. Res.* **87**, 311-321.

Pharr, G.M. and Ashby, M.F. (1983) On creep enhanced by a liquid phase, *Acta Metallurgica* **31**, 129-138.

Phillips, W.J. (1974) The dynamic emplacement of cone sheets, *Tectonophysics* **24**, 69-84.

Philpotts, A.R. and Asher, P.M. (1994) Magmatic flow-direction indicators in a giant diabase feeder dike, Connecticut, *Geology* **22**, 363-366.

Pichavant, M., Boher, M., Stenger, J.F., Aissa, M., and Charoy, B. (1987) Relations de phases des granites de Beauvoir entre 1 et 3 kb en condition de saturation en H_2O, *Géologie de la France* **2-3**, 77-86.

Pichavant, M., Kontak, D.J., Valencia Herrera, J., and Clark, A.H. (1988) The Miocene-Pliocene Macusani Volcanics, southeast Peru. I.: Mineralogy and magmatic evolution of a two-mica aluminosilicate-bearing ignimbrite suite, *Contr. Mineral. Petrol.* **100**, 300-324.

Pinkerton H. and Stevenson, R.J. (1992) Methods of determining the rheological properties of magmas at sub-liquidus temperatures, *J. Volc. Geotherm. Res.* **53**, 47-66.

Pitcher, W.S. (1979) The nature, ascent and emplacement of granitic magma, *J. Geol. Soc. Lond.* ***136***, 627-662.

Pitcher, W.S. (1993) The nature and origin of granite, *Blackie Acad. Prof. London*, 322 p.

Platt, J.P. and Vissers, R.L.M. (1980) Extensional structures in anisotropic rocks, *J. Struct. Geol.* **2**, 397-410.

Pons, J. (1970) Relations entre la structure et la pétrofrabrique des roches éruptives de la bordure méridionale du massif granitique du Quérigut (Ariège, France), *C. R. Acad. Sci. Paris* **271**, D, 1665-1668.

Pons, J., Oudin, C., and Valero, J. (1992) Kinematics of large scale synorogenic intrusions: example of the lower Proterozoic Saraya batholith (Eastern Senegal), *Geol. Rundschau* **81/2**, 473-486.

Pons, J., Barbey, P., Dupuis, D., and Legar, J.M. (1995) Mechanism of pluton emplacement and structural evolution of a 2.1 Ga juvenile continental crust: the Birimian of southwest Niger, *Precamb. Res.* **70**, 281-301.

Rahaman, M.A. (1988) Recent advances in the study of the basement complex of Nigeria, in *Geol. Surv. Nigeria.*, Precambrian geology of Nigeria, 11-43.

Ramberg, H. (1981) *Gravity, deformation and the Earth's crust*, Academic Press, 452 p.

Ramberg, H. and Ghosh, S. K. (1977) Rotation and strain of linear and planar structures in three-dimensional progressive deformation, *Tectonophysics* **40**, 309-337.

Ramsay, J.G. (1958) Superimposed folding at Loch Monar, Inverness-shire and Ross-shire, *J. Geol. Soc.* London **113**, 271-307.

Ramsay, J.G. (1967) Folding and fracturing of rocks, Mc. Graw-Hill, New York, 568 p.

Ramsay, J.G. (1989) Emplacement kinematics of a granite diapir: the Chindamora batholith, Zimbabwe, *J. Struct. Geol.* **11**, 191-209.

Raposo, M.I.B. and Ernesto, M. (1995) Anisotropy of magnetic susceptibility in the Ponta Grossa dyke swarm (Brazil) and its relationship with magma flow direction, *Phys. Earth Planet. Interiors* **87**, 183-196.

Read, H.H. (1957) *The Granite Controversy*, Thomas Murby & Co., 430 pp.

Reed, L. and Tryggvason, E. (1974) Preferred orientations of rigid particles in a viscous matrix deformed by pure shear and simple shear, *Tectonophysics* **24**, 85-98.

Regan, R.D. and Hinze, W.J. (1976) The effect of finite data length in the spectral analysis of ideal gravity anomalies, *Geophysics* **42**, 44-55.

Ribeiro, A. (1990) Central Iberian Zone, introduction in R.D. Dallmeyer and E. Martínez García (eds.), *Pre-Mesozoic Geology of Iberia*, Springer-Verlag, Berlin, pp. 143-144.

Richter, C., van der Pluijm, B.A., and Housen, B.A. (1993) The quantification of crystallographic preferred orientation using magnetic anisotropy, *J. Struct. Geol.* **15**, 113-116.

Riegger, O.K. and van Vlack, L.H. (1960) Wetting angle measurement, *AIME Trans.* **218**, 933-935.

Riklin, K., (1983) Contact metamorphism of the Permian red sandstones in the Adamello area, *Memorie della Società Geologica Italiana* **26**, 159-169.

Riley, G.N., Jr. and Kohlstedt, D.L. (1991) Kinetics of melt migration in upper mantle-type rocks, *Earth Planet. Sci. Lett.* **105**, 500-521.

Riley, G.N., Jr., Kohlstedt, D.L., and Richter, F. M. (1990) Melt migration in a silicate liquid-olivine system: an experimental test of compaction theory, *Geophys. Res. Lett.* **17**, 2101-2104.

Ringwood, A.E. (1974) The petrological evolution of island arc system, *J. Geol. Soc. London* **130**, 183-204.

Robin, P.Y.F. (1979) Theory of metamorphic segregation and related processes, *Geochim. Cosmochim. Acta*, **43**, 1587-1600.

Rochette, P. (1987) Magnetic susceptibility of the rock matrix related to magnetic fabric studies, *J. Struct. Geol.* **9**, 1015-1020.

Rochette, P., Jackson, M., and Aubourg, C. (1992) Rock magnetism and the interpretation of anisotropy of magnetic susceptibility, *Rev. Geophys.* **30**, 209-226.

Rochette, P., Scaillet, B., Guillot, S., Pêcher, A., and Le Fort, P. (1994) Magnetic mineralogy of the High Himalayan leucogranites: structural implications, *Earth Planet. Sci. Letters* **126**, 217-234.

Román Berdiel, T., Gapais, D., and Brun, J.P. (1995) Analogue models of laccolith formation, *J. Struct. Geol.* **17**, 1337-1346.

Roman-Berdiel, T., Pueyo-Morer, E.L., and Casas-Sainz, A.M. (1995) Granite emplacement during contemporary shortening and normal faulting : structural and magnetic study of the Veiga Massif (NW Spain), *J. Struct. Ceol.* **17**, 1689-1706.

Rönnlund, P. (1991) Diapiric walls and stocks from layer and line sources, *Uppsala Univ. Dept. Mineral. Petrol. Research Report* **68**, 43 p.

Rosenberg, C.L., Berger, A., and Schmidt, S.M. (1995) Observations from the floor of a granitoid pluton: inferences on the driving force of final emplacement, *Geology* **23**, 443-446.

Royse, K. and Tarney, J. (1995) Emplacement mechanisms of Nain (Labrador) anorthosites: evidence from deformed margins bearing on their petrogenesis, *Terra Abstracts, Terra Nova* **7**, 142.

Rubie, D.C. and Brearley, A.J. (1991) A model for rates of disequilibrium melting during metamorphism, in J.R. Ashworth and M. Brown (eds.), *High Temperature Metamorphism and Crustal Anatexis*, Unwin Hyman, London, pp. 57-86.

Rubin, A.M. (1995) Getting granite dikes out of the source region, *J. Geophys. Res.* **100**, 5911-5929.

Rushmer, T. (1995) An experimental deformation study of partially molten amphibolite: application to low-melt fraction segregation, *J. Geophys. Res.* **100**, 15681-15695.

Rutter, E.H. and Neumann, D.H.K. (1995) Experimental deformation of partially molten Westerly granite under fluid-absent conditions, with implications for the extraction of granitic magmas, *J. Geophys. Res.* **100 B8**, 15697-15715.
Saint Blanquat (de), M. (1993) EXAMS, an Excel macro for automatic computing of AMS data. *Unpublished program Univ. Toulouse.*
Saint Blanquat (de), M., Olivier, P., and Tikoff, B. (1993), Magnetic/magmatic structures in the Mono Creek pluton (Sierra Nevada batholith, California, USA), *Terra Abstracts* **5**, 111.
Saint Blanquat (de), M., and Tikoff B. (1995) Passive and forceful emplacement of the Mono Creek pluton (Sierra Nevada, California) during regional transpressive deformation, *Terra Abstracts* **7**, 138.
Saint Blanquat, M (de) and Tikoff, B. (1996) Development of magmatic to solid-state fabrics during syntectonic emplacement of the Mono Creek granite, Sierra Nevada batholith, in J.L. Bouchez, D.H.W. Hutton and E. Stephens (eds.), *Granite: from segregation of melt to emplacement fabrics*, Kluwer Academic Publishers, Dordrecht, pp. 231-252.
Sakuyama, M. (1983) Petrology of arc volcanic rocks and their origin by mantle diapirs, *J. Volc. Geotherm. Res.* **18**, 297-320.
Sander, B. (1948) Einführung in die Gefügekunde der geologischen Körper, Springer Verlag, Vienna, vol. I, 215 p.; vol II (1950), 409 p.
Sanderson, D.J. and Meneilly, A.W. (1981) Analysis of three-dimensional strain modified uniform distributions: andalusite fabrics from a granite aureole, *J. Struct. Geol.* **3**, 109-116.
Sanderson, D.J. and Marchini, W.R.D. (1984) Transpression, *J. Struct. Geol.* **6**, 449-458.
Sato, M. (1978) Oxygen fugacity of basaltic magmas and the role of gas-forming elements, *Geophys. Res. Lett.* **5**, 447-9.
Sato, T. and Nagahama, H. (1979) Geology of the Yakushima-Seinanbu district, *Geol. Surv. Japan*, Quadrangle Series, Scale 1 : 50,000, Tanegashima **16**, 9 (in Japanese).
Savage, W.Z. (1978) The development of residual stress in cooling rock bodies, *Geophys. Res. Lett.* **5**, 633-636.
Sawyer, E.W. (1991) Disequilibrium melting and the rate of melt-residuum separation during migmatization of mafic rocks from the Grenville front, Quebec, *J. Petrol. 32*, 701-783.
Sawyer, E.W. (1994) Melt segregation in the continental crust, *Geology* **22**, 1019-1022.
Sawyer, E.W. (1996) Melt segregation and magma flow in migmatites: implications for the generation of granite magmas, *Trans. Roy. Soc. Edinburgh* **87**, 85-94.
Sawyer, E.W. and Robin, P.Y. (1986) The sub-solidus segregation of layer-parallel quartz-feldspar veins in greenschist to upper amphibolite facies metasediments, *J. Metam. Geol.* **4**, 237-260.
Scaillet, B. and Pichavant, M. (1994) Phase relations of amphibole-biotite bearing granites. II: Effects of fO_2, *Terra Nova* **6**, 42.
Scaillet, B., Pêcher, A., Rochette, P., and Champenois, M. (1995) The Gangotri granite (Garhwal Himalaya): laccolithic emplacement in an extending collisional belt, *J. Geophys. Res.* **100/B1**, 585-607
Scaillet, B., Pichavant M., and Roux, J. (1995) Experimental crystallization of leucogranite magmas, *J. Petrol.* **36**, 664-706.
Scaillet, B., Holtz, F., Pichavant, M., and Schmidt, M. (1996) The viscosity of Himalayan leucogranites: implications for mechanisms of granitic magma ascent, *J. Geophys. Res.* (in press).
Scheidegger, A.E. (1965) The analysis of finite strain using lines with an initial random orientation, *Geol. Soc. Am. Bull.* **4**, 1519-1528.
Schmeling, H., Cruden, A.R., and Marquart, G. (1988) Finite deformation in and around a fluid sphere moving through a viscous medium: implications for diapiric ascent, *Tectonophysics* **149**, 17-34.
Schrödinger, E. (1944) *What is Life*? Cambridge University Press, UK., 200 pp.
Schulmann, K. and Venera, Z. (1995) Fabric study, flow kinematics and emplacement of ultra potassic intrusion in a mid crustal level, the Bohemian Massif, *Terra Abstracts, Terra Nova* **7**, 138.
Schulmann, K., Jesek, J., and Venera, Z. (1997) Perpendicular linerar fabrics in granite: markers of combined simple shear and pure shear flows? in J.L. Bouchez, D.H.W. Hutton and W.E. Stephens (eds.), *Granite: from segregation of melt to emplacement fabrics*, Kluwer Academic Publishers, Dordrecht, pp. 159-176.
Schwarz, (1992) Magnetic expressions of intrusions including magnetic aureoles, *Tectonophysics* **192**, 191-200.
Segall, P., McKee, E.H., Martel, S.J., and Turrin, B.D. (1990) Late Cretaceous age of fractures in the Sierra Nevada batholith, California, *Geology* **18**, 1248-1251.
Shannon, C.E. (1948) A mathematical theory of communications, *Tec. J. Bell. System* **27**, 379-423.
Sharp, W.D, Tobisch, O., and Renne, P.R. (1993) Late Cretaceous (85-80 Ma), syn-arc cleavage development in metamorphic rocks of the Ritter Range, Central Sierra Nevada, California, *Geol. Soc. Am. Abstracts*, Cordilleran and Rocky Mtn Sect., 145.
Shaw, H.R. (1963) The four-phase curve sanidine-quartz-liquid-gas between 500 and 4000 bars, *Amer. Mineral.* **48**, 883-896.
Shaw, H.R. (1969) Rheology of basalt in the melting range, *J. Petrol.* **10**, 510-535.
Shaw, H.R. (1972) Viscosities of magmatic silicate liquid: an empirical method of prediction, *Am. J. Sci.* **272**, 870-893.
Shaw, H.R. (1980) Fracture mechanisms of magma transport from the mantle to the surface, in R.B. Hargraves (ed.), *Physics of Magmatic Processes*, Princeton University Press, Princeton, N.J., pp. 201-264.

Shemenda, A.I. (1994) *Subduction: insights from physical modeling*, Kluwer Academic Publishers, Dordrecht, 215 p.
Shibata, K. (1978) Contemporaneity of Tertiary granites in the Outer Zone of Southwest Japan, *Bull. Geol. Surv. Japan* **29**, 551-554 (in Japanese).
Shibata, K. and Ishihara, S. (1979) Initial $^{87}Sr/^{86}Sr$ ratios of plutonic rocks from Japan, *Contrib. Mineral. Petrol.* **70**, 381-390.
Shigeno, H. (1980) Tension joint system prevailing in the Okinoshima granitic rocks, Shikoku, Southwest Japan: on the regional stress field of the Shimanto belt during the late Neogene, *Bull. Inst. Sci., Kyushu Univ.* **13**, 145-153 (in Japanese).
Shimamoto, T. (1993) Rheology of rocks and plate tectonics, in E. T. Brown and J. A. Hudson (eds.), *Comprehensive rock engineering, I: Fundamentals*, Pergamon Press, Oxford, pp. 93-109.
Shimamura, H. and Asada, T. (1975) T waves from deep earthquakes generated exactly at the bottom of deep-sea trenches, *Earth Planet. Sci. Lett.* **27**, 137-142.
Shiono, K., Mikumo, T., and Ishikawa, Y. (1980) Tectonics of the Kyushu-Ryukyu arc as evidenced from seismicity and focal mechanism of shallow to intermediate-depth earthquakes, *J. Phys. Earth* **28**, 17-43.
Siegesmund, S., Ullemeyer, K., and Dahms, M. (1995) Control of magnetic rock fabrics by mica preferred orientation: a quantitative approach, *J. Struct. Geol.* **17**, 1601-1613.
Simpson, C. and Schmid, S.M. (1983) An evaluation of criteria to deduce the sense of movement in sheared rocks, *Geol. Soc. Am. Bull.* **94**, 1281-1288.
Smith, C.S. (1964) Some elementary principles of polycrystalline microstructure, *Metall. Rev.* **9**, 1-48.
Spanner, B.G. and Kruhl, J.H. (1996) Varying mechanisms of magma emplacement due to a change of tectonic setting during the Brasiliano orogeny: the Carmo and Sindacta Pluton (SE Brazil). *J. Struct. Geol.* (in press).
Sparks, R.S.J. and Marshall, L.A. (1986) Thermal and mechanical constraints on mixing between mafic and silicic magmas, *J. Volc. Geotherm. Res.* **29**, 99-124.
Speer, J.A., McSween, H.Y., and Gates, A.E. (1994) Generation, segregation, ascent and emplacement of Alleghanian plutons in the southern Applachians, *J. Geol.* **102**, 249-267.
Spera, F.J. (1980) Aspects of magma transport, in R.B. Hargraves (ed.), *Physics of magmatic processes*, Princeton Univ. Press, Princeton (New Jersey), pp. 265-323.
Spiegelman, M. and McKenzie, D. (1987) Simple 2-D models for melt extraction at mid-ocean ridges and island arcs, *Earth Planet. Sci. Lett.* **83**, 137-152.
Spry, A. (1969) *Metamorphic textures*, Pergamon, Oxford, 350 pp.
Stacey, F.D. (1960) Magnetic anisotropy of igneous rocks, *J. Geoph. Res.* **65**, 2429-2442.
Stanley, J.M. and Green, R. (1976) Gravity gradients and the interpretation of the truncated plate, *Geophysics* **41**, 1370-1376.
Stein, G., Charvet, J., Lapierre, H., and Fabbri, O. (1994) Geodynamic setting of volcano-plutonic rocks in so-called "paleo-accretionary prism": fore-arc activity or post-collisional magmatism? The Shimanto belt as a case study, *Lithos* **33**, 85-107.
Stephansson, O. (1975) Polydiapirism of granitic rocks in the Svecofennian of Central Sweden, *Precamb. Res.* **2**,189-214.
Stephenson, A. (1994) Distribution anisotropy: two simple models for magnetic lineation and foliation, *Phys. Earth Planet. Inter.* **82**, 49-53.
Stern, T.W., Bateman, P.C., Morgan, B.A., Newell, M.F., and Peck, D.L. (1981) Isotopic U-Pb ages of zircon from granitoids of the Central Sierra Nevada, California, *U.S. Geol. Survey Prof. Paper*, 1185.
Stevenson D.J. (1989) Spontaneous small-scale melt segregation in partial melts undergoing deformation, *Geophys. Res. Letters* **16**, 1067-1070.
Stewart, D.B. (1967) Four-phase curve in the system $CaAl_2Si_2O_8$-SiO_2-H_2O between 1 and 10 kb, *Schweiz. Miner. Petrogr. Mitt.* **47**, 35-59.
Sykes, M.L. and Holloway, J.R. (1987) Evolution of granitic magmas during ascent: a phase equilibrium model, in B. Mysen (ed.), *Magmatic processes: physicochemical principles*, The Geochemical Society, Spec. Publ. n° 1, University Park, pp. 447-461.
Taira, A. and Ogawa, Y. (1991) Cretaceous to Holocene fore-arc evolution in Japan and its implication to crustal dynamics, *Episodes* **14**, 205-212.
Takahashi, M. (1986) Anatomy of a middle Miocene Valles-type caldera cluster: geology of the Okueyama volcano-plutonic complex, southwest Japan, *J. Volc. Geotherm. Res.* **29**, 33-70.
Talbot, C.J. (1977) Inclined and asymmetric upward-moving gravity structures, *Tectonophysics* **42**, 159-181.
Talbot, C.J. and Jackson, M.P.A. (1987) Internal kinematics of salt diapirs, *Amer. Assoc. Petroleum Geol. Bull.* **71**, 1068-1093.
Taneda, S. and Kinoshita, K. (1972) An alkaline rock body in Tanegashima island, South Kyushu, Japan, *Kazan 2* **17**, 88-97 (in Japanese).
Tarling, D.H. and Hrouda, F. (1993) *The magnetic anisotropy of rocks*, Chapman and Hall, London, 217 pp.
Tasch, P. (1965) Communication theory and the fossil record of invertebrates, *Trans. Kansas. Ac. Sci. 68*, 322-329.
Tatsumi, Y., Hamilton, D.L., and Nesbitt, R.W. (1986) Chemical characteristics of fluid phase released from a subducted lithosphere and origin of arc magmas: evidence from high-pressure experiments and natural rocks, *J. Volc. Geotherm. Res.* **29**, 293-309.
Tatsumi, Y. and Eggins, S. (1995) *Subduction zone magmatism*, Blackwell, 211 p.

Telford, W.M., Geldart, L.P., Sheriff, R.E., and Keys, D.A. (1990)*Applied geophysics*, Cambridge University Press, 860 pp.
Tepper, J.H., Nelson, B.K., Bergantz, G.W., and Irving, A.J. (1993) Petrology of the Chilliwack batholith, North Cascades, Washington: generation of calc-alkaline granitoids by melting of mafic lower crust with variable water fugacity. *Contr. Mineral. Petrol.* **113**, 333-351.
Terzaghi, K. (1936) The shearing resistance of saturated soils, *Proceedings of the 1st International conference on Soil Mechanics*, Cambridge Massachuetts **1**, 54-56.
Terzaghi, K. and Peck, R.B. (1967) *Soil Mechanics in engineering practice*, Wiley, 729 pp.
Tikoff (1994), Transpression: strain theory and application to the emplacement and deformation of granite, Sierra Nevada, California, Ph.D. dissertation, *University of Minnesota*, 277 pp.
Tikoff, B. and Teyssier, C. (1991) Dextral shearing in the east-central Sierra Nevada magmatic arc: implications for late Cretaceous tectonics and passive emplacement of granite, *Geol. Soc. Am. Abstr. Programs* **23**, 5, A176.
Tikoff, B. and Teyssier, C. (1992) Crustal-scale, en échelon "P-shear" tensional bridges: a possible solution to the batholithic room problem, *Geology* **20**, 927-930.
Tikoff, B. and Greene, D. (1994) Transpressional deformation within the Sierra Crest shear zone system, Sierra Nevada, California (92-80 Ma): vertical and horizontal stretching lineations within a single shear zone, *Geol. Soc. Am. Abstr. Programs* **26**, 7, 385.
Tikoff, B. and Teyssier, C. (1994) Strain and fabric analysis based on porphyroclast interaction, *J. Struct. Geol.* **16**, 477-491.
Tobisch, O.T. and Paterson, S.R. (1990) The Yarra granite: an intradeformational pluton associated with ductile thrusting, Lachlan Fold belt, southeastern Australia, *Geol. Soc. Am. Bull.* **102**, 693-703.
Tobisch, O.T., Renne, P.R., and Saleeby, J.B. (1993) Deformation resulting from extension during pluton ascent and emplacement, central Sierra Nevada, California, *J. Struct. Geol.* **15**, 609-628.
Tobisch, O.T., Saleeby, J.B., Renne, P.R., McNulty, B. and Tong, W. (1995) Variations in deformation fields during development of a large volume magmatic arc, central Sierra Nevada, California, *Geol. Soc. Am. Bull.* **107/2**, 148-166.
Toksöz, M.N. and Hsui, A.T. (1978) Numerical studies of back-arc convection and the formation of marginal basins, *Tectonophysics* **50**, 177-196.
Tommasi, A., Vauchez, A., Fernandes, L.A.D., and Porcher, C.C. (1994) Magma-assisted strain localization in an orogen-parallel transcurrent shear zone of southern Brazil, *Tectonics* **13**, 421-437.
Tommasi, A., Vauchez, A., and Daudré, B. (1995) Initiation and propagation of shear zones in a heterogeneous continental lithosphere, *J. Geophys. Res.* **100**, 22083-22101.
Toteu, S.F., Van Schmus, W.R., Penaye, J., and Nyobe, J.B. (1994) U-Pb and Sm-Nd evidence for Eburnian and Pan-African high-grade metamorphism in cratonic rocks of southern Cameroon, *Precamb. Res.* **67**, 321-347.
Treagus, S.H. and Sokoutis, D. (1992) Laboratory modelling of strain variations across rheological boundaries, *J. Struct. Geol.* **14**, 405-424.
Tribe, I.R., Strachan, R.A., and D'Lemos, R.S. (1996) Neoproterozoic shear zone tectonics within the Icartian basement of Guernsey and Sark, Channel Islands, *Geol. Mag.* **133**, 177-192.
Tubosun, I.A., Lancelot, J.R., Rahaman, M.A., and Ocan, O. (1984) U-Pb Panafrican ages of two charnockite-granite association from southwestern Nigeria, *Contr. Mineral. Petrol.* **88**, 188-195.
Tullis, T.E. (1976) Experiments on the origin of slaty cleavage and schistosity, *Geol. Soc. Am. Bull.* **87**, 745-753.
Turner, D.C. (1983) Upper Proterozoic Schist Belts in the Nigerian sector of the Panafrican province of West Africa, *Precamb. Res.* **21**, 55-79.
Turner, D.C. (1986) Magma distribution and crustal extension in the Nigerian Younger Granite province: evidence from the Wase area. *J. Afr. Earth Sci.* **5**, 243-247.
Ulmer, P. (1986) Basische und ultrabasiche Gesteine des Adamello (Provinzen Brescia und Trento, Norditalien, Eidgenössischen Technischen Hochschule, Ph.D. Dissertation 8105, Zürich 274 pp.
Utsu, T. (1967) Anomalies in seismic wave velocity and attenuation associated with a deep earthquake zone (1), *J. Fac. Sci. Hokkaido Univ.* **3**, 1-25.
Utsu, T. and Okada, H. (1968) Anomalies in seismic wave velocity and attenuation associated with a deep earthquake zone (2), *J. Fac. Sci. Hokkaido Univ.* **3**, 65-84.
Uyeda, S. (1982) Subduction zones: an introduction to comparative subductology, *Tectonophysics* **81**, 133-159.
Uyeda, S. (1991) The Japanese island arc and the subduction process, *Episodes* **14**, 190-198.
Uyeda, S., Fuller, M.D., Belshe, J.C., and Girdler, R.W. (1963) Anisotropy of magnetic susceptibility of rocks and minerals, *J. Geophys. Res.* **68**, 279-291.
van Breemen, D., Pidgeon, R.T., and Bowden, P. (1977). Age and isotopic studies of some Panafrican granites from northcentral Nigeria, *Precamb. Res.*, **4**, 307-319.
van der Molen, I. and Paterson, M.S. (1979) Experimental deformation of partially melted granite, *Contr. Mineral. Petrol.* **70**, 299-318.
van der Voo, R. and Klootwijk, C.T. (1972) Paleomagnetic reconnaissance study of the Flamanville granite with special reference to the anisotropy of its susceptibility, *Geol. Mijnb.* **51**, 609-617.
Vance, J.A. (1962) Zoning in igneous plagioclases: normal and oscillatory zoning, *Amer. J. Sci.* **260**, 746-760.

Vauchez, A., Pacheco Neves, S., and Tommasi, A. (1997) Transcurrent shear zones and magma emplacement in neoproterozoic belts of Brazil, in J.L. Bouchez, D.H.W. Hutton and W.E. Stephens (eds.), *Granite: from segregation of melt to emplacement fabrics*, Kluwer Academic Publishers, Dordrecht, pp. 275-293.
Vauchez, A. and Egydio-Silva, M. (1992) Termination of a continental-scale strike-slip fault in partially melted crust: the West-Pernambuco shear zone, northeast Brazil, *Geology* **20**, 1007-1010.
Vauchez, A., Neves, S.P., Caby, R., Corsini, M., Egydio-Silva, M., Arthaud, M., and Amaro, V. (1995) The Borborema shear zone system, *J. S. Am. Earth Sci.* **8**, 247-266.
Vergne, R. and Fernandez, A. (1990) Modèle d'anisotropie de susceptibilité magnétique induite par orientation préférentielle de forme de marqueurs paramagnétiques anisotropes dans une roche, *Rev. Phys. Appl.* **25**, 1049-1093.
Verhaeren, M., Claes, S, and Hertogen, J. (1995) Emplacement history of the Kagenfels granite (Late Hercynian, northern Vosges, France), *Terra Abstracts, Terra Nova* **7**, 141.
Vernon, R.H. (1968) Microstructures of high-grade metamorphic rocks at Broken Hill, Australia, *J. Petrol.* **9**, 1-22.
Vernon, R.H. (1990) Crystallization and hybridization in microgranitoid enclave magmas: microstructural evidence, *J. Geophys. Res.* **95**, 17849-17859.
Vernon, R.H., Etheridge, M.A., and Wall, V.J. (1988) Shape and microstructure of microgranitoid enclaves: indicators of magma mingling and flow, *Lithos* **22**, 1-11.
Vernon, R.H. and Paterson, S.R. (1993) The Ardara pluton, Ireland: deflating an expanded intrusion, *Lithos* **31**, 17-32.
Vicenzi, E.P., Rapp, R.P., and Watson, E.B. (1988) Crystal/melt wetting characteristics in partially-molten amphibolite, *EOS* **69**, 482.
Vigneresse, J.L. (1983) Enracinement des granites armoricains estimé d'après la gravimétrie, *Bull. Soc. géol. Min. Bretagne* **15C**, 1-15.
Vigneresse, J.L. (1990) Use and misuse of geophysical data to determine the shape at depth of granitic intrusions, *Geol. J.* **25**, 248-260.
Vigneresse, J.L. (1995) Control of granite emplacement by regional deformation, *Tectonophysics* **249**, 173-186.
Vigneresse, J.L. and Cannat, M. (1987) Mesure des paramètres physiques dans le sondage d'Echassières (vitesse sismique, porosité, densité), *Géologie de la France* **2-3**, 145-148.
Vigneresse, J.L., Barbey, P., and Cuney, M. (1996) An approach to rheological transitions during partial melting and during crystallization with application to melt segregation and transfer, *J. Petrol.* in press.
Von Bargen, N. and Waff, H.S. (1986) Permeabilities, interfacial areas and curvatures of partially molten systems: results of numerical computations of equilibrium microstructures, *J. Geophys. Res.* **91**, 9261-9276.
Waff, H.S. and Faul, U.H. (1992) Effects of crystalline anisotropy on fluid distribution in ultramafic partial melts, *J. Geophys. Res.* **97**, 9003-9014.
Wallbrecher (1986) Gefuegeanalytische Arbeitsweisen, Encke Verlag, Stuttgart, 220 pp.
Watson, E.B. (1982) Melt infiltration and magma evolution, *Geology* **10**, 236-240.
Watson, E.B. (1991) Diffusion in fluid-bearing and slightly melted rocks: experimental and numerical approaches illustrated by iron transport in dunite, *Contr. Mineral. Petrol.* **107**, 417-434.
Watson, E.B. (1994) Diffusion in volatile-bearing magmas, in M. Carroll and J.R. Holloway (eds.), *Volatiles in Magmas, Rev. Mineral.* **30**, 371-411.
Watson, E.B. and Brenan, J.M. (1987) Fluids in the lithosphere, 1. Experimentally determined wetting characteristics of CO_2-H_2O fluids and their implications for fluid transport, host-rock physical properties and fluid inclusion formation, *Earth Planet. Sci. Lett.* **85**, 497-515.
Watson, E.B. and Lupulescu, A. (1993) Aqueous fluid connectivity and chemical transport in clinopyroxene-rich rocks, *Earth Planet. Sci. Lett.* **117**, 279-294.
Watson, E.B., Brenan, J.M., and Baker, D.R. (1990) Distribution of fluids in the continental mantle, in M.A. Menzies (ed.), *Continental Mantle*, Clarendon, Oxford, pp. 111-125.
Webster, J.D., Holloway, J.R., and Hervig, R.L. (1987) Phase equilibria of a Be, U and F-enriched vitrophyre from Spor Mountain, Utah, *Geochim. Cosmochim. Acta.* **51**, 389-402.
Weijermars, R. (1986) Finite strain of laminar flows can be visualized in SGM36-polymer, *Naturwissenschaften* **73**, S. 33.
Weijermars, R. (1986) Flow behaviour and physical chemistry of bouncing putties and related polymers in view of tectonic laboratory applications, *Tectonophysics* **124**, 325-358.
Weinberg, R.F. (1992) Internal circulation in a buoyant two-fluid Newtonian sphere: implications for composed magmatic diapirs, *Earth Planet. Sci. Lett.* **110**, 77-94.
Weinberg, R.F. (1993) Drops, surface tension and diapir models, *J. Struct. Geol.* **15**, 227-232.
Weinberg, R.F. (1994) Re-examining pluton emplacement process: discussion, *J. Struct. Geol.* **16**, 743-746.
Weinberg, R.F. and Schmeling, H. (1992) Polydiapirs: multiwavelength gravity structures, *J. Struct. Geol.* **14**, 425-436.
Weinberg, R. F. and Podladchikov, Y. (1994) Diapiric ascent of magmas through power law crust and mantle, *J. Geophys. Res.* **99**, 9543-9559.
Whitney, J. (1975) The effects of pressure, temperature and X_{H_2O} on phase assemblage in four synthetic rock compositions, *J. Geol.* **83**, 1-31.

Whitney, J. (1988) The origin of granite: the role and source of water in the evolution of granitic magmas, *Geol. Soc. Amer. Bull.* **100**, 1886-1897.
Wickham, S.M. (1987) The segregation and emplacement of granitic magmas, *J. Geol. Soc. London* **144**, 281-297.
Williams, M.L., Hanmer, S., Kopf, C., and Darrach, M. (1995) Syntectonic generation and segregation of tonalitic melts from amphibolite dikes in the lower crust, Striding-Athabasca mylonite zone, northern Saskatchewan, *J. Geophys. Res.* **100**, 15717-15737.
Williams, Q. and Tobisch, O.T. (1994) Microgranitic enclave shapes and magmatic strain histories: constraints from drop deformation theory, *J. Geophys. Res.* **99 (B12)**, 24359-24368.
Willis, D.G. (1977) A kinematic model of preferred orientation, *Geol. Soc. Am. Bull.* **88**, 883-894.
Wilson, J. (1995) A sequential emplacement model for the Las Tazas pluton, Chile, *Terra Abstracts, Terra Nova* **7**, 138.
Woakes, M., Rahaman, M.A., and Ajibade, A.C. (1987). Some metallogenic features of the Nigerian basement, *J. African Earth Sciences* **6**, 655-664.
Wolf, M.B. and Wyllie, P.J. (1991) Dehydration-melting of solid amphibolite at 10 kb: textural development, liquid interconnectivity and application to the segregation of magmas, *Mineral. Petrol.* **44**, 151-179.
Woodcock, N.H. (1977) Specification of fabric shapes using an eigenvalue method, *Geol. Soc. Am. Bull.* **88**, 1231-1236.
Wortis, M. (1988) Equilibrium crystal shapes and interfacial phase transitions, in R. Vanselow and R.F. Howe (eds.), *Chemistry and physics of solid surfaces, vol. VIII*, Springer-Verlag, Heidelberg, pp. 367-405.
Wray, P.J. (1976) The geometry of two-phase aggregates in which the shape of the second phase is determined by its wetting angle, *Acta Metall.* **24**, 125-135.
Wright, E.P. (1971) Basement Complex: the geology of the Jos Plateau, *Geol. Surv. Bull. Nigeria* **32**, 12-47.
Wyllie, P. J. (1977) Crustal anatexis: an experimental review, *Tectonophysics* **43**, 41-71.
Yamaoka, K., Fukao, Y., and Kumazawa, M. (1986) Spherical shell tectonics: effects of sphericity and inextensibility on the geometry of the descending lithosphere, *Rev. Geophys.* **24**, 27-53.
Yoshii, T. (1979) A detailed cross-section of the deep seismic zone beneath northeastern Honshu, Japan, *Tectonophysics* **55**, 349-360.
Zapletal, K. (1990) Low-field susceptibility anisotropy of some biotite crystals, *Phys. Earth Plan. Inter.* **63**, 85-97
Zhong, S. and Gurnis, M. (1995) Mantle convection with plates and mobile, faulted plate margins, *Science* **2a 7**, 838-848.

INDEX

MIX
Papier aus verantwortungsvollen Quellen
Paper from responsible sources
FSC® C105338

If you have any concerns about our products,
you can contact us on
ProductSafety@springernature.com

In case Publisher is established outside the EU,
the EU authorized representative is:
Springer Nature Customer Service Center GmbH
Europaplatz 3, 69115 Heidelberg, Germany

Printed by Libri Plureos GmbH
in Hamburg, Germany